D0535080

Resource Management Information Systems

Resource Management Information Systems
Process and Practice

Keith R. McCloy

UK Taylor & Francis Ltd, 4 John St, London WC1N 2ET
USA Taylor & Francis Inc., 1900 Frost Road, Suite 101, Bristol PA 19007

British Library Cataloguing in Publication Data

A catalogue record for this book is available from the British Library

ISBN 0 7484 0119 9
0 7484 0120 2

Library of Congress Cataloging in Publication Data are available

Cover design by Amanda Barragry

Typeset and printed by Graphicraft Typesetters Ltd., Hong Kong

To Robyn
Sarah, Sophia and Simon

Contents

Preface

The purpose of this book is to enable the reader to design, build, implement, operate and use Resource Management Information Systems (RMISs) as decision support tools in the management of resources.

RMISs contain spatial information, predictive models and other tools designed to support the resource manager in making decisions.

Because many resources are dynamic, i.e. they change, sometimes quite rapidly, with time and space, the RMIS must have up-to-date information for it to be fully useful for many management purposes. In consequence, often the sources of information for a RMIS are derived from remotely sensed data, although other sources can include maps and field observations. This book covers the acquisition of remotely sensed data and the extraction of information from that data, as well as the derivation of other forms of information.

The management of resources requires the consideration of many types of information, with the mix of information required depending on the decision to be made, the skill and experience of the manager, as well as to whom, and how the decision has to be conveyed. Integration of the different types of information so as to create information that can be used in making decisions, or management information, occurs within a Geographic Information System (GIS). There are different types of GISs, each with its own strengths and weaknesses, so that some are better for some purposes than are others. This book covers the characteristics and uses of GISs in the management of resources.

An essential goal of all resource management decisions must be to strive towards the development of sustainable practices, i.e. those that:

- do not cause degradation of renewable resources beyond the capacity of the environment to replace them;
- use non-renewable resources at a rate which ensures that their cost is not prohibitive by the time society has developed alternatives; and
- do not create wastes beyond the capacity of the environment to absorb them.

Information systems are an essential component in any such strategy, and an essential component in those systems will be to predict the likely outcomes of proposed actions. Predictive modelling is thus an essential component of RMISs.

The importance of information in the management of all resources means that all disciplines concerned in any way with the management of resources should be familiar with RMISs as a decision support tool. This text is thus directed at all resource managers, and at students at both undergraduate and postgraduate levels who wish to manage resources, including students of agriculture, forestry, marine and aquatic resources, environmental science, landuse planning, valuation, engineering and geography.

A number of people have assisted in the preparation of this book. The assistance of these people is gratefully acknowledged, in particular: Drs Tony Milne, Bruce Forster and Kevin McDougall for proofreading of chapters of the text; colleagues Bill Gates and Sue Moffatt for critical comment; Connie Julian, Gerald Galgana and Serg Abad in the preparation of many of the colour prints; and Mary-Jo Julian for typing some of the text.

Special appreciation must go to my wife, Robyn, our children, Sarah, Sophia and Simon, and a friend, Robert Ridgway for their support and encouragement during the preparation of the text.

Keith R. McCloy
Sydney, Australia

Photographic section

(Note: captions are situated at the end of this section.)

Photograph 1.1

Photograph 1.2

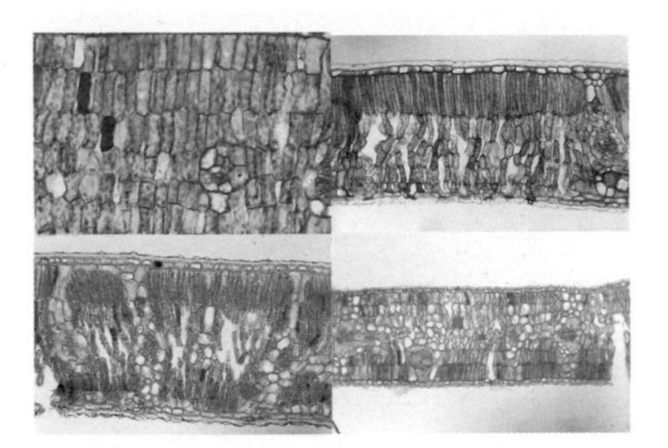

Photograph 2.1

Photograph 2.2

Photograph 2.3

Photograph 2.4

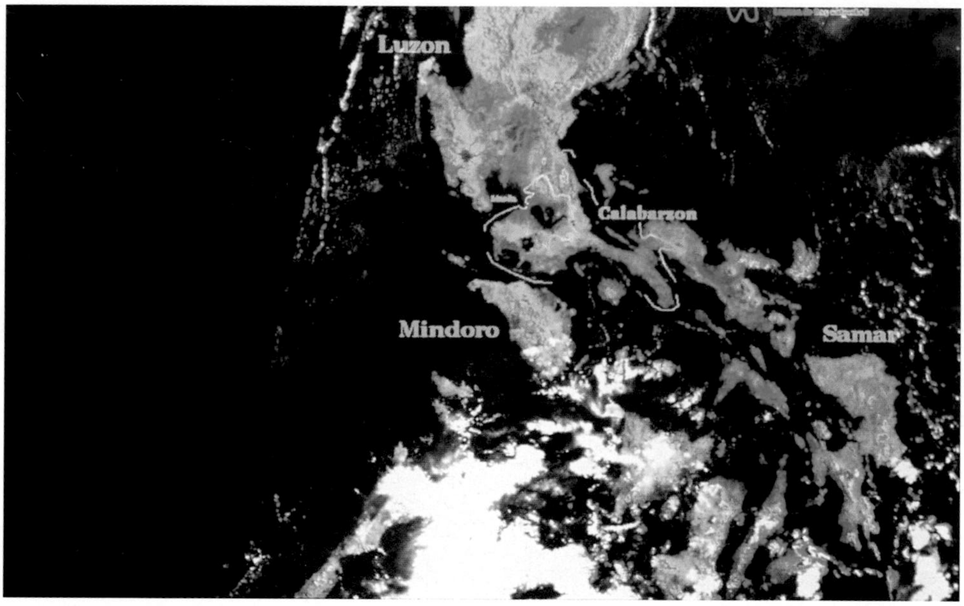

Photograph 2.5

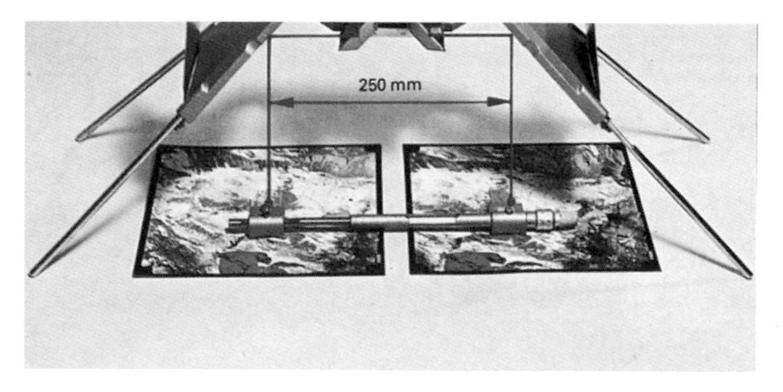

Photograph 3.1

Photograph 3.2

Photograph 3.3

Photograph 3.4

Photograph 3.5

Photograph 4.1

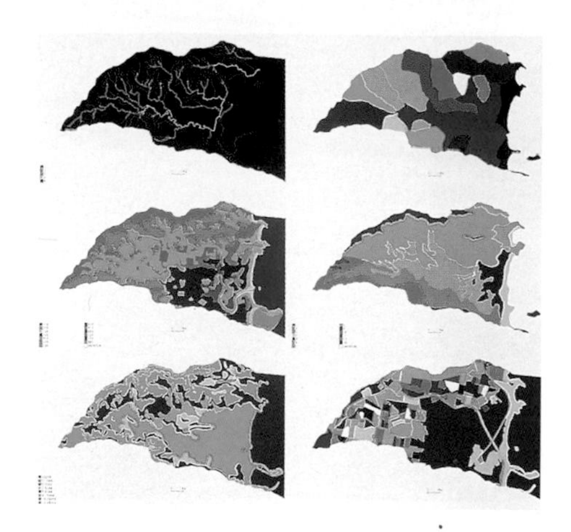

Photograph 4.2

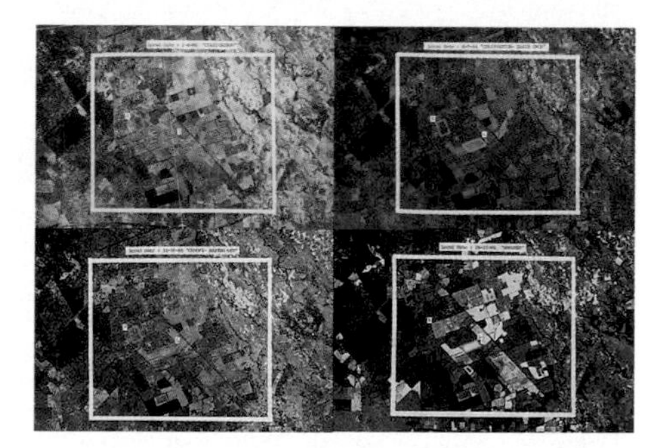

Photograph 4.3

Photograph 4.4

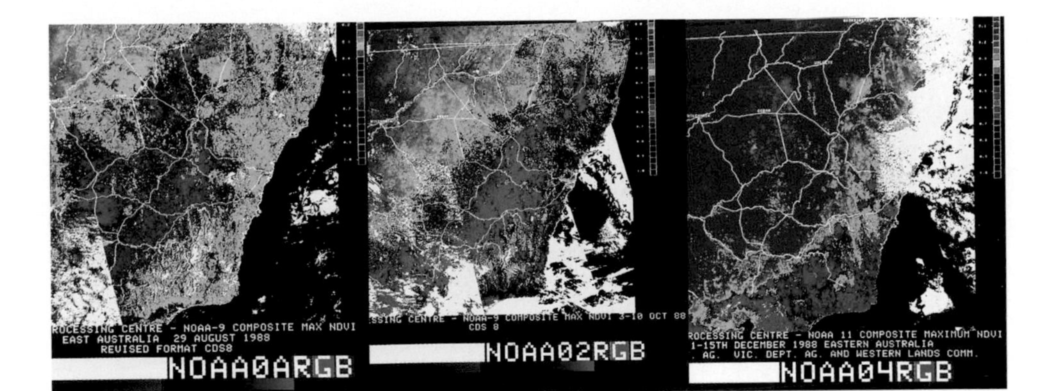

Photograph 4.5

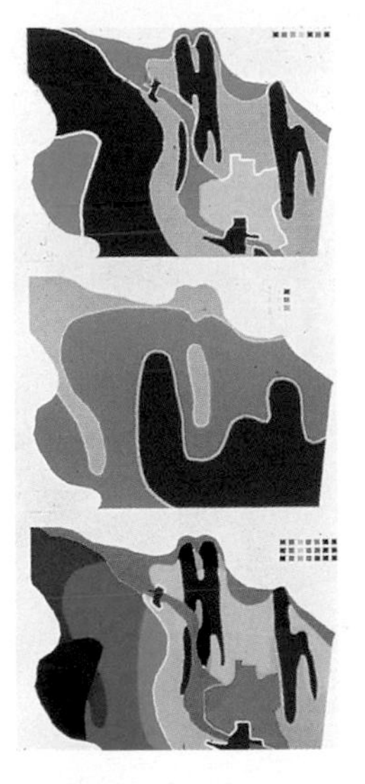

Photograph 4.6

Photograph 4.7

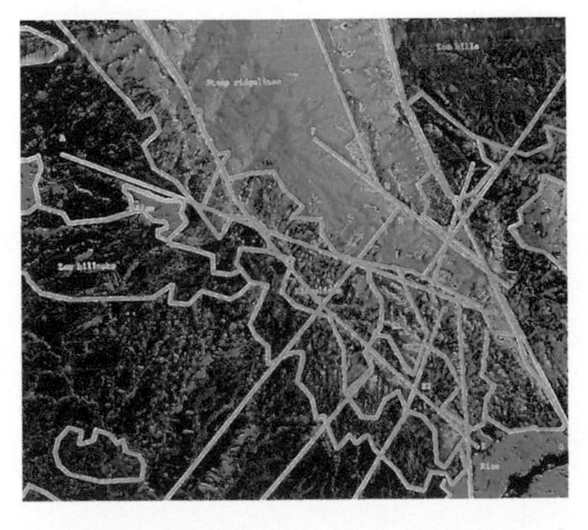

Photograph 4.8

Photograph 5.1

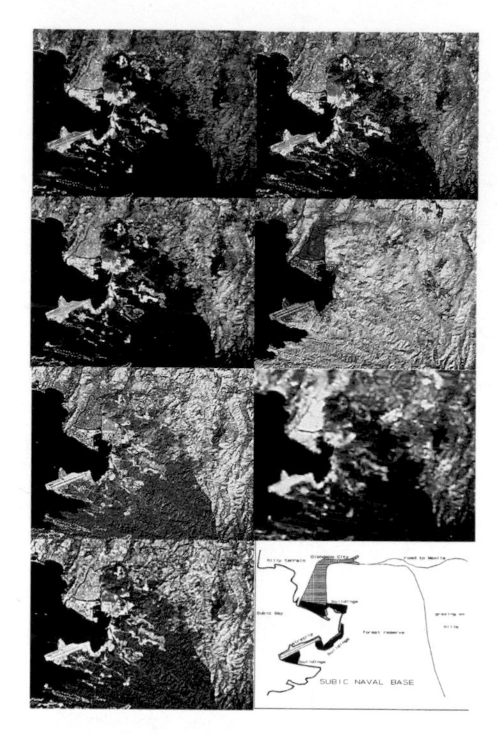

Photograph 5.2

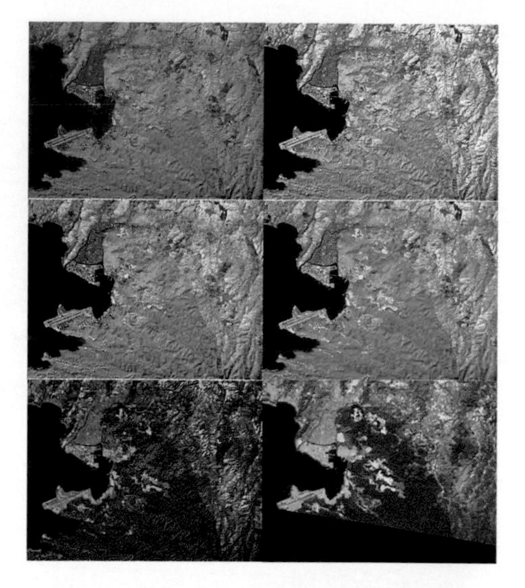

Photograph 5.3

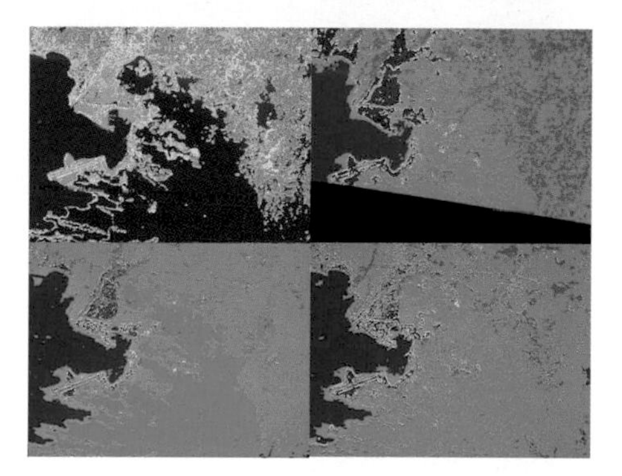

Photograph 5.5

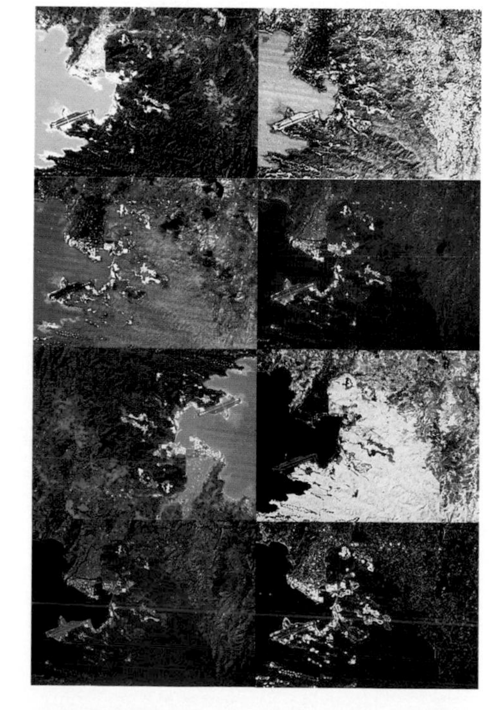

Photograph 5.4

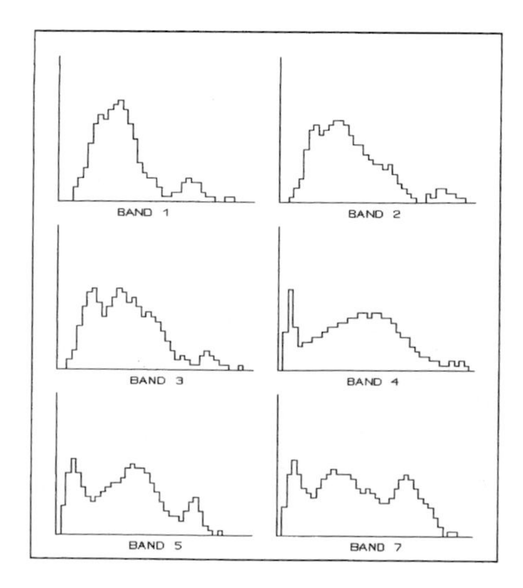

Photograph 5.6

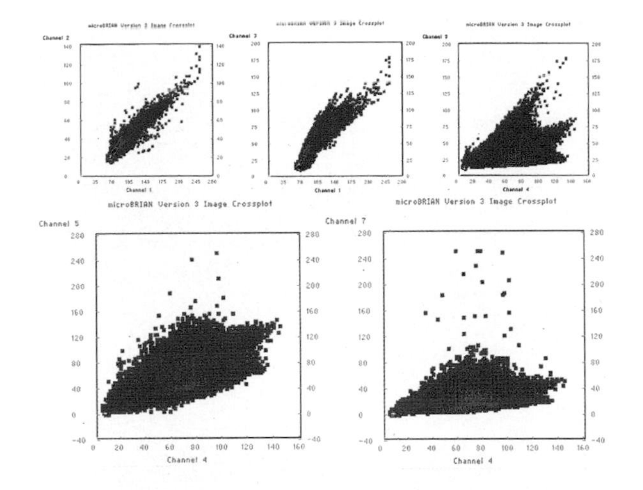

Photograph 5.7

Photograph 5.8

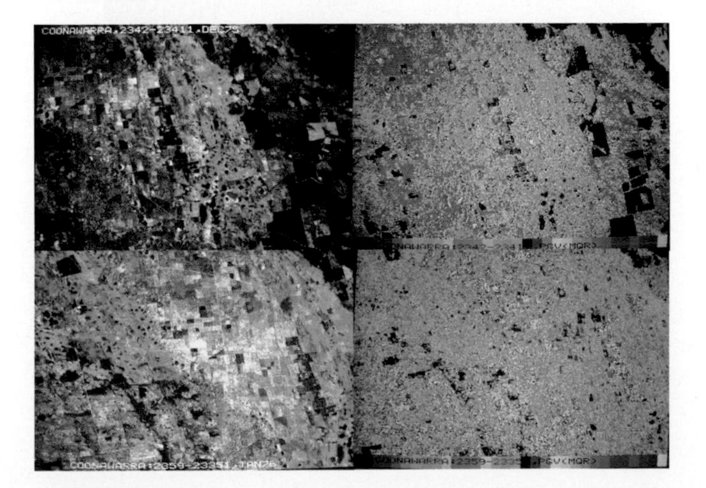

Photograph 5.9

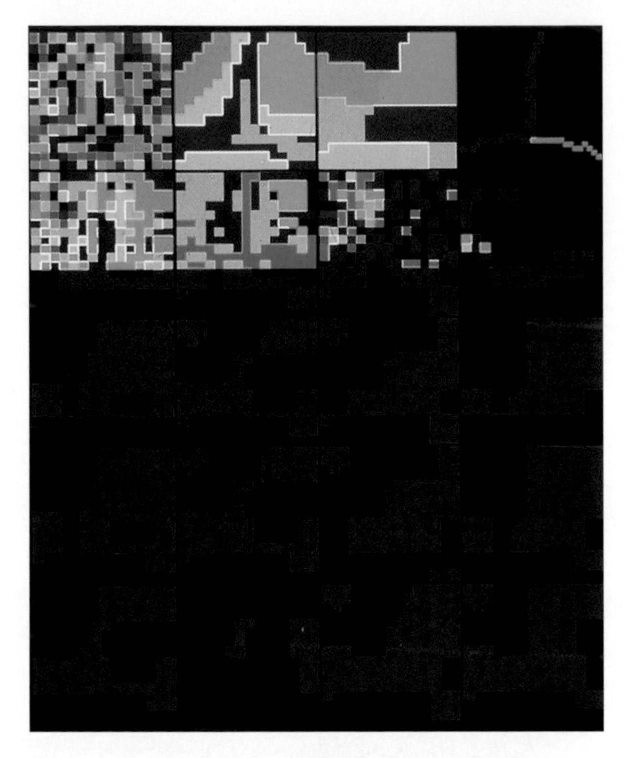

Photograph 7.1

Photograph 7.2

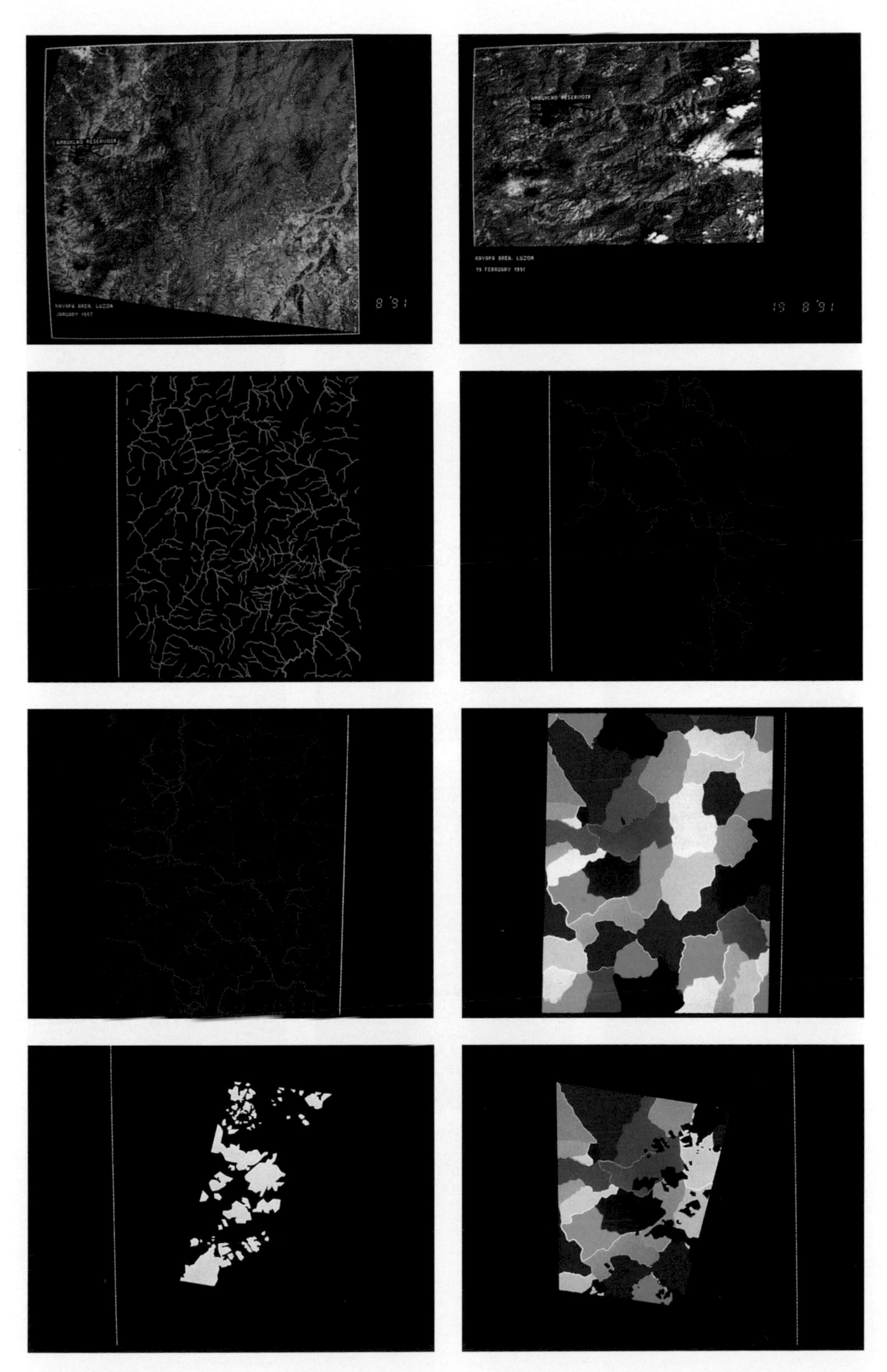

Photograph 9.1

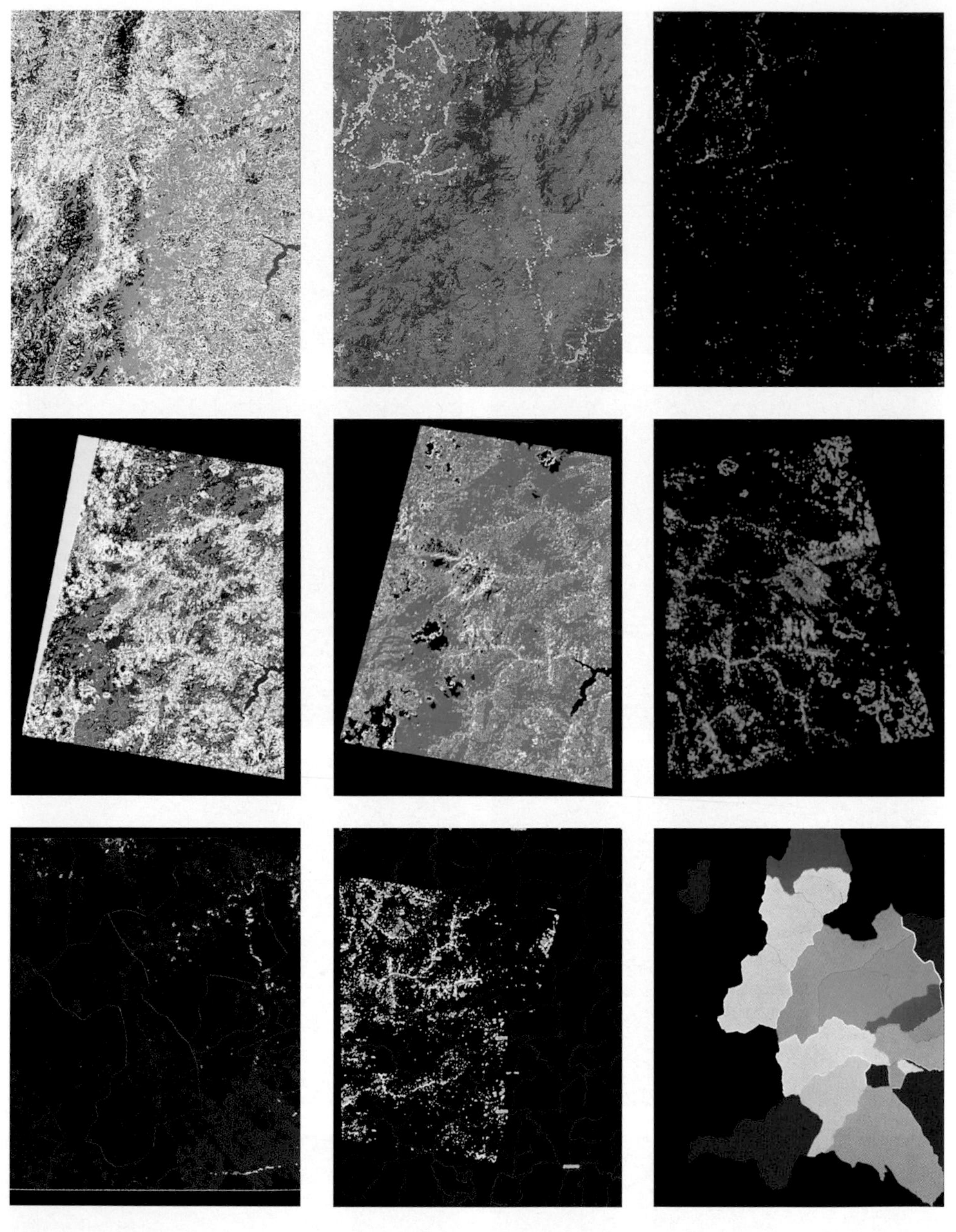

Photograph 9.2

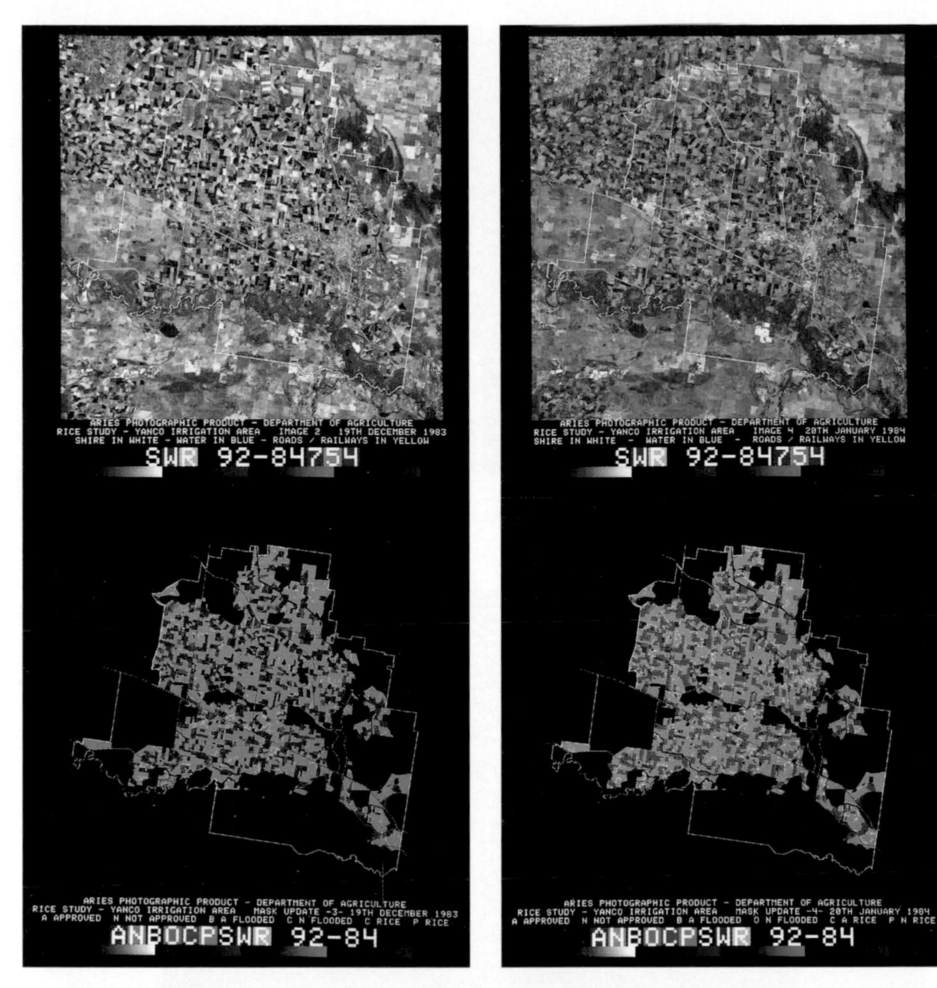

Photograph 9.3

Photograph 9.4

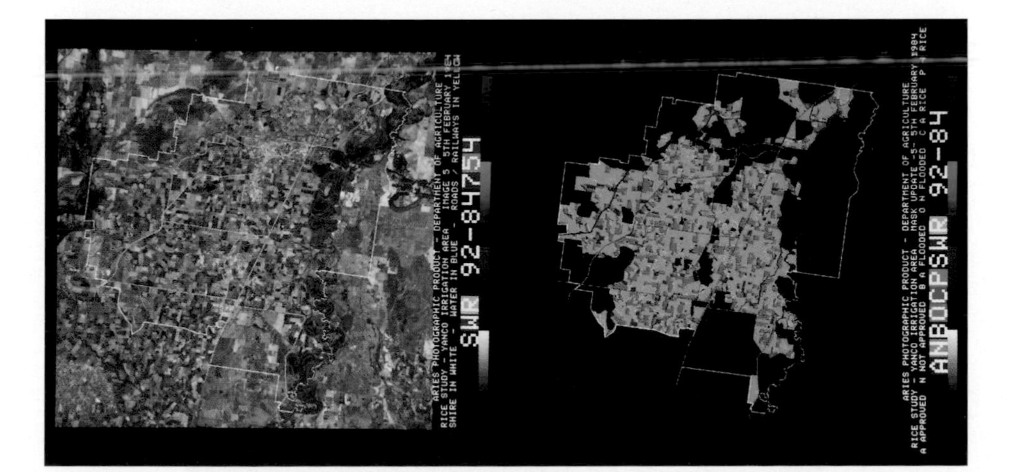

Photograph 9.5

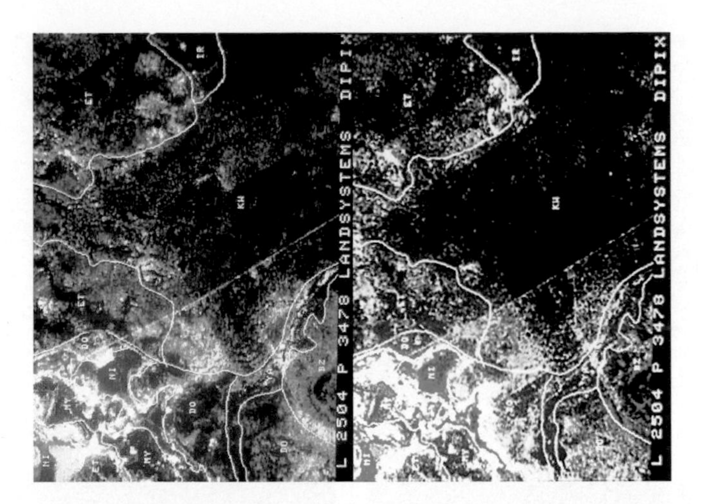

Photograph 9.6

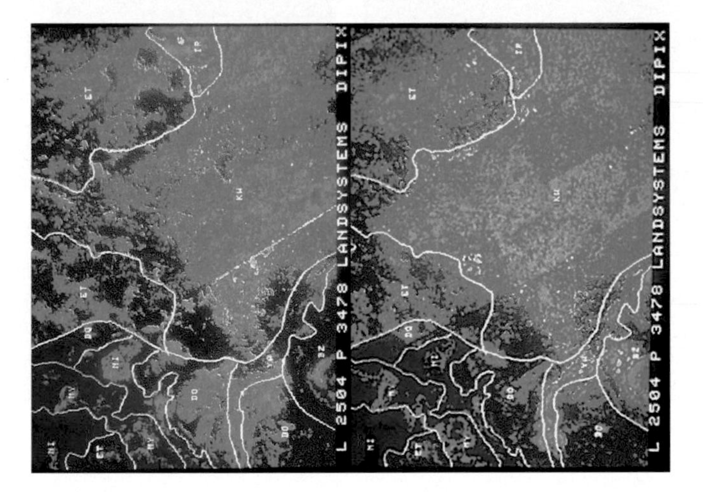

Photograph 9.7

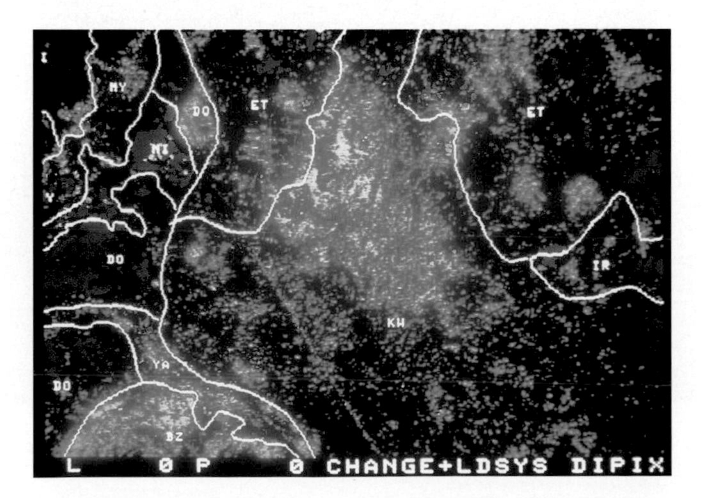

Photograph 9.8

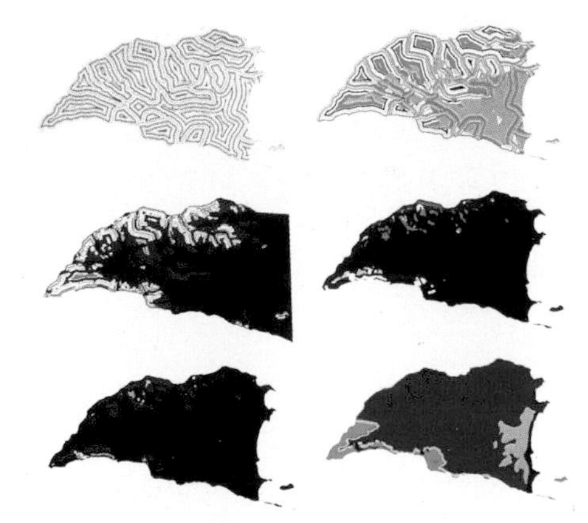

Photograph 9.9

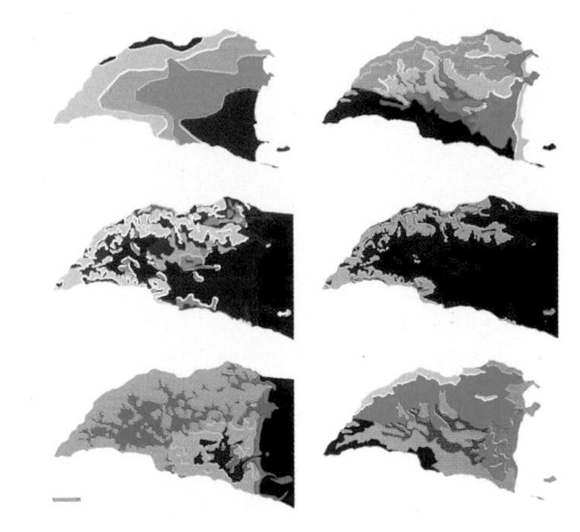

Photograph 9.10

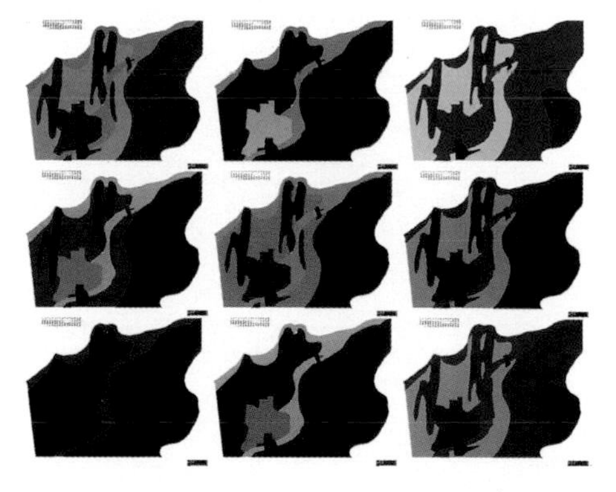

Photograph 9.11

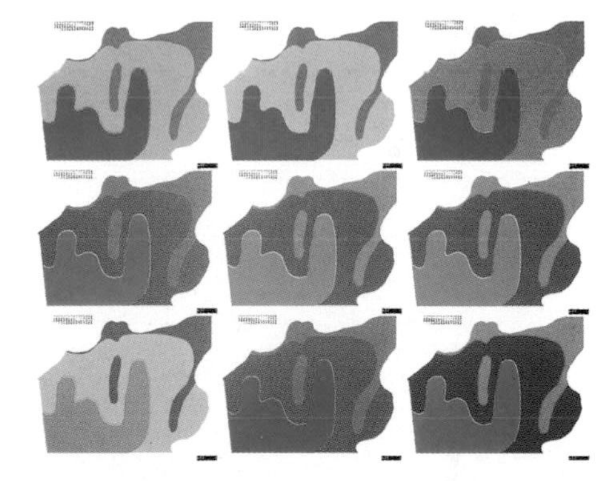

Photograph 9.12

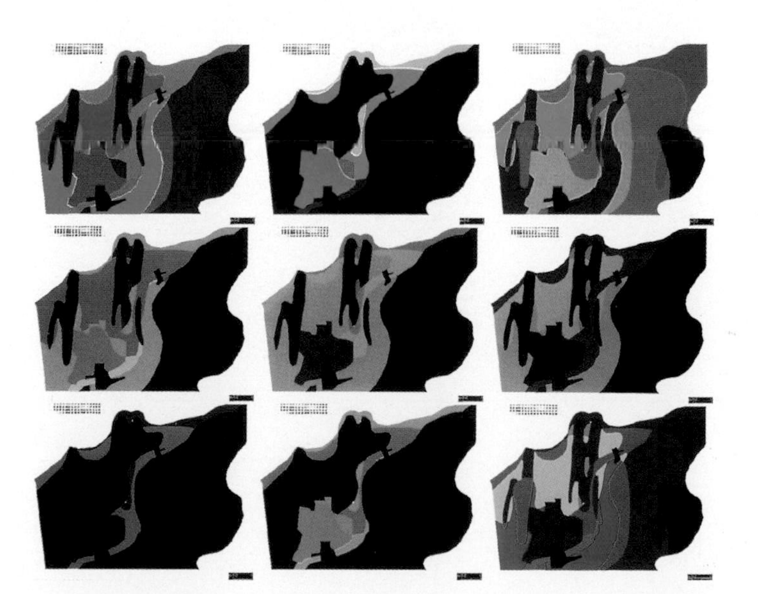

Photograph 9.13

Captions for photographic section

Photograph 1.1 *Soil erosion in Australia due to surface runoff. The photograph illustrates two forms of resource degradation in the rangelands of Australia: soil erosion and woody weed infestation.*

Photograph 1.2 *No tillage cultivation as practised in New South Wales. Photograph provided courtesy of Mr Warwick Felton, New South Wales Department of Agriculture.*

Photograph 2.1 *Dyed sections through leaves of* E. sieberi, E. pilularis, E. saligna *and* E. aglomorata (O'Niell 1990).

Photograph 2.2 *A typical camera used to acquire large format (23 cm by 23 cm) images that are suitable for topographic mapping, geologic and many other interpretation tasks.*

Photograph 2.3 *A typical video camera.*

Photograph 2.4 *Boom set on vehicle to hold a pair of spectrometers for field observations.*

Photograph 2.5 *A NOAA AVHRR image of the island of Luzon in the Philippines acquired in April 1992 in which band 1, band 2 and band 4 are displayed in blue, green and red respectively. The mountain ranges in the island are very obvious, being depicted in yellow colours, due to the cooler areas that are more reflective in the NIR due to vegetation. The warmer lower areas are shown as green, indicating vigorous green vegetative growth. The water areas are depicted as reddish in colour, indicating negligible reflectance in bands 1 and 2, but some radiance in the thermal infra-red. The image also shows some interesting cloud shapes around the island.*

Photograph 3.1 *A typical mirror stereoscope and parallax bar.*

Photograph 3.2 *A colour aerial photograph of part of the Gold Coast of Queensland. The high rise buildings exhibit considerable displacements due to their heights, displacements that can be used to locate the nadir point and to calculate the heights of the buildings.*

Photograph 3.3 *A stereo-triplet of an urban area.*

Photograph 3.4 *A temporal sequence of images of an area of cropping in NSW, Australia, taken in June (cultivation), August (immature canopy), November (mature canopy) and December (harvest), showing the seasonal changes that are recorded in such image sets.*

Photograph 3.5 *A side-looking airborne radar image in C band acquired of part of the Philippines in 1991. The image includes portion of the active Philippines Fault and the city of San José.*

Photograph 4.1 *A stereogram of aerial stereo photographs of the Coffs Harbour and Harbour Creek watershed area.*

Photograph 4.2 *Maps of six parameters extracted from analysis of either aerial photographs or maps of the Harbour Creek watershed. The maps are drainage, sub-watershed areas, land-use, terrain classification, slope classes and properties in the area.*

Photograph 4.3 *Landsat satellite images for cropping showing the images at cultivation (June), canopy development (September), mature canopy (October) and harvest (December).*

Photograph 4.4 *Interpreted analysis of the June, September and December images to identify crop areas, with the properties of interest in this analysis. The colours are coded as red – early crop identified in June, orange – late crop in June, green – early crops identified in September, brown – late crops in September and dark green for late crops identified in December.*

Photograph 4.5 *Images of a vegetation index that indicates the greenness in the vegetation, from red – negligible green vegetation, yellow – some green vegetation, green – significant green vegetation. The three images are composites of daily images acquired over the periods 29 August, 3–10 October and 11–15 December 1988 by the NOAA AVHRR sensor.*

Photograph 4.6 *Three layers of information derived from the visual interpretation of satellite image data. Image 1 shows the land-use intensity classes extracted from Landsat data, with the classes: red – Schist Ridges, green – alluvial floodplain, grey – fertile slopes, yellow – irrigated area, magenta – low hills; cyan – steep hills and purple – urban areas. Image 2 depicts the strata used in Forrest Shire to sample for crop yield from the vegetation index images depicted in Photo 4.5, with the yield strata colour coded as: red – significant green vegetation, green – some green vegetation, grey – negligible green vegetation. Image 3 shows the polygons created in Forrest Shire from the combination of the land-using and yield strata. The colours of the polygons, in terms of the land-using and yield strata, are given in the key in the photograph. The area of each polygon is given in Table 4.12.*

Photograph 4.7 *False colour composite image of an integrated set of channels of remotely sensed data of the Philippine Fault area in Luzon in the Philippines. The channels are depicted as red – radar, green – SPOT Ch3(NIR) and blue – SPOT Ch2(red).*

Photograph 4.8 False colour composite image of derived, transformed channels of data from an integrated set of channels of remotely sensed data of the Philippine Fault area in Luzon in the Philippines. The displayed channels are depicted as red – principal components Ch2, green – normalized difference vegetation index and blue – principal components Ch4, from the integrated radar and SPOT dataset. The image contains some interpretation of landforms and geologic shear formations from interpretation of both images.

Photograph 5.1 A low cost image processing system running both image processing and GIS software.

Photograph 5.2 Linearly enhanced Landsat TM bands 1, 2, 3, 4, 5, 6, 7 and a location map of the Subic Bay (Philippines) area. Subic Bay is dark on all images, but there are some shallow water details in bands 1 and 2. The forested areas are dark in bands 3, 5 and 7, to a lesser extent in bands 1 and 2 and highly reflective in band 4. The urban commercial areas of Olongapo City are distinctly darker in bands 4 and 5 than the adjacent naval industrial areas. All built-up areas have low response in bands 4 and 5, compared with exposed soil surfaces.

Photograph 5.3 Linearly enhanced TM colour composites of (a) bands 1, 2 and 3 as blue, green and red, (b) bands 3, 4 and 5, (c) bands 3, 4 and 7, (d) bands 1, 4 and 3; (e) bands 2, 3 and 4, and (f) MOS–1 bands 1, 2 and 4. Images (a), (b), (c) and (d) provide an approximation to true colour images but in general they display less information than that portrayed in the more conventional false colour composites depicted in (e) and (f). Differences in the resolution of the MOS–1 and the TM data are obvious in the images.

Photograph 5.4 Image enhancements. The first four principal components in (a), (b), (c) and (d); (e) colour composite of PCA–1 (blue), PCA–2 (green) and PCA–3 (red); (f) normalized difference vegetation index (NDVI); (g) average (smoothing) filter and (h) edge detector filter.

Photograph 5.5 Classification of image data showing (a) the spectral classes derived from a mixed supervised/unsupervised, minimum distance classification of the TM image data of the Subic Bay area, and (b), (c) and (d) the agglomeration of spectral classes into informational classes, being urban/commercial (red), ocean (blue), forest (green), coastal edge (cyan), exposed rock and cement (pink), agricultural lands (light green) and bare soil (olive brown) for MOS–1, TM and SPOT XS images of the same area, over a period of two years.

Photograph 5.6 Histograms of the seven TM wave bands for the Subic Bay Image depicted in Photo 5.2.

Photograph 5.7 Two-dimensional scattergrams of the band pair combinations from the TM data: (a) band 1/2; (b) band 1/3; (c) band 4/3; (d) band 4/5 and (e) band 4/7.

Photograph 5.8 *Enhancement of image data for Landsat MSS data acquired in December 1975 for south eastern Australia showing (a) ratio vegetation index, (b) normalized difference vegetation index, (c) greenness vegetation index and (d) Kauth image of GVI and SBI.*

Photograph 5.9 *Estimation of pasture conditions using image data. Images acquired on 29 December 1975 and 16 January 1976 – (a) and (b) – have been transformed into estimates of the proportion of green vegetation in the canopy – images (c) and (d) – using a regression model developed using sample areas from the processing of the image data and corresponding field data collected during overpass.*

Photograph 7.1 *A set of geo-coded data on elevation, soil classes, land-ownership, drainage, slope, aspect, woody cover density, soil erosion and cropping in 1980–1991. The values in these data sets are given in Diagram 7.13.*

Photograph 7.2 *A set of information derived from the data in Photo 7.1 by processing in a GIS. The derived information includes slope (3° ranges), woody cover (10 per cent ranges), property borders, crop sequences, crop sum, properties with rotation causing most erosion, factors for predicting erosion risk, predicted erosion, comparison of predicted with actual erosion and this comparison with property boundaries.*

Photograph 9.1 *AA – rectified SPOT multispectral image of the Kayapa–Bokod area of the Philippines acquired in February 1987, AB – rectified SPOT multispectral image of January 1991, BA – creeks, BB – roads, CA – administrative data, CB – watersheds for the study area, DA – cloud cover in 1991 and DB – watersheds in 1991 with cloud areas masked out.*

Photograph 9.2 *AA – classification of the 1987 (pre-earthquake image) into 25 spectral classes, AB – eight information classes from AA, AC – classes of interest from AB, BA – spectral classification of the 1991 (post-earthquake image) into 43 spectral classes, BB – ten information classes from BA, BC – classes of interest from BB, CA – change in landslides and alluvium/colluvium in the study area between 1987 and 1991, CB – Similar to CA, with the sub-watershed boundaries, CC – sub-watershed damage priorities as function of the density of landslides in the 1987–1991 period.*

Photograph 9.3 *The Landsat MSS image of 19 December 1983 and updated current mask after this image with red and yellow being areas that 'could grow rice', black being areas that 'cannot grow rice' and blue representing 'potentially rice', i.e. those areas that 'looked like rice' on the first image.*

Photograph 9.4 *The Landsat MSS image of 20 January 1984 and updated current mask after this image with similar colours as for Photo 9.3, as well as green representing 'rice', i.e. areas that 'looked like rice' on the first two images.*

Photograph 9.5 *The Landsat MSS image of 5 February 1984 and updated current mask after this image with similar colours as for Photograph 9.4, as well as light green representing 'rice' – i.e. areas that 'looked like rice' on the first three images.*

Photograph 9.6 *The 1975 and 1987 Landsat MSS images of the rangelands area of Australia, showing the red soil country that has a significant shrub invasion problem as greenish in colour and the grey soils of the Darling River as grey.*

Photograph 9.7 *Classification of the images in Photo 9.6 by the vector classifier and transformation of the derived parameters using a regression equation to show estimates of shrub cover as increasing intensities of green. Red indicates decreasing densities of shrub cover and the blue indicates areas not included in the class by the classifier.*

Photograph 9.8 *Change in woody cover between 1975 and 1987 depicted as red for increase in cover and green for a decrease in cover and an aerial photograph acquired in 1987 and used to estimated woody cover so as to develop the regression equation to estimate cover from the vector classification of the image data.*

Photograph 9.9 *GIS layers used to estimate erosion loss using the universal soil loss equation. The images represent, from top left to bottom right, (a) rainfall erosivity contours, (b) soil factor from terrain classification, (c) per cent slope from slope classes, (d) slope factor, (e) cover factor and (f) acceptable levels of soil loss.*

Photograph 9.10 *GIS layers used to estimate erosion loss using the USLE. The images are (a) shortest distance from watershed boundary to cell, (b) distance in (a) to power* (m + *1), (c) the LS factor, (d) predicted annual soil loss from the catchment using the USLE model, (e) actual soil erosion, exceeding acceptable levels and (f) acceptable levels of soil loss exceed predicted levels of soil loss.*

Photograph 9.11 *Crop densities for the nine winter grain crops given in Table 9.21. Each map represents one crop. Top left – bottom right: wheat (varieties Cane, Albert and Dune), oats (varieties Dora, Dora2 and Spain), barley (varieties Ibis and Ibis2) and Triticale. The colours in the images represent the same scale on all images, indicated by the key in the images, from 0 at the top left to 63 in the bottom right as percentage cover of that crop.*

Photograph 9.12 *Yields for the nine winter grain crops given in Table 9.22. Each map represents one crop. Top left – bottom right: wheat (varieties Cane, Albert and Dune), oats (varieties Dora, Dora2 and Spain), barley (varieties Ibis and Ibis2) and triticale. The colours in the images are represented by the key in the images, from 0 at the top left to 63 in the bottom right as (yield* $\times$ *10) tonnes* ha^{-1} *for that crop.*

Photograph 9.13 *Production estimates derived as the product of the crop area and yield for the nine winter grain crops shown in Photo 9.1 and 9.12. Each map represents one crop. Top left – bottom right: wheat (varieties Cane, Albert and Dune), oats (varieties Dora, Dora2 and Spain), barley (varieties Ibis and Ibis2) and triticale. The colours in the images are represented by the key, from 0 at the top left to 63 in the bottom right as (production* $\times$ *1000) tonnes from that cell for that crop.*

1
Introduction

1.1 Major macroscale issues in resource management

The term 'Resource Management Information System (RMIS)' is used to convey the dominant theme of this text: the role of information systems in supporting the proper management of physical environmental resources. Information conventionally available to resource managers includes market reports, weather predictions, product information and professional advice from various sources. It rarely includes information on landuse – both current and historical – landuse trends and landcover conditions. Yet all physical resources exist within a spatial context, so that spatial information is an essential component of an information base designed to support the proper management of environmental resources.

Integration of spatial information with these other types of information for the routine management of resources has implications for the whole resource management process: the mix and value placed on the different items of information; how management will operate to use the new mix; how management structures adapt to take advantage of the opportunities that arise from the use of a new mix of information; how the resources are actually managed; and finally, how the resource manager relates to the wider community that has identified its investment in the maintenance of environmental resources. All of these facets of resource management will be affected by changes in the information base available for resource managers. This book is concerned with all of these facets, and therefore it considers not only the technologies that constitute resource information systems and their scientific underpinnings, but also resource management practices and how they need to adapt to these technologies.

All information has a price and a manager has to assess the relevance of specific information for making specific decisions. If the provision of information brings benefits that outweigh the costs, then that information may be seen as worth using; those benefits might be improved productivity or resource maintenance, an improvement in community quality of life or a better understanding of the environment within which the community lives. Achieving the different benefits often involves choices; sometimes these benefits are complementary and sometimes they compete. Benefits are often measured in different units making it difficult to compare them and to arrive at a balance when a choice has to be made. One criteria for assessing benefits in relation to physical resources is in terms of the criticality of the different environmental units; less critical or over abundant units tend to be considered to have a lower unit value. In this type of assessment, spatial information is essential in determining the relative abundance of environmental units and their criticality in environmental processes.

The resource manager also wants to make decisions on the smallest information set, so as to simplify the decision-making process. An important challenge to those deriving information sets is to search for either subsets or simplifying transformations of the information sets, that are just as useful to the resource manager in making resource management decisions.

Provision of this geographic dimension to the information available to resource managers comes from the use of remotely sensed data, geographic information systems and modelling. Before considering these technologies, it is important to evaluate the need for 'more' information so as to make proper decisions from the perspectives of both the resource manager, and the community.

The community is, and many individual resource managers are, concerned with two issues: the management of commercial resources to improve productivity; and with the maintenance of environmental resources. Despite this interest, currently most management systems are designed for the short-term management of commercial resources, with the aim of maintaining or improving productivity. Existing information systems of market reports, weather information and predictions, product information and advice, all focus on improving commercial profitability. Management decisions in relation to these resources are relatively straightforward given this simple objective; a manager has to make an acceptable profit in the near future, otherwise he will go out of business.

However, these management methods appear to be quite inadequate when addressing the long-term maintenance of resources. There are many well documented examples of resource depletion and degradation in Australia, the US and elsewhere, which is threatening the viability of many of the industries that depend on those resources – clear evidence that the management of those resources is inadequate, even irresponsible, when taking a longer term perspective.

Consider the most immediate and obvious form of degradation or soil erosion by water transport (Photo 1.1). Man-induced soil erosion usually starts with the first clearing of the land, often for agricultural purposes. This means that soil erosion in Australia would have significantly increased in the early- to mid-19th century; yet soil erosion was not acknowledged as a serious problem in Australia until the 1930s, a delay of at least 80 years. The rate of erosion may have been reasonably constant over that time, but it took 80 or more years for the accumulated effects to reach sufficient magnitude, for the problem to receive the community attention that it deserved. Even today, more than 100 years after commencement of the problem, agricultural management causes erosion, with the management techniques currently recommended by Departments of Agriculture and their equivalents, minimizing but not eliminating this form of degradation. The initial response to the problem of erosion in the 1930s – of contour banks and better cultivation practices – do not reduce microscale erosion, but they do reduce its accumulation to larger scales and hence do reduce its magnitude. Clearly these responses are better than doing nothing about the problem, but our community objective should be to eliminate soil erosion as a form of resource degradation. It has only been in the 1970s that the next step, of introducing new cropping and other landusing practices that significantly reduce or eliminate actual erosion as it occurs, have been developed. These practices, of no or minimum tillage (Photo 1.2), are still not widely adopted because the costs of herbicide can make them uneconomic using short-term or one-seasonal criteria, whilst ignoring the longer term benefit of retaining more of the topsoil. The longer term benefits of using herbicides instead of cultivation must also be assessed, relative to the environmental costs that are incurred.

By the time that erosion had been recognized as a major problem by the community, it had affected large tracts of country and caused the removal and transport of thousands of tons of soil. It will take thousands of years to rebuild the soil losses that have occurred over decades in Australia.

A second disadvantage of current management techniques is the inability of managers to discriminate long-term trends from the short-term fluctuations that are a natural part of all environmental systems. In Australia, there has been an argument for over 30 years that agricultural practices are causing desertification in the marginal cropping belt. The protagonists are arguing, that the area is becoming more arid through agricultural practices, whilst the other view is that the area is naturally arid and that the current aridity is part of the natural cycle of events. The real problem is the paucity of adequate temporal, geographic data that could be used to draw substantiated conclusions as to the actual processes and causes that are at work.

Another problem is the lack of predictive power available to current management. A manager who is unaware of the long-term impact of a decision can be excused for ignoring these long-term implications. Even when some members of the community are aware of the implications, this awareness is often subjective, lacking in rigorous quantitative definition. Clearly, under these conditions, there is considerable potential for conflict in making resource management decisions. The long-term maintenance of resources requires management decisions that are based on the long-term implications of those decisions, i.e. they need to be based, in part, on estimates of effects derived from prediction. Predictive models, or computer-based models designed to develop predictions, require rigorously established and maintained, extensive and accurate quantitative spatial databases of relevant resource parameters, if they are to be useful. These databases are essential to the development of many models, and provide essential data to drive all models. Only by means of remote sensing can these databases of parameters, many of which change relatively rapidly over time, be developed and maintained.

Yet another limitation of current management practices is the nature by which they are taken and implemented. Generally, current management decisions are discipline-oriented; agricultural managers make agricultural decisions, foresters make forestry decisions and so on, with little consideration of the implications for other facets of the environment. Yet the interdependence of the various components of the environment means that single discipline-based decisions may well have an adverse effect on other aspects of the environment. Decisions need to be taken in the context of the region in which they are made, where that region needs to be of sufficient size and environmental composition to take into account the main environmental interactions that may be affected by the decision. The environmental impact of decisions varies between environments and forms of degradation, so the definition of a region must allow these regions to vary, depending on the issue and degradation that is being considered. For example, regions that are impacted by water-borne erosion are catchment or subcatchment areas; regions that are impacted by weeds or soil trace element depletion may use environmental regions that are primarily dependant on climatic and soil factors; regions that are impacted by wind erosion will be quite different. GISs provide the technology that can allow resource managers to acquire resource information and to partition that information readily into regions appropriate for the analysis of specific issues.

In summary, current management methods are inadequate for the long-term maintenance of all resources because they:

1. lack adequate quantitative databases to identify long-term trends and develop predictive models;
2. lack predictive models;
3. are discipline-based rather than regionally based; and
4. cannot readily partition available information into relevant and appropriate management units.

So, although current management methods may be adequate for the short-term management of resources, they are totally inadequate for the long-term maintenance of resources. Because short-term management of resources involves spatial considerations, for many purposes this management will be improved by the proper use of spatial information as can be best supplied by RMISs. Funding of RMISs will be most readily available if those systems are of use for commercially driven applications. A major theme of this book is thus that RMISs should be designed so as to make them cost effective for the short-term management of resources. An important benefit of this will be that the developed systems could provide the information necessary to manage environmental resources at little extra cost.

1.2 The nature of geographic information

Data are defined as the recorded values of an attribute or parameter. They are the raw measurements, often made by instruments or collected by field staff, that form the basis on which information is created. Generally, they do not constitute information of themselves. Most data provide an objective historical record of specific attributes of the surface being observed. For example, the data acquired by aerial photographs are related to the energy emanating from the surface in the wavebands being sensed by the film in the camera at an instant in the past. The data used in RMISs are often of values that are not of much direct interest to managers in their form of reflectance or radiance data or individual point measurements. They have to be transformed, or converted, into something more meaningful and useful in terms of the information needs of the resource manager, and less constrained to the historical perspective of the source data. This transformation or conversion creates 'raw' information.

Information is defined as specific factual material that describes the nature or character of the surface, or of aspects of that surface, which are important to the resource manager. For a variety of reasons, information is often processed to an intermediate stage, called raw or base information from the data. This raw information can be used directly by the resource manager, but more usually the manager requires information that is an integration of different sets of raw information each of which may be derived from a different source. This integrated information, uniquely required by a particular manager to address a specific management issue, is called 'management' information. For example, an aerial photograph may be used to identify crop areas. Such raw information can be used by various managers, but they frequently require more complex information for making management decisions. If the manager wants to know the area of crop on a particular soil type or property, integration of the raw cropping information with either property or soil information would provide the required management information.

Knowledge is defined as an understanding of the characteristics and processes that occur in an object or surface. Knowledge has two dimensions: a conceptual, strategic or big-picture dimension; and a detailed, or tactical, dimension. Generally, resource managers build their conceptual models through education, learning and discussion; indeed many people in society have conceptual models of many environmental processes. Thus many people have a conceptual understanding of the processes involved in soil erosion. Conceptual models of conditions or processes are essential to the making of decisions, but they are not sufficient. The resource manager also needs a detailed or tactical model that provides details on the status of the object or surface of interest, or how the actual object or surface is acting or responding to the surrounding conditions. Thus, whilst most farmers can tell how soil erosion occurs, many continue to have erosion occurring on their property, partly because they do not have adequate tactical models of the process itself and its expression on their property. The tactical models required to make decisions are constructed using information to define the parameters for quantified versions of the conceptual models. Once this tactical model has been constructed then the manager should be in a position to make decisions in relation to that specific issue. Clearly, the information needed by the resource manager is that which will allow construction of an adequate tactical knowledge base on which to make the relevant decisions; other decisions by other managers will require different tactical models and hence other information. The direct use of information is thus to construct the conceptual and tactical knowledge necessary to make specific decisions.

A secondary use of information is to identify limitations in the currently used conceptual models, so as to modify or improve on those models. New technologies often reveal limitations in existing conceptual or tactical models, since these technologies will depict the real world in ways that may not have been explored previously. Since remote sensing is a new technology, there will be considerable scope for remote sensing to be used to renovate or renew our conceptual models of environmental processes.

Geographic data or information are defined as those forms of data or information that incorporate spatial, temporal and attribute information about the features or resources of interest. Some types of geographic data and information have the same temporal information value for each attribute value in the data set, as occurs with an aerial photograph or a remotely sensed image, since the image is acquired at the one time. Other types of geographic data have one spatial coordinate but many temporal coordinates such as data recorded on a logging device.

A map is the classic form of geographic information. Each symbol on the map represents one type of feature, indicating that the feature so represented occurs at that location at some specified level of accuracy. The distribution of symbols on a map indicates the relative locations of the different features in the area covered, to another level of accuracy. The absolute accuracy of location of the first feature is usually lower than the relative accuracy between features when the map is drawn from information derived from remotely sensed images, including aerial photographs, due to the characteristics of those images. Again, maps depict information that is correct at a temporal coordinate value (or time) prior to the date of publication of the map. Since this temporal information is important to users, it is usually printed on the map.

The accuracy of a map must be considered in spatial, informational and temporal terms. Most maps are derived from remotely sensed images and thus have accuracy characteristics typical of these data sources, while maps that are constructed from other data sources, such as point observations, will have other accuracy characteristics.

Accuracy can be defined as the closeness of the estimated values of an attribute to the true values in that estimate. Modern physics shows us that it is not possible ever to measure the 'true' value of any physical quantity, so measures of accuracy are estimates of closeness of the estimated value to the best estimate that can be made of that value. Accuracies are usually measured in different ways for different attributes, and when considering different perspectives in using the information. Thus, a user of cropping information may be more concerned about having either good estimates of the area of crop, or he may be concerned with identifying all of the paddocks that were cropped. Both questions are legitimate to different resource managers, and they require different approaches to accuracy assessment.

The **spatial accuracy** of a map depends on many factors: the map scale; the accuracy of plotting detail on the map; the size of the symbols used relative to the scaled size of the feature; the density of features depicted on the map, particularly if their symbols are larger than the actual features at the map scale; and the age of the map. **Informational accuracy** also refers to the likelihood of features being correctly depicted on the map, that they are in the locations indicated, that all of the features are shown and that no extraneous features are wrongly depicted. **Temporal accuracy** is affected by the changes that occurred between acquisition of the imagery and the printing of the final map. The actual temporal accuracy is also affected by the changes that have occurred between the printing of the map and its use by the reader. These aspects of accuracy will vary from feature to feature because of differences in accuracy in identifying the different features and because of the frequency of changes in them over time.

A simple example is to consider the case of houses and roads. Roads are easily seen and well known by the community – they will rarely be wrongly depicted on a map. Road positions do not change frequently, so that they will tend to hold their accuracy over time. Depending on the scale of the map, groups of houses may be depicted by the one house symbol; it is more difficult to identify houses and so there will be a larger error level in marking houses on a map. Finally, there are always changes to the stock of houses in an area so the accuracy of the map in depicting houses will decrease with time more quickly than will the accuracy associated with roads.

Historically, maps and all other forms of geographic data including diagrams, plans and records, have been kept in an analogue, or hard-copy, form e.g. on paper. However, all forms of analogue data and information can now be stored as digital data and information, suitable for analysis in a computer.

Whilst analogue data are appropriate for visual assessment and interpretation, this form of storage contains high overhead costs, including costs of reproduction, retrieval for use, storage, reformatting for different applications and integration with other data. These costs have meant that map information has never been used to its full potential, a limitation that will become worse as labour costs increase relative to other costs.

Both forms of data – analogue and digital – have a part to play in deriving information, and in the management of resources. They have different strengths and weaknesses, related to both the characteristics of their forms of storage and display, as well as to how they can be used in analysis. In deriving information, the computer has strengths in conducting processes that involve considerable amounts of simple numerical or logical number or character processing. However, computers are not capable of making intuitive decisions. A human interpreter, on the other hand, while slower in conducting detailed numerical or logical processing, is better at integrative and intuitive logic. At present, computers are also very limited in their capacity to identify complex figures, shapes or patterns, whereas the human is very good at this type of activity. The relative power of the computer and the human interpreter will change as the technology evolves, but each will (no doubt) retain strengths relative to the other. In consequence, the most effective systems are likely to be those that incorporate both digital analysis and visual interpretation in deriving further information.

Other types of geographic data include aerial photographs, satellite images and field data which include data on the time and location of the observations. Thus soil tests that include data on the location and time of acquisition are geographic data.

Geographic data and information are derived and used respectively, in systems designed for that purpose, from the very simple to the very complex, expert systems. The simplest systems, which may include a map and the observer, or an aerial photograph and the interpreter, are suitable for many tasks, and should be used when appropriate. However, when the information required cannot be extracted, using these simple systems, with sufficient accuracy, or when the derivation of the required information requires multiple data sets, more sophisticated or complex processing systems are needed. In general, the more complex the system, the more vulnerable it is to errors arising from unanticipated sources, unanticipated complexities interfering with its operation, and the more expensive it is likely to be to operate. The systems used to derive geographic information should therefore be as simple as possible, commensurate with deriving the information required to an acceptable level of accuracy.

In developing RMISs it is therefore important not to be deluded by the perceived glamour of highly automated or digital systems, if development of these systems involves overkill or is not cost-effective. The KIS (Keep It Simple) principle is a very good one to abide by in the development of RMISs.

The most critical factor in successfully implementing RMISs is the proper design of such systems. Remote sensing, forming a major component of such systems, is full of case histories of poor system design and the consequent development of uneconomic systems. The US invested more than $400 m in the LACIE and AgRISTARS programs in the 1970s, a major component of which was the development of a system to determine the in-season area, and to estimate the in-season production of crops, focusing on winter cereals. The system has not yet been implemented operationally because it is not cost effective relative to existing techniques. A very important factor in this failure is not that remote sensing is an inappropriate technology, but that the system design was wrong.

Generally speaking, remote sensing is not as yet being widely used in operational programs in an extensive way, not because the technology is inappropriate *per se*, but probably because the system design to implement remote sensing is inappropriate. The main purpose of this text is to impart the skills, knowledge and philosophical approaches necessary to design, develop, evaluate and implement successful, cost effective RMISs.

1.3 The nature of RMISs

RMISs are defined as systems that use geographic data and information, with associated ancillary data and information, to derive information required by resource managers in a cost effective manner. Before analysing we first review the role or purpose of an RMIS in the management of resources.

An RMIS is intended to provide information to resource managers, presenting that information in a way that allows the manager to understand the information and its implications so as to construct a tactical knowledge model before making a management decision. An RMIS is therefore a decision support system and does not in any way take the actual decision-making process away from the resource manager, nor his responsibilities in the management of resources. Because the RMIS is providing a comprehensive set of relevant information to the manager, it should facilitate the implementation of more timely decisions, and should be accompanied by an improvement in the quality of the decisions made by the manager. However an RMIS is not a panacea that will relieve the manager of his responsibilities, nor the challenges, in making management decisions.

Because an RMIS is a management decision support system, it is essential that the system interface to the manager be easy for the manager to learn, use and understand. Clearly the more that this language mimics the language of the manager, the more convenient it will be, and so it is important to phase questions in the RMIS in terms used by the resource manager. The current trend in computer interfaces, using the concepts of windows and pointers or a mouse, improves the convenience of the interface, requiring minimum training by those not familiar with computers. The RMIS must also communicate its information to the manager in a manner that reflects their concerns and priorities if it is to be of most economic use to those managers. An RMIS is often based on sophisticated technology; it often requires sophisticated technological skills to implement; in consequence the workshop component of an RMIS may contain staff with highly specialized technical skills. Whilst this workshop component must be accessible to the manager, it will not be through the workshop that the manager will communicate with the RMIS. It is more likely that he will communicate directly with the RMIS. In consequence the interface in the RMIS to the manager must be suitable for use by the manager.

An RMIS will comprise at least four parts:

1. input
2. analysis
3. estimation and prediction
4. decision support

Input includes acceptance by the system of all of the data and information used by the system to derive management information, e.g. aerial photographs, field data, maps and remotely sensed images. The maps can include topographic, property or cadastral (land ownership), geologic, soil, landuse and other relevant maps in either analogue or digital forms. If the system is to use analogue data then the input will include the facilities necessary to convert analogue data into digital data, as well as the facilities necessary to store both forms of data.

Input must also consider error checking and validation of the data that is being entered. The adage, 'garbage in, garbage out', is quite correct. Users will be more nervous about digital data than analogue data because they cannot 'see' the digital data and hence they cannot assess the quality of the information by themselves, as they can with an analogue document. In consequence, if they discover errors in the information given by an RMIS then they are likely to be less forgiving of the system; they will place less reliance on the system when making management decisions. The ongoing viability of an RMIS depends on its credibility with its users, and this in turn depends on users understanding the accuracies that they can both expect and are actually getting, in the information provided by the system.

Analysis involves the derivation of raw information from the input data and information, as well as the derivation of management information from the raw information. The derivation of raw information is usually done from the analysis of remotely sensed images, derived from maps or from field data. Whilst this raw information can be used on its own, it is more frequently used in conjunction with other information. The processing of sets of raw information to derive management information is usually done within a Geographic Information System (GIS).

Typical types of management information that might be derived using a GIS could be to answer questions of the form, 'What fields in County Alpha have been cropped in each of the last four years and what is the average yield of these fields relative to the average for the county?', or 'Show me all areas that have been cropped at least twice over the last five years and are within x km of railhead Beta. What is the area involved in cropping each year and the average production?'

The raw information for these questions is the cropping history, yield information and land administration information. The cropping history information can be readily taken from remotely sensed images. The yield information may come from remotely sensed data or from field data in the form of annual returns. The administrative information will have come from digitization of maps that may be maintained as a digital database. Each layer of raw information can be used on its own, but they become capable of being used to provide much more powerful and useful information when they are combined with other layers of raw information. These same layers can be used to address many other management questions, either on their own or in conjunction with other layers of raw information such as soil or slope information. The manager is the person in the best position to define clearly the questions that he wants addressed; he will also want the system to be responsive to his needs, and to provide the requested information when required. This component of an RMIS must therefore be very accessible to, and useable by, the resource manager.

Estimation or prediction are defined as the extrapolation from data or information on specific resource attributes to derive estimates of those or other attributes by making assumptions about the behaviour of parameters that affect the estimation or prediction. Estimation is extrapolation from a specific set of attributes to another set at the same time. It thus involves models that estimate the second set of parameters from values given to the model for the parameters in the first set. Prediction is extrapolation over time and thus normally includes assumptions about climatic events over the period of time of extrapolation. Both estimation and prediction are modelling tasks, and as such similar considerations are involved in the design and development of both types of models. But the method of use of both types of models can be quite different within an RMIS. For example, predictive models are usually implemented to address 'what-if' type questions – 'what would happen to my pasture condition if my property receives *x* mm of rainfall over the next *y* days?' or 'how long will I have adequate fodder without rain?' – as a precursor to making a management decision. Estimation models are most often used to derive estimates of physical parameters of the environment such as above ground vegetative mass or biomass, leaf area index (LAI) or other parameters of interest to the resource manager.

Decision support is the interface with the resource manager; its design must consider not only what information is required by managers, but what they want the information for, how the manager will use the RMIS to derive that information and how they will use the information in the decision-making process. The method of use affects the presentation of information, e.g. as a map, as statistics, as digital data for a computer or video suitable for television. The knowledge levels of users, the facilities available to them and their attitude to the technology will influence the training and ongoing support required if they are to gain the maximum benefit from an RMIS.

All information in an RMIS contains errors. The way the manager uses the information in the RMIS must both acknowledge the existence of these errors and protect the resources, the community and the manager from errors of judgement that arise due to errors in the information base. This can be done in a variety of ways as will be discussed later in the text.

Thus RMISs, structured in four components, must deal with the technologies of computer science, remote sensing, geographic information systems and modelling. These technologies could be considered to be the tools and techniques upon which an RMIS depends. However, it should also be clear that a very critical component upon which a successful RMIS depends is the good design of that RMIS. These systems can fulfil either a mapping or a monitoring function in the management of resources. Whichever role they fulfil, but particularly if it is a monitoring function, then the RMIS must take cognizance of the environmental conditions that exist within the area of interest and within which the RMIS must provide management information. The reasons for this will soon become clearer, but they can be illustrated here in a rather superficial way. The frequency of occurrence of cloud cover can be a significant factor in the design of an RMIS information acquisition component, and hence the way that the RMIS is used, simply because of the impact of cloud cover on the acquisition of remotely sensed imagery. For example, the likelihood of cloud cover may force the system designer to opt for daily AVHRR imagery rather than use Landsat at 16-day intervals. The differences in resolution (1.1 km versus 30 m spatial resolution) between the two will affect the information that can be gained from the data. Another alternative would be to develop a system that integrates the two, affecting the complexity and the costs of the program.

1.4 Structure of the book

The major objective of this text is to impart the knowledge, skills and attitudes necessary to design, develop, evaluate and implement operational RMISs or components of such systems. To achieve this objective, it is necessary to impart knowledge on and skills in the tools and techniques available so as to use properly and to gain the most from those tools and techniques. It is also important to impart an understanding of the need to appreciate the human and natural environments in which the RMIS will work, as these environments will affect the design and use of the implemented RMIS. This text cannot provide the information necessary on all of these human and physical environments as this task is well addressed in other disciplines, but will address the need for this information. This text will cover the principles of and methods used in the design and implementation of an RMIS. Finally it is important for the reader to appreciate how this information can be used in the actual management of resources.

Whilst this text cannot address the many other disciplines concerned with the physical environment, it is assumed that students using this text have an adequate appreciation of the elements of these environments that might influence their implementation or use of an RMIS. All students using this text must expect to apply their knowledge of these other disciplines during the course of their study. The text will however, deal with the three facets of RMIS in detail namely – tools and techniques, principles of design and use and applications.

These scientific fields of endeavour are quite new. Many of the basic paradigms or rules that control the use or application of the tools and techniques are still being developed and evaluated. As the process of development continues, so existing tools and techniques will evolve, and others will be developed. In consequence, all students of this science must accept that they cannot just learn about RMISs and leave it at that. They will need to continue to learn and explore as these systems evolve.

The newness of this science also means that all working hypotheses and conventional wisdoms should be treated with considerable scepticism as many of them will be either extensively modified or replaced as the science develops. Whilst similar statements can be made about all sciences, it is important to emphasize them in this case because the immaturity of the science leaves it particularly vulnerable to errors of judgement. It is also important to recognize that this science, by bridging across from the physical to the mathematical and biological sciences in quite integrated ways, creates opportunities for both growth and errors of judgement.

The author is well aware of the pitfalls and risks associated with the conduct of research and development of this science, having participated in remote sensing and geographic information system research, development and implementation since 1973. Everything in this text is given in good faith and the belief that it is the best information and advice currently available. Nevertheless some of the material may prove to be wrong in the fullness of time, as our knowledge and experience in using this science evolves. It is therefore important that all readers of this text do so with care, caution and I would ask with constructive criticism in mind. I welcome constructive comments and discussion with readers on the contents and presentation of this text.

Having said all of this, I wish all readers a rewarding study of the scientific disciplines of remote sensing, geographic information systems and modelling and hope that this text contributes to improving your understanding, skills and competence in developing and using Resource Management Information Systems.

2
The physical principles of remote sensing

2.1 Introduction

Remote sensing is defined as the acquisition of data using a remotely located sensing device, and the extraction of information from that data. The eye and the brain could be considered as a remote sensing system: light entering the eye is converted by the retina into electrical signals that are transmitted to the brain; on receipt of these signals the brain conducts a process of analysis, or signal processing, to extract information from the data, in this case about the world about us. In practice, the term remote sensing is restricted to data acquired by man-made sensing devices, such as a camera and to electromagnetic radiation, such as light. The acquisition of photography using a remotely located camera, and then the interpretation of that data to extract information by viewing the photographs, is the simplest form of remote sensing.

The definition of remote sensing shows that there are two main subdivisions: **data acquisition** and **information extraction**. The theoretical basis for data acquisition lies primarily within the domain of the physical and mathematical sciences whereas information extraction is based primarily on the mathematical, statistical, biological and psychological sciences. However, the nature of the data used in that analysis depends upon the scientific basis and technological characteristics of the sensor and acquisition conditions.

A user, requiring the acquisition of specific data, is most likely to arrange for a specialist group to acquire the data due to the specialized, highly technical and capital intensive nature of the work. These groups have the expertise to provide advice on the acquisition of data. However, the characteristics of the imagery must be thoroughly understood if it is to provide accurate, reliable and economical information. The user must therefore have a good understanding of the characteristics of remotely sensed imagery, but not a detailed understanding of the technical aspects of acquisition itself. There are exceptions to this position, specifically in relation to those aspects of data acquisition that the user has some control over, including films, filters and flight planning. These aspects will be dealt with in more detail than the remainder of data acquisition.

Diagram 2.1 illustrates the progression from energy source through data acquisition, information extraction to management decision. All of the components depicted need to be discussed and understood if the student is to be effective in the design and implementation of a remote sensing project, so all are covered in later chapters.

The aim of any operational remote sensing program is to provide information as an input to management information systems. The nature of the required information will specify the constraints in establishing the remote sensing program. It follows that proper design of a program is dependant upon accurately specifying the information to be provided by the program. The information required by management can often be acquired by other means, but remote sensing is selected when this method provides either better, more cost effective, or new but necessary information. The user is interested in determining whether remote sensing is the best available method of getting the information that he requires and he can do this by asking the following three questions.

1. Does the information that is required about the physical environment change with spatial position in that environment?

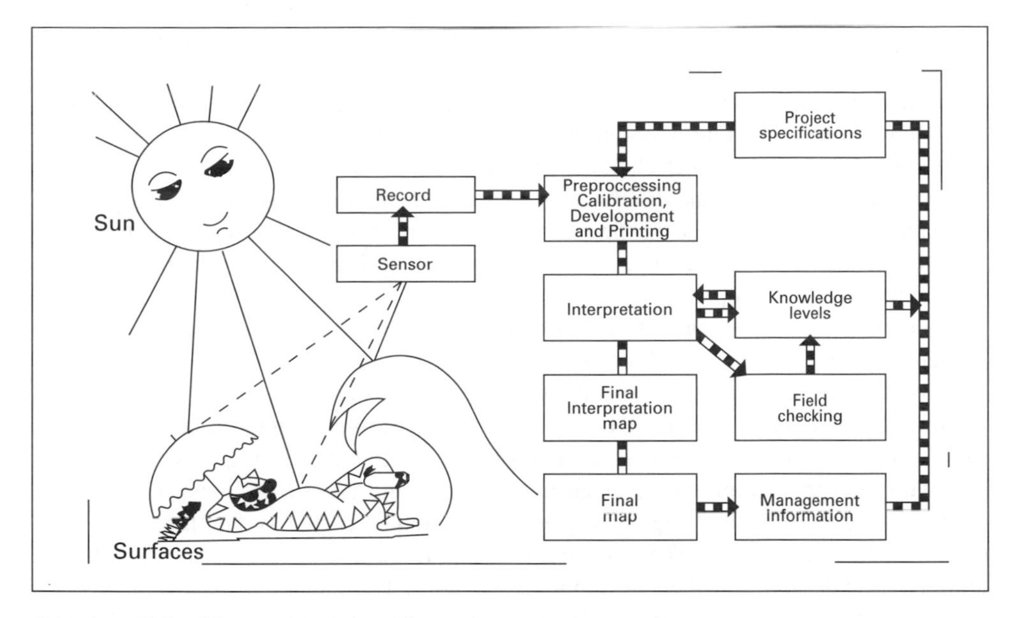

Diagram 2.1 The components of a remote sensing project.

2. Is the information required dependent on time?
3. Does the collection of the required information either affect the resource being monitored, cause undue time delays, or create unacceptable administrative or cost structures to be created for its collection?

If the answer to question 1 is 'yes', and particularly if this spatial location is important then remote sensing might be the best way of obtaining the required information. There are other methods of acquiring this sort of information, but they all depend upon point observations by human observers, with interpolation between those point observations to create a continuous map of the information type. They suffer a tyranny that high accuracy requires a dense network of points, achieved at great expense, but the greater difficulty is the time required to create the map under many conditions where the information is actually changing whilst the map is being made. Remotely sensed data provides a permanent record of the whole of the area of interest at a point in time, so that the extraction of information does not involve interpolation between points (but does involve other types of interpolation and extrapolation that will be discussed later). Information derived from remotely sensed data is also consistent across the image area at one instant in time.

Remotely sensed data has the advantage that the information derived is temporally and spatially consistent. Many types of important information in the management of renewable resources do vary with time, and therefore require this consistency. Just as important though, the information may need to be collected at a regular interval to see how it changes with time. Remote sensing, by providing a permanent record at intervals of time, means that not only can the information be derived at each image time, but that the different information sets can be compared much more precisely, and on a point by point basis, than can be done when the information is derived by alternative methods. With point observations by human observers it is unlikely that the point observations will be at the same locations, and there is a degree of error associated with the interpolation between the points that precludes accurate comparison. These criticisms of alternatives to remote sensing on technical grounds are quite separate to the very great cost that would be associated with repetitive mapping of an area using observers, a cost that has prohibited the collection of this sort of information on renewable resources up to now.

Remote sensing is thus a tool in the armoury of the manager of renewable resources, to be used when its particular characteristics make it suitable to the task at hand. It has its own set of advantages and disadvantages relative to other methods of collecting spatial information. However, it is the first time a tool with the characteristics particular to remote sensing has been available; it will take a while to learn to understand and exploit these characteristics properly.

The historical development of the science of remote sensing starts with the development of photography and **photo-interpretation**. It was not until more sophisticated sensors that could sense in wavebands other than the visible, and digital analysis techniques were developed that the activity, known as photo-interpretation, became too big to be accurately described by that term and the name remote sensing was coined.

The first photographs were taken by Daguerre and Niepce in about 1822. These photographs, taken by the Daguerreotype process, required clumsy facilities. The cameras were large and bulky. The emulsion film had to be laid on glass plates just before taking the exposure, and then developed soon after. Despite these disadvantages, the French demonstrated the potential of the tool in topographic mapping. Laussedat took extensive terrestrial photographs for mapping purposes, demonstrating that the concept of stereoscopy works with photography, significantly increasing the value of the photographs for mapping. Tournachon demonstrated that photographs could be taken from a balloon in 1852.

The next important step was the development of gelatin with light-sensitive silver halide salts as the means of recording an image. This process meant that development did not have to occur immediately the photograph was taken, and was much simpler and easier to handle than the Daguerreotype process cameras. Many photographs were taken from balloons and kites once the technique was developed. However, it was not until the aircraft was developed that aerial photography could really take off as this was the only stable platform that would allow acquisition of photographs of large areas. The first aerial photographs were taken by Wilbur Wright in 1909.

Hauron developed the first colour separation film in 1895. From this, and the work of others, evolved Kodachrome in 1935. Colour film was first used from an aircraft in 1935, but it took a number of years to develop reliable methods of acquiring colour aerial photography, due to haze, vibration, and the need for good navigation facilities.

The interpretation of aerial photographs was first appreciated by the military in World War I. The skills acquired then were applied and extended for topographic and geologic mapping between the two world wars, with the art of photo-interpretation reaching full maturity in World War II. Photo-interpretation has continued to be used extensively, and extended to many other applications since.

The idea of using **thermal sensing** to detect men and machines was first proposed during World War II. Devices to detect thermal radiation gradually led to the development of scanners sensing in the visible and infra-red wavebands. These instruments had evolved to a reasonably satisfactory level by the 1950s, and received considerable impetus after this time because of the use of scanners in satellites, particularly the TIROS and NIMBUS meteorological satellites. Unlike photography, the data acquired by a scanner can be readily converted into a signal suitable for radio transmission, much more convenient for obtaining data from a satellite than canisters containing film. Satellite-borne scanners were first used for earth resource related purposes with the launch of Earth Resources Technology Satellite (ERTS) 1 in 1972. ERTS-1, renamed Landsat 1, was the

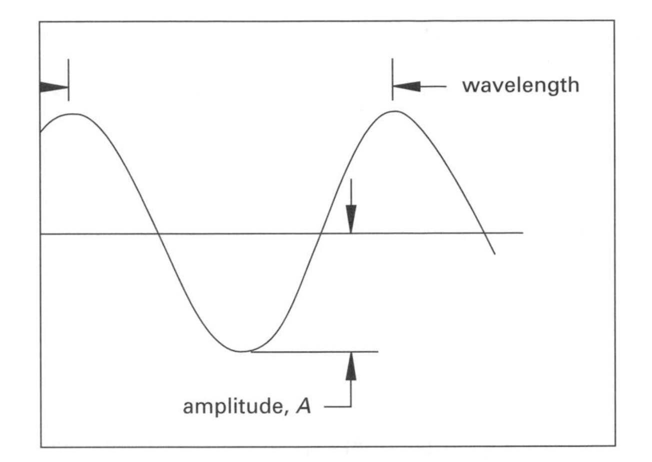

Diagram 2.2 The sine wave and its parameters of wavelength, λ and amplitude, A.

first of a series of experimental earth monitoring satellites launched by the US. Up until 1989 five Landsat satellites had been launched, with some variation in sensors on the different satellites. Landsat 1 contained a successfully used **multi spectral scanner (MSS)** with 80 m resolution data and four video cameras that proved to be unsuccessful. The video cameras were not included in the payloads of Landsats 2 and 3. Landsats 4 and 5 included a **thematic mapper (TM)** scanner with 30 m resolution as well as the MSS at 80 m resolution. In 1986, France launched SPOT One, a new scanner design that contained fewer moving parts than the Landsat scanners. SPOT One could also be pointed off track, so that it represented a significant development on the Landsat technology.

The first sensors to be developed that depended on their own energy source to illuminate the target surface was RADAR (radio detection and ranging) developed in the UK during World War II. The first imaging radars were developed in the US in the 1950s. These imaging radars were sidelooking airborne radar (SLAR) using real aperture antennae. A major difficulty of these radars was the long antennae required to achieve a reasonable resolution at the wavelengths being sensed. This problem was solved with the development of synthetic aperture radar in the 1960s in the US. The first satellite-borne radar accessible to the civilian community was Seasat A in 1976.

2.2 Electromagnetic radiation

2.2.1 The nature of electromagnetic radiation

Electromagnetic radiation can be considered to be transmitted through space as an oscillating sine wave of the form (Diagram 2.2). The parameters that define the shape of a sine wave are its wavelength, λ, (the distance between identical points on adjacent symmetrical wave segments) and its amplitude, A (the magnitude of the oscillation from the mean value).

The frequency of electromagnetic radiation is the number of oscillations or wave segments that pass a point in a given time. All electromagnetic radiation travels at nearly the speed of light, 300 000 000 m s^{-1} at sea level, so that

$$f_c \times \lambda = c = 300\,000\,000$$

i.e. the frequency of radiation (f_c) is inversely proportional to the wavelength (λ).

Table 2.1 Metric units of distance

Unit name	Abbreviation	Relationship to metre
Angstrom	Å	1Å = 10^{-10} m
nanometre	nm	1 nm = 10^{-9} m
micrometre	µm	1 µm = 10^{-6} m
millimetre	mm	1 mm = 10^{-3} m
centimetre	cm	1 cm = 10^{-2} m
kilometre	km	1 km = 10^{3} m

Electromagnetic radiation varies from very short to very long wavelengths. Only a subset of these are of use in remote sensing because absorption or scattering makes the atmosphere opaque to the others (section 2.2.4). The metric units of measurement used to define wavelength are given in Table 2.1.

All physical bodies, at temperatures above absolute zero (0 K or –273°C) radiate electromagnetic radiation. If the body is a perfect absorber and emitter of radiation then it is called a blackbody; as a perfect absorber it will look absolutely black to the human eye. The energy (in watts m^{-2} Å^{-1}) that it emits obeys Planck's Law:

$$E_\lambda = \frac{2\pi hc^2}{\lambda^5\{\exp(hc/\lambda kt) - 1\}}$$

where

h = 6.626196 × 10^{-34} (Planck's constant, in J sec)
c = 300 000 000 (speed of light in a vacuum in m s^{-1})
k = 1.38 × 10^{-23} J deg^{-1} (Boltzmann constant)
λ = wavelength metres
T = temperature in Kelvin

Typical energy radiation curves, for blackbodies of various temperatures (Diagram 2.3), show the variation in radiated energy with wavelength. The curves depicted illustrate two characteristics of the distribution of energy radiated from a surface.

1. As the temperature of the source increases, so the wavelength of the peak level of radiated energy shortens. Thus surfaces at a temperature of 300 K will have peak radiation levels at longer wavelengths than surfaces at temperatures of 6000 K. To the human eye, as the temperature of a source increases, so the colour changes from dark to bright red, yellow, white then blue, where these colours match changes to shorter wavelengths with higher energy levels.
2. As the temperature of a source increases, so the amount of radiated energy increases. The total amount of energy radiated from a surface is given by integrating the area under the energy curve for that surface. This area is larger for the surface radiating at 6000 K than it is for the surface at temperature 300 K.

The wavelength of the peak energy radiated by a body is given by Wein's Displacement Law:

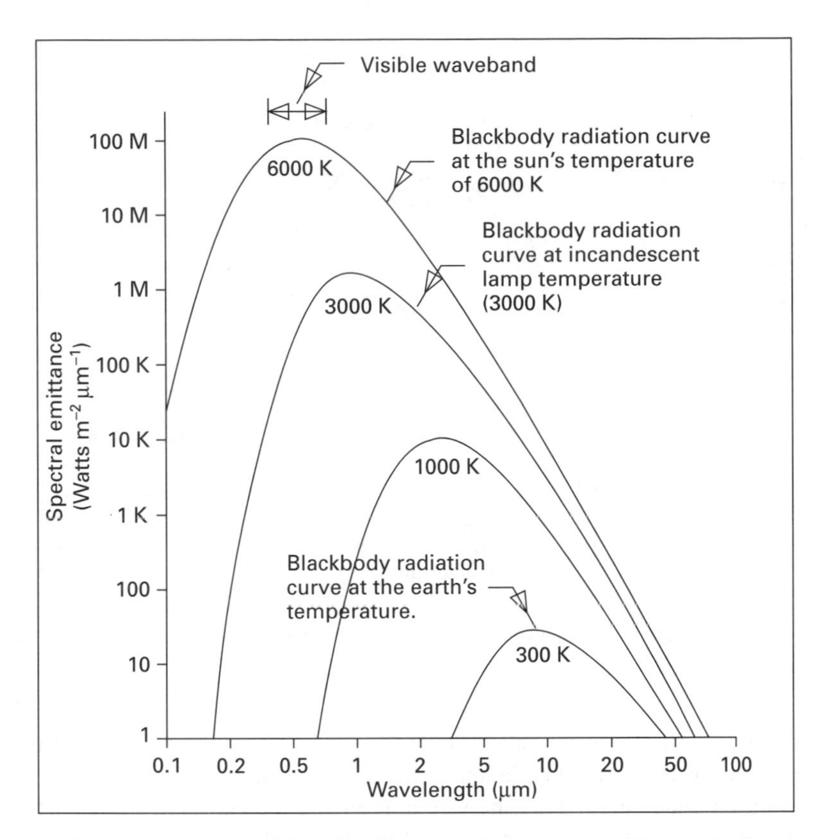

Diagram 2.3 Energy emitted by blackbody surfaces at the different body temperatures of 300 K, 1000 K, 3000 K and 6000 K.

$$\lambda_p = \frac{a}{T} \tag{2.1}$$

where

λ_p = peak wavelength (in mm)
a = constant (= 2.898 mm K)

The equations for E_λ and λ_p (Diagram 2.4) mean that the energy emitted by a surface rises rapidly to a peak, and then tails off much more slowly as the wavelength increases.

Most surfaces are not perfect emitters or radiators of energy, i.e. they are not perfect black bodies. The efficiency of a surface as an emitter or radiator is termed the emissivity of the surface:

$$E_\lambda = e_\lambda \times E_{b,\lambda}$$

where

e_λ = emissivity at wavelength λ
E_λ = energy emitted by surface at wavelength λ
$E_{b,\lambda}$ = energy emitted by blackbody at wavelength λ

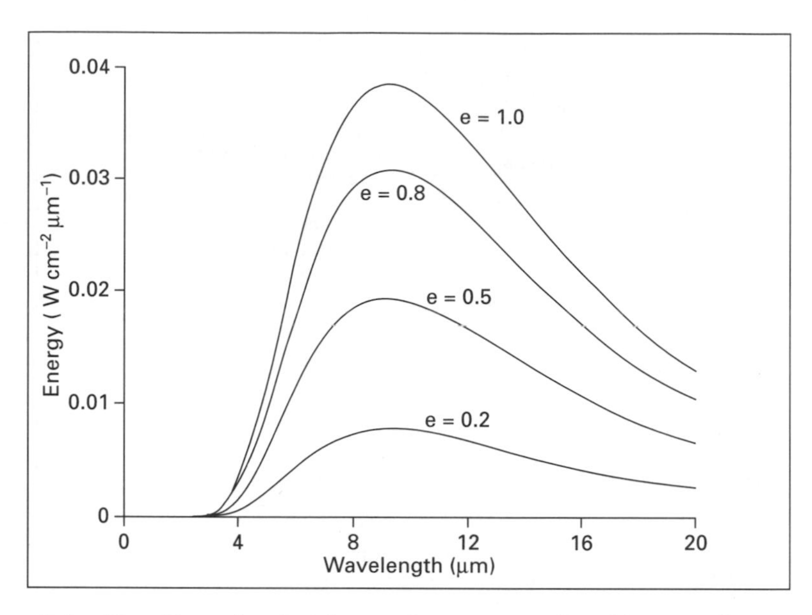

Diagram 2.4 The effect of emissivity on the energy radiated by a surface, at a temperature of T = 313 K.

All surfaces with $e_\lambda < 1$ are called greybodies. The effect of variations in the emissivity of a surface on the emitted radiation is shown in Diagram 2.4, in which e_λ is assumed to be constant for each curve. This shows that reducing the emissivity of a surface does not change the wavelength of peak radiation from that surface, or the wavelength distribution of energy from the surface; what it does is reduce the amount of energy radiated at each wavelength.

Changes in the temperature of a surface affect both the wavelength of peak radiation, as per Wein's Displacement Law, as well as the amount of radiation at each wavelength. A variation in emissivity affects the amount of energy emitted, but does not affect the wavelength of peak radiation. If the energy being radiated from a surface is only recorded in the one waveband, then the actual energy levels received at the sensor will be affected by both the temperature and emissivity of the surface.

2.2.2 Radiometric terms and definitions

Radiometric energy, Q, is a measure of the capacity of radiation to do work, e.g. to heat a surface, move an object or cause a change in the state of the object. The unit of measurement of radiant energy is the joule (J) or kilowatt-hour (kW hr).

Radiant flux, O, is the rate of application of energy per unit time. It can be considered to be like the flow of energy past a point per unit time, or conceptually similar to the flow of water in an ocean current past a point in unit time. **Radiant flux density**, E or M, is the radiant flux that flows through unit area of a surface. In the previous analogy of the ocean, it is the flow of water per unit cross-sectional area. Just as the flow of water will change from place to place, so the radiant flux density can change from surface element to surface element. Radiant flux density incident on a surface is called the **irradiance**, E, and the flux density that is emitted by a surface is called the **exitance**, M. Both E and M are measured in units of W m^{-2}.

Radiant intensity, I, is the radiant flux per unit solid angle leaving a point source (Diagram 2.5). The radiant energy from a point source, being radiated in all directions,

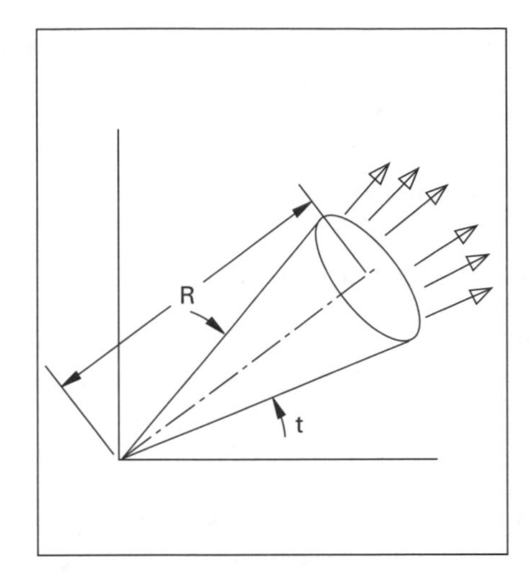

Diagram 2.5 Radiant intensity from a point source.

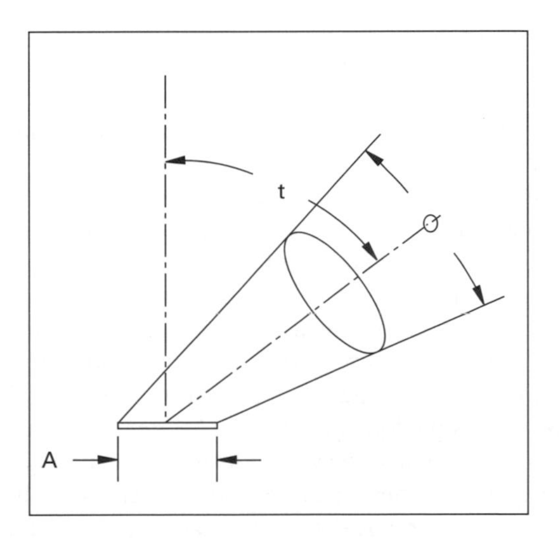

Diagram 2.6 Radiance into an extended source area, A.

has a radiant flux density that decreases as the surface moves further away from the source. The intensity, however, does not decrease with increasing distance from the source as it is a function of the cone angle.

Usually the point source is being observed by a sensing device with a fixed aperture area, A (Diagram 2.6). As the distance between the point source and the aperture A increases, (i.e. R increases in Diagram 2.5), so the solid angle decreases and the flux passing through the aperture is reduced. The decrease is proportional to the square of the distance from the source, since the area of the aperture is a function of the square of the radius of the aperture. **Radiance**, L, is the radiant flux per unit solid angle leaving an extended source in a given direction per unit projected source area in that direction. With an extended source being observed by a sensor with a cone angle of O, as R increases so the area on the extended source will increase so that the radiance incident on the detector area A will be constant, for constant t, as long as the flux density is constant across the extended source.

All of the terms defined above are wavelength dependent, and this can be indicated by the use of the term 'spectral' in front of the term, the term then applying to a unit of wavelength such as a nanometer or micrometer, e.g.:

$$\text{spectral radiance} = L \text{ W m}^{-2} \text{ nm}^{-1}$$

2.2.3 Energy radiated by the Sun and the Earth

The energy distributed by the Sun has a spectral distribution similar to that of a blackbody at a temperature of 6000 K. The temperature of the Earth's surface varies, but is often at about 30°C or 300 K. The peak radiance from the Sun is at about 520 nm, coinciding with the part of the spectrum visible to the human eye as green light. The human eye can detect radiation from about 390 nm, or blue in colour, through the green part of the spectrum at about 520 nm, to the red part of the spectrum up to about 700 nm. Thus the human eye is designed to exploit the peak radiation wavelengths that are radiated by the Sun.

2.2.4 Effects of the atmosphere

The atmosphere is transparent to electromagnetic radiation at some wavelengths, opaque at others, and partially transparent at most. This is due to scattering of the radiation by atmospheric molecules or particulate matter, absorption of energy by the constituents of the atmosphere, often as molecular resonance absorption, and emission of radiation by matter in the atmosphere where the contribution of each depends upon atmosphere conditions and the wavelength being considered.

The effect of this absorption is to reduce the amount of energy that penetrates the earth's atmosphere to the surface, and then is reflected back to a sensor. The wavelength regions for which the atmosphere is relatively transparent are called windows, the main windows of interest in remote sensing being shown in Diagram 2.7. Scattering of electromagnetic radiation occurs predominantly at the shorter wavelengths due to molecular or Rayleigh scattering, and at longer wavelengths due to scattering by particulate matter in the atmosphere. Rayleigh scattering causes the characteristic blueness of the sky.

The effects of scattering and absorption are to attenuate, or reduce, the direct energy that will be incident on a surface by the transmissivity of the atmosphere, t_λ, and to add a constant radiance level, $E_{A,\lambda}$ due to scattering from the atmosphere. In the normal practice of acquiring imagery under cloud-free conditions, the contribution of the atmosphere is small compared to the attenuated contribution from the Sun, for the windows that are normally used.

$$E_{i,\lambda} = t_\lambda \times E_{s,\lambda} \times \cos\theta_s + E_{A,\lambda} \qquad (2.2)$$

where

t_λ = average atmospheric spectral transmittance for the ray path through the atmosphere
$E_{s,\lambda}$ = solar radiance at the top of the atmosphere
θ_s = solar zenith angle
$E_{A,\lambda}$ = atmospheric spectral radiance
λ = wavelength (in nm)

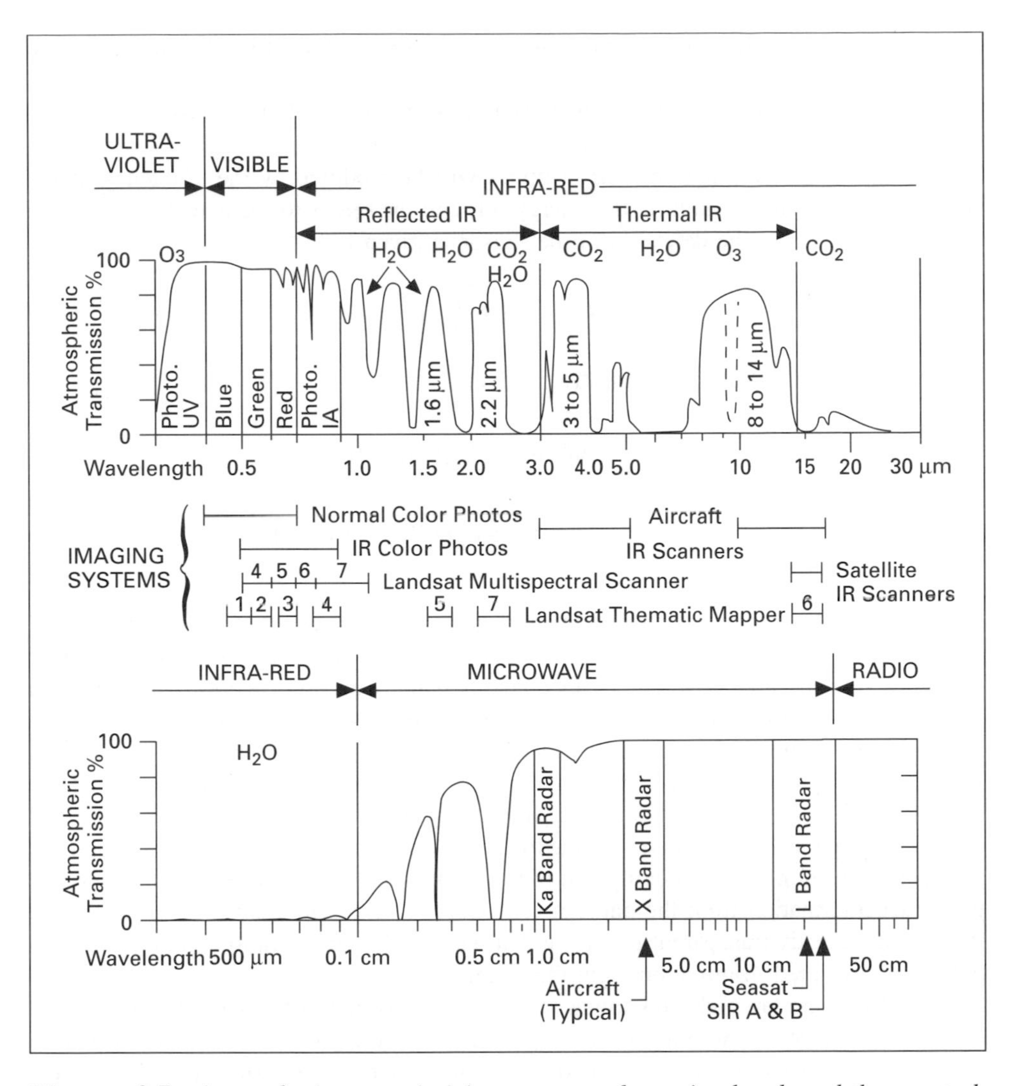

Diagram 2.7 Atmospheric transmissivity, common absorption bands and the spectral ranges of selected remote sensing systems, from Sabins (1987).

If the spectral reflectance, r_λ, at the surface is defined as the ratio of the reflected to incident energy, then the spectral radiance, L_λ, reflected from the surface due to the incident irradiance, $E_{i,\lambda}$ is given by

$$L_\lambda = \frac{r_\lambda E_{i,\lambda}}{\pi} \tag{2.3}$$

The radiance from the surface is further attenuated during transmission to the sensor due to atmospheric transmissivity, and gains an additive component due to atmospheric scattering:

$$E_{O,\lambda} = t_{\lambda'} \times L_\lambda + E_{A,\lambda'} \tag{2.4}$$

where

$t_{\lambda'}$ = average atmospheric spectral transmittance for ray path from the surface to the sensor
$E_{A,\lambda'}$ = atmospheric radiance incident on sensor from the direction of surface

The energy, $E_{o,\lambda}$ that enters the sensor optics will be modified by the efficiency of the sensor optics, e_λ, in transmitting the energy through the optics to the detector units. The energy that activates the detectors in the sensor is given by

$$E_{d,\lambda} = e_\lambda \times E_{O,\lambda} \tag{2.5}$$

Substitute (2.2), (2.3) and (2.4) into (2.5) gives

$$E_{d,\lambda} = e_\lambda\{t_{\lambda'}[r_\lambda(t_\lambda E_{S,\lambda}\cos\theta_S + E_{A,\lambda})/\pi] + E_{A,\lambda'}$$

$$= \frac{e_\lambda}{\pi}\left\{r_\lambda t_\lambda t_{\lambda'} E_{s,\lambda}\cos\theta_s + r_\lambda t_{\lambda'} E_{A,\lambda} + \pi E_{A,\lambda'}\right\} \tag{2.6}$$

Atmospheric conditions change from place to place as well as from day to day. In consequence, atmospheric conditions will change from image to image, and indeed will change across the area covered by an image. Changes in atmospheric conditions can cause changes in atmospheric transmissivity and atmospheric path radiance or skylight. Increasing levels of atmospheric path radiance are normally accompanied by a reduction in transmissivity, but molecular absorption means that reduced transmissivity need not be accompanied by increased atmospheric path radiance.

Lowering transmissivity and raising the atmospheric path radiance cause an increase in the noise level in the data compared to the signal value. This makes it more difficult to extract information from the data. In consequence, it is important to use wavebands that are sufficiently transparent to ensure that the signal-to-noise ratio is high enough to extract the required information from the acquired data.

In (2.6) there are eight unknowns, of which one is due to the sensor (e_λ), four are due to either atmospheric transmissivity ($t_{\lambda'}$, t_λ) or radiance ($E_{A,\lambda'}$, $E_{A,\lambda}$), two are due to the incident solar radiation ($E_{s,\lambda}$, θ_s) and only one is due to the surface (r_λ). In practice the equation is often simplified as follows.

1. Replacing radiance at the top of the atmosphere with incident irradiance at the surface as measured in the field, and substituting (2.3) and (2.4) into (2.5) to give

$$E_{d,\lambda} = e_\lambda\{t_{\lambda'} r_\lambda E_{i,\lambda}/ \quad + E_{A,\lambda'}\} \tag{2.7}$$

2. Assuming perfectly efficient sensor optics, that is make $e_\lambda = 1$. The resulting equation is then

$$E_{d,\lambda} = \{t_{\lambda'} r_\lambda E_{i,\lambda} + \quad E_{A,\lambda'}\}/\pi$$

so that

$$r_\lambda = \frac{\pi}{t_{\lambda'} E_{i,\lambda}}\left\{E_{d,\lambda} - E_{A,\lambda'}\right\} \tag{2.8}$$

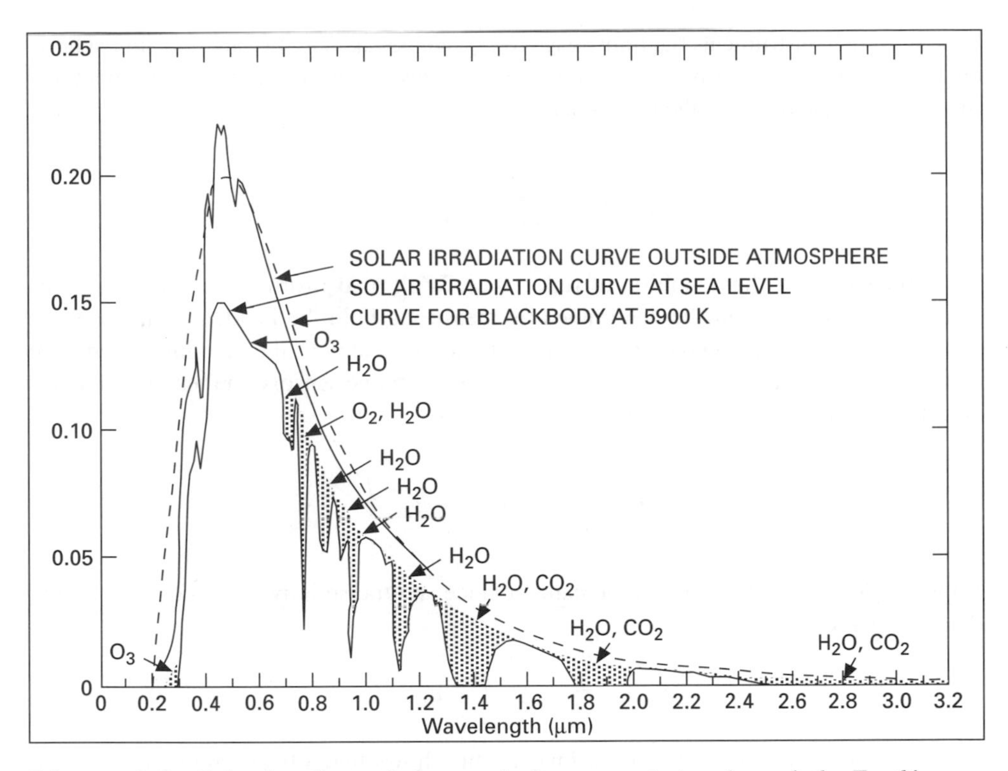

Diagram 2.8 Solar irradiance before and after transmission through the Earth's atmosphere, with atmospheric absorption, for the Sun at the zenith, from Lintz and Simonett (1976).

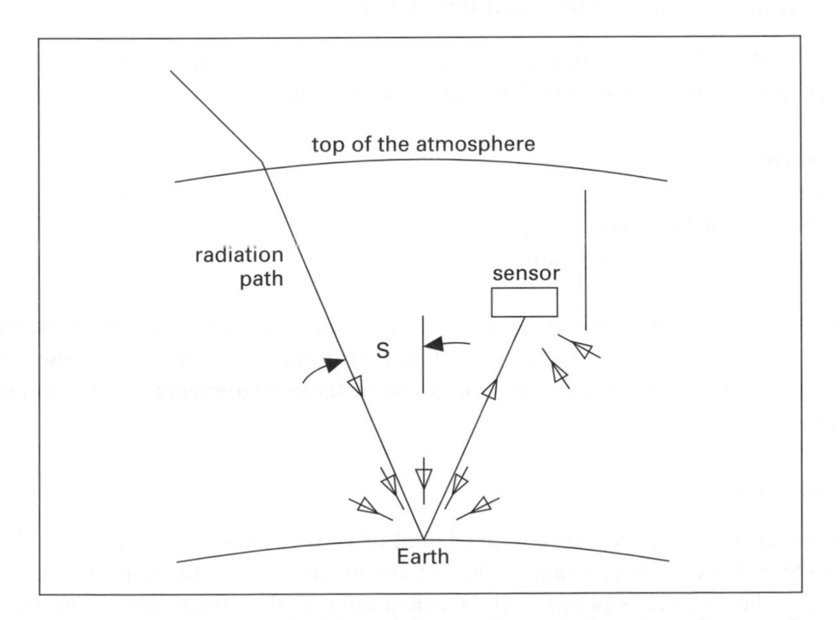

Diagram 2.9 Transmission of solar irradiance through the atmosphere to the sensor.

With field instruments the transmissivity, $t_{\lambda'}$ can be taken as 1.0, and the atmospheric component, $E_{A,\lambda'}$ as zero because of the short distance between the field instrument and the target. Equation (2.8) then becomes

$$r_{\lambda} = \frac{\pi E_{d,\lambda}}{E_{i,\lambda}} = \frac{L_{d,\lambda}}{E_{i,\lambda}}$$

in accordance with the definition of reflectance. With field instruments, the reflectance from the surface can thus be determined by measuring both the incident irradiance, the reflected radiance and computing the ratio of the two, as long as the field instruments are calibrated so as to allow for e_{λ}. For airborne and spaceborne sensors, the sensor efficiency is normally taken as unity. Equation (2.7) becomes

$$r_{\lambda} = \frac{\pi}{t_{\lambda'} E_{i,\lambda}} \left\{ E_{d,\lambda} - E_{A,\lambda'} \right\}$$

If the incident radiance, atmospheric path radiance and transmissivity are constant during data acquisition, then

$$r_{\lambda} = c_1 E_{d,\lambda} + c_2$$

where c_1 and c_2 are constants for the image. This shows that a linear model is a simplification of the relationship between reflectance, incident irradiance and radiance incident on a sensor.

2.2.5 Measurement of radiance and irradiance

Radiance and irradiance are measured using instruments called **spectrometers** (Diagram 2.10). A spectrometer consists of four main components:

1. the collecting optics;
2. filter unit;
3. detector or detectors; and
4. the recording device or display.

Versions of these instruments are also called spectroradiometers or spectrophotometers, the latter usually being restricted either to that part of the spectrum that is visible to the human eye, or to reflected solar radiation. Some of these instruments contain an internal calibration source.

Collecting optics

The collecting optics define a cone-shaped **field of view** (FOV). For all spectrometers the solid angle of this cone is constant at the instrument, so that the area sensed on the target depends on the distance between, and the orientation of, the sensor relative to the target surface.

The collecting optics collect the energy within the instruments FOV and direct this energy through the filter unit and onto the detectors in the sensor. The energy incident on the detectors is recorded or sampled at specific times, the area actually sensed at that

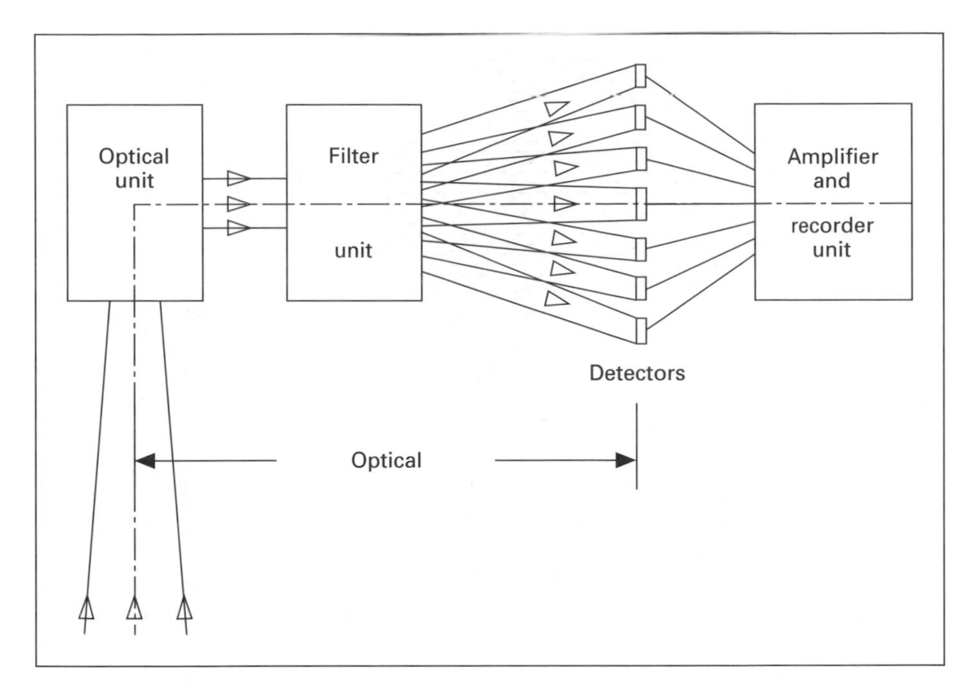

Diagram 2.10 Schematic diagram of the construction of a spectrometer, consisting of an optical system, filter unit, detectors and recorder.

instant being called the instrument's **instantaneous field of view** (IFOV). The instrument IFOV is thus the actual area sensed at the time that the sensor was sampled.

Filter unit

This unit filters the required wavebands for projection onto the detector units. It can do this in a number of ways:

1. split the incoming beam into separate wavebands using a prism or similar filter unit;
2. split the incoming beam into a number of subbeams that are individually filtered using either gelatin or metal oxide filters;
3. filter the required waveband from the incident beam using a diffraction grating.

Gelatin filters A gelatin filter allows a specific set of wavebands to be transmitted through the filter, with different types of filters allowing different wavebands to be transmitted. Gelatin filters are relatively cheap, but they cannot provide narrow transmission spectra and hence cannot be used when the user wants to detect narrow wavebands. It is also often difficult to achieve a sharp cut-off in transmission on the edges of the sensed band (Diagram 2.11).

Metal oxide filters These can be used to provide either much narrower or sharper transmission curves than gelatin filters. They operate by combining the absorption characteristics of various metallic oxides to create the desired transmission spectra. The selected metallic oxides are sprayed as a fine coating onto a glass base and then welded to the base using heat. Metal oxide filters are very stable.

Gelatin or metal oxide filters have the advantage that they are relatively cheap, but they have the disadvantage that the incident beam has to be split before filtering; there

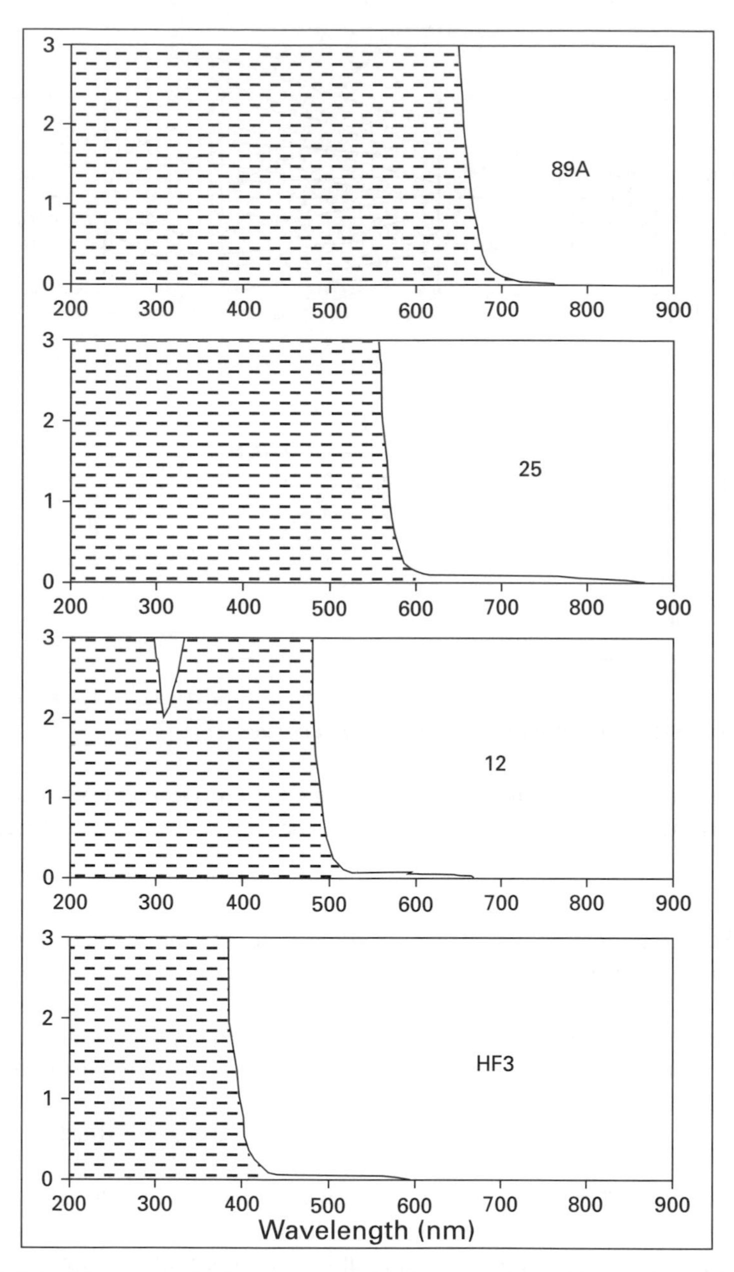

Diagram 2.11 The transmission spectra of selected gelatin filters, Haze Filter 3, 12 – yellow, 25 – red and 89A – deep red (from Kodak, with permission).

is less of the total energy available to create each waveband, and total amounts of energy are often a significant issue in sensor design.

Diffraction grating This allows light of a specific wavelength to be transmitted through the grating, where this wavelength depends on the grating interval and the angle of incidence of the light to the grating. A grating can be used in field instruments where there is time to change the orientation of the grating between measurements, but they are not usually suitable for sensors on moving platforms such as an aircraft or satellite.

Detectors

The detectors contain light sensitive chemicals that respond to incident electromagnetic radiation by converting that radiation into an electrical charge. The charge is taken, or sampled, by electronic circuitry at a regular interval, and reset to zero after sampling. The charge accumulates due to the incident radiation between each sampling. Each sample charge reading is then amplified and transmitted to the storage device. The magnitude of the charge is a function of the amount of incident radiation accumulated between samples, the sensitivity of the detector and the detector response speed. The latter of these three factors causes a lag between change in incident radiation and detector response. This lag will influence the values that are recorded from the detector.

The response of a detector is not linearly related to the incident radiation, so that it is usually necessary to calibrate the electrical signal before use. In addition, as detectors age, or change their temperature, so their response might change. It is therefore essential to calibrate instruments at regular intervals to ensure that the results being obtained are correct. The response of different detectors will be quite different in different parts of the electromagnetic spectrum so that different detectors are often used to record the energy levels that are incident in different wavebands.

Output device

The simplest output device is a meter attached to the output of the detectors, and which can be read by the operator. In many systems this output is sampled electronically, and recorded automatically onto a suitable recording device, such as a cassette tape, a disc unit or optical disc. This approach has significant benefits as it allows the more efficient collection of field data. Creation of a digital record also means that the raw data can be easily calibrated, allowing faster and more convenient analysis.

Spectrometers can be grouped into three main categories. The first category records the radiance or irradiance in set wavebands, usually matching the wavebands of a particular scanning instrument. This type are usually cheaper, they can be more robust and are usually simpler to use than the other types of spectrometer. Their disadvantage is that the results obtained are only of use for those specific wavebands sensed. The second category of spectrometer provides high resolution data over an extensive range of wavebands, using very precise and delicate optical and filter systems. Whilst they are usually more expensive and may be more difficult to use, they have the advantage that the acquired data can be integrated over any broader bands of interest to the analyst. They can thus provide data that can match existing or proposed wavebands in either current or potential sensing systems. They provide a more powerful data set than that collected using the first type of spectrometer.

The third category of spectrometers are very precise laboratory instruments that measure the reflectance of elements of the surface of interest, often for areas of 1 cm^2 or smaller. Often, these instruments are also capable of measuring the transmittance of the sample, thereby providing the analyst with information on the reflectance, absorptance and transmittance of the surface type. The biggest disadvantage of these instruments is that they provide information on the components of the surface, and not the reflectance of the surface itself. They are, however very useful in providing the reflectance (r_λ), absorptance (a_λ) and transmittance (t_λ) for the elements that make up the surface as is necessary for modelling of the reflectance from a surface.

The use of spectrometers for measuring the reflectance of surfaces in the field is discussed in Chapter 6 under the collection of field data.

2.3 Interaction of radiation with matter

2.3.1 The nature of reflectance

Solar energy is attenuated and scattered by the atmosphere and its aerosols prior to being intercepted by the ground surface or the vegetative canopy. At the surface, the energy is either absorbed, transmitted or reflected in accordance with the Law of the Conservation of Energy:

$$E_{i,\lambda} = E_{a,\lambda} + E_{r,\lambda} + E_{t,\lambda}$$

$$= E_{i,\lambda}(a_\lambda + r_\lambda + t_\lambda)$$

$$\Rightarrow \qquad a_\lambda + r_\lambda + t_\lambda = 1$$

where E represents the incident, absorbed, reflected and transmitted spectral energy and a, r and t are the spectral absorptance, reflectance and transmittance of the surface.

The transmitted energy is then available for further absorption, reflection or transmission within the surface. Some of this energy is eventually reflected out of the surface to contribute to the reflected component from the surface. The total amount of energy reflected from a surface is an integration of the myriad interactions that occur at and within the surface and within the sensor field of view. This energy is further attenuated and scattered by the atmosphere before entering the sensor optics.

Of these various factors that influence the final calibrated radiance values in the sensor, the impact of the surface on radiance is of most interest to users, because it is the component that contains information on the surface features. The other components are of concern to users because they interfere with our capacity to extract information from the data. Thus from (2.6), the parameter of most interest is r_λ, the surface reflectance.

Reflectance at surface interface

Radiation incident on a surface is either reflected, absorbed or transmitted at the interface of the surface. The nature of the reflection at that interface depends upon:

1. the angle of incidence of the energy;
2. surface roughness as a function of wavelength; and
3. The materials on either side of the interface, as they will affect the angle of refraction and the percentages that are reflected, absorbed and transmitted at the interface.

The smoother the surface, the more specular or mirror-like the reflectance is likely to be, i.e. the greater the proportion of the energy will be reflected in accordance with Snell's Law (which states that the angle or reflection equals the angle of incidence). The rougher the surface the more scattered in direction the reflection is likely to be. This reflectance is more lambertian in form, where lambertian reflectance is defined as reflectance that scatters or reflects the energy equally in all directions, independent of the angle of incidence of the incident energy (Diagram 2.12).

With specular surfaces, the closer the incident ray is to the normal from the surface, or the smaller the angle of incidence, the lower the level of reflectance that is likely to occur. As the angle of incidence increases so the reflectance is likely to increase, particularly when the ray becomes a grazing ray, or the angle of incidence approaches 90°. It is for this reason that water is more reflectant at large angles of incidence.

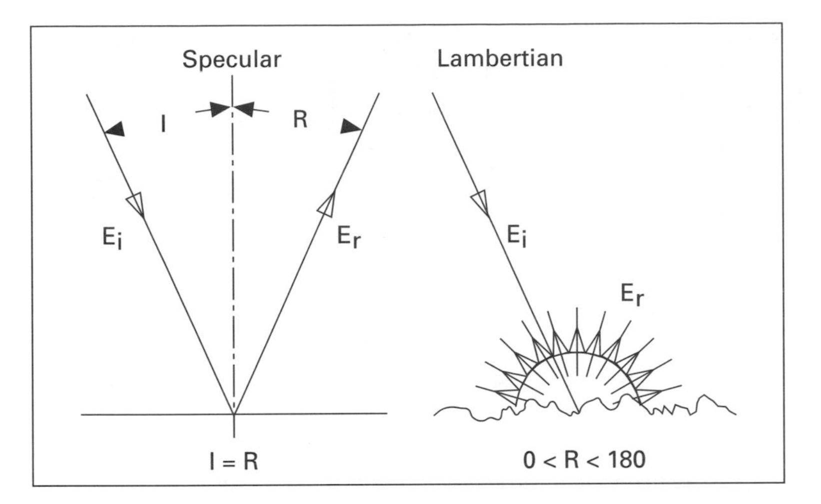

Diagram 2.12 Specular reflection in accordance with Snell's Law and lambertian reflectance.

The reflectance of specular surfaces is highly dependent on the geometric relationship between the source, usually the Sun, the surface elements and the sensor. As surfaces become more lambertian their reflectance becomes less dependant on this geometric relationship. Since few surfaces are perfectly lambertian reflectors, so few have reflectance characteristics that are completely independent of this geometric relationship. It will be seen that the closer surfaces are to the lambertian ideal, the better they are from the remote sensing perspective. Whatever the nature of the surfaces, their reflectance characteristics are an important consideration in remote sensing.

Reflectance within a surface

Energy transmitted into a surface at the interface is then available for further absorption, reflection or transmission within the surface. During transmission through the surface, the energy is selectively absorbed by the chemical constituents of the surface, and is reflected or scattered by surface elements within that surface. For example, energy transmitted into a green leaf will undergo selective absorption by the pigments in the leaf, such as the chlorophyll pigments and be reflected from the cell walls within the leaf (Photo 2.1).

Selective absorption means that the energy eventually emerging from the surface will be deficient in those wavelengths that have been selectively absorbed. The reflection from surface elements within the surface will tend to scatter the energy in all directions and so the reflected energy emerging from the surface will tend to be lambertian in form. The closeness of this reflectance to the lambertian ideal depends on the nature of the reflections within the surface. If these are from surface elements distributed equally at all orientations, then the resulting reflectance will be a close approximation of a lambertian surface, as is the case with many soil surfaces. If the internal surface elements have strong correlation with certain orientations then the resulting reflectance may not be a good fit to the lambertian ideal.

Reflection that occurs within a surface is called **body reflection**, in contrast to surface reflection that occurs at the surface of objects. These characteristics of body reflection, often leading to near lambertian reflectance, and selective absorption, are critical to the detection of surfaces in remote sensing, since remote sensing assumes the existence of these characteristics. Consider how remote sensing would operate if the majority of

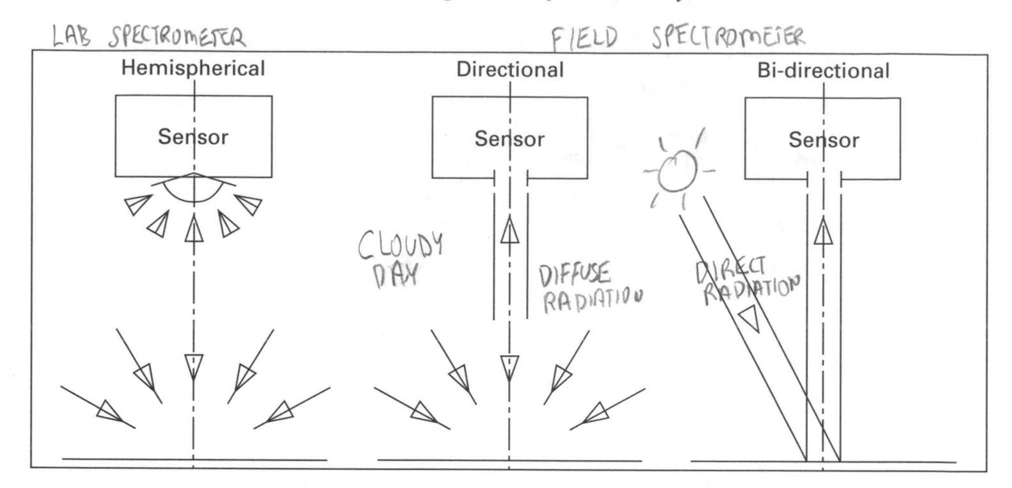

Diagram 2.13 Hemispherical, directional and bi-directional reflectance.

surfaces were specular in nature. Images of these surfaces would be predominantly black, when reflection is away from the sensor, or white when reflection is towards the sensor, exhibiting very little gradation between these limits. Those elements of the surface that were reflecting towards the sensor would change dramatically as the sensor moved, so that each image would appear quite different, and indeed sequential images, including sequential aerial photographs would be quite different to each other. If there was no selective absorption then reflection would have more contrast than with selective absorption, leading to images of maximum, or minimum contrast for all elements within the image area. The result would be images that could not be interpreted very easily, if at all, so that specular reflectance would mean that remote sensing would probably be of little use in resource management.

Both the reflected and incident radiation at a surface can be measured in three different ways (Diagram 2.13): bi-directional, directional and hemispherical reflectance.

Bi-directional reflectance If an observer is looking at a surface that is illuminated by torchlight then the source of energy comes from one direction, and the surface is being observed at another direction. The ratio of these two energies is called the bi-directional reflectance of the surface, being dependent on the two directions involved. Illumination of surfaces by the Sun, involving predominantly illumination from the one direction, and observation at a sensor is a very close approximation to bi-directional reflectance. Because bi-directional reflectance is dependant on the orientation of the source and the sensor relative to the surface, it will change as this geometry changes, if the surface is not a perfectly lambertian reflector. In consequence, the reflectance of some surfaces change to some degree, from image to image, due to this cause, when those surfaces are illuminated by the Sun. Most forms of active sensors, or sensors that have their own energy source to illuminate the target, use a point source and sense in one direction so that they represent perfect bi-directional reflectance.

Directional reflectance If the observer views the surface under overcast conditions when the main source of illumination is scattered skylight, then the illumination is approximately equal from all directions, but the sensor is in one direction relative to the surface. Reflectance of this form is called directional reflectance.

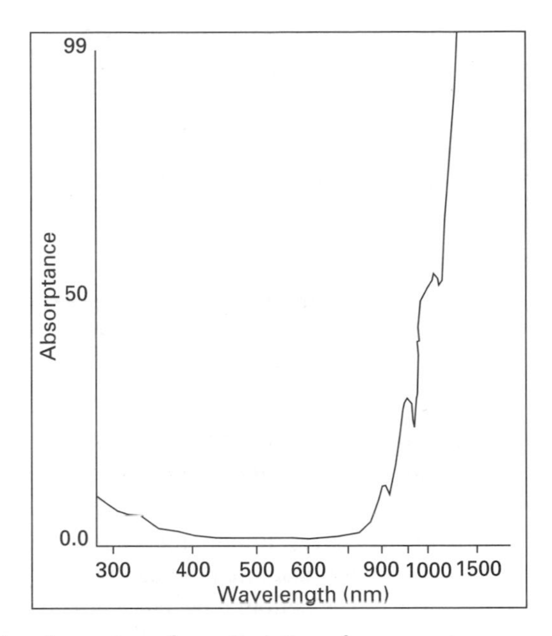

Diagram 2.14 The absorption characteristics of pure water.

Hemispherical reflectance Hemispherical reflectance occurs when both the incident and reflected energy is measured over the whole hemisphere. Hemispherical reflectance is of little concern in relation to sensors, but is of great interest in the modelling of light interactions within the surface and in assessing the impact of reflectance on aerial photography and other imagery that is taken with sensors that have a wide field of view. Hemispherical reflectance should be identical to bi-directional reflectance for lambertian surfaces, and provides an average reflectance value for specular surfaces. Comparison of hemispherical and bi-directional reflectance values can thus indicate how close surfaces are to the lambertian model.

Laboratory spectrometers usually measure hemispherical reflectance, and field spectrometers usually measure an approximation to bi-directional reflectance, or directional reflectance under cloudy conditions.

2.3.2 The reflectance of water surfaces

Energy incident on a water surface is primarily reflected from or transmitted into the surface, where the proportions of each depend upon the angle of incidence between the incident energy and the surface. The reflected energy is specular in character, usually as grazing rays from wave elements. Most of the energy is transmitted into the water surface where it is absorbed by the chemical constituents of the water body, scattered by particulate matter in the water or reflected from the floor of the water body. The absorptance of water increases with increasing wavelength from a minimum absorptance at about 420 nm (Diagram 2.14) until it is greater than 90 per cent in the reflected infra-red portion of the spectrum. Most scattering and reflection from water is therefore in the blue–green part of the spectrum, particularly for relatively pure water.

Both chlorophyll and suspended soil sediments significantly affect the reflectance of water, with both of them absorbing energy in accordance with their individual absorptance characteristics (Diagram 2.15). The effects of sediments depends on the reflectance characteristics of the sediment particles (section 2.3.3), the density of the particles in the

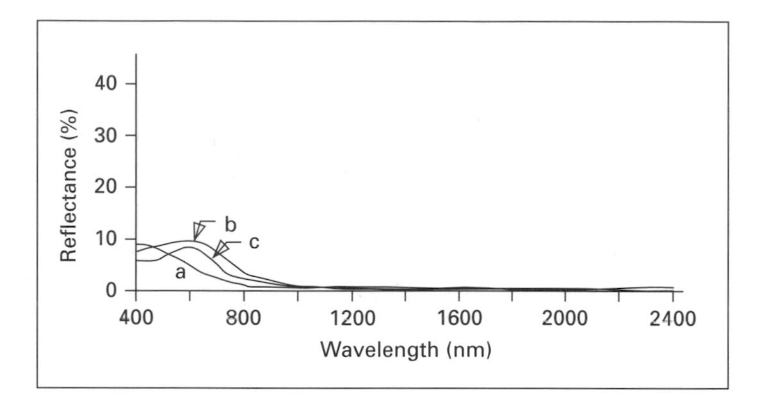

Diagram 2.15 Typical effects of chlorophyll and sediments on water reflectance: (a) Ocean water; (b) turbid water and (c) water with chlorophyll.

water, and the depth of the particulate layer. Bottom reflectance affects the reflectance of a water body when sufficient energy is transmitted through the body to be reflected from the floor and returned to the surface. The bottom reflectance affects the amount of energy reflected and hence available for transmission through the water body.

Patterns in water bodies are thus due primarily to variations in suspended sediment, chlorophyll concentrations, variations in depth to the floor and bottom reflectance. Because all of these factors can have similar effects on reflectance, it is often necessary to use auxiliary data, such as field visits and bathymetric data, to identify the actual cause in each situation.

2.3.3 The reflectance characteristics of soils

Soil reflectance occurs in the surface layers of the soil, with a significant proportion occurring as surface reflectance from the soil grains themselves. Several factors affect the reflectance of soils: chemical composition, humus content, surface soil moisture and surface roughness.

Chemical composition The chemical composition of the soil affects soil colour due to selective absorption. Different chemicals have different absorption bands and in consequence the existence of different chemicals in the soil will affect the selective absorption of energy during transmission through the soil. The most obvious effect of this type is the redness of soils with a high ferric iron content, due to selective absorption of blue–green radiation.

Humus content Humus, the organic products derived from the breakdown or decomposition of vegetable and animal matter, usually exhibits absorption at all wavelengths in the visible region, but with slightly higher absorption in the blue–green part of the spectrum due to the antho-cyanin compounds created by breakdown of chlorophyll in leaf elements. The level of humus in soils varies over time so that the reflectance of soils can also vary over time.

The decompositional state of plant litter significantly affects the reflectance curves of organic soils. Fully decomposed vegetative material gives a concave curve as illustrated in curve (c) of Diagram 2.16. Soil with a significant component of preserved fibres have a curve of type (b), or an organically affected reflectance curve. Soils with minimally decomposed litter exhibit a concave curve up to 750 nm and then quite high

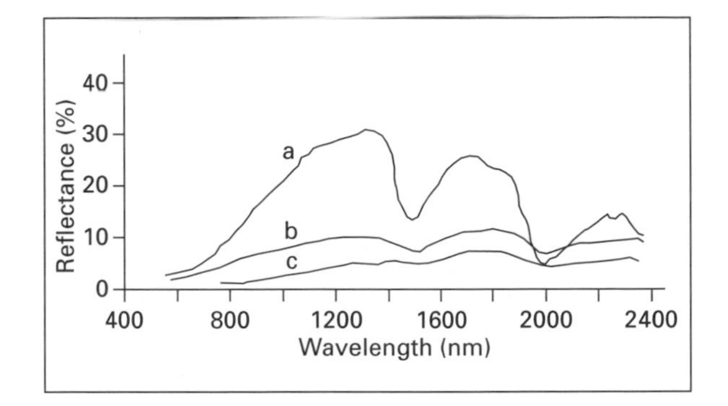

Diagram 2.16 Representative reflectance of typical soils with (a) minimally altered (fibric), (b) partially altered (ferric) and (c) fully decomposed (sapric) organic matter after Stoner and Baumgardner (1981).

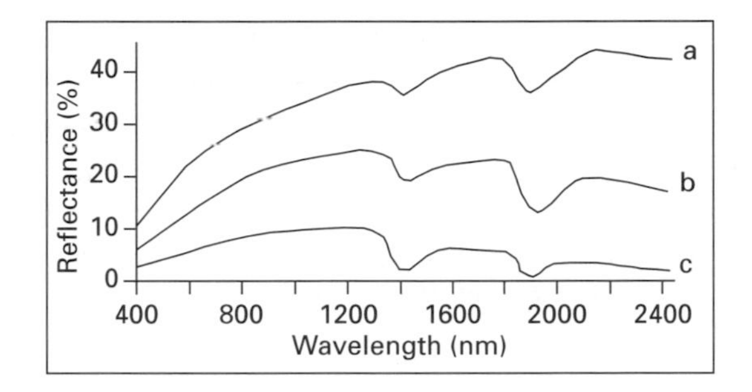

Diagram 2.17 Changes in reflectance spectra for organic soil with changes in moisture content: (a) 5 per cent (b) 20 per cent (c) 40 per cent.

reflectance beyond 900 nm. Curves of this type (type (a) in Diagram 2.16) are called **fibric curves** and are similar to the curves of standing dead or brown vegetation.

Surface soil moisture Water selectively absorbs at all wavelengths, but with increasing absorptance at longer wavelengths as previously discussed. Because most of the reflectance of soils occurs in the top layers of soil particles, only the water in these top layers will affect the reflectance of the soil surface. If these layers are moist then the water will have a strong influence on reflectance. However, this layer dries out quite rapidly, so normally the influence of surface water on soil reflectance is quite minimal. The influence of three levels of moisture content on the reflectance of one soil type are shown in Diagram 2.17.

Surface roughness The combination of surface roughness and solar elevation affects the amount of shadowing that occurs in the soil and in consequence the soil reflectance. Surface roughness can be due to the soil grain size for soils and sands that do not form clods, or the soil texture due to the clod size for other soils. For example, claypans or flat surfaces with a very finely textured surface are more highly reflective than rougher sandy surfaces even though sand itself is often more reflective than the particles of clay in the claypan.

Surface roughness can also be due to cultivation or other human activities on the soil surface that break up the surface and either increase or decrease its texture. Cultivation

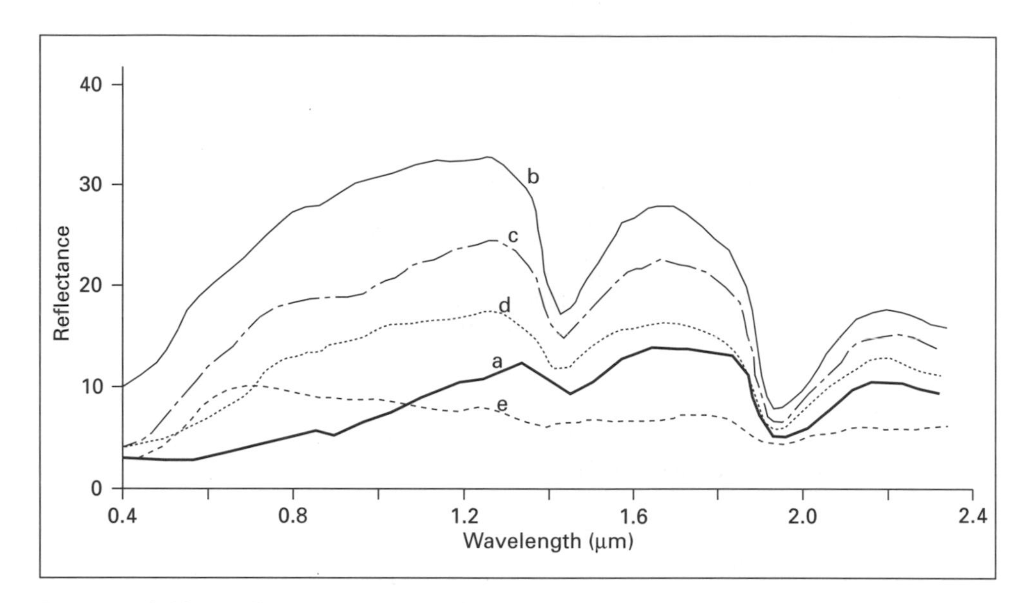

Diagram 2.18 Reflectance spectra of surface samples of five mineral soils: (a) organic dominated (high organic content, moderately fine texture); (b) minimally altered (low organic and iron content); (c) iron altered (low organic and medium iron content); (d) organic affected (high organic content and moderately coarse texture); and (e) iron dominated (high iron content with fine texture) from Stoner and Baumgardner (1981).

increases the surface clod size immediately after cultivation, thereby increasing the texture and reducing the reflectance due to increased shadowing by the clods. The marked reduction in the reflectance of cultivated soils is primarily due to this cause. As the cultivation ages so the clods are smoothed, the texture decreases and the reflectance increases. Images of areas under cultivation will thus exhibit variations in soil colour that are a function of the period since cultivation.

Work by Condit (1970) and Stoner and Baumgardner (1981) indicates that the reflectance curves for soils of the US can be grouped into five distinctive shapes (Diagram 2.18).

1. **Organic dominated** has a low overall reflectance with characteristic concave shape from about 500 nm to 1300 nm. Water absorption bands are noticeable at 1450 nm and 1950 nm, the broadness of the bands indicating that the absorption is due to both chemically bound and free water.
2. **Minimally altered** is high reflectance at all wavelengths with characteristic convex shape in the range 500–1300 nm. The water absorption bands are usually noticeable at 1450 nm and 1950 nm unless there is negligible bound water as occurs with sandy soils.
3. **Iron affected** where soils have lower reflectance than type 2 soils, with slight absorption evident at the ferric iron absorption bands of 450 nm and 900 nm. The water absorption bands are usually present.
4. **Organic affected** has higher overall reflectance than type 1 soils, maintains the concave shape in the 500–700 nm region and has a convex shape at longer wavelengths. It usually exhibits the water absorption bands at 1450 nm and 1950 nm.
5. **Iron dominated** has a unique spectral curve shape with reflectance peaking in the red

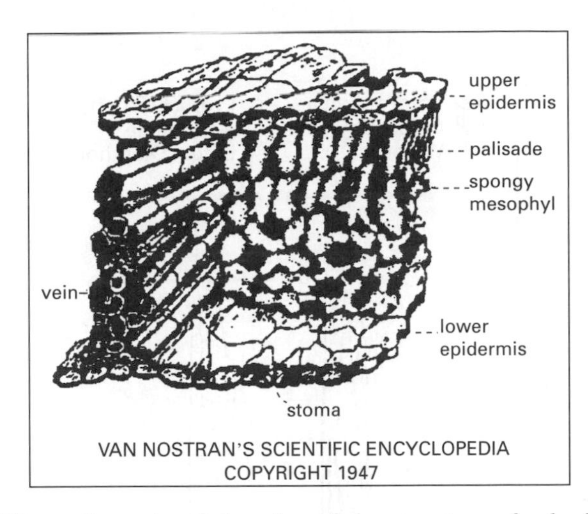

Diagram 2.19 Three-dimensional drawing of the structure of a leaf, similar to a citrus leaf, from Weigand et al. *(1972).*

at about 720 nm, and then decreasing due to iron absorption. The iron absorption dominates at the longer wavelengths so that the water absorption bands are often obliterated.

2.3.4 The reflectance of vegetation

Our concern in remote sensing is to understand better the relationship between reflectance from a plant canopy and the physical attributes of that canopy so as to extract more accurate and reliable information on the physical status of the canopy. Although the structure of leaves varies with species and environmental conditions, most leaves contain mesophyll cells enclosed by their upper and lower epidermis. The mesophyll cells usually consist of palisade cells adjacent to the upper epidermis and inner spongy mesophyll cells (Diagram 2.19 and Photo 2.1). Both the upper and lower epidermal cells diffuse and transmit the majority of the incident radiation. The long narrow palisade cells contain chloroplasts, with chlorophyll pigments, that selectively absorb radiation in the visible part of the spectrum. The spongy mesophyll cells contain many inter-cellular interstices that are usually filled with air and water vapour. Leaf cells are large in size in comparison with the wavelength of solar radiation, in comparison with the chloroplasts that are of similar dimensions to the wavelength of solar radiation.

Light entering a green leaf is subjected to multiple reflections at the cell walls within the leaf, scattering within the epidermis and chloroplasts of the palisade cells and absorption by the pigments within the palisade cells, the water in the leaf cells and by the constituent chemicals of the leaf (Diagram 2.26).

The multiple reflections and scattering mean that light reflected from most leaves is approximately lambertian in character; only leaves with a smooth glossy or waxy surface differ significantly from this general rule. The absorption in the leaf means that the light reflected out of the leaf is deficient in those wavelengths that have been selectively absorbed.

The absorption spectra of the principal plant pigments and water are shown in Diagram 2.20; the pigments are shown to have high, but different levels of absorption in the blue part of the spectrum (375–495 nm), and the chlorophyll A and B pigments also absorb in the red part of the spectrum (600–700 nm). Leaf water exhibits increasing

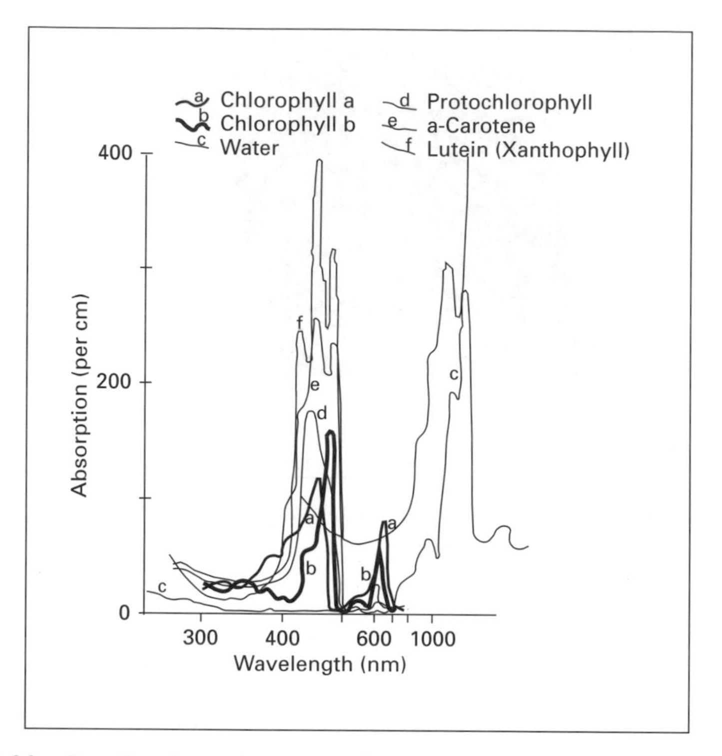

Diagram 2.20 Specific absorption versus frequency and wavelength of the principal plant pigments and liquid water from Gates (1965).

absorption with increasing wavelength, with peaks coinciding with the moisture absorption bands at about 990 nm, 1200 nm, 1550 nm, 2000 nm, 2.8 and 3.5 μm.

The importance of selective absorption in green leaves is indicated in Diagrams 2.21 and 2.22. In Diagram 2.21 the spectral absorption of green and albino *Hedera helix* leaves from the same plant show significantly greater absorption for the green leaf due to the absorption characteristics of the leaf pigments that are absent from the albino leaves.

Diagram 2.22 shows the reflectance, transmittance and absorptance curves in the range 400–4000 nm for three different types of green leaves, and four desert succulent plants. The diagram shows that green vegetation is highly absorbent in the visible part of the spectrum due to the plant pigments, but with varying amounts of absorption of green light (500–580 nm). In all cases there is negligible transmission in the visible part of the spectrum. At about 700 nm, the 'red edge', changes the curve in a most dramatic way; the absorption due to the plant pigments ceases, and absorption by the plant becomes negligible. All species reflect and transmit highly in the near infra-red (NIR) (760–1300 nm), but the extreme thickness of succulent plant tissue means that the transmitted component of the energy is either reflected or absorbed by these plants. Beyond about 1200 nm, the absorption characteristics of water begin to dominate. Beyond about 2000 nm, leaves become efficient absorbers and emitters, a characteristic that also allows them to act as efficient radiators of energy at these wavelengths so as to expel accumulated thermal energy.

In summary, plants absorb energy efficiently in the chlorophyll absorption bands where they use the energy for photosynthesis, absorb poorly in the NIR where the energy is of no use to their photosynthetic activity and would only accumulate as heat, and

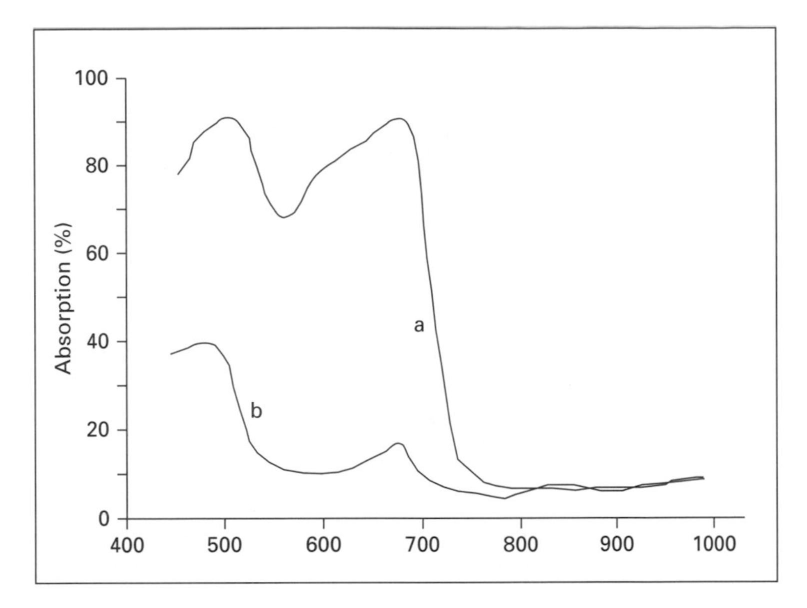

Diagram 2.21 The spectral absorption of (a) green and (b) albino leaves of a Hedera helix *plant from Gates (1965).*

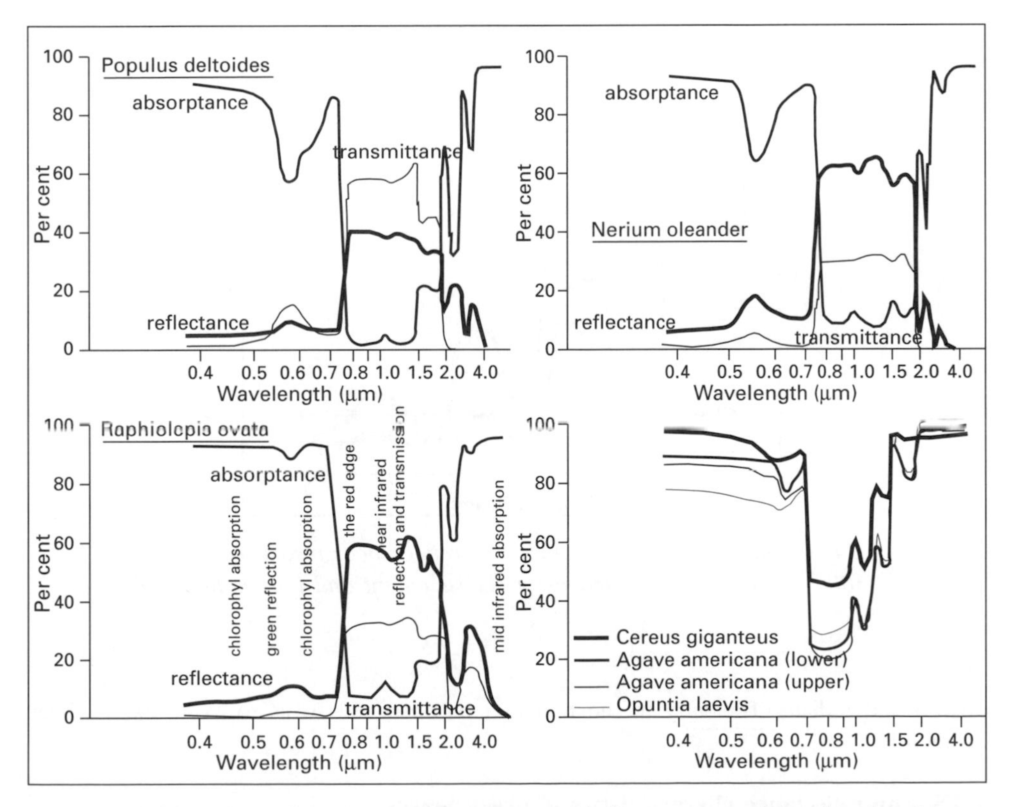

Diagram 2.22 The reflectance, transmittance and absorptance of Populus deltoides *leaves, being moderately thin and light coloured,* Nerium oleander, *a thick green leaf,* Raphiolepis ovata, *a thick dark green leaf and four selected desert succulent plants from Gates (1965).*

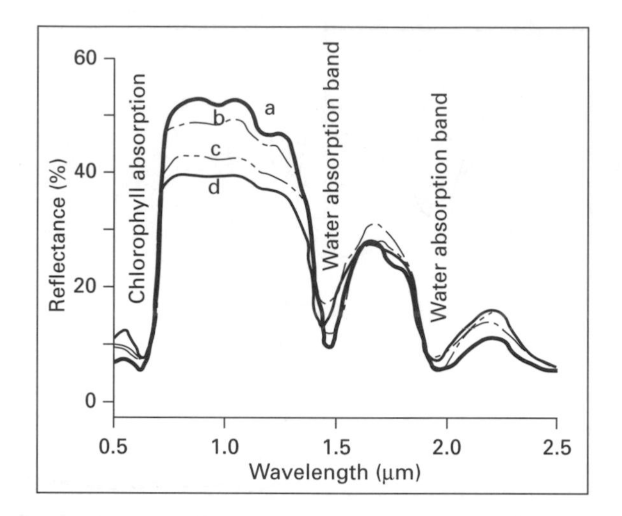

Diagram 2.23 Reflectance spectra of four plant genera differing in plant leaf structural or mesophyl arrangement, being: dorsiventral – (a) oleander and (b) hyacinth; isolateral – (c) eucalyptus; and compact – (d) corn. From Gausmann et al. *(1970).*

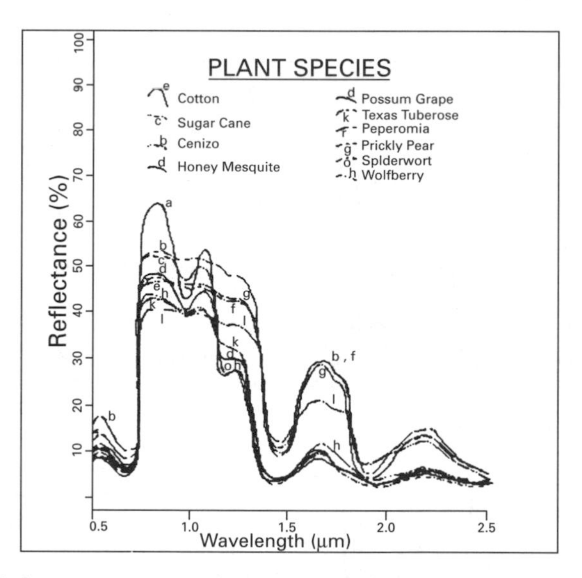

Diagram 2.24 Laboratory spectrophotometric reflectances measured over the waveband 500 – 2500 nm for the leaves of six succulent and four non-succulent species.

absorb and radiate efficiently in the mid and far infra-red so as to expel excess heat efficiently.

2.3.5 The reflectance characteristics of green leaves

The reflectance spectra of the leaves of a variety of vegetation species are shown in Diagrams 2.23 and 2.24. These diagrams show that all plants respond generally in the same way to incident radiation, and wish to reflect in similar ways. There are small and

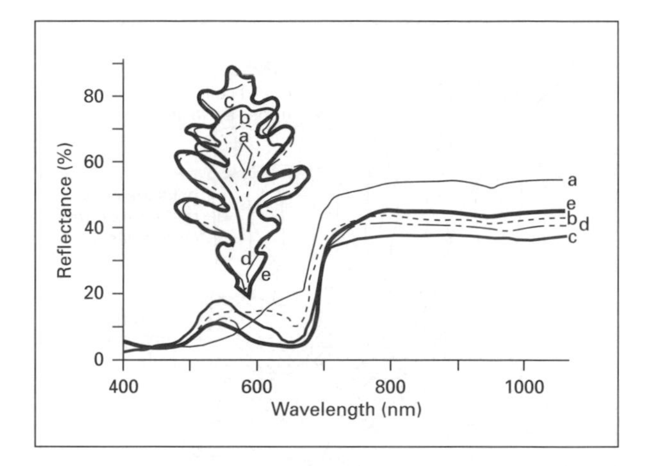

Diagram 2.25 Evolution of the spectral reflectance of a Quercus alba *leaf during the growing season for: (a) 17 April 1964; (b) 22 April 1954; (c) 5 May 1954; (d) 11 May 1954; and (e) 18 May 1964, from Gates (1965).*

important differences in the reflectance of different species that may eventually be used to obtain better information out of remotely sensed data.

The evolution of the spectral reflectance of plant leaves during the life of a leaf is illustrated for *Quercus alba* in Diagram 2.25. Initially the young leaves exhibited the absorption characteristics of proto-chlorophyll; as the leaves develop chlorophyll A and B so there is an increasing absorption in the red part of the spectrum. The increasing concentration of chlorophyll causes a deepening of the red absorption band and a narrowing of the green reflectance band. In early maturity the leaf is lighter with higher green reflectance. The leaf darkens with age, this ageing coinciding with increasing intercellular air spaces and a corresponding increase in NIR reflectance due to lower water absorption. The 'red edge' shifts from about 700 nm to about 725 nm as the leaf matures.

Green leaf elements, as shown in Diagram 2.19, are not very transmissive in the visible portion of the spectrum, but are highly transmissive in the NIR. The effect of this on a plant canopy can be deduced from Diagram 2.27. Increasing the stack thickness by adding one leaf at a time to the stack up to a thickness of about six leaves, causes the reflectance to increase in the NIR but shows negligible change in the visible part of the spectrum.

This means that the reflectance of green plant canopies will change in the NIR up to a green leaf area index (GLAI) of about six, but will show little change above this. Changes in crop canopy reflectance as crop Leaf Area and soil cover increase, are shown in Diagram 2.28. The extent of these changes are greater in the NIR than they are in the visible part of the spectrum, as would be expected from the previous discussion.

The reflectance of green leaves will also change with moisture stress. As leaves dehydrate they lose the ability to maintain a temperature balance through transpiration, thereby inducing a rise in temperature in the leaves. Simultaneously the leaves lose turgidity and start to wilt, affecting the canopy geometry and hence the canopy reflectance. Individual leaf reflectance will start to change due to the lower absorption caused by the reduced moisture content in the leaves. The leaves will, however, maintain their reflectance component due to the leaf pigments until these pigments start to decompose, which occurs after the onset of moisture stress and wilting.

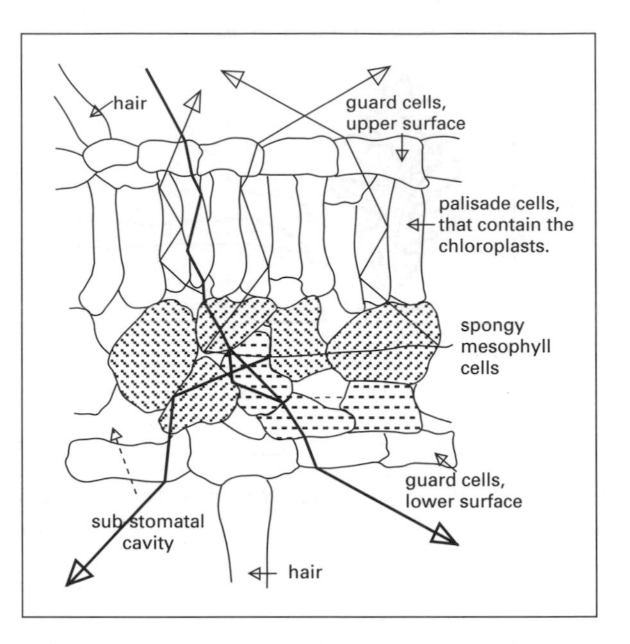

Diagram 2.26 Reflection, transmission and absorption in a leaf.

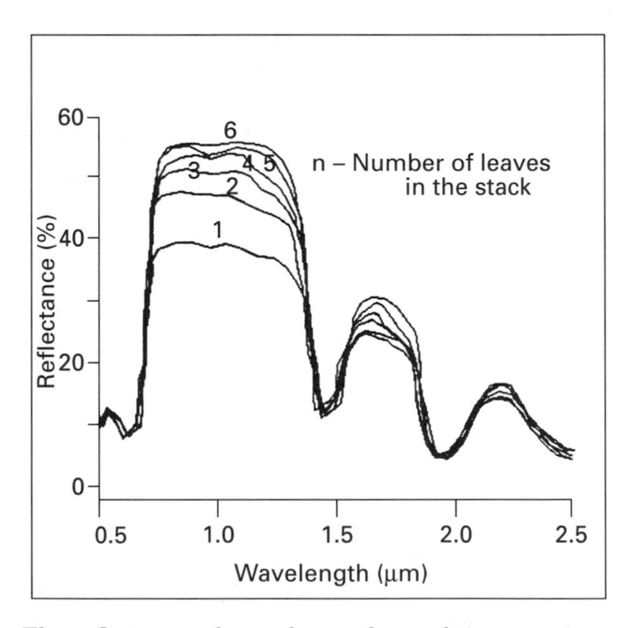

Diagram 2.27 The reflectance of corn leaves for each increase in stack thickness by one leaf, up to a thickness of six leaves.

2.3.6 The reflectance characteristics of dead leaves

There is negligible transmission through dead leaves and vegetation at all wavelengths as shown in Diagram 2.29, in contrast to the situation with green leaves. The reflectance spectra of dead leaves are similar for all species, with significant absorption in the blue-green waveband, and higher reflectance than green vegetation at longer wavelengths. The

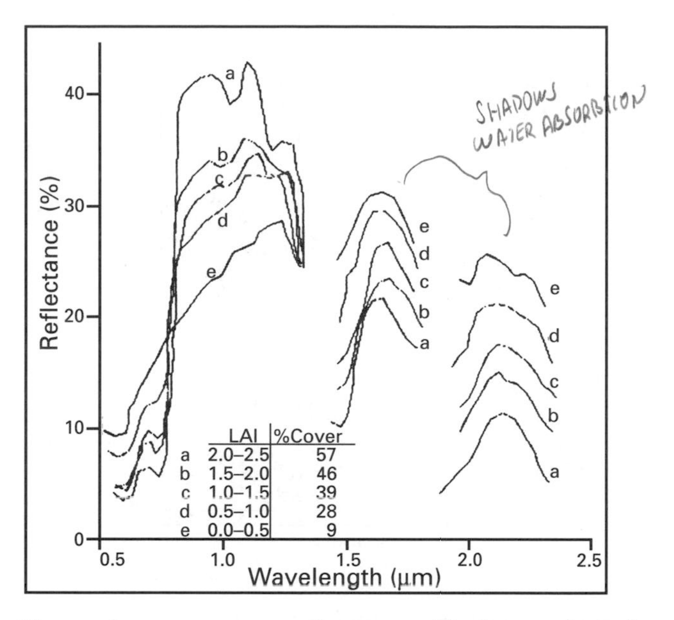

Diagram 2.28 Change in crop canopy reflectance with changes in leaf area index (LAI) and per cent cover.

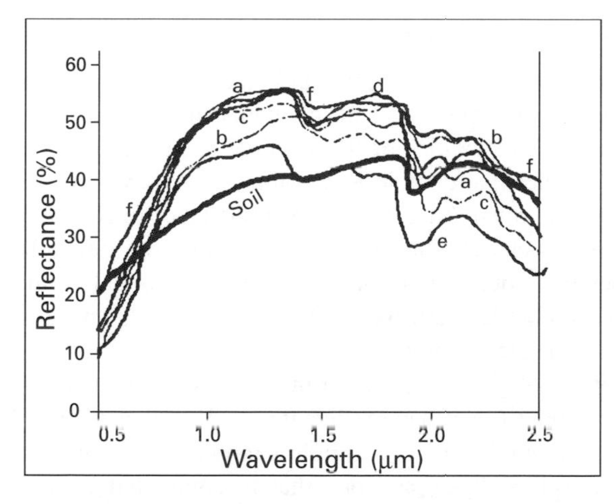

Diagram 2.29 Laboratory spectrophotometrically measured reflectances of dead leaves and bare soils for six crop species: (a) avocado; (b) sugarcane; (c) citrus; (d) corn; (e) cotton; and (f) sorghum. From Gausman et al. *(1976).*

high reflectance of the brown leaf elements is most likely due to both the hardening of the cell walls, making them more reflective, and the reduction in cell water, reducing absorption in the moisture absorption bands.

As brown vegetation decomposes the absorption levels increase, particularly at shorter wavelengths due primarily to the absorption by antho-cyanin compounds that are formed as the breakdown chemicals from decomposition of the chlorophyll plant pigments. In consequence, dead vegetation tends to be very reflective soon after it has browned off, and to become darker with age.

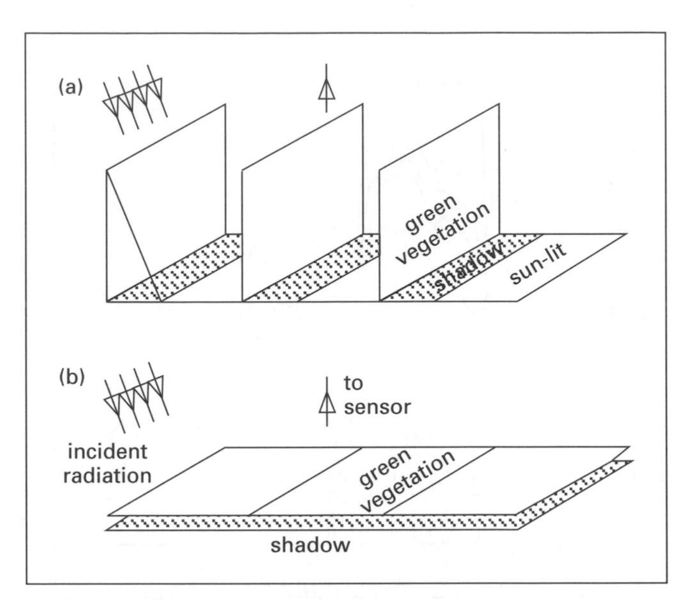

Diagram 2.30 The geometry of energy interactions in (a) vertical erectophile and (b) horizontal planophile vegetation canopies.

2.3.7 Vegetative canopy reflectance

The reflectance from a vegetative canopy is affected by the location, orientation, reflectance, transmittance and absorptance of all elements that constitute the canopy, the distribution (orientation and elevation) of the incident radiation and the reflectance of the soil background. Canopies that contain leaves that are glossy and thus have a high level of specular reflection will often cause the canopy reflectance to diverge from the lambertian ideal, i.e. these surfaces can exhibit directional variation in their reflectance. This is in contrast to canopies that contain matt or highly lambertian leaf elements as these canopies tend to exhibit lambertian reflectance. Wilting of leaves changes the geometry and hence the reflectance of the canopy.

The transmissivity of green vegetation in the NIR means that canopies up to a leaf area index (LAI) of about 6 can have reflectance values that include some contribution from the soil. Consider models of two simple canopies: one with a dominance of horizontal leaf elements (planophile); and one that is dominated by vertical leaf elements (erectophile), even when the number of leaf elements in each are similar (Diagram 2.30). Both canopies have a LAI of one, i.e. the ratio of the area of green leaves above the corresponding soil area is unity. Assume that the solar elevation is 60° and that there is negligible skylight contributing to scene illumination.

In the vertical (erectophile) canopy, the incident radiation will predominantly follow one of four paths:

1. transmission through a leaf then reflectance at the soil interface;
2. soil reflectance;
3. reflectance from the leaf and then reflectance from the soil interface; or
4. reflectance from the leaf elements towards the sensor.

The relative contribution by each of these depends upon the elevation of the sun, the direction of the sensor and the sensor field of view relative to the canopy size.

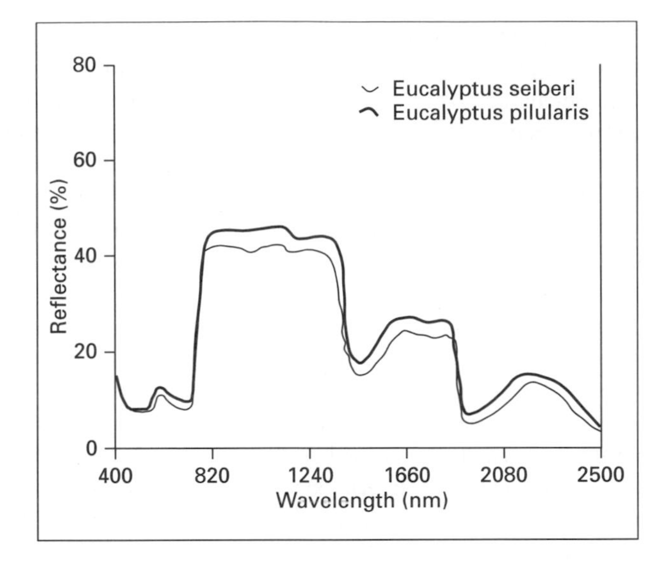

Diagram 2.31 The reflectance of selected Eucalyptus sp *plants from O'Niell (1990).*

In the horizontal (planophile) canopy, the energy will predominantly be:

1. reflected from the vegetation surface; or
2. transmitted through leaf elements before reflection from lower leaf elements and subsequent transmission to the surface; or
3. by transmission from all of the leaves, reflection at the soil interface and transmission through the leaves to the surface.

In considering these two simple canopies, quite a few assumptions have been made concerning canopy geometry and the reflectance characteristics of leaf elements. In practice, the canopy is much more complex than this, however the modelling shows that:

1. the reflectance from a canopy dominated by erectophile components is much more likely to show changes in reflectance with changes in solar elevation and angle to the sensor than will a canopy with predominantly planophile components;
2. changes in the proportions of vertical and horizontal leaf components can effect canopy reflectance; and
3. the size of the sensor IFOV relative to the scene elements can affect the reflectances recorded in the data.

Consider how this might assist in explaining the reflectance characteristic of different canopies. The canopies of *Eucalyptus sp* have much lower reflectance values than the canopies of many deciduous and tropical trees, even though the reflectance of the individual leaf elements are similar as shown in Diagram 2.31. *Eucalyptus sp* trees have leaf elements that hang vertically or much more so than the predominantly horizontal orientation of the leaves of most deciduous and tropical trees. In consequence, a Eucalyptus canopy is quite open when viewed (sensed) directly from above. When the sun is at low angles of elevation the leaves create considerable shadow, so that vertical imagery will contain significant shadow and soil reflectance. When the sun is at higher elevations the shadowing is reduced, but the contribution of soil reflectance increases.

All plants use the same pigments for photosynthesis, and the same chemical materials for plant building. For plants occupying similar environmental regimes, they all have to handle similar energy budgets. To compete successfully with other species, a plant will need to develop unique characteristics that will allow it to occupy an ecological niche or niches. Plants do this in many ways: changing the mix of pigments used in photosynthesis; modifying their canopy geometry so as to optimize their ability to use energy or to minimize environmental impact on them; adapting their growth patterns; and modifying their leaf structure. All of these techniques affect the energy balance of the plant, and all of them affect the reflectance from the canopy in some way during their phenological life cycle. Thus analysis of diurnal and seasonal changes in canopy reflectance is an important strategy in attempting to discriminate between different plant communities.

Consider winter and summer growing herbage or grasses that occupy different ecological niches by growing at different seasons. The two communities will look quite different on the one image if the imagery is chosen to ensure that the phenological differences are maximized. Agricultural crops have a different phenological cycle to most of the other landcovers due to physical interference by man, and most obvious at cultivation and at harvest time when the crop status changes from a mature crop to litter and straw. Both of these activities are quite unique to crops and thus allow better discrimination of crops from other landcovers. The changes at harvest time are often more useful because they are usually more dramatic in their effect in the imagery, the period of harvest covers a shorter period than cultivation and sowing, and there is a greater probability of cloud-free conditions suited to the acquisition of image data.

2.4 Passive sensing systems

Sensing systems collect energy from the surface, and convert that energy into an analogue or digital record. There are important distinguishing characteristics of this image data.

1. The relative spatial locations of all data elements are known to a high degree of accuracy, which is essential if the data are to be used to create an image that accurately depicts the imaged surface. The relative spatial locations are known because the geometric model of the sensor are known. The closer the sensor fits the geometric model, the more spatially accurate will be the image. With analogue sensors, such as the camera, the data are preserved in their correct relative position in the recording medium or film. With digital sensors, the data are stored without preservation of spatial organizational information attached to the individual image element, but the data are preserved in such a way that allows the reconstruction of the data into an image.
2. The absolute spatial locations can be determined. These coordinates depend on the sensor geometry as well as the coordinates and orientations of the sensor relative to the object coordinate system. External control points, with known absolute and image coordinates, are normally used to determine the absolute coordinates of all image points. Because absolute coordinates depend upon both internal and external geometries, the accuracy of establishing absolute coordinates of image data is usually inferior to the internal accuracy of the data. Techniques to rectify image data, or create a new image that is a close fit to an absolute coordinate system, is discussed in Chapter 5.
3. An image depicts conditions at an instant of time; images are acquired within fractions of a second from an aircraft, and fractions of a minute from a satellite. If this is fast, relative to the speed of change on the surface, then the image can be considered to

be acquired at an instant in time. A set of temporal images record conditions at different times. The use of temporal data sets requires comparison of the images and hence requires good absolute accuracy between the images.
4. Image data is a record of specific energy radiations from the surface, not of the surface features themselves. Unlike field data, that usually measures the surface features of interest, remotely sensed data records the effect of the surface on electromagnetic radiation under specific illumination and sensing conditions. Extrapolation is necessary to derive estimates of physical parameters from this data, using techniques discussed in Chapters 3 and 5.

Remote sensing devices are either passive systems, using existing sources of energy to illuminate the object, or active systems that generate their own illumination. Passive systems are usually simpler and cheaper, but they depend on the characteristics of the natural source of energy that they use. Active systems are independent of the normally used naturally occurring energy sources such as the Sun and the Earth itself.

The most commonly used passive sensor systems in remote sensing are the camera, and the scanner.

2.4.1 The camera

A camera forms an image by focusing the incoming electromagnetic radiation from the scene onto the film in the camera in accordance with specific rules.

1. There is a one-to-one correspondence between the object, the perspective centre (point O in Diagram 2.32) and the image point. These three points all lie on the one line, so that objects at different spatial locations must be imaged at different positions in the image.
2. Straight lines in the object space must be imaged as straight lines in the image space. Curved lines in the image space are due to distortions in the camera lens that limit their use for quantitative remote sensing tasks.
3. The light entering the camera must be controlled in both duration and intensity to ensure proper exposure of the film.

Light from the object is focused by the lens cone (Digram 2.33) onto the film in accordance with these principles. Start to read the **focal length** (f_c in Diagram 2.32) is the length of the normal from the perspective centre, on the rear nodal plane of the lens, to the image plane when the camera is focused on infinity or the incoming radiation from a point is nearly parallel across the lens aperture. The **principal point** (P) is at the foot of this normal on the image plane (Diagram 2.32), and is defined by the intersection of the lines joining the fiducial marks on an aerial photograph.

The lens of a camera must strive to maintain close fidelity to the geometric model that is the basis of the camera's construction, so as to maintain geometric accuracy, and to provide as fast an exposure as possible so as to minimize image blur. These requirements place competing demands on the construction of the camera since smaller lens apertures come closer to the geometric model of a camera in which the perspective centre is defined as a point, whilst the larger the aperture the more light that will enter the camera in a given time, allowing shorter exposures. These competing demands cannot be solved perfectly resulting in camera distortions, e.g. chromatic aberration, spherical aberration and coma, or astigmatism.

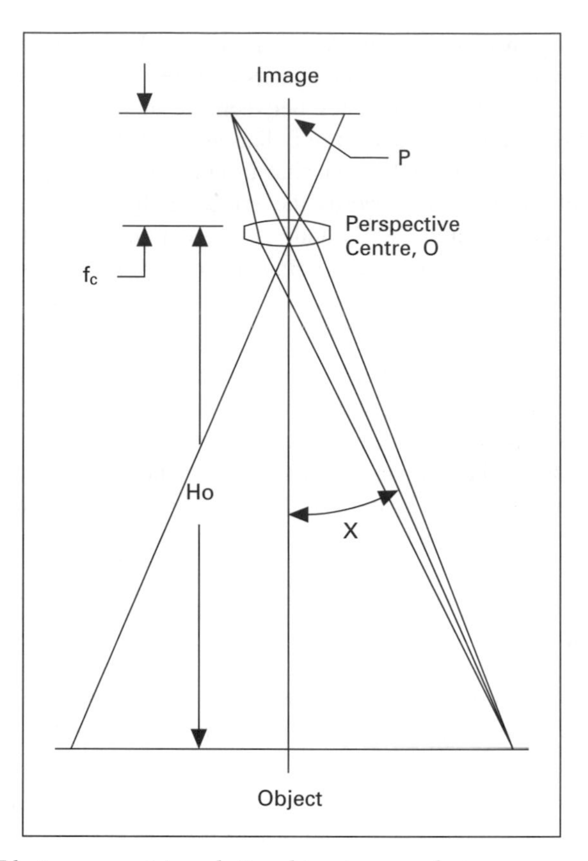

Diagram 2.32 Photogrammetric relationships in a perfect camera.

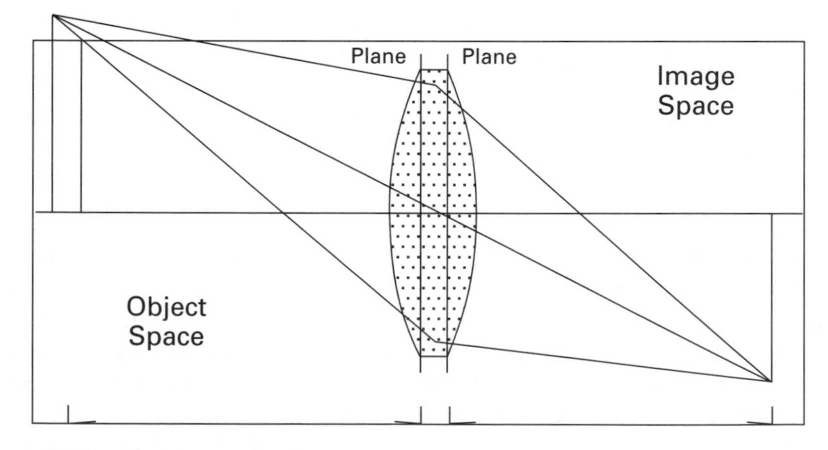

Diagram 2.33 Geometry of a lens cone.

Chromatic aberration If light of different wavelengths is refracted by different amounts in the lens then it will be focused at different distances from the lens. This means that some wavelengths may not be in focus on the film plane, since the plane is at a constant distance from the lens.

Spherical aberration and coma Refraction at the edges of the lens causes more bending of the rays than at the centre of the lens. There are effectively different focal

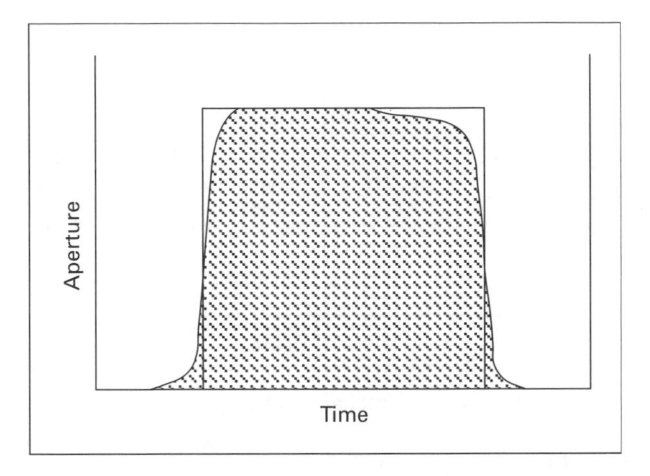

Diagram 2.34 Shutter efficiency. The rectangular outline represents ideal shutter action: open in zero time, stay completely open, close in zero time. The shaded area shows the actual shutter aperture.

distances at different distances out from the central axis of the lens causing image blur.

Astigmatism The curvature of the lens is different in planes through the lens axis to the curvature on chords that do not pass through the axis. This difference in curvature creates differences in the focusing distance on the plane through the centre of the lens to the plane through the chord. The result is that a point may be in focus in one direction, but be out of focus at right angles.

The lens of most cameras contain the diaphragm and shutter. The **diaphragm** controls the amount of light that can enter the camera in a unit period of time. The larger the diaphragm aperture, the more light that can enter the camera in a unit of time, but the larger the geometric errors that will occur. The intensity of the light in the camera decreases as the square of the distance travelled from the lens (because the same amount of energy must cover a larger area). The F number, or ratio of area of aperture to focal length, indicates equivalent light intensities on the image plane of cameras of any focal length, and is therefore used as the standard measure of exposure per unit time. The larger the F number the smaller the aperture relative to the focal length, and hence the less light that can enter the camera in a given time and vice versa. The range of F numbers used in most cameras is:

F number	Relative aperture size
32	Very small
16	Medium
2.8	Large

The **camera shutter** is designed to control the length of time that light is allowed to enter the camera, and hence control the total exposure. Shutters do not operate with perfect precision, taking a period of time to open, and a period to close, during which time the shutter does not allow the full amount of light into the camera. Diagram 2.34 illustrates the typical efficiency achieved with many modern camera shutters.

The **magazine** (Diagram 2.35) holds the film such that it can be supported in the

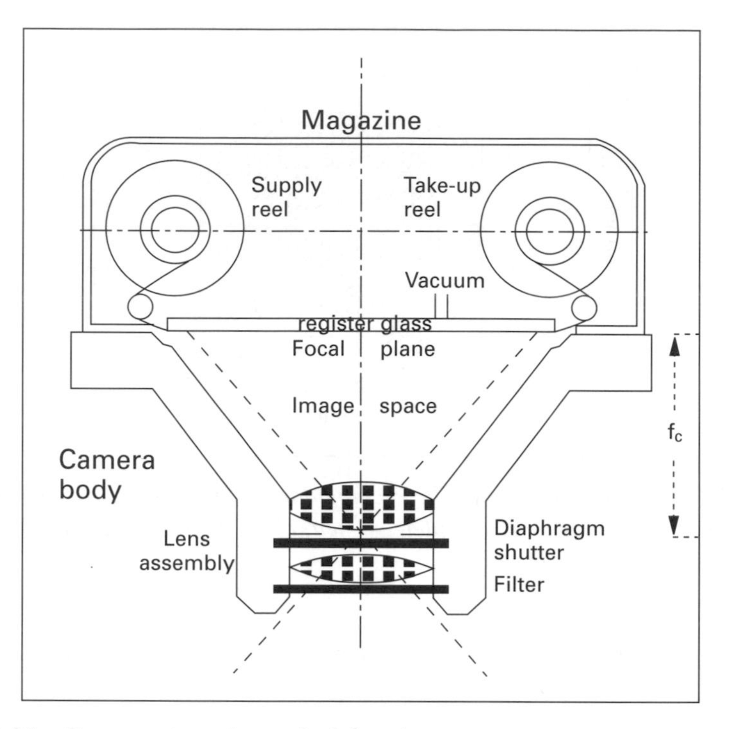

Diagram 2.35 Construction of a typical framing camera.

image plane during exposure, and wound on to prepare for the next exposure. The film is held so that it is sealed from light other than that allowed to enter to expose the film by the shutter and aperture. The magazine also has counters to record the exposures taken or still available, and facilities to record critical information about the film on each exposure such as the aircraft altitude, time of exposure, date, flight or sortie number and the exposure number.

The **camera body** holds the lens cone in the correct position relative to the image plane and the magazine so as to minimize distortions in the images taken. It is a light tight compartment, sealing the film from stray light, and includes the register glass, a flat ground glass plane just behind the image plane and against which the film is pressed during exposure.

The **suspension mount** secures the camera to the platform, and usually contains mechanisms to damp vibrations from the platform being transmitted through to the camera, as well as levelling and orientation adjustments. The level and orientation of the camera will vary with aircraft payload, the wind direction and velocity relative to aircraft velocity and direction.

In a camera, unlike other sensors, the sensing and recording functions are undertaken by the same elements of the sensor, the emulsion surface on the film. This characteristic of cameras simplifies their construction and hence reduces their cost relative to other sensors, as well as eliminating a number of sources of geometric error. Because of the fixed exposure and the development steps that occur in creating a photograph, it is difficult to relate densities on a photograph with radiances that are incident on the camera. Exposure panels, of known or constant reflectance characteristics can be used to help control and calibrate the density values in the film.

The most common type of camera is the framing camera (Diagram 2.35) or a camera that simultaneously exposes the whole of the image area for each fixed exposure. Typical framing cameras are all 35 mm cameras, various 70 mm cameras, and 23 cm format mapping cameras. Mapping cameras are designed to minimize distortions so as to produce high precision negatives, 23 cm by 23 cm in size (Photo 2.2), are flown in aircraft properly set up for the purpose, with properly installed mounts, intervalometers to ensure the proper interval between exposures and navigation sights. Suitable 70 mm cameras can be used for taking aerial photography from a light aircraft, but with better resolution and quality than can be achieved with most 35 mm format cameras. The smaller format cameras can be set up in banks to take simultaneous photography, e.g. simultaneous colour and colour infra-red photography.

With aerial cameras the focus is set at infinity so that the distance from the rear nodal plane of the lens to the image plane is set at the focal length of the lens. It is essential to use high quality framing cameras for accurate measurement work as in the production of topographic maps. Other types of framing cameras of lower geometric quality can often be used where accurate geometric information is not being sought from the photographic data.

Often photo-interpretation does not require accurate geometric measurements and can therefore use other, or cheaper cameras in relation to geometric quality. However, any work involving detailed analysis of images requires photographs that are of the highest possible image quality, in terms of resolution, focus, contrast and colour rendition. For this reason photo-interpretation requires cameras that are of high image-making quality, but can be somewhat inferior to mapping cameras in geometric quality.

There are four main **types of film** used in remote sensing, identified by their sensitivity to light of different wavelengths, and how they record this radiation. The light sensitive layer depends on the reaction of silver halide salts to electromagnetic radiation. Silver halide grains are naturally sensitive to light in the blue and ultraviolet parts of the electromagnetic spectrum, but their sensitivity is extended into the remainder of the visible part of the spectrum, and into the near infra-red, by the use of dyes called **colour sensitizers**. The grains are still sensitive to the shorter wavelengths after they are sensitized, and so this radiation needs to be filtered out to stop it entering the camera and exposing the film. Without filters, this shorter wavelength radiation would contribute to, and often dominate the exposure, masking the contribution due to longer wavelength radiation. There are four common types of film emulsion (Diagrams 2.36 and 2.37).

1. **Panchromatic** – a one-layer film that creates black and white photographs, recording the radiation in the visible part of the spectrum, from about 350 nm to 700 nm, in accordance with the spectral sensitivity of the film.
2. **Black and white infra-red** – a one-layer film that creates black and white photographs recording the radiation from the blue part of the spectrum through into the NIR, at about 900 nm in accordance with the spectral sensitivity of the film. These films are usually used to record the incident energy in the NIR, and so a filter must be used to stop energy below about 700 nm being transmitted to the film.
3. **Colour** – a three-layer emulsion film constructed similarly to that shown in Diagram 2.38. The layers are sensitized to record energy up to the blue, green and red wavebands respectively. A yellow filter is inserted between the green and red sensitive layers to filter out blue and green light from activating the red sensitive layer. Each layer contains a dye, usually of yellow, magenta and cyan respectively so that exposure and development of that layer creates an image in that colour.
4. **Colour infra-red** – a three-layer emulsion film with the layers sensitized to the green, red and NIR wavebands. It is of similar construction to that of colour film, but without

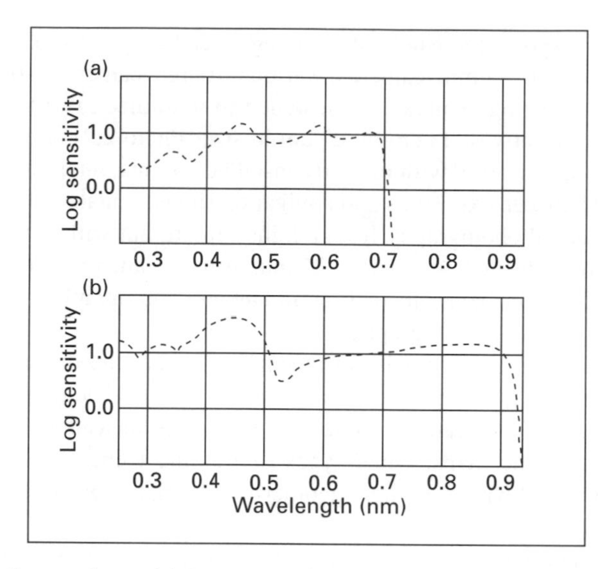

Diagram 2.36 Spectral sensitivity curves for two common one-layer emulsion films: (a) panchromatic; and (b) b&w infra-red.

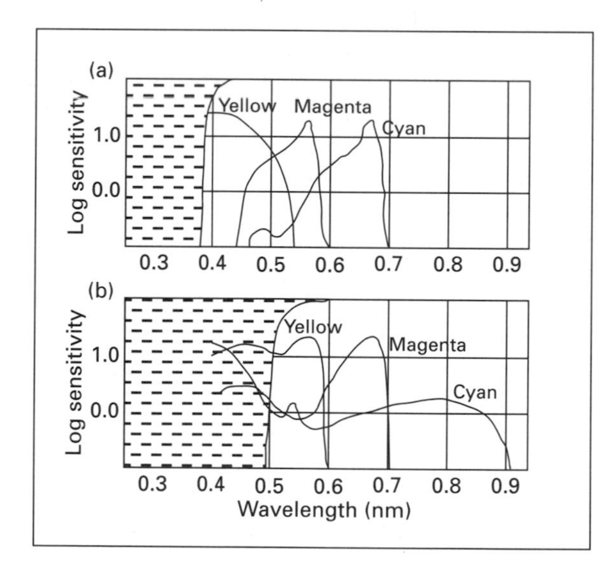

Diagram 2.37 Spectral sensitivity curves for (a) three-layer colour and (b) colour IR films.

the yellow filter. All of the layers of this film retain their sensitivity to blue light, and so a filter must be used to stop the blue light being transmitted onto the film.

Yellow or neutral density filters are used to filter out some of the blue light and hence correct for the effect of haze. A yellow filter with cut-off at about 500 nm is normally used with colour infra-red film, and a deep red (Wratten 25) filter is normally used with black and white IR film to record incident NIR radiation on the film. Filters are in two forms (Diagram 2.39):

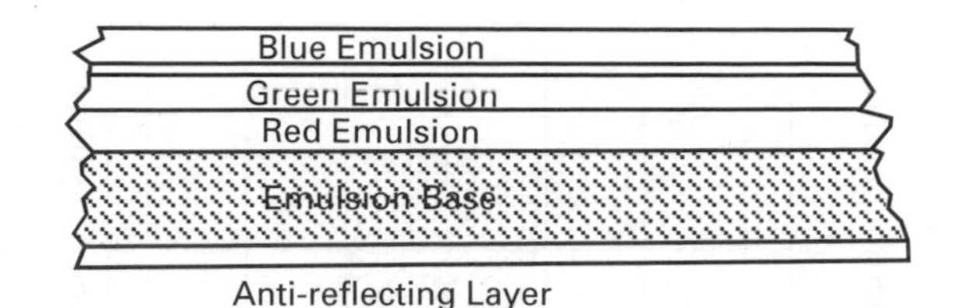

Diagram 2.38 Cross-section through a colour film emulsion showing the emulsion layers.

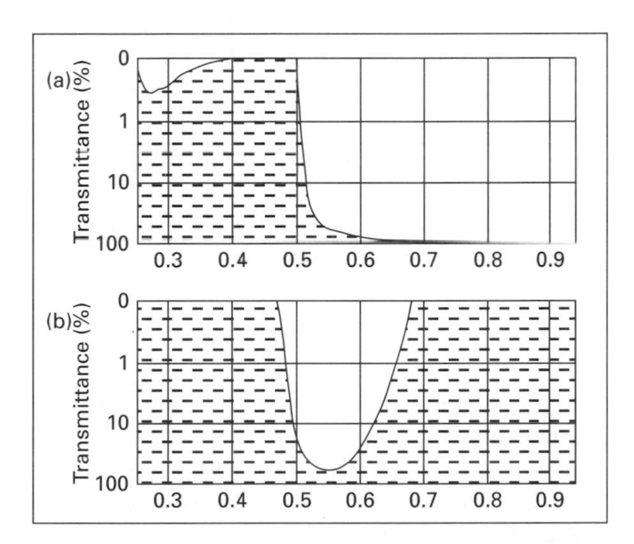

Diagram 2.39 Typical transmission curves for (a) cut-off and (b) band filters.

1. **cut-off filters** which have either a lower or upper cut-off, only transmitting radiation above, or below the cut-off waveband–yellow filters are typical cut-off filters; and
2. **band filters** which transmit energy within a selected waveband.

Filters only transmit part of the radiation to which the film is sensitive. Sometimes the manufacturer expects the user to utilize the filter, and publishes the film speed as if the film will be used with the filter. However, sometimes the user may wish to record radiation in only part of the normally used waveband. Thus manufacturers of black and white IR film expect users to utilize a deep red filter with the film and so the film speed is set accordingly. Now, panchromatic film is normally used to record radiation in the whole of the visible part of the spectrum, so if the user wishes to record only part of that spectrum, such as green light, then the green filter is used to reduce the amount of radiation available to activate the film, relative to the amount anticipated by the manufacturer in determining the speed of the film. This restriction effectively makes the film slower, requiring a longer exposure. This change in exposure due to use of a filter is known as the **filter factor** (FF) for that film and filter combination. A yellow filter that has a cut-off at 500 nm, restricting the transmission of blue light, would normally have a filter factor of 1.5 or 2 for use with panchromatic film; but a filter factor of 1 with colour IR film as the manufacturer assumes the use of this filter with this film.

The filter factor can be calculated by plotting the filter transmission curve on the film

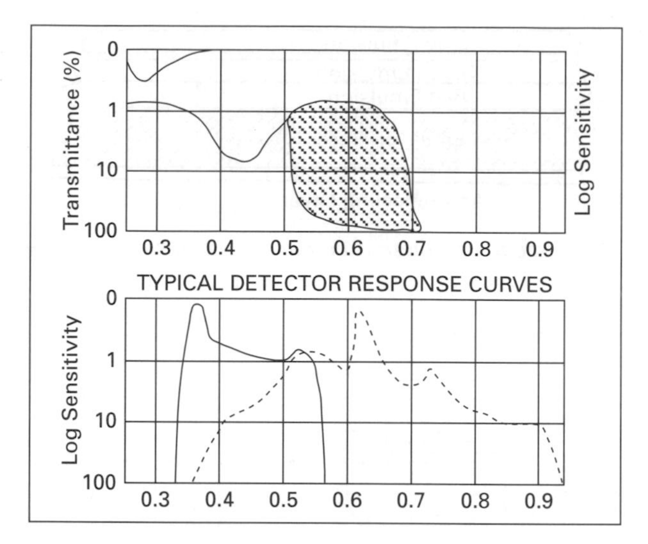

Diagram 2.40 Determination of the filter factor and the responsivity versus wavelength for several typical photon detectors.

sensitivity curve (Diagram 2.40); the filter factor is the ratio of the area bounded by the film sensitivity curve in normal usage over the area bounded by the film sensitivity and filter transmission curves when transmission is restricted by the filter – normally a multiplicative factor greater than one. The filter factor can also be determined by photographing a target without the filter, then with the filter at a series of exposures. Comparing the film densities and selecting the filtered exposure that gives a density closest to the unfiltered density – the ratio of the two exposures is then an estimate of the filter factor.

Resolution of detail in an image is of critical importance in visual interpretation. Historically, this has been measured as the number of lines per mm that can be resolved in a photograph of a bar chart which is a number of sets of parallel black and white panels, with the panels being of equal width in each set, but different between the sets, so that the narrowest panels will be blurred in the photograph and the widest will be clearly seen. The use of bar targets to determine resolution is easily understood, but it does not show the loss of resolution as the detail becomes finer, nor does it indicate the differences in resolution that occur when contrast is reduced. These difficulties are overcome by the use of the **optical transfer function** (OTF). If a bar chart with intensities that varied in a sinusoidal fashion is photographed, then the resulting image densities also vary sinusoidally (Diagram 2.41). As the frequency of the bars increases (or the wavelength of the sine curve is reduced) the photograph exhibits a reduction in the amplitude in the image. At long wavelengths, the reduction in contrast is negligible, but as the wavelength decreases, so the reduction in contrast increases until eventually the sine wave cannot be discriminated.

The OTF indicates the percentage of the information that will be recorded for different sized objects, at maximum contrast. The level of contrast between objects and their surroundings is determined as a proportion of this maximum contrast. The product of this proportion and the percentage recorded in the OTF gives the percentage that will be recorded at the actual contrast. Clearly the higher the contrast in the original scene, the better the resolution in the information that can be extracted.

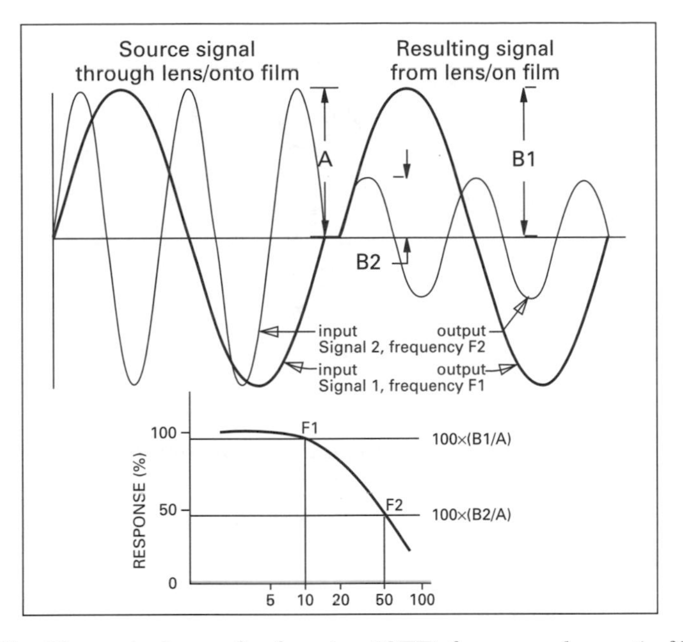

Diagram 2.41 The optical transfer function (OTF) for a panchromatic film.

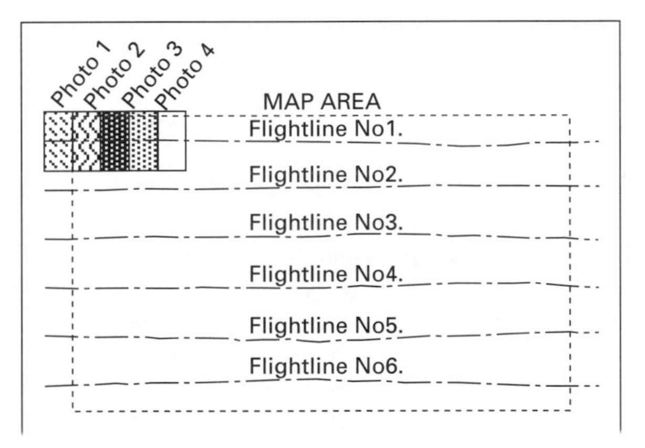

Diagram 2.42 Layout of an aerial sortie.

2.4.2 Acquisition of aerial photography with a framing camera

Aerial photography is usually required of a larger area than can be covered with the one exposure of the camera. This problem is usually solved by covering the area with a number of photographs, arranged in sets or strips along flightlines of the aircraft (Diagram 2.42). Within each strip the photographs are taken at a regular interval designed to ensure that there is 10 per cent or more overlap between the adjacent photographs. Usually the overlap is either 60 or 80 per cent, to provide sets of photographs in which all points are photographed on at least two adjacent photographs. The reason for this is to provide stereoscopic coverage of the area of interest, as will be discussed in Chapter 3. About 15 per cent of the coverage of one strip is also covered on the adjacent strip, to ensure no gaps between adjacent strips of photography.

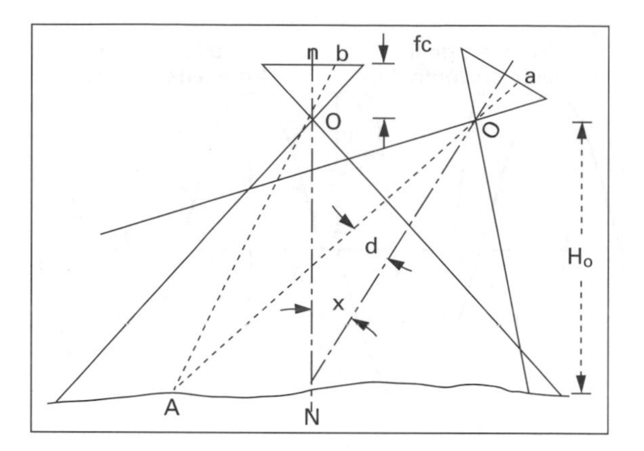

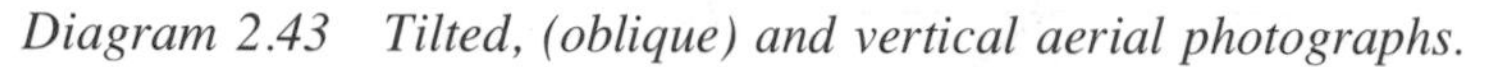
Diagram 2.43 Tilted, (oblique) and vertical aerial photographs.

An aerial photograph contains not only the image, but also essential information about the image, including:

- agency that took the photograph;
- flight or sortie number;
- photograph number, within the flight or sortie;
- focal length of the camera;
- altimeter reading at the time of exposure of the photograph;
- time and date of the photograph; and
- bubble to indicate the photograph tilt at the time of exposure.

In the centre of each side of an aerial photograph (Photo 3.2) are the fiducial marks. The intersection of the lines joining the fiducial marks on opposite sides of the photograph indicates the position of the principal point of the photograph.

The scale (and hence the scale factor, x) of an aerial photograph, as in a map, is defined as the ratio of one unit of distance on the photograph, or map, to the number of units of distance to cover the equivalent distance on the ground (x). Thus if an aerial photograph is at a scale of 1:48 600 then 1 mm on the aerial photograph represents 48 600 mm (or 48.6 metres) on the ground.

The scale of a vertical aerial photograph can be determined by measuring the distance between two features on the aerial photograph, and measuring the equivalent ground distance, i.e. the product of the map distance and the map scale factor. The scale of a vertical aerial photograph can also be calculated from the flying height above ground level and the focal length. In Diagram 2.43 the scale at point b on the image is given by

$$Ob{:}OA \quad \text{or} \quad 1{:}\frac{OA}{Ob}$$

By similar triangles

$$\frac{OA}{ON} = \frac{Ob}{f_c} \quad \text{so that} \quad \frac{OA}{Ob} = \frac{ON}{f_c}$$

If $ON : H_o$ the flying height above ground level then the scale of a vertical aerial photograph is:

$$1{:}\frac{ON}{f_c} = 1{:}\frac{H_o}{f_c}$$

For flat, horizontal ground surfaces a vertical aerial photograph has constant scale at all points in the photograph. Height differences introduce changes in H_o and hence introduce changes in scale in the photograph. Tilts in a photograph introduce changes in scale across the photograph. Under normal conditions where there are variations in topographic height and small tilts in the photograph, calculation of the scale using the flying height and focal length will give only an approximate value for the scale.

Example

The distance between two points A and B is measured as 22 mm and 38 mm respectively on an aerial photograph and a map. If the map is at a scale of 1:25 000 then what is the scale of the aerial photograph?

$$\text{Ground distance} = 38 \times 25\,000 \text{ mm}$$

$$\text{Scale} = 22{:}38 \times 25\,000 \simeq 1{:}43\,000$$

Because all aerial photographs contain small tilts, and some differences in elevation, there will be some variation in scale across the photographs. It is therefore a good idea to calculate the scale of the aerial photographs using at least three distances scattered across the photograph, and to then adopt the average.

Photographs that are taken with large tilts are known as oblique photographs. High oblique photographs contain the horizon whilst low oblique photographs do not. Consider the tilted photograph in Diagram 2.43. Scale at A is $1{:}OA/Oa$. Now,

$$OA = H_o \sec(X + d)$$

and

$$Oa = \frac{f_c}{\cos(d)}$$

so the scale at A is

$$1{:}\frac{H_o \cos(d)}{f_c \cos(X + d)}$$

The scale of an oblique aerial photograph is dependent on both the magnitude of the tilt and the position of the point in the photograph, relative to the principal point, in the direction of tilt. The scale of an oblique aerial photograph is continuously changing in the direction of tilt, but is constant at right angles to the direction of tilt. If there are variations in altitude in the terrain then these variations will also affect the scale.

An advantage of oblique aerial photographs is that very large areas can be recorded in an oblique photograph. The disadvantage is that it is very difficult to take quantitative measurements from such photographs. In consequence, oblique photographs can be valuable when the user wishes to conduct some interpretation without measurement, or where a permanent record is required and the change in scale is not a serious disadvantage to that record.

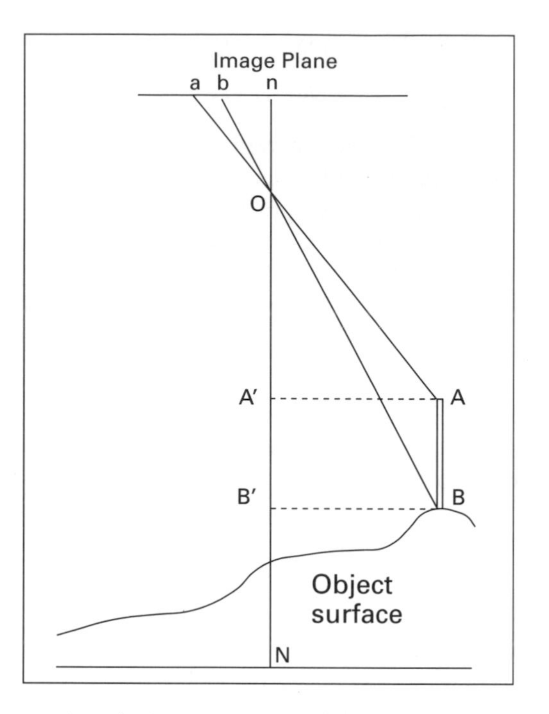

Diagram 2.44 Geometric relations in an aerial photograph.

The effects of height differences on an aerial photograph

A height difference, *AB*, (Diagram 2.44) introduces a displacement, ab, in the aerial photograph as well as introducing a change of scale between a and b. The displacement ab is radial from the **nadir point** in the aerial photograph, where the nadir point is defined as the point on the ground surface vertically below the front nodal plane (principal point) of the lens cone, and its corresponding point on the image plane. In a vertical aerial photograph the nadir and principal points coincide so that height displacements are radial from the principal point. The greater the height difference, and the further out the point is from the nadir, the greater the height displacement. The effects of height differences on displacements in a photograph are clearly shown on Photograph 3.2. Consider the height displacement ab in Diagram 2.44:

$$\frac{AA'}{OA'} = \frac{na}{f_c}$$

$$\Rightarrow \qquad OA' = \frac{AA' f_c}{na}$$

Also

$$\frac{BB'}{OB'} = \frac{nb}{f_c}$$

$$\Rightarrow \qquad OB' = \frac{BB' f_c}{nb}$$

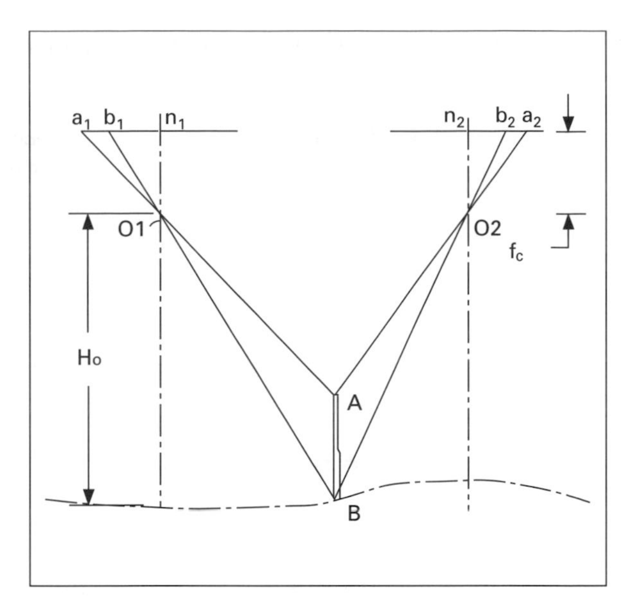

Diagram 2.45 Height displacements for a point in an overlapping pair of aerial photographs.

The height difference AB, ΔH_{AB} is given by

$$\Delta H_{AB} = OB' - OA'$$

$$= f_c\left(\frac{BB'}{nb} - \frac{AA'}{na}\right)$$

Now, $BB' = AA'$ by definition so

$$\Delta H_{AB} = BB'f_c\frac{(na - nb)}{na \times nb}$$

Let OB' equal H_o the flying height, then

$$\Delta H_{AB} = H_o f_c \frac{ab}{na}$$

When a point is recorded in two photographs then the height displacements in the two photographs will be of different magnitude, and will be in opposite directions as is shown in Diagram 2.45.

Types of lens cones

Aerial cameras are normally designated as being normal, wide angle or superwide angle depending on the field of view of the camera, in accordance with Table 2.2. As the field of view increases, so the focal length decreases for a constant image size and the flying height would decrease for images of the same scale as shown in Diagram 2.46. Constant flying height would increase the coverage and reduce the scale as the focal length is reduced.

Diagram 2.46 also shows that, for a given scale photography, as the focal length

Table 2.2 Relationship between lens cone type, field of view and focal length for aerial cameras

Class	Field of view (Cone angle)	Focal length (mm) (23 cm × 23 cm format)
Normal	< 60°	150
Wide angle	90°	115
Superwide angle	120°	90

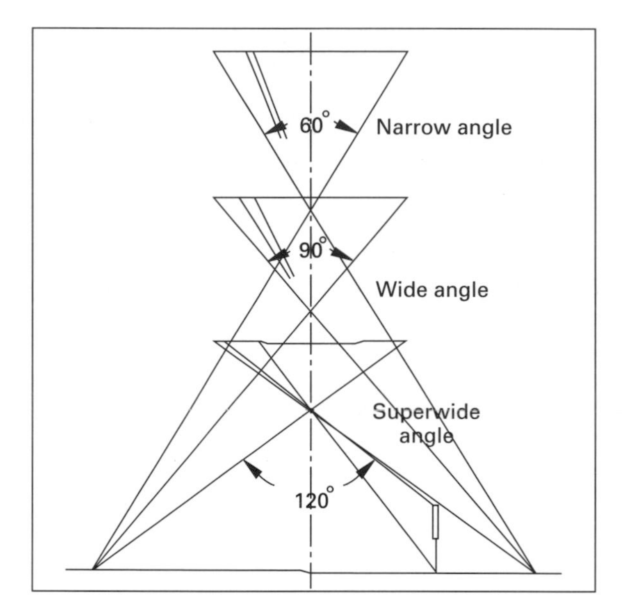

Diagram 2.46 Relationship between cameras of different cone angles.

decreases so the height displacement increases for a given height difference because the distance between the camera and the object decreases. In consequence, superwide angle photography will contain the largest displacements due to height differences and will thus be best for measurement of height differences. This same characteristic makes superwide angle photography the least suitable when displacements due to height differences interfere with the task being conducted, such as the construction of a mosaic or measurement of planimetric distances.

2.4.3 The scanner

In scanning sensors, the reflected radiance from the surface is optically scanned and electronically recorded. The reflected radiance is focused onto light-sensitive detecting devices that convert the electromagnetic energy into an electrical signal which is then converted into either a photographic image, or more usually into a digital signal for use in computer analysis. There are three commonly used forms of scanners – television, the moving mirror scanner and pushbroom scanners.

2.4.4 Television

The incident electromagnetic energy is focused by means of a lens cone onto a light-sensitive surface in the image plane, as occurs in a camera. The surface is usually either

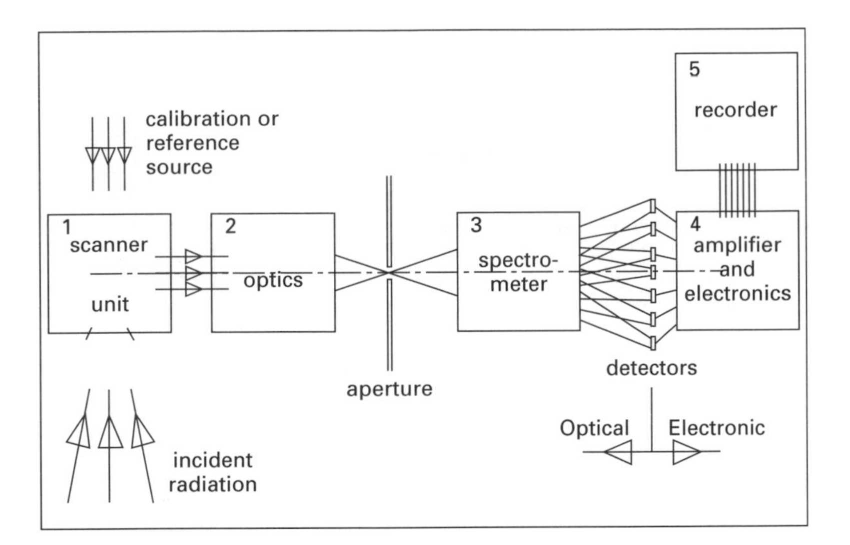

Diagram 2.47 The components of a moving mirror scanner.

a phosphor coating that is activated by electromagnetic radiation or a two-dimensional array of very small light-sensitive detectors, the most common being charged couple devices (CCDs) from their method of activation. (Photo 2.3 shows a typical video camera.)

The phosphor coating may be charged electrically so as to amplify the energy levels prior to converting them into an electrical potential. The surface is scanned as a series of parallel scanlines by an electron beam or gun that responds to differences in potential on the surface by creating differences in potential at the gun. As the surface is scanned by the gun, the electrical potential at the surface is reset to zero. The differences in potential on the gun are used to modulate an electrical signal that is then amplified before being stored as an analogue signal on a videotape recorder. This analogue signal can be converted into a digital signal for storage and use in a computer.

The CCD array elements accumulate a charge when activated by electromagnetic radiation that is then sensed by sequentially scanning all of the elements of the array. The magnitude of this charge is related to the intensity of the incident radiation.

2.4.5 The moving mirror scanner

A moving mirror scanner uses a rotating or oscillating mirror to scan the surface, and to direct a narrow beam of energy onto a small set of detectors, after passing through a filter unit. The moving mirror scanner consists of five main components: the scanning element; the objective lens; the spectrometer or filter unit; detectors; and the recorder (Diagram 2.47).

The scanning element A mechanically driven mirror scans across a strip approximately at right angles to the direction of movement of the platform.

The **instantaneous field of view (IFOV)** of the scanner is the field of view of the mirror at the instant that the energy is sensed on the detector. The IFOV controls the area on the surface that is being sensed at any instant by the mirror of the scanner, where this area depends upon the angular field of view of the mirror, the distance between and the angle of the sensor to the surface. The area of the mirror controls the total amount of energy that can come into the scanner, and thus larger mirrors are required if either narrow wavebands or low light levels are expected. The scanning unit may also collect energy from a reference source to calibrate the detector and electronics.

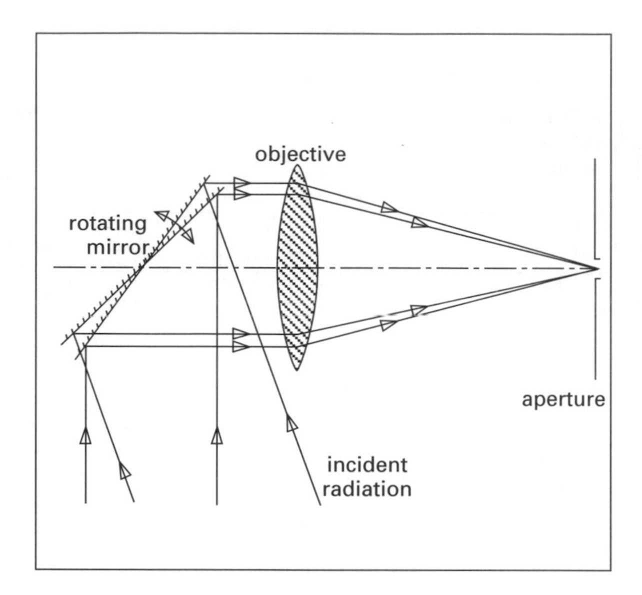

Diagram 2.48 The object space scanning unit.

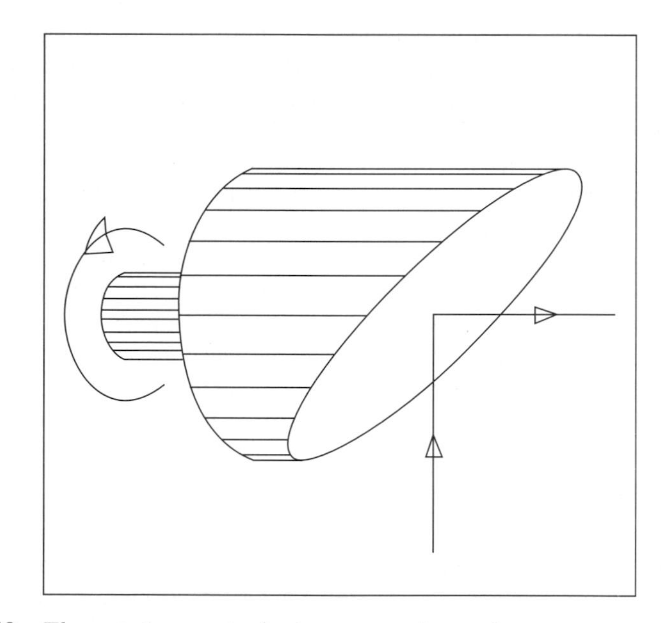

Diagram 2.49 The rotating conical mirror scanning unit.

The scanning unit can be of two types: object space or image space scanning units. **Object space scanners** form the majority of mechanical scanners because they always generate an image on the axis of the objective lens, and the optical aberrations that affect the signal are minimized. For these scanners the scanning mirror is either an oscillating (Diagram 2.48) or conical mirror (Diagram 2.49). The conical mirror is rarely used because mirror imbalance is a difficult problem to eliminate. The oscillating mirror generates a linear scan, unlike the conical rotating mirror – a most desirable characteristic. However, it is extremely difficult to keep the velocity of the mirror constant across the whole of the sweep, so that positional errors are common.

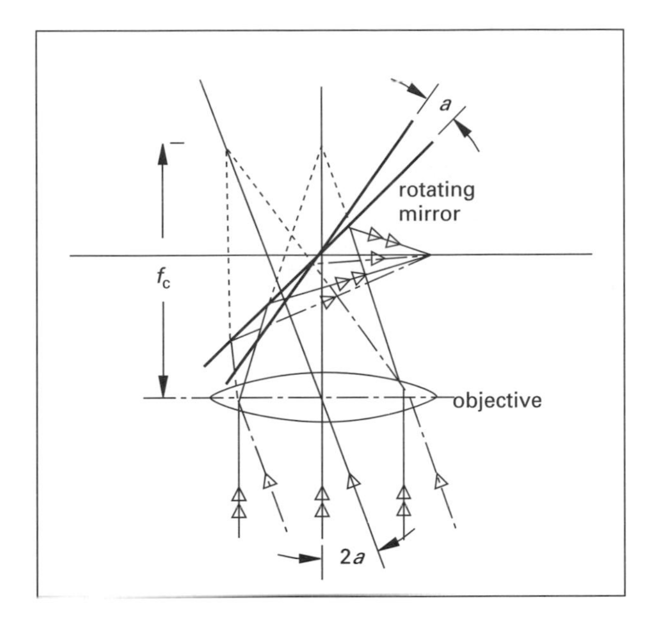

Diagram 2.50 An image space scanner.

In the object space scanners the scanning unit comes before the objective lens of the scanner whereas in **image space scanners** the objective lens precedes the scanning unit (Diagram 2.50). As a consequence the objective lens has to accept all of the energy coming in from the whole of the area to be imaged, unlike the object space scanners in which the objective lens only has to handle a parallel beam the size of the IFOV of the mirror.

The objective lens The objective lens focuses the incoming radiation onto the spectrometer or filter unit. The objective must be optically corrected to minimize radiometric and geometric distortions. Because of the wider field of view of the image space scanners, the objective lens is more susceptible to distortions and hence is more expensive.

Spectrometer or filter unit The purpose of this unit is to split the beam into a number of specified wavebands. This can be done in a number of ways. The beam can be split into a number of paths, and a separate filter be used on each path. This approach is relatively simple and can be used satisfactorily for a few wavebands when the amount of available energy is not a limitation. Alternatively, the whole of the incident beam is projected through a filter unit that splits the energy into separate wavebands. These wavebands are then projected onto separate detector units. A typical solution in this type of scanner is to use a dichroic grating to separate longer wavelength infra-red radiation from the visible radiation, and a prism that separates the visible spectrum into a number of wavebands (Diagram 2.51).

Detectors These convert the incident electromagnetic radiation into an electrical signal. The sensitivity of a detector varies as a function of wavelength, and can also vary with age. It is important to calibrate detectors regularly, even if the scanner containing the detectors contains a calibration source, as this can also deteriorate with age.

The recorder The electrical signal is normally amplified before it is used either to create an analogue record either on film or in a videotape recorder, or to be converted into a

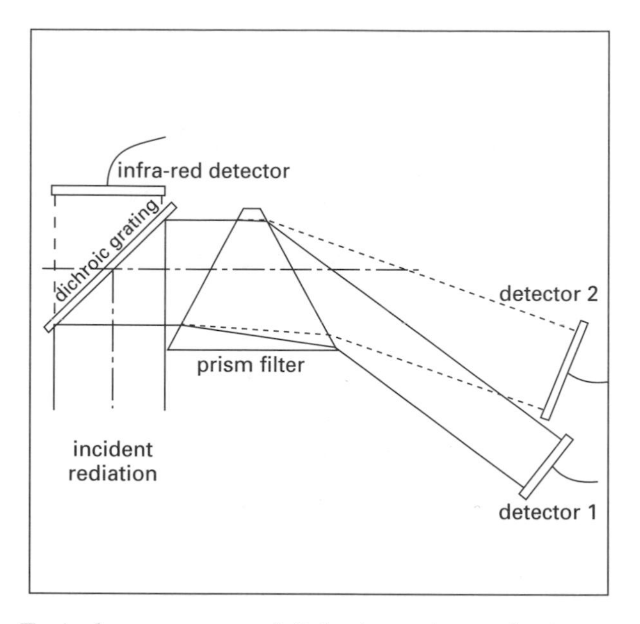

Diagram 2.51 Typical arrangement of dichroic grating and prism to split incoming radiation into a number of wavebands.

digital signal for subsequent analysis by computer. The continuous electrical signal can be used to create a photographic image by illuminating a light source that exposes a segment of emulsion in a strip across the film. The whole of the strip is exposed in synchronization with the mirror, the strip representing the swathe sensed in a scanline by the sensor. The film is stepped forward at the completion of each scanline so as to then record the next strip. For digital output the electrical signal is sampled at a regular interval by means of an analogue-to-digital convertor (ADC) to create digital values for the IFOV at a regular interval across the scanline.

The resolution of scanner data

The scanner records incident radiation as a series of scanlines at about right angles to the flight line of the platform. Within each scanline there is a set of recorded values called picture elements or **pixels**, with each pixel being the same size as the IFOV. The pixel is thus one measure of the spatial resolution limit of scanner data. The digital values normally range between 0, representing negligible radiance into the sensor, and $(2^n - 1)$ at the full scale or saturation response of the detector where each data value is stored in n bits. The values normally have a range of 2^n so as to allow the maximum number of values or shades for a given storage capacity. The radiometric resolution of the data is the difference in radiance between that recording a zero digital value and that just recording a maximum digital value, divided by the number of digital values or 2^n. The value of n is usually set at 8 or 16.

The number of wavebands being sensed, and their spectral range, specifies the spectral sensitivity or **spectral resolution** of the scanner. The frequency of acquisition, particularly of satellite systems, defines the **temporal resolution** of the data.

The actual resolution of the information that can be derived depends on a complex mix of these four resolutions: spatial, spectral, radiometric and temporal – as well as the discriminability of that information from other types of information in the data.

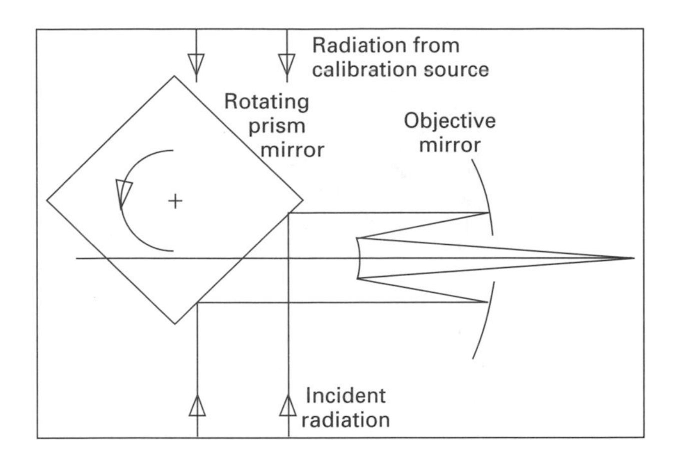

Diagram 2.52 The construction of a typical thermal scanner unit.

Thermal scanner data

Scanners used to acquire thermal data have to address some unique problems. All units in the scanner, including the detector, radiate energy as a function of their temperature in accordance with Planck's Law (section 2.2.1). The energy radiated by the sensor will activate the detector, thereby creating a large noise or dark current signal from the scanner that will usually swamp the signal coming from the target. The solution is to cool the detector, and all other elements of the scanner that may introduce a significant signal, sufficiently to reduce this noise to an acceptable level. This cooling is normally done by immersing the detector in liquid nitrogen with a boiling point of –120°C.

The other difficulty is in fabricating optics that will transmit the radiation. The common solution is to use a reflecting mirror as the objective optics as shown in Diagram 2.52).

Sources of error in oscillating mirror scanner imagery

As a scanner sweeps across the object, it builds up the image as a series of scanlines, recording a set of pixel values across each scanline. Because of the serial nature of data acquisition, changes in the altitude or orientation of the platform during or between scans will affect the location of pixels and scanlines on the surface of the object. The more common sources of geometric error in scanner data are as follows:

Changes in elevation or orientation of the platform The IFOV and angular interval between pixels remains constant, independent of the altitude of the scanner. Changes in elevation will change both the size of, and the ground interval between, pixels. None of the current methods of monitoring variations in altitude or orientation are to a sufficient accuracy to calculate the ground coordinates of image data to the precision required for most purposes. Image rectification is required for this purpose.

Changes in pixel size with sweep angle As the mirror rotates out from the nadir, so the distance between the object and the sensor increases. From Diagram 2.53, the geometric relations are:

$$\begin{aligned}\text{Pixel size at nadir} &= 2\text{MN} \\ &= 2H_o \tan\left(\frac{x}{2}\right)\end{aligned}$$

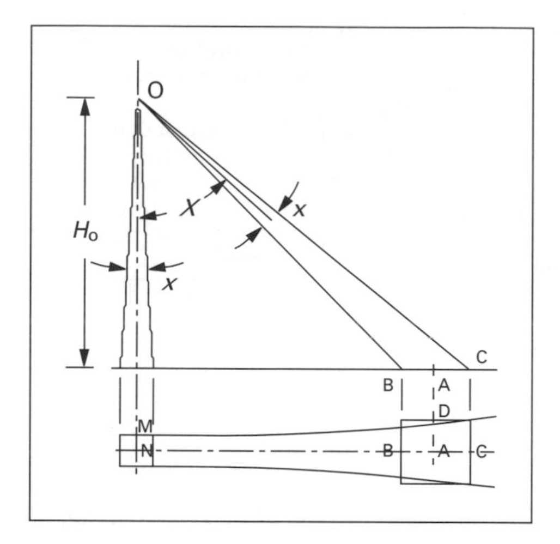

Diagram 2.53 Geometric relations in a scanline.

Pixel size at inclination X in scanline direction

$$= NC - NB$$

$$= H_o\left[\tan\left(X + \frac{x}{2}\right) - \tan\left(X - \frac{x}{2}\right)\right]$$

$$= H_o\frac{\left[2\tan\frac{x}{2}\left(1 + \tan^2 X\right)\right]}{\left[1 - \tan^2 X \tan^2\frac{x}{2}\right]}$$

Pixel size at right angles

$$= 2AD$$

$$= 2H_o\sec(X)\tan\left(\frac{x}{2}\right)$$

Overlap between adjacent scanlines increases as a function of sec (X). Since the sampling interval or angle along the scanline is constant at $x/2$, the pixel size and interval will both increase with increasing X. In consequence there will be negligible overlap between pixels along the scanline.

Finite scan time A finite scan time will introduce a shift or displacement across a scanline in the same direction as the direction of travel of the platform. The size of this displacement, D, depends on the time from the start of the scan to the pixel, t_i seconds, and the platform forward velocity, V according to the equation $D = Vt_i$ at pixel i.

Variations in mirror velocity Rotating mirror scanners can be kept to a uniform velocity across the scanline. With oscillating mirror scanners it is very difficult to maintain constant mirror velocity, due to acceleration at the start of the scan and deceleration at the end. This change in velocity creates errors in position of the pixels across the scanline if the pixels are assumed to be at a constant separation. The scanner mirror velocity can be calibrated, and corrections applied for this source of error.

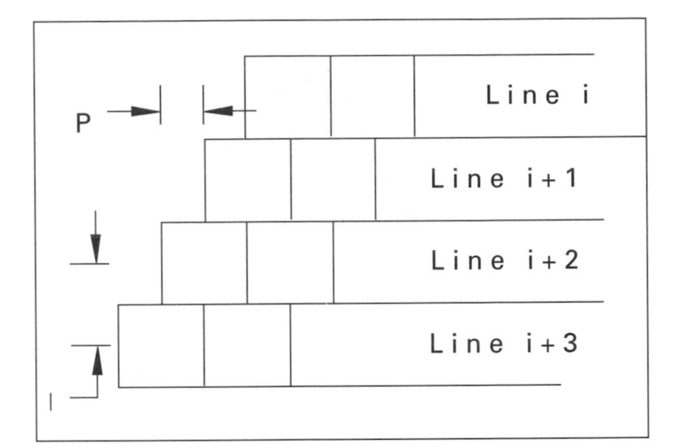

Diagram 2.54 Stepping of satellite scanner data due to the earth's rotation.

Earth rotation Introduces displacements due to the relative velocities of the earth and the platform, ignoring the forward velocity due to the platform. With aircraft there is little relative velocity and so there is little error from this source. However, there is considerable relative velocity between a satellite and the earth and in consequence large displacements are incurred. If the earth radius is given as $R = 6\,370\,000$ m then in time t seconds a point on the earth's surface at the equator moves an angular distance of d seconds of arc. The earth completes one rotation of $360 \times 60 \times 60$ seconds of arc in $24 \times 60 \times 60$ seconds of time. In time t:

$$d = \frac{360 \times 60 \times 60t}{24 \times 60 \times 60}$$
$$= 13.84t$$

Thus the angular distance moved is $1.098 \times 10^{-4}t$ radians.

At latitude θ the effective radius $R' = R\cos(\theta)$ so that the distance covered (in radians) is.

$$D = dR'$$
$$= 1.098tR\cos(\theta) \times 10^{-4}$$

This creates a step between adjacent scans as shown in Diagram 2.54.

Earth curvature and height differences Curvature of the earth introduces displacements in the position sampled by a scanner as shown in Diagram 2.55. With both earth curvature and heights, the errors increase as the angle of the mirror increases out from the nadir.

Radiometric errors are another major source of error in scanner data. The detectors in a scanner change and decay with time. There is electronic noise in the detector circuitry that can affect the signal attributed to the detector. Whilst these errors do not affect the geometric quality of the data, they do affect the validity of the data received, and hence

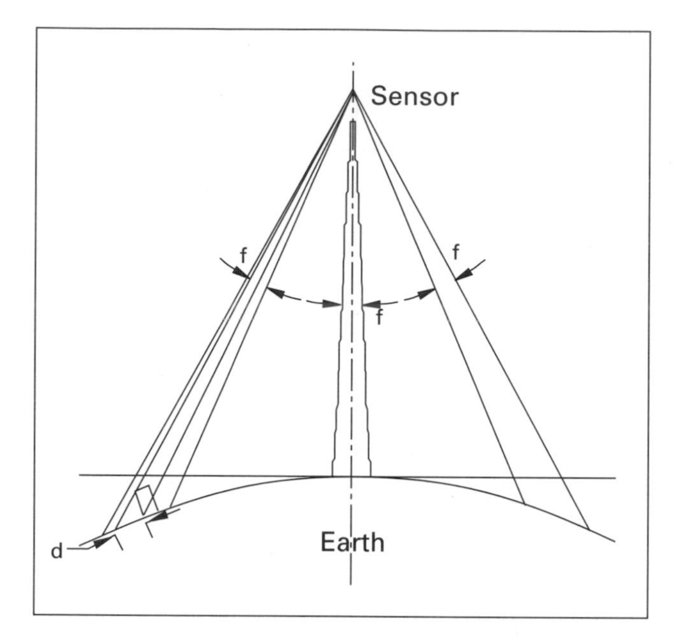

Diagram 2.55 Displacements due to earth curvature and height displacements.

the accuracy of any derived information. The major sources of radiometric errors are as follows.

- **Changes in detector response** with time or environmental conditions.
- **Detector lag** which is a finite lag between the time when energy falls on the detector, and the detector responds. Similarly once a detector is responding then there is a lag in the response after the incident radiation ceases. If the incident radiation is intense and saturates the detector then the lag time can be increased.
- **Sensor electronic noise** – all detectors and circuitry generate electronic noise or dark current. The dark current can change with age of the circuitry in the scanner.

2.4.6 Pushbroom scanners

This sensor uses a wide angle optical system that focuses a strip across the whole of the scene onto a linear array of detectors. The signal from each detector is sampled to create a record for the pixels in that line. The detector array is sampled again when the platform has moved forward by the size of the pixels on the earth's surface. In this way the sensor can be visualized as a broom that sweeps out a swathe of data.

If only one band is to be imaged then a single array of detectors is adequate. When multiple wavebands are to be recorded than a linear array of detectors is required for each band, with appropriate filters set in front of each array. The arrays are arranged so that they image adjacent scanlines.

The main disadvantages of the pushbroom scanner are the number of detectors and their calibration. The major advantage is that the scanner does not contain a movable mirror and so a major source of malfunction or wear, particularly in a satellite, is removed. The other advantage is the long dwell time between sampling the detectors compared to the time available in mechanical scanners. This dwell time can be used to accumulate a larger signal and hence used to either sense in narrower wavebands or develop a better signal-to-noise ratio.

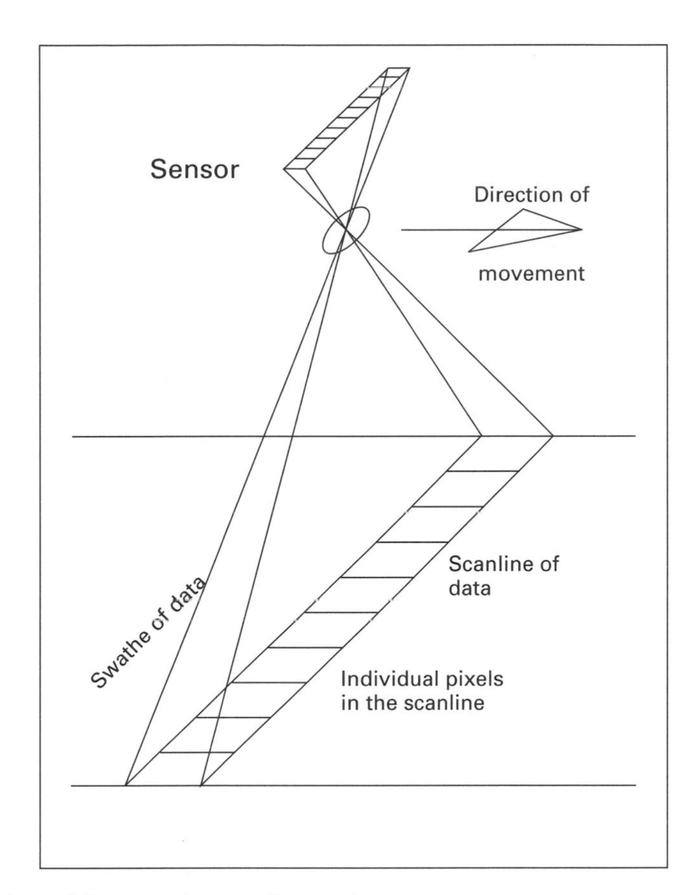

Diagram 2.56 Schematic layout of a pushbroom scanner.

2.5 Active sensing systems

2.5.1 Introduction

Active sensing systems illuminate the target with their own source and then sense the reflected radiation. This is in contrast to the passive systems as previously discussed that depend upon independent, usually naturally occurring, sources such as the Sun or the Earth itself. This makes active systems much more independent of natural events than passive systems.

The most common type of active sensing system is radar (*ra*dio *d*etection *a*nd *r*anging) which was developed during the Second World War. The name comes from the early use to which radar was put in detecting and estimating the distance to aircraft. This type of radar depicts the reflecting objects as bright spots on a black screen. In contrast remote sensing uses radars that create an image of the surface. The image is depicted as variations in density or greyscale on a photograph, or variations in digital values across the area imaged.

The theoretical resolution limit of radars are coarser than that of imagery in the visible region because the wavelengths of the radiation used is much longer in the radio portion of the spectrum than in the visible and NIR portion. Another constraint is the energy required to generate the illumination signal. The power requirements of radars is many times greater than the energy requirements of passive systems. Generally this does

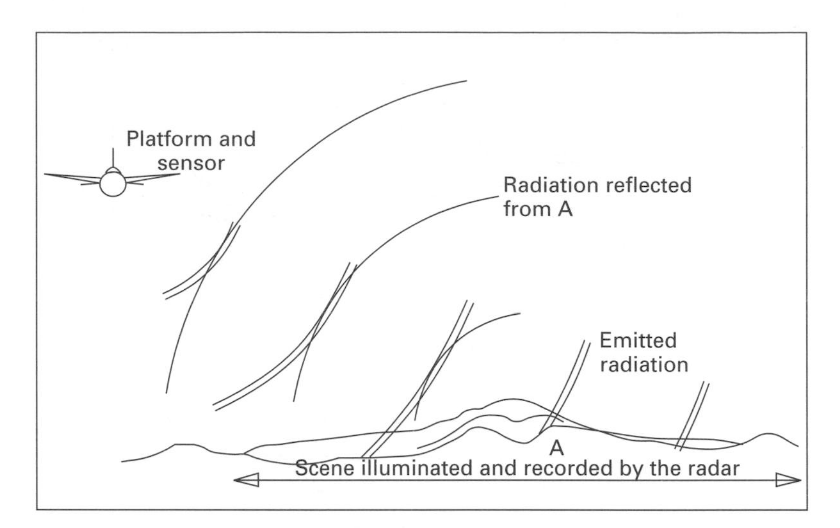

Diagram 2.57 Geometry of a side-looking imaging radar.

not inhibit the acquisition of airborne radar imagery, but it may limit the capability of satellite borne sensors.

2.5.2 The geometry of radar systems

Radars used in remote sensing are side-looking airborne radars (SLAR) (Diagram 2.57). The radar antennas transmit a pulse of energy on a slant angle from the side of the aircraft down to the ground surface. Some of the signal is reflected back from the surface towards the radar antennas. The radar antennas accepts the return energy, converts the travel time into a slant distance that is converted into a ground distance, and compares the output with the return signal strength. The radar creates an image that is proportional to the strength of the return relative to the output signal and corrected for the path length.

Radars are categorized by the wavelength of the radio waves transmitted by the instrument. The wavelength of the radar affects the way that the radar images the surface. It is important to select the most appropriate radars for a task, and to be aware of the radar wavelength in the interpretation of radar imagery. The radars are grouped into categories as a function of wavelength as shown in Table 2.3.

Resolution of radar data

A pulse of the radar beam illuminates area A on Diagram 2.58(a) at a particular instant. The duration of the pulse defines the resolution of the data in the cross-track direction. The shorter the pulse the higher the resolution, but the weaker the accumulated return signal. This resolution is constant at all distances from the sensor, since the pulse length is constant.

The pulse beam width is shown as area B, so that the width of B defines the resolution in the along-track direction. The along-track resolution decreases with distance from the aircraft due to the increasing beam width. The angular width of the beam, B, is inversely proportional to the length of the antennas so that longer antennas will reduce the angular beam width. The intersection of areas A and B indicates the resolution of the data. The reflected return signal is sampled at a regular interval, equivalent to a strip of pixels out from the aircraft. The strength of the return signal is dependent upon the path length and the reflectance or backscatter from the surface. The return signal slant distance

Table 2.3 Radar bands and their wavelength ranges

Band	Frequency (mHz)	Wavelength (cm)	Common wavelengths of imaging radars (cm)
P	220–390	133.3–76.9	
L	390–1 550	76.9–19.3	23.5 and 25
S	1 550–3 900	19.3– 7.7	
C	3 900–6 200	7.7– 4.8	6
X	6 200–10 900	4.8– 2.7	3.0 and 3.2
K	10 900–36 000	2.7– 0.8	
K_U		2.4– 1.7	
K		1.7– 1.1	
K_a		1.1– 0.8	
Q	36 000–46 000	0.8– 0.6	
V	46 000–56 000	0.6– 0.5	

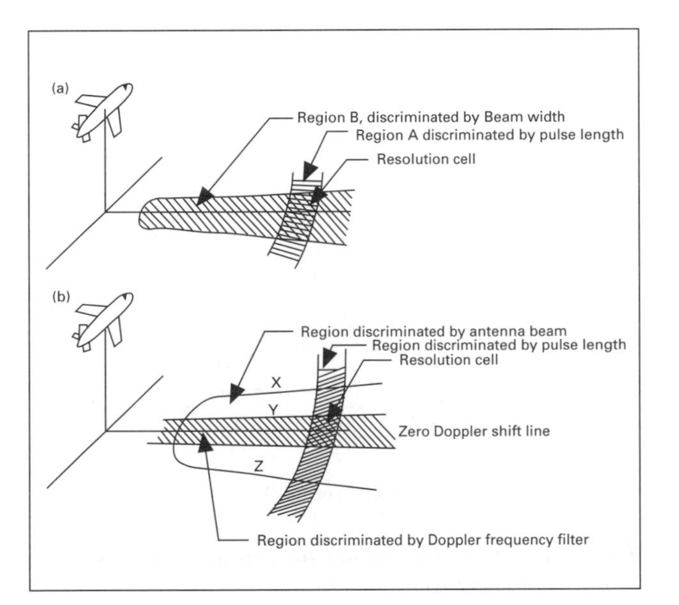

Diagram 2.58 Resolution constraints on radar imagery: (a) real aperture radar with beam width and pulse constraints; and (b) synthetic aperture radar with filter and pulse constraints.

is converted into a horizontal distance and the signal is corrected for atmospheric attenuation to create the image data. This type of radar is known as **real aperture radar** (**RAR**). Its coarse along-track resolution will always be constrained by the largest antennae that can be carried on the platform.

Another way of narrowing the beam width is by the use of **synthetic aperture radar (SAR)**. Instead of using one long antenna, a short antenna is used at a number of points as if it is part of a long antenna. The values received at the set of antenna positions have to be combined to derive an image of the area. Another way of perceiving synthetic aperture radar is in terms of the Doppler shift. The points X,Y and Z in Diagram 2.58(b) at different locations in area A will exhibit different frequency shifts in the radar signal

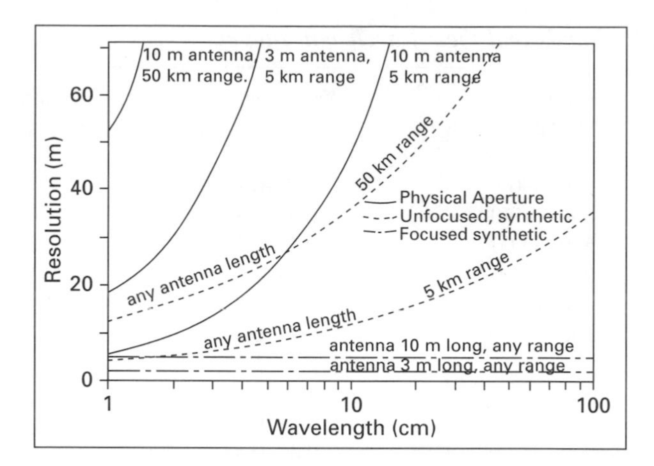

Diagram 2.59 Design trade-off information for SARs: the effects of changes in wavelength on resolution for different types of SAR.

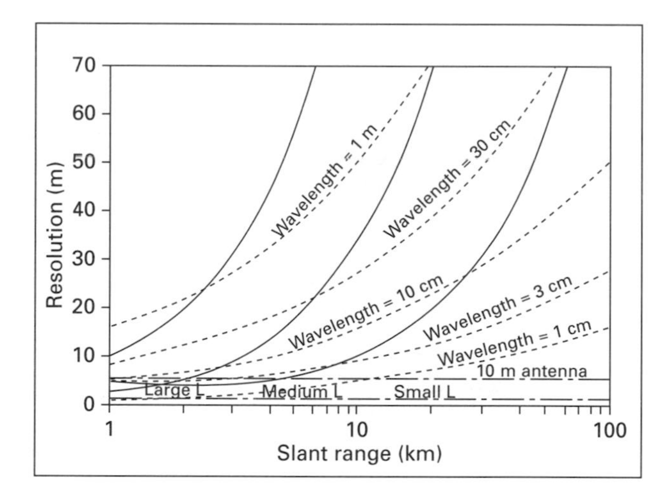

Diagram 2.60 Design trade-off information for SARs: the effects of changes in slant range on resolution for different types of SAR.

transmitted from and returned to the radar. The frequency shift is due to the Doppler shift; when a signal is transmitted from a target to an object, the frequency of the signal is affected by the relative speeds of the target and platform. As the points X,Y and Z will have different speeds relative to the platform, both in velocity and direction, so the radar signal will be affected by Doppler shift. These differences are quantized by use of a narrow bandwidth filter which allows only a small zone, C, through this filter, creating smaller pixels than previously and hence improving the resolution of the data.

Raw synthetic aperture radar data is a record of response and frequency distributions that have to be integrated in a computationally expensive way to derive the final image data. However, the advantage of synthetic aperture data is that much higher resolutions can be achieved. Indeed the resolutions of SAR data are independent of the distance between the sensor and the surface as shown in Diagrams 2.59 and 2.60.

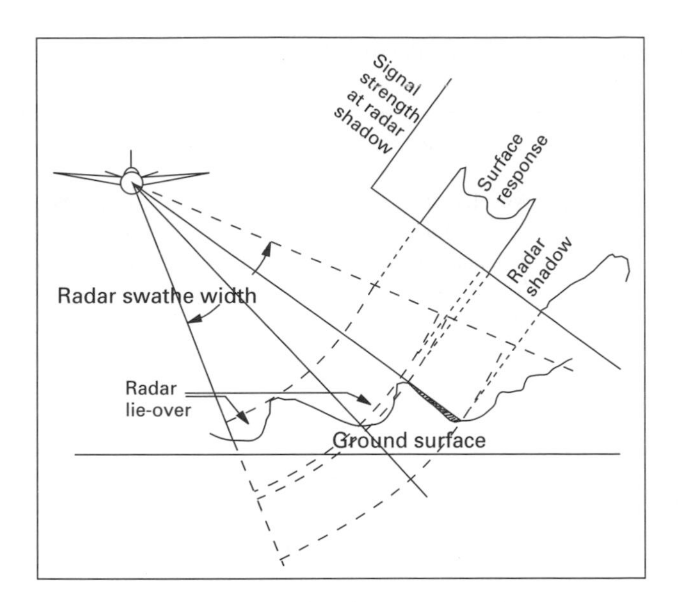

Diagram 2.61 The effects of topographic heights on radar imagery.

Effect of height displacements

The slope of the radar signal can cause **radar shadows** and **radar lieover** (Diagram 2.61). Radar shadows occur when the slope of the terrain is at steeper gradients away from the sensor than the radar signal. Radar lieover occurs when the top of a hill is closer to the sensor than the bottom, and so is imaged first.

Radar images contain displacements due to height differences, much like aerial photographs. These displacements can be used in pairs of radar images to construct stereoscopic images. Suitable sets of radar imagery can therefore be used stereoscopically to visually analyse landforms.

2.5.3 The information content of radar imagery

The radar equation is the fundamental equation showing the amount of energy received by a radar system from a target:

$$P_r = \frac{P_t G_t}{4\pi R^2} \times \sigma \times \frac{A}{4\pi R^2}$$

where

P_r = received power
P_t = transmitted power
G_t = transmitter gain
R = slant range
σ = effective backscattering cross-section
A = effective aperture of the receiving antennas

The first component of the equation indicates the power per unit area at the target; the last quantity indicates the capacity of the antennas to accept reflected radiation. These

Table 2.4 Factors that affect the effective backscatter cross-section area per unit area

	Source	Target
Properties	Wavelength	Roughness
	Polarization	Slope
	Incidence angle	Inhomogeneity
	Azimuth look angle	Permittivity and conductivity
	Resolution	Resonant sized objects

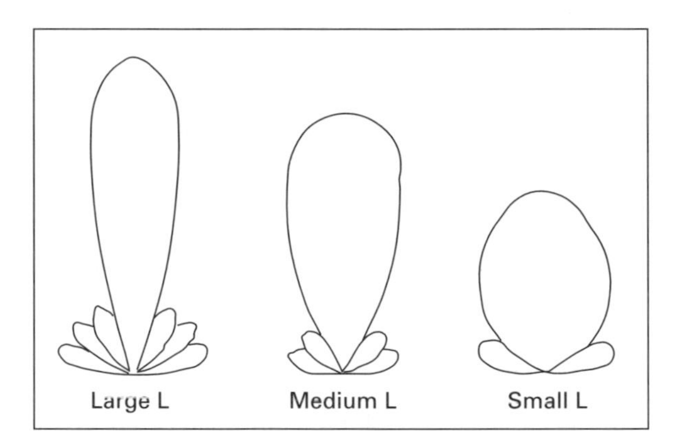

Diagram 2.62 Backscatter cross-section area angular distribution for uniform normally incident radiation as a function of facet size, L.

two components indicate the influence of system components on the received power at the antennas.

The effective backscatter cross-section, σ, indicates the effect of the surface on the signal. It is defined as the cross-section area that would reflect 100 per cent of the signal, yet have the same response as the actual signal. Since the actual signal will be due to surfaces that are imperfect reflectors, so σ must be less than the actual cross-section. The effective backscatter cross-section is thus a function of both the reflective ability of the surface and the area being sensed. To eliminate the effects of the variable area, the effective backscatter cross-section per unit area or σ_o is used. Then, σ_o is independent of the spatial resolution of the data, and the factors that do affect σ_o are as given in Table 2.4.

Surface roughness

Most surfaces are neither perfect lambertian nor specular reflectors to radar signals, but act somewhere between these limits. One theory of radar reflection considers that a surface consists of many facets. If the facets are small relative to the radar wavelength, the reflection due to normally incident radiation would be approximately lambertian (Diagram 2.62). As the facet size increases, so the reflection becomes more specular until very large facets, relative to wavelength, are specular reflectors.

A surface is considered to be smooth in radar if the height of the features in the surface, Δh are given by the equation,

$$\Delta h < \left(\frac{1}{8}\right)\lambda \cos \theta$$

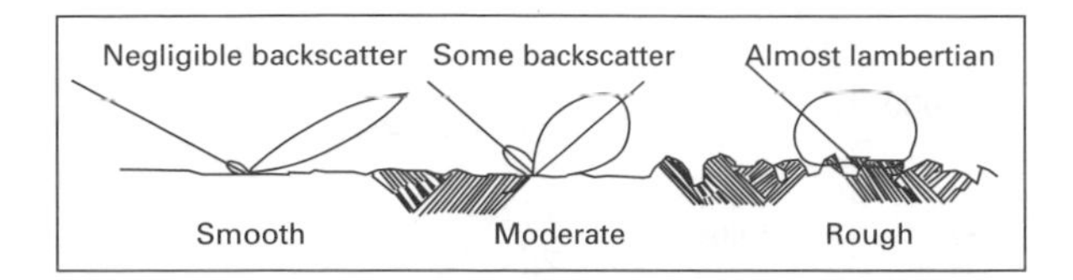

Diagram 2.63 Backscattering cross-section angular distribution for incident radiation at angle of incidence, θ, as a function of facet size, L.

Surface roughness is thus a function of both the wavelength of the radar, λ and its angle of incidence, θ.

If a surface is smooth relative to wavelength then most of the energy is reflected away from the antennas and the surface appears dark in the image (Diagram 2.63). This is the typical situation for smooth water, airport runways, claypans, etc. that are smooth to most radars. As the surface becomes rougher, so the backscatter towards the receiving antennas increases, and the signal increases. Ploughed or cultivated fields appear rough to shorter wavelength radars and will thus reflect strongly back towards the sensor.

The effect of surface slope on radar response can also be explained in terms of the facet theory. The slope affects the direction of specular reflection from the surface. In consequence, slopes facing the radar will, in general, show a strong return whereas those slopes facing away from the radar will show a weak return. The effect of the slope is moderated by surface roughness.

Inhomogeneous targets are those that are not spatially and/or vertically homogeneous in those characteristics that affect a radar signal. These targets will give different returns when illuminated from different directions. Consider a sloping, planar surface. If it faces towards the radar sensor then the return signal will be much greater than if the surface is sloping at right angles to the sensor. The signal from a surface that is sloping away from the sensor will be much smaller again. Consider the more subtle case of a cultivated field where the cultivation rows follow the rectangular sides of the field. The radar response will be much higher from the sides that have cultivation rows parallel to the flightline because of the reflective characteristics of these rows, than from the sides where the rows are perpendicular to the flightline when the radar sees a much smoother surface. These are two of many cases of inhomogeneity. The effect of these inhomogeneities is that radar data do not exhibit a consistency in their returns, sometimes features may be there, other times not.

Because of the wide beam width of radar, particularly with synthetic aperture radar, inhomogeneities may cause the signal return from a target to be different in different parts of the beam. In real aperture radar these inconsistencies are averaged out in creating the final signal. In SAR the same resolution element can exhibit significantly different returns at different positions in the beam. If the number of samples taken of a resolution element are sufficiently large then these differences are satisfactorily averaged out. If, however the number of samples is small then the variation may not be averaged out resulting in speckle in the data, an artifact that can be a major problem during analysis.

Permittivity and conductivity

The permittivity of a surface is its dielectric constant, given as a complex term, $e = e' - e''j$, in which the imaginary part, e'', is the electrical conductivity of the surface. Sample values of e'' for fresh, sea and pure water are shown in Diagram 2.64.

The dielectrical properties of various materials are listed in Table 2.5 for one wave-

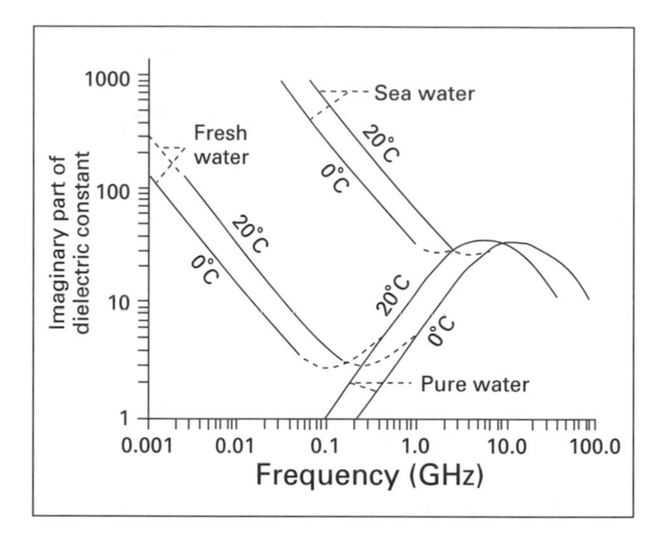

Diagram 2.64 Conductivity of pure, fresh and sea water at different frequencies and at two temperatures, 0°C and 20°C, from Paris (1969).

Table 2.5 The dielectric properties of sample materials in the microwave region (most at 3 MHz) from Lintz and Simonett (1976)

Material	e'	e''/e'
Sandy, dry soil	2.55	0.006
Loamy, dry soil	2.44	0.001
Freshly fallen snow	1.20	0.0003
Distilled water	77	0.157
Mahogany wood	1.9	0.025

length since the dielectric constant of surface materials changes with the frequency of the radar. Permittivity measures the ability of a surface to absorb energy, the higher the permittivity the higher the proportion of the energy that is absorbed. Conductivity is a measure of the ability of a medium to conduct electrical energy away from the surface. High conductivity indicates a good conductor.

All of the permittivities listed in Table 2.5 are quite small, except for distilled water. The presence of water in the materials of a surface, whether they be vegetation or soil, will dramatically affect the permittivity of the surface. The higher the permittivity and conductivity, the less energy will be available after the signal intersects the surface to be reflected back to the receiver. Thus moist surfaces will not exhibit high reflectances in comparison with similar but dry surfaces (Diagrams 2.65–7).

The ratio e''/e' also indicates the depth of the material that might be penetrated by the radar as a function of wavelength, and still return a signal to the antennas. If $e''/e' < 0.16$ then the distance at which the penetrating wave is reduced to 37 per cent of its original value exceeds a wavelength in the material, assuming 100 per cent reflection in the material. Because the signal has to return the same distance, a value of 0.08 should be used. Thus for dry soils and dry snow, radar should penetrate several wavelengths into the surface.

Under some conditions resonant sized objects can exhibit a high backscatter

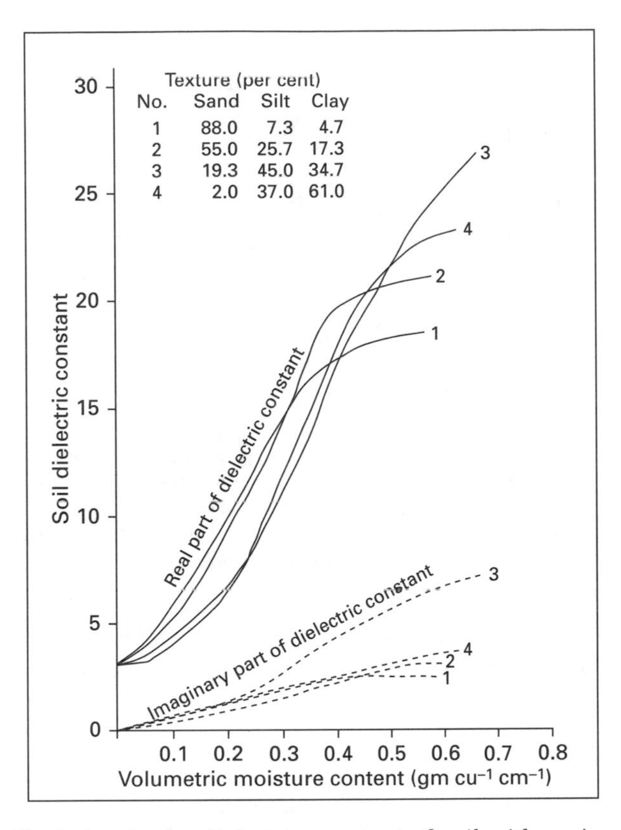

Diagram 2.65 Variation in the dielectric constant of soil with various volumetric moisture contents for four soils at 5 GHz frequency, from Wang and Schmugge (1980).

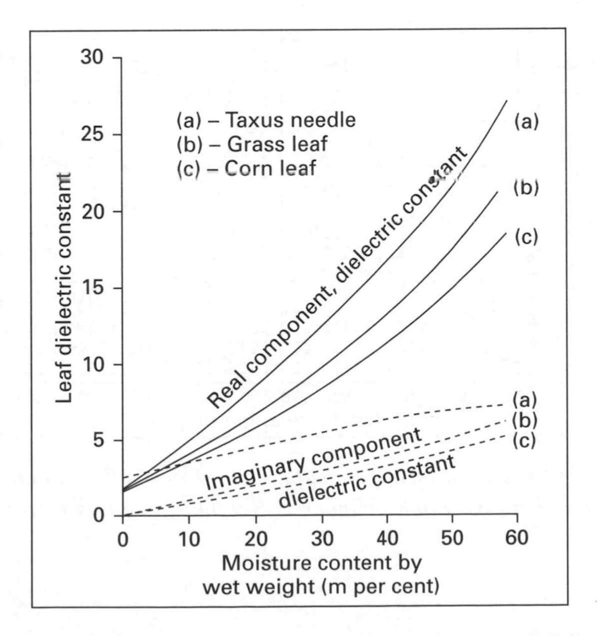

Diagram 2.66 Variation in the dielectric constant for leaves of three species as a function of leaf moisture content at 8.5 GHz frequency, from Carlson (1967).

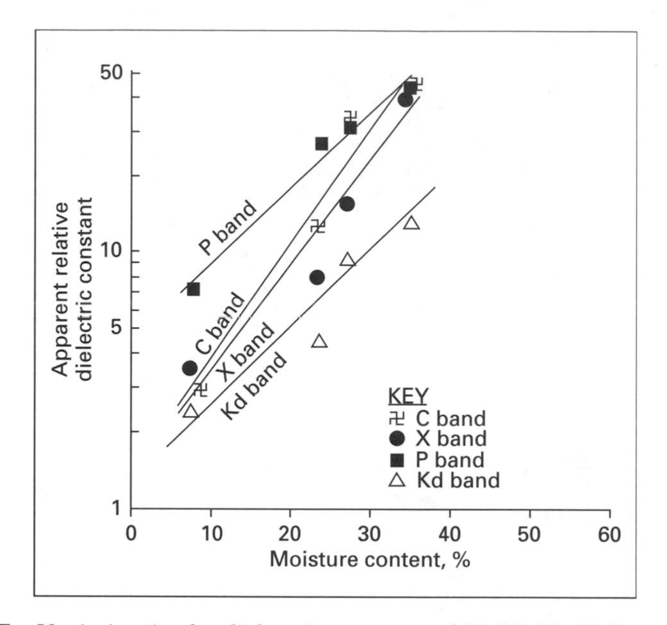

Diagram 2.67 Variation in the dielectric constant of Richfield silt loam as a function of moisture content (per cent) for four different radar frequencies, from Lundien (1966).

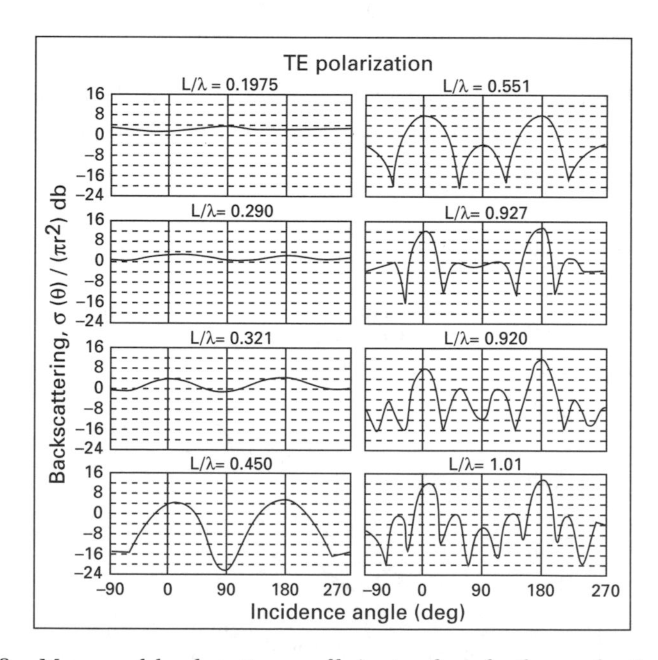

Diagram 2.68 Measured backscatter coefficients of perfectly conducting thin circular cylinders of various lengths. Each cylinder is 3.9 mm in radius (after Carswell, 1965).

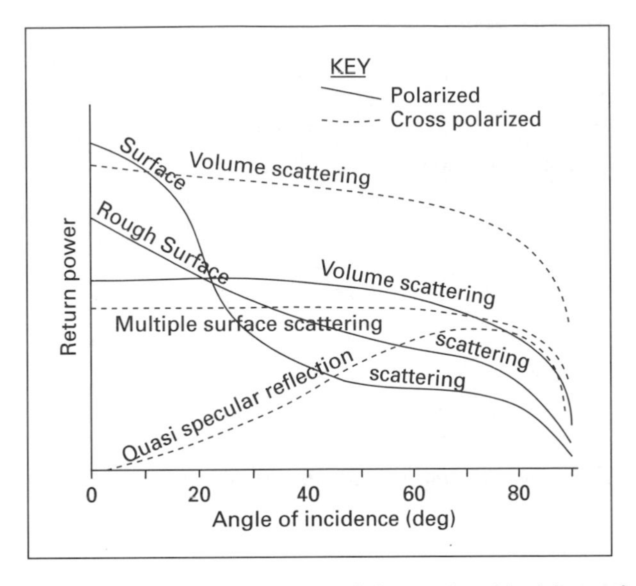

Diagram 2.69 Return power as a function of the angle of incidence for like- and cross-polarized conditions.

coefficient. When the size of an object, L is very small relative to the radar wavelength, λ, then the backscatter tends to be small and constant at all angles of incidence. As L/λ approactes 1.0, then the backscatter coefficient can be very high at some angles of incidence and low at others. The effect of angles of incidence on backscatter for a variety of surface types are shown in Diagram 2.69.

The change in backscatter coefficient is caused by resonance of the objects when they are about the same size as the wavelength of the radiation. The effect of the objects resonating is to reinforce the radiation when the frequency of resonation matches that of the incident radiation, and to reduce the radiation when they are out of synchronization. Resonant sized objects often act as inhomogeneous surfaces due to the nature of the objects.

Radars can be polarized for transmission, and filtered to accept polarized signals on reception. They are usually polarized for combinations of horizontal (H) and vertical (V) polarization, being called **like-polarized** when they are either HH or VV and **cross-polarized** for HV or VH. Surface and volume scattering affects the way radar signals are reflected and hence affects the loss of polarization on reflection of the signal at the surface. Significant loss of polarization usually occurs with multiple reflections, as in volume scattering. Targets that have high volume scattering and high loss of polarization are likely to have a larger backscatter coefficient in cross-polarized images than they will in like-polarized images. Surface scattering targets are likely to have a relatively higher backscatter coefficient than other targets in like-polarized images.

2.6 Platforms

Platforms are the means of supporting and moving a sensor so as to acquire data or imagery as required. Different platforms have characteristics that make them suitable for some tasks and not for others. An appreciation of these characteristics allows a user to select the platform most appropriate for the task.

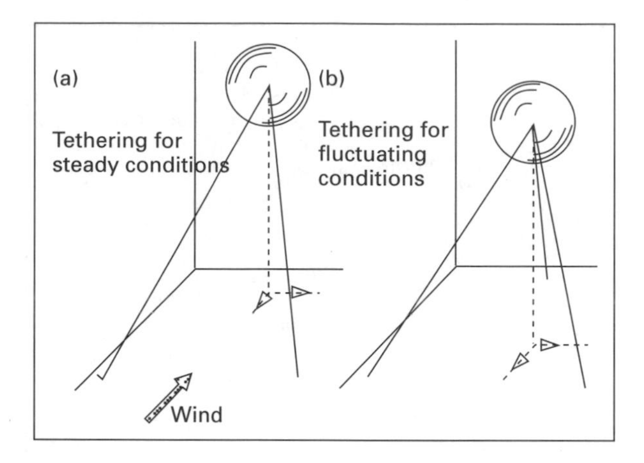

Diagram 2.70 Tethering of a balloon under (a) steady and (b) fluctuating conditions.

2.6.1 Terrestrial platforms

Terrestrial platforms are those rigidly secured to the Earth's surface, including the tripod, boom (Photo 2.4), tower and cherry picker. They are suitable for the acquisition of large scale photography, for temporal data sets and for observations with spectrometers. The advantage of terrestrial platforms is that the precise position and pointing of the sensor can be known relative to the surface. The other advantage of terrestrial platforms is the ability to precisely re-establish the position of the sensor relative to the surface. This means that sets of images or spectral observations can be taken over a period of time for the same area. To ensure precise re-establishment it is necessary to establish permanent marks that can be used to position and point the sensor.

2.6.2 The balloon

Balloons can be used to lift a payload of sensors above an area of study. Balloons would normally be tethered using either two or three lightweight but high tensile lines (Diagram 2.70). The sensors on the balloon should be activated by a radio controlled device so as to reduce the weight being carried by the balloon. Balloons move about in the air and so the sensor should be located in a gimbal mount to ensure that it is pointing vertically down.

The advantage of the balloon is the ability to establish a sensor in a fixed position for relatively long periods so as to monitor specific events. The disadvantages include the relatively high cost of launching the balloon, and the difficulty of replacing magazines and correcting faults.

2.6.3 The helicopter or boat

Both are highly manœuverable platforms that can also be held relatively stationary over the object being sensed and can be stopped so as to collect field data. However they sacrifice stability, the helicopter in particular having high vibration levels that can seriously interfere with the acquisition of data.

2.6.4 Aircraft

The aircraft provides a platform that is fairly stable and moves at a relatively constant high velocity in a specified direction. It is used to acquire data covering large areas as strips or runs of photographs along each flightline, with a number of parallel flight lines covering the whole area (Diagram 2.43).

Within each flightline photographs are usually taken so as to have 60 per cent overlap between adjacent photographs in the run, and at least 10 per cent sidelap between adjacent strips. The exposures along the flightline are triggered by means of an intervalometer. The time between exposure can be calculated in either of two ways. The simplest is to use the aircraft ground speed and the distance between adjacent photograph centres to compute the time between exposures. However this technique suffers from the difficulty of estimating the aircraft ground speed. A better method is by the use of a viewing telescope which displays the ground under the aircraft, and includes a moving line projected onto the field of view. The projected line is adjusted so as to move across the screen at such a rate that it stays stationary relative to the ground features. When it does this it provides a very good estimate of the aircraft ground velocity. A navigation sight is used to maintain the aircraft heading.

The sensor may be placed in a gyroscopic mount to minimize the effects of aircraft movement during flight. These movements may not create serious distortions in photography because of the relatively short exposure time. Significant movements of the platform will introduce errors in scanner data which is continuously acquiring data, and the distortions are hard to correct. The motions that are of interest are set out below.

Forward motion This can cause image blur since the object moves relative to the camera during exposure. With aircraft velocity, V (m s^{-1}), and exposure E (s), the distance (in mm) moved on an image, I, of scale factor S will be given by

$$I = \frac{1000\,VE}{S}$$

If the resolution of the film, R, is set in lines per mm then the maximum exposure that will maintain this resolution can be calculated from:

$$R = 1/I \leq \frac{S}{1000\,VE}$$

$$\text{so that} \quad E \leq \frac{S}{1000\,VR}$$

The maximum exposure time can be increased by increasing the scale factor, reducing the aircraft speed or reducing the acceptable resolution in the photography.

Sideways motion Side wind causes the direction of heading to be different to the direction of flight. Modern camera and scanner mounts can be rotated and tilted to allow for differences in both heading and nose-to-tail tilt. If the mount does not allow for it then each photograph will be rotated relative to its neighbours as shown in Diagram 2.71.

Rotations of pitch and roll With good aircraft platforms these rotations are kept in the range 2–5°.

With cameras these rotations introduce constant tilts across each photograph which can be corrected using appropriate facilities and/or techniques. With a scanner, where the imagery is acquired continuously during the flight, these rotations will introduce significant displacements into the imagery, as shown in Diagram 2.72, that can be difficult to eradicate.

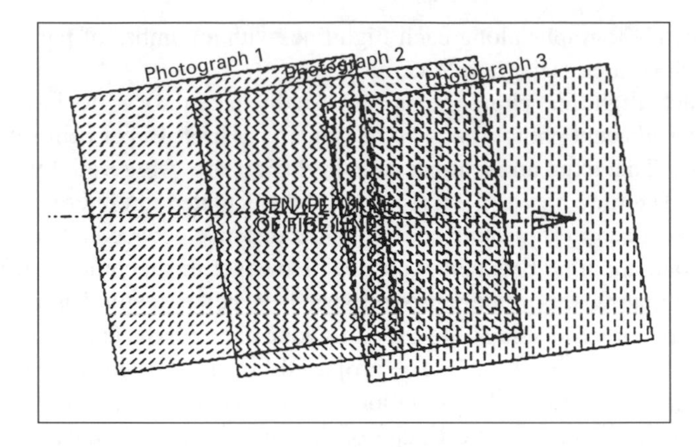

Diagram 2.71 The effect on photography of differences between aircraft heading and direction of flight.

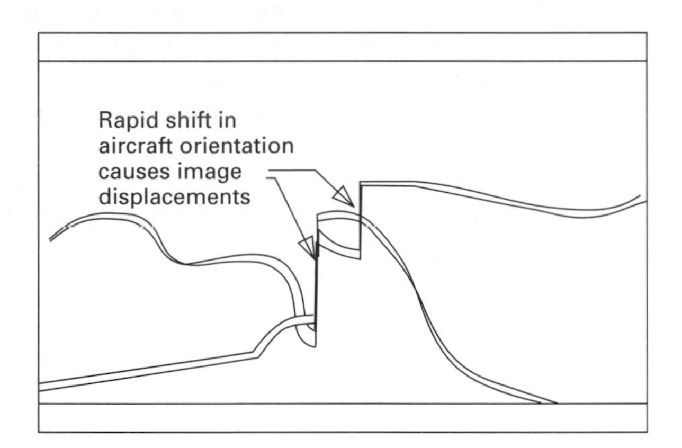

Diagram 2.72 The effect of sudden changes in roll rotation on scanner imagery.

Hotspots

Hotspots are areas on a photograph that appear as a brightening on the photograph in a way that is not related to the detail on the photograph. Hotspots are caused by either cloud reflection, as depicted in Diagram 2.73 or refraction around aircraft, causing a hotspot at a point on the Earth's surface directly below the line from the Sun through the aircraft. At low altitudes the aircraft shadow dominates; at higher altitudes refraction of the Sun's rays causes brightening the area on the photography. It is a problem when the altitude of the Sun is high enough for this point to occur on the photography.

2.6.5 Planning an aerial sortie

Prior to planning an aerial sortie the following information must be known:

1. scale and resolution of the required photography;
2. film and filters required;
3. area of coverage and overlap requirements (60 per cent for stereoscopy) – does the information sought from the imagery influence the direction of the flight lines?;
4. the essential or preferred time of day or season for acquisition of the photography; and

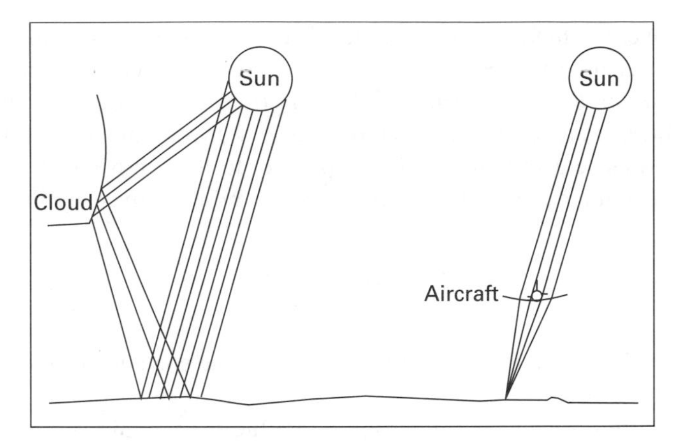

Diagram 2.73 Creation of hotspots due either to reflection from a cloud or to refraction around an aircraft.

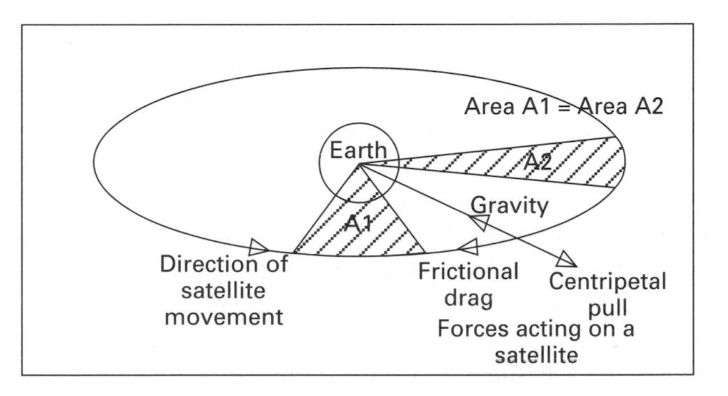

Diagram 2.74 The forces acting on a satellite.

5. any specific geometric requirements of the photography – e.g. the use of normal angle camera versus superwide angle.

Given this information the flight planner will have to ascertain:

1. whether the available aircraft, camera exposure range with the required film type will provide the required resolution;
2. whether hotspots are likely to occur;
3. that the available equipment is suitable and that no undue limitations exist on flying over the area; and
4. the best flight lines for the sortie to achieve user needs at lowest cost, the number of exposures per flightline and total number of exposures so as to prepare adequate film for the sortie.

2.6.6 The satellite platform

Orbiting satellites are under the influence of three forces: gravity (g) or attraction to the earth mass; centripetal (c) or the outwards force due to the angular change in momentum of the satellite; and drag (d) or the resistance of matter in the orbit (Diagram 2.74).

These forces must be in balance if the satellite is to stay in orbit for any length of time. A satellite orbit is normally elliptical, with the satellite moving much faster when it is closer to the earth (higher gravity pull needs to be counteracted by higher centripetal throw), and slower as it moves away from the earth. However this would cause great difficulty with sensor systems because of the variations in altitude and velocity. In consequence satellites designed to carry remote sensors are designed to be launched into orbits that are as close as possible to being parallel to the Earth's surface.

The period, T hours, of a satellite in a circular orbit is given by the expression:

$$T = 1.4\left[1 + \left(\frac{E}{R}\right)\right]^{1.5} \tag{2.9}$$

where R is the radius of the Earth, (6370 km) and E is the elevation of the satellite above the Earth's surface (in km).

The lowest orbit capable of supporting a satellite for a significant period of time is about 400 km above the Earth due to atmospheric drag. The highest altitude probably of interest for polar orbiting satellites is about 1000 km, whilst geostationary satellites, at orbits of about 36 000 km are used to provide high frequency repetitive coverage at coarse resolutions.

Polar orbiting satellites have their orbit in a plane that passes close to the poles. A polar orbit can be designed to precess about the globe, thereby achieving full coverage of the globe. If this plane is set to precess so that the plane retains the same angle to the Sun, then imagery will always be acquired at the same solar time. A Sun synchronous orbit will not exhibit changes in reflectance due to changes in solar elevation except that due to seasonal variations in solar elevation. The disadvantage is that the sensor cannot acquire imagery at other times during the day.

Consider the requirements of a satellite system that must image the full globe once every day. The Earth completes one rotation about the Sun once every 365 days, and so each day the plane of the satellite orbit needs to precess or rotate by 1/365th of 24 hours. Since this annual rotation is in addition to the 365 rotations, in effect the Earth completes 366 rotations of the Sun in 365 days. Thus the satellite needs to pass over the same location on the Earth's surface every (24 – 24/365 = 23.93424) hours. Let the satellite complete 14 orbits in this time, of 1.70959 hours per orbit. From (2.9) a period of 1.70959 hours would mean a satellite altitude of 907 km. The coverage per orbit is (360/14 degrees =) 25° 42′ 51″. To ensure full coverage, the swathe for an orbit could be set at about 27°, or 3000 km. It can be seen that this is a very large area to cover, with considerable distortion out from the edges of the imagery due to the Earth's curvature. Alternatively the satellite can be in lower Earth orbit, have more orbits and acquire less coverage per orbit.

Another option is to acquire full coverage every n days, completing partial coverage every day in N orbits such that the Nth orbit, 23.934 hours after the first orbit, is stepped by the swathe of the scanner relative to the first orbit. These steps are designed so that the swathes acquired each day progressively complete coverage of the Earth's surface between the swathes acquired on the first day, in n days. Once coverage is complete then the cycle is repeated.

The advantage of this approach is that the swathe width per orbit can be significantly reduced. Consider the previous case where the satellite, acquiring 14 orbits every 23.934 hours has to sweep out 25° 42′ 51″ per orbit. If the sensor is to have a swathe width of 4° 30′ (27°/6) then the satellite will need to cover 364° 30′ in the 14 orbits. Now the plane of the satellite precesses 360° in 23.934 hours, so that the duration of the 14 orbits will need to be 23.934 × 364.5/360 hours if the 14th orbit is to be at the same solar time. This requires a satellite orbit period of 1.7309 hours.

Whilst a satellite provides a very stable platform it is still subject to rotations about its three axes. Satellites use a reference to determine their orientation and to decide when corrections need to be made. The reference is usually by fix on a star; as the satellite oscillates so the direction to the star changes and retro rockets are used to re-establish orientation.

Because of the cost of loading and recovering film from a satellite, most sensors on satellites will be electronic, either video, scanners or radar. The data acquired by these systems is converted into digital data to minimize the effects of atmospheric attenuation on the data, before transmission to Earth.

The forward velocity of the satellite is used to provide the advance during acquisition of data. Both the rotation velocity of the Earth relative to the satellite and the forward velocity of the satellite introduce significant errors during sampling of the data. Other difficulties that can arise are geometric distortions due to the Earth's curvature and radiometric variations in the returned signal due to variations in path length.

Despite these limitations satellite data are very suitable for the ongoing monitoring of an area because of the repeatability of the data, the small errors due to rotations in the data, and the small effects of height displacements on the data.

2.7 Satellite sensor systems

Satellite-borne sensors have been launched by many countries, including the US, Russia, France, India and Japan. There are now a wide variety of satellite-borne sensors available to many users, with the characteristics varying widely between the different sensors. Diagram 2.75 shows the spectral sensing characteristics of some of the main sensors currently available on either airborne or spaceborne platforms. Of all of these, the Landsat, SPOT and AVHRR systems are most accessible to most people, and most used. They will be discussed in the next sections as examples of complementary sensor systems.

2.7.1 The Landsat satellite system

The platform

Five Landsat satellites have been launched since 1972. The first three had similar characteristics that differed from those of the subsequent Landsat satellites (Landsats 4 and 5 with Landsat 5 depicted in Diagram 2.76). The characteristics of each satellite are given in Table 2.6. Each Landsat satellite is in a Sun-synchronous, near polar orbit at an inclination of 99°, passing over the equator at about 10 a.m. local time.

The orbit precession completes ($24 \times 60/103.3$) orbits per day and progressively fills in the swathes over the cycle repeat time of 18 (Landsats 1–3) or 16 days (Landsats 4–5). Diagram 2.77 shows the orbit characteristics for Landsats 4–5. There is some overlap between the adjacent swathes of image data, with this side lap being least at the equator, and greatest near the poles. The amount of sidelap is shown in Diagram 2.78.

Geometric characteristics of the multispectral scanner (MSS)

The MSS uses an oscillating flat mirror to scan a six scanline swathe on each east-to-west mirror sweep (Diagram 2.79). The mirror sweep is perpendicular to the spacecraft heading. In the time it takes to scan across the swathe and have the mirror return to its start position, the sub satellite point has advanced 474 m, approximately the width of the six scanlines.

The oscillating mirror reflects the incident radiation onto a rectangular array of 24 optical fibres. The fibres are configured as a 4×6 matrix to provide input to the six

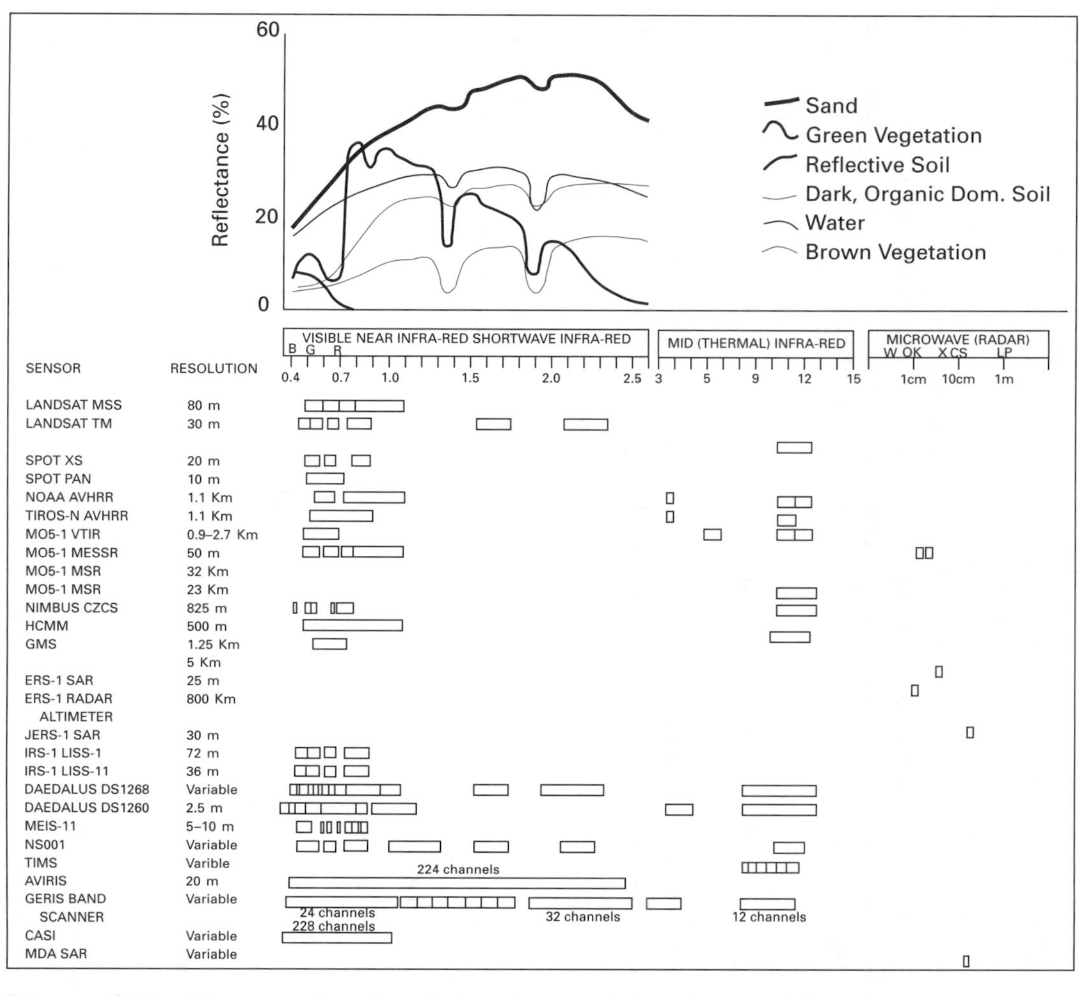

Diagram 2.75 The spectral and resolution characteristics of some of the more common satellite and airborne sensing systems.

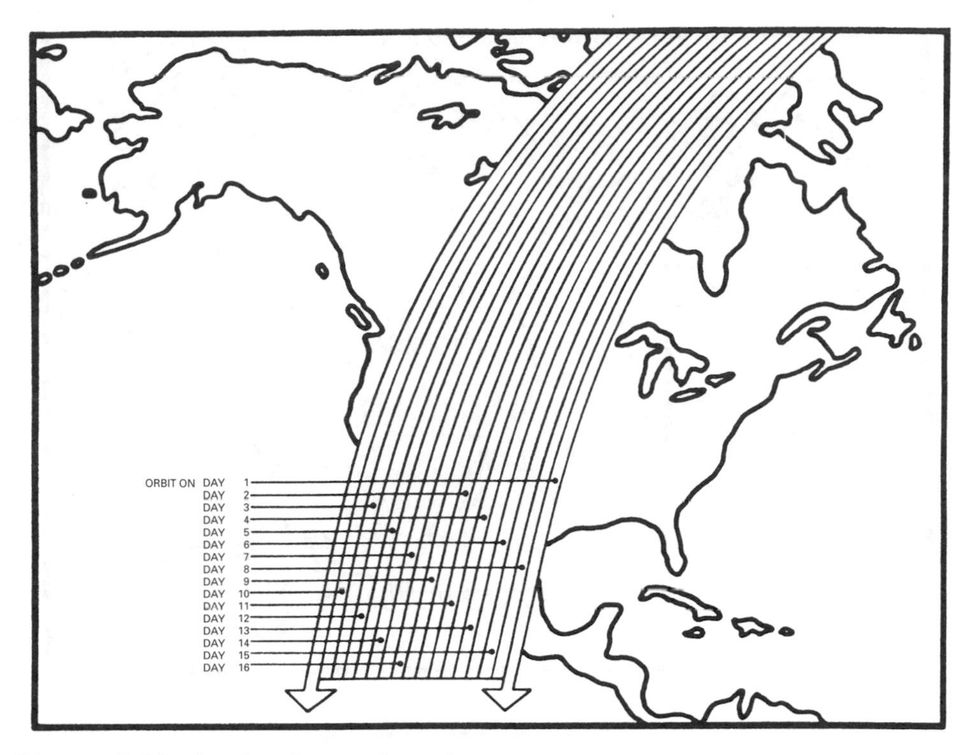

Diagram 2.76 Landsat 4 ground swathe pattern.

Table 2.6 Characteristics of the different Landsat satellites

Landsat	Launch date	Orbit period	Time of coverage Altitude	Acq.	cycle	Sensors MSS	VIDEO	TM
1	23 July 1973	103.3	950 km	0942	18 days	Y	Y	N
2	22 Jan 1975	103.3	950 km	0942	18 days	Y	N	N
3	5 Mar 1978	103.3	950 km	0942	18 days	Y	N	N
4	16 July 1982	102.1	870 km	0935	16 days	Y	N	Y
5	1 Mar 1984	102.1	870 km	0935	16 days	Y	N	Y

scanlines in each of the four wavebands. The light from each fibre is passed through the appropriate filter onto a detector. The 24 detectors are sampled every 9 microseconds in a serial fashion. The IFOV of the MSS creates pixels of about 79 m in size on the ground. However, the sampling rate is such that only about 57 m has been covered before the detectors are resampled, representing the separation between pixels along the scanline. Each scanline contains about 3200 pixels covering a ground distance of about 185 km in contrast to the 2340 scanlines that cover about the same distance in the along-track direction (Diagram 2.80).

Geometric distortions in the MSS scanner

The following geometric distortions exist in MSS data.

- Non-linearity in mirror velocity (Diagram 2.81) that approximates a sine curve. The correction for this error is of the form:

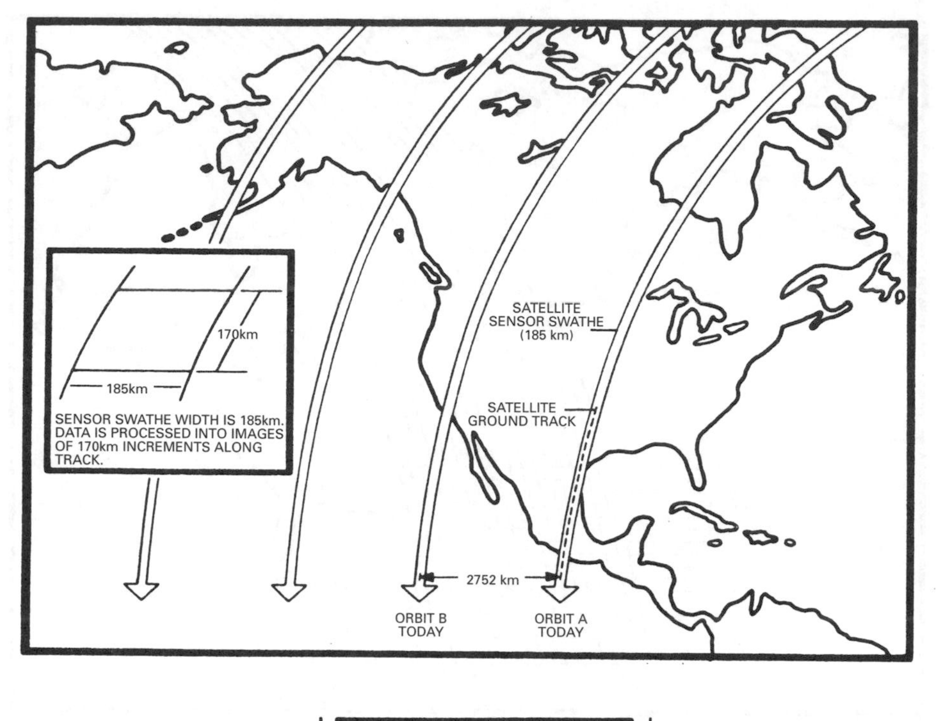

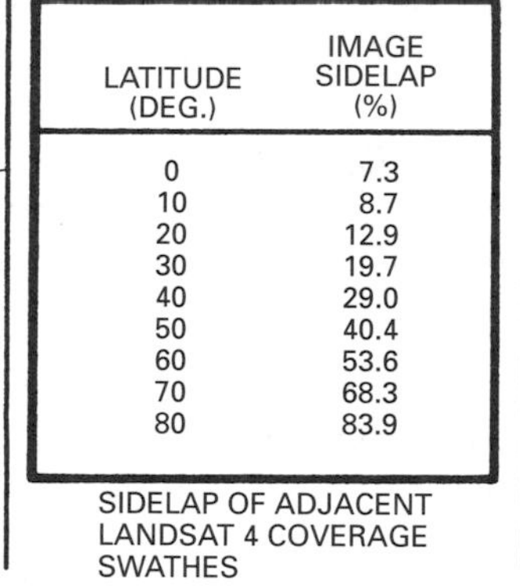

LATITUDE (DEG.)	IMAGE SIDELAP (%)
0	7.3
10	8.7
20	12.9
30	19.7
40	29.0
50	40.4
60	53.6
70	68.3
80	83.9

SIDELAP OF ADJACENT LANDSAT 4 COVERAGE SWATHES

Diagram 2.77 Landsat 4 sensor swathe and orbit characteristics.

$$\text{correction (in m)}, C = 323\sin\left(1718 + \left(\text{pixel} \times \frac{2\pi}{3245}\right)\right) - 73$$

- A random cross scan jitter giving random errors of about 5 m on the ground.
- A scan to scan repeatability error, giving a maximum error of about 7.5 m on the ground.
- A scan start and end variation giving errors of about 8 m.
- A systematic detector alignment error of about 8 m.
- A sensor delay distortion because the detectors are sampled sequentially rather than simultaneously. This gives an error of 22 m displacement of line six compared to line one in each scan. The displacement for line L in the six line scan is given by

$$\text{Correction (in m)}, C = \frac{22}{5} \times (L - 1)$$

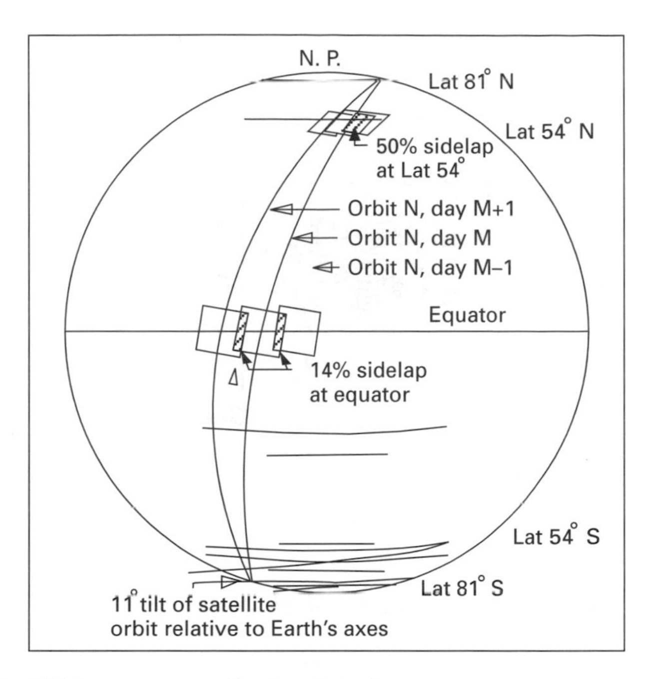

Diagram 2.78 Sidelap coverage for Landsat data.

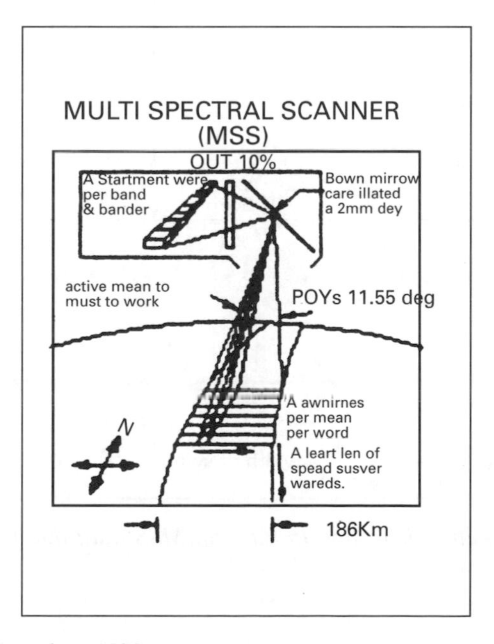

Diagram 2.79 The Landsat MSS scanning arrangement.

Distortions due to the satellite orbit

The satellite's altitude, pitch, roll and yaw, as well as changes in these during acquisition of the image, all affect the position a pixel will occupy on the Earth's surface. The satellite cannot maintain an orbit that is perfectly parallel to the Earth's curvature so some variation in altitude must occur. The main effect of changes in altitude is to change the scale of the image data. However, the distance between the sixth row of a single mirror

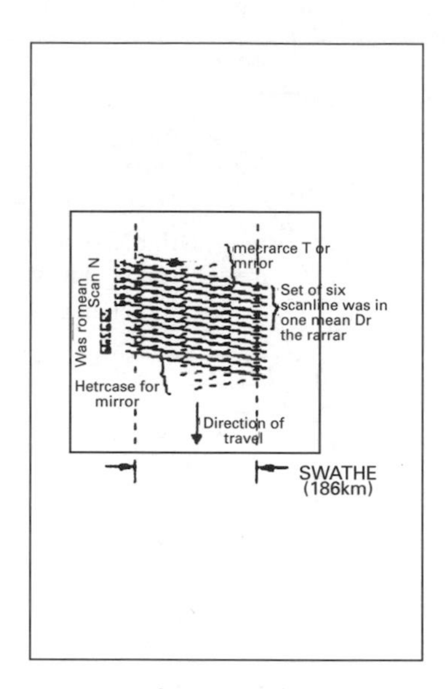

Diagram 2.80 Ground scan pattern for a single MSS detector.

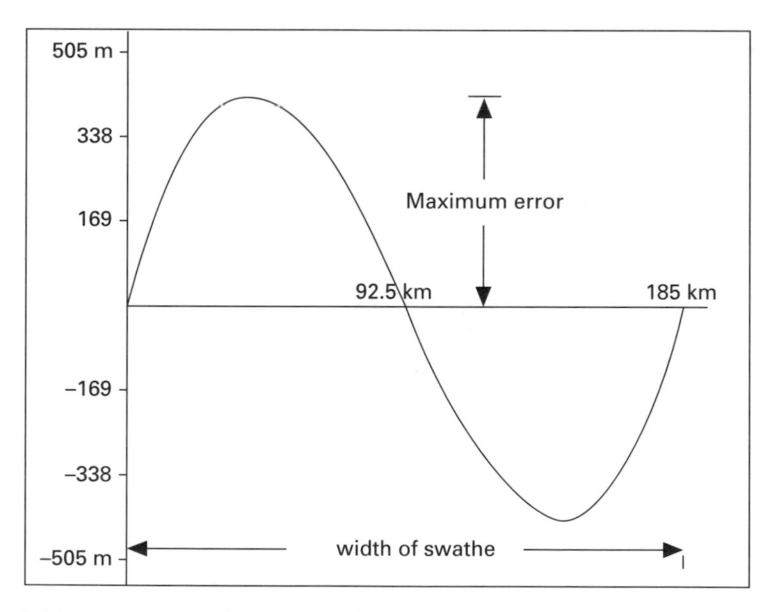

Diagram 2.81 Geometric distortion in Landsat MSS data due to non-linear mirror velocity.

sweep, and the first row of the next sweep, is affected by variations in the satellite's altitude and velocity. Since the mirror has a finite scan time, and the satellite is moving during acquisition of the data in a scan, then each scan will be skewed (Diagram 2.82) in the along-track direction.

Earth's rotation

The Earth's rotation has to be treated in two different stages. The first is to consider the rate of rotation while the mirror is in its active west-to-east sweep, the second is to

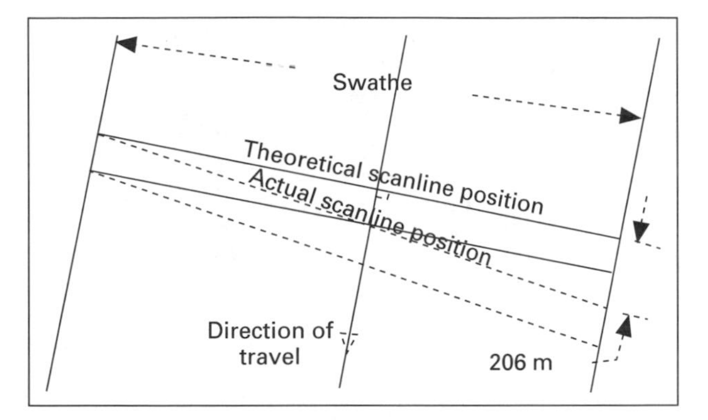

Diagram 2.82 Geometric error or skew introduced due to the platform velocity.

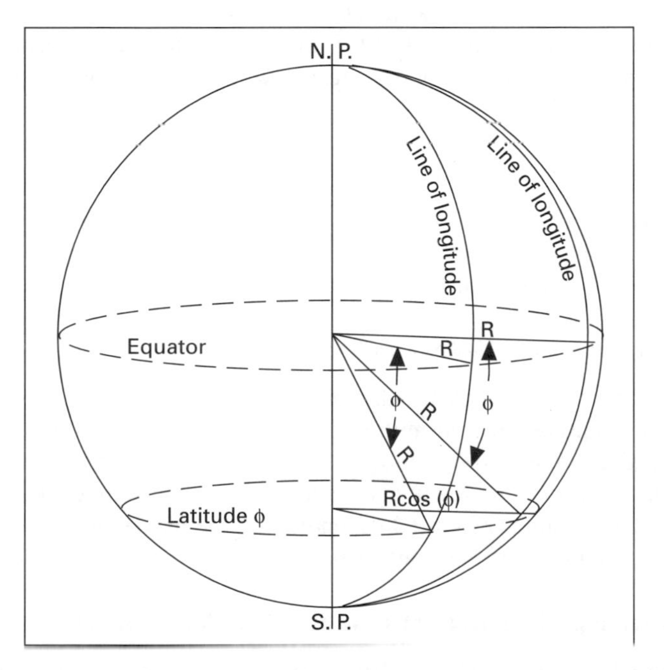

Diagram 2.83 Distance from the Earth's rotation axis as a function of latitude.

consider the Earth's rotation during the retrace of the mirror. The Earth rotates one complete revolution in 24 hours. The velocity of the surface of the Earth is the distance travelled by a point on the surface in that period. The distance around the globe (Diagram 2.83), is $2\pi R$ cos (latitude). The velocity in m s^{-1} at the surface of the earth can now be calculated to be:

$$V = \frac{2\pi R \cos(\text{lat}) \times 1000}{24 \times 60 \times 60}$$

where R is the radius of the Earth (= 6370 km) and lat is the latitude of the point. Hence,

$$V = 463 \cos(\text{lat})$$

Table 2.7 Landsat MSS wavebands, quantization levels at transmission from the satellite and after calibration and radiometric constants

Band No.	Wavelength range	Quantization levels	Calibrated levels
4	500–600	64	256
5	600–700	64	256
6	700–800	64	256
7	800–1100	64	64

The interval between sampling pixels is 9 microseconds, representing 0.004 m per pixel at the equator. The total mirror scan and retrace takes 73×10^{-3} seconds. The Earth's rotation in this period can be calculated to be about 34 m at the equator, so that each scan of six scanlines is stepped 34 m west of the immediately preceding scan, at the equator. The errors due to the Earth's rotation decrease away from the equator as a function of cos (lat).

Effects of height differences

Height differences introduce displacements in the images, just as they do into aerial photographs. Generally the displacements that are introduced are very small, usually being ignored in the analysis of Landsat data.

The radiometric characteristics of Landsat MSS data

The Landsat MSS records reflected radiation in four wavebands, and converts the readings into digital values before transmitting the data to receiving facilities, where the data is then calibrated for radiometric distortions and quantized into 256 levels. Zero count in the digital data is meant to represent zero radiance from the source, and the radiance at the maximum digital value (256 or 64 depending on the band) represents the maximum radiance before saturation of the detector occurs.

2.7.2 The systeme probatoire d' observation de la terre (SPOT)

The SPOT-1 satellite was launched in 1986 and SPOT-2 in 1989. The satellite (Diagram 2.84) is in a near polar (inclination of 98.7°), Sun-synchronous orbit at an altitude of 832 km, orbiting above the same location on the Earth's surface once every 26 days. The satellite has an orbit period of 101 minutes, crossing the equator at about 10.30 a.m. local solar time. The satellite carries two multispectral sensors (HRV) that operate using the pushbroom principle (Diagram 2.56). These scanners (Diagram 2.85) focus the incoming radiation onto a linear array of charged couple devices (CCDs). The linear array of CCDs is electronically scanned to create a record corresponding to a scanline of data. Successive lines are created by the satellite movement across the surface causing a new area to be focused onto the CCDs, and the CCDs being resampled to create a record for the second scan. A dichroic mirror splits the incoming radiation into three beams that are then focused through filters onto separate CCD linear arrays for each of the three wavebands as shown in Table 2.8.

The HRV sensors can operate in two modes: a panchromatic mode, corresponding to a broad spectral band (similar to panchromatic or black and white photographs); and a multispectral mode, corresponding to observation in three narrower spectral bands

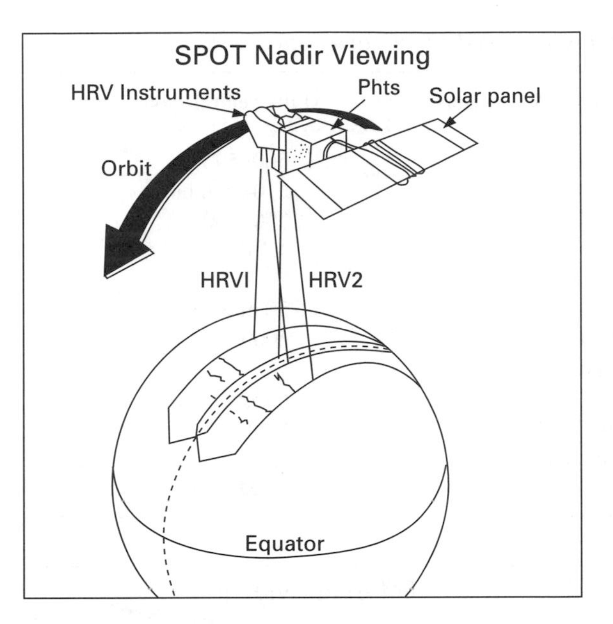

Diagram 2.84 SPOT HRV1 and HRV2 instrument nadir viewing to provide coverage of 117 km swathe.

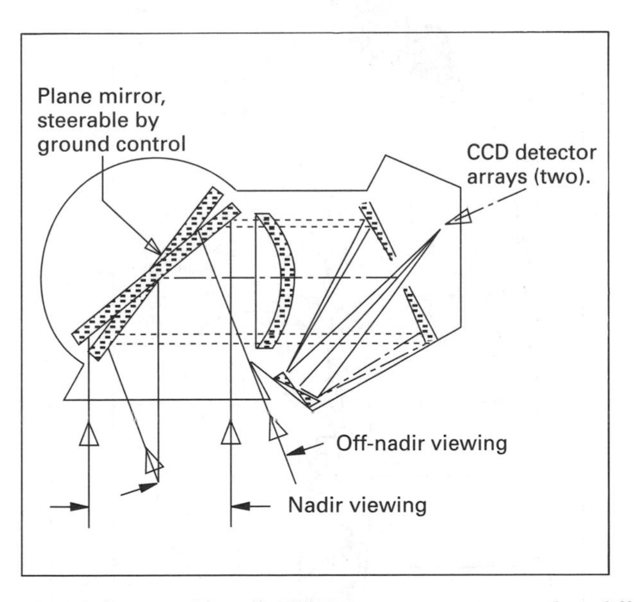

Diagram 2.85 The ability to point the HRV instrument means that different swathe segments can be imaged within an orbit and that the same ground area can be imaged on various orbits to provide more frequent cover.

given in Table 2.8. The ground spatial resolution is 10 m × 10 m in the first mode and 20 m × 20 m in the second, when the instruments are viewing directly below the satellite. Radiant energy reflected from the terrain enters the HRV via a plane mirror and is then projected onto the charge-coupled detector (CCD) arrays. Each array consists of 6000 detectors arranged linearly to form a scanline of data across the direction of orbit of the satellite. When looking directly at the terrain beneath the sensor system, the two HRV

Table 2.8 SPOT sensor system characteristics

Characteristics of the HRV sensors	Multispectral mode	Panchromatic mode
Spectral bands	500–590 nm 610–690 nm 790–900 nm	500–900 nm
Instrument IFOV	4.13°	4.13°
Ground sampling area (at nadir)	20 m × 20 m	10 m × 10 m
Number of pixels per line	3000	6000
Ground swathe width (at nadir)	60 km	60 km
Pixel quantization (compressed from 8 bits)	8 bits	6 bits
Image data bit rate	25 Mbits s^{-1}	25 Mbits s^{-1}

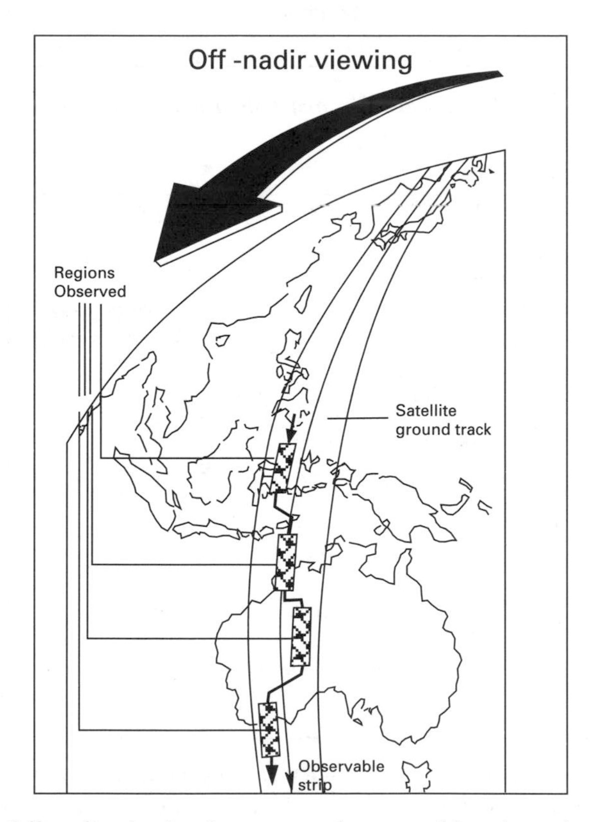

Diagram 2.86 Off-nadir viewing by means of a steerable mirror in the HRV instruments provides flexibility in the acquisition of data.

instruments may be pointed to cover adjacent areas, each 60 km by 60 km in size. In this configuration the total swathe width is 117 km, giving 3 km overlap in the coverage by the two instruments. Global coverage can be achieved with this setting over 26 days.

A tiltable mirror outside of the scanner lens allows the scanner to sense swathes that are off-axis (Diagram 2.86). The mirror can point to either side of the axis, so images can

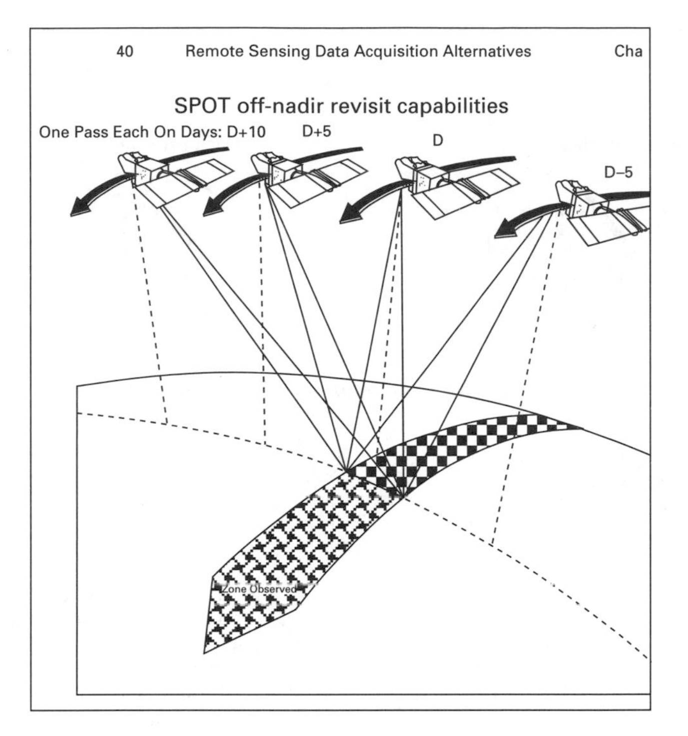

Diagram 2.87 Off-nadir viewing allows more frequent data to be acquired of areas that need to be monitored and increases the chance of acquiring data of areas that are frequently covered by cloud.

be taken that are off-axis on either side of the orbit. The amount that the image is off-axis can also be controlled by varying the tilt on the mirror. The benefits of this are twofold.

1. Any point on the Earth's surface can be imaged at more frequent intervals than would otherwise be the case. If the HRV instruments were only capable of nadir viewing, the revisit frequency for any given region of the world would be 26 days. Using the tilting mirror, any point could be observed on seven different passes if it were on the equator, and on 11 occasions if at a latitude of 45°. A given region may be revisited on dates separated alternatively by 1 and 4 (or occasionally 5) days (Diagram 2.87).
2. The second advantage is that images taken at different tilts will introduce different displacements in the images due to height differences, and hence pairs of these images can be used to provide stereoscopic pairs of the common area of coverage (Diagram 2.88). The greater the difference in the tilts, up to a maximum of 27.8° either side of the central axis, the larger the base to height ratio and the larger the difference in the displacements due to height differences, making the pairs better for stereoscopic analysis.

2.7.3 The advanced very high resolution radiometer (AVHRR) sensor

The National Oceanographic and Atmospheric Administration (NOAA) operates a series of meteorological satellites that have sun synchronous, near-polar orbits at a nominal altitude of 833 km. The advanced very high resolution radiometer (AVHRR) sensor on board these satellites has a field of view of about 1.4 milliradians (Table 2.9) and an interval between samples (or pixels) of 0.95 milliradians. This means that the sensor has

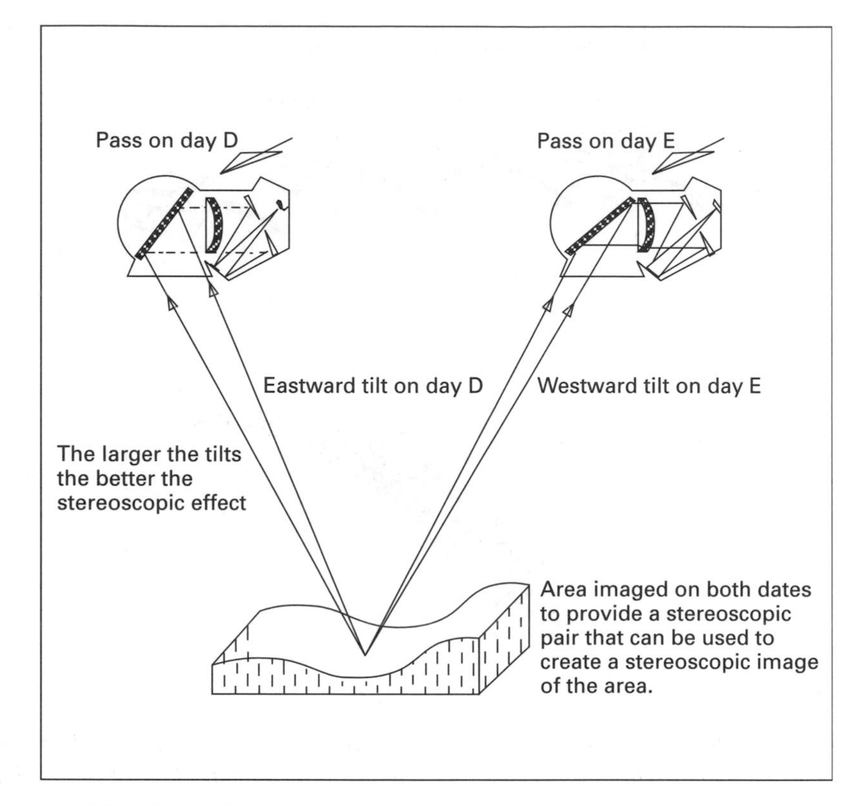

Diagram 2.88 The ability to point the HRV instruments provides the capability to acquire stereoscopic pairs of images.

Table 2.9 The main characteristics of the NOAA AVHRR sensor

Parameter	Channel 1	Channel 2	Channel 3	Channel 4	Channel 5
Spectral range (nm)	0.55–0.68	0.73–1.1	3.55–3.95	10.5–11.5	11.5–12.5
Field of view (milliradians)	1.39	1.41	1.51	1.41	1.3
Resolution (km) at nadir	1.1	1.1	1.1	1.1	1.1
Data precision (bits)	10	10	10	10	10
	0–1023	0–1023	0–1023	0–1023	0–1023

a ground resolution of about 1.1 km at the nadir point from the satellite, and over samples, that is there is some overlap between adjacent pixel areas, along the scanline.

The sensor acquires 2048 samples per scanline, so that the mirror sweeps out an arc of 55.4° either side of the nadir. At the extremity of the sweep, the distance from the earth to the satellite is about 1761 km, more than twice the distance at the nadir. At this point the size of the field of view, normal to the direction to the satellite, is 2290 m, which has a footprint on the Earth's surface of 6263 m. Thus at the extremity of the swathe the pixels are about six times the size of the pixels at the nadir point. The direction from the ground surface to the satellite is about 13° above the horizon at the extremity of each

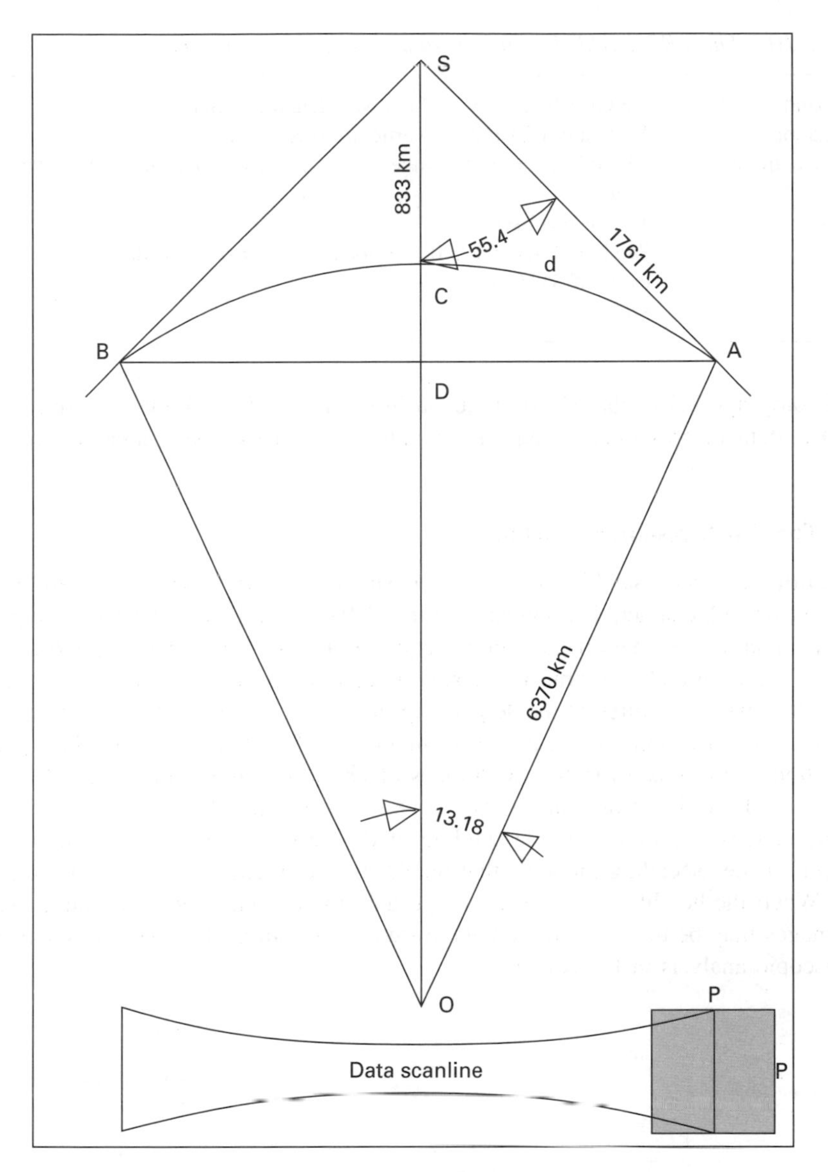

Diagram 2.89 Geometry of the NOAA AVHRR scanner, to illustrate the distortions that occur at the extremities of the scans in the data.

scanline, so changes in elevation will introduce distortions in the data (Diagram 2.89). A scanline supports an angle of 13.16° either side of the nadir, at the centre of the earth.

The satellite orbit is 102.12 minutes, representing 25.53° at the Earth's centre. There is thus some overlap between the swathes of data acquired from adjacent orbits, but separated in time by 102.12 minutes. The satellite orbit means that it does not exactly replicate its orbit paths every day, so that the location of the satellite changes relative to the ground, from day to day.

The extremely wide swathe relative to the altitude of the satellite creates geometric and radiometric distortions in the data. Radiometric distortions are introduced due to variations in atmospheric scattering and absorption as a function of path length and variations in atmospheric conditions across the area of the image.

Table 2.10 The ERS-1 SAR instrument and satellite characteristics

Waveband	5.3 GHz frequency; 5.7 cm wavelength; C Band
Polarization	Vertically transmitted, vertically received or V-V
Incidence angle	23° off axis at mid swathe, creating a 100 km wide swathe, 250 km of nadir
Resolution	12.5 m pixel size
Orbit	785 km altitude; 10.30 a.m. local time crossing of the equator; repeat cycle of 35 days.
Inclination	98.5°

Photograph 2.5 of the island of Luzon in the Philippines shows the advantages of AVHRR data in providing a frequent synoptic view of extensive areas of the earth's surface.

2.7.4 The Earth resources satellite–1

The Earth Resources Satellite–1, or ERS–1, supports a sidelooking, synthetic aperture radar sensor with characteristics given in Table 2.10. The ability of the sensor to generate its own signal means that images can be acquired on both the descending and ascending nodes of the orbit and under most atmospheric conditions. The satellite has a near-polar orbit, with images acquired in the descending mode at a bearing of 98.5°, whilst the data acquired in the ascending mode is at a bearing of 262.5°. This means that there is a mis-orientation of the images in the two modes of about 17°, requiring rectification of the image data if the different data sets are to be used together. This geometry also means that the images acquired in the descending mode are in the opposite direction to those acquired in the ascending mode so that height displacements will be in opposite directions. When the height differences are large then the displacements introduced into the two images may be too large to view them stereoscopically. The imagery is suitable for stereoscopic analysis in flat country.

3
Visual interpretation

3.1 Overview

3.1.1 Remotely sensed data and visual interpretation

Remotely sensed data can be in the form of either visual or digital images. Different types of visual images include photographs and other forms of hard-copy. Digital images are stored on computer disks, tapes or cassettes as spatially related digital data and will be discussed in subsequent chapters.

Visual interpretation of an image or photograph is one of two important ways of extracting information from remotely sensed data. It uses skills much more readily available to the human interpreter than through computer processing of digital image data. The most important skill is the interpreter's ability to conduct analytical reasoning and to identify and explain complex patterns.

Digital image processing is the second method and will be discussed in Chapter 5. Digital processing and extraction of information is assuming increasing importance, particularly when used in conjunction with visual interpretation. Currently it is used for the more routine numerical tasks associated with image enhancement needed to provide better images for later visual interpretation. Increasingly, it will be used in the more complex tasks of mapping and monitoring which may not require the skills of the human interpreter.

3.1.2 The effects of height differences on remotely sensed images

In analysing photographs, it is important to recognize that all images are affected to some degree by height differences in the terrain. Many aerial photographs are taken to exploit this characteristic so as to estimate such height differences. This is the basis of most topographic mapping where a major objective is to map elevations in an area by means of contours.

These height differences can be estimated:

- with great precision using photogrammetric instruments with aerial photographs;
- with less precision when replacing the aerial photographs with satellite images acquired for this purpose; and
- using simple techniques.

Introduction of displacements due to height differences means that single images will not be spatially accurate. The introduction of these displacements can be ignored when they are negligible, either because the terrain is flat or the geometry of the sensor introduces negligible displacements at that position. These displacements will be interpreted as differences in spatial location if only a single photograph is used unless:

- the image has been corrected for their effects; or
- the process of transferring the information from the image to a map makes corrections for the displacements.

Photogrammetric processes can map spatial position from images with great precision. There are also simple techniques that can map spatial position, but within limited areas due to height displacements.

The first section of this chapter studies how height displacements occur and how they can be used to estimate differences in elevation, and how to map planimetric detail. The chapter then deals with the techniques of photo-interpretation and how to conduct an interpretation task. Photo-interpretation certainly benefits from innate skills in observation and analysis, but just as important is the attitude of the analyst to the task. Interest, a desire to learn, confidence and a preparedness to work are prerequisites to successful interpretation. Competence in interpretation also comes from practice, so it is important to consolidate study of this chapter with practical experience in the techniques and methods discussed.

3.2 Stereoscopy

3.2.1 Introduction

Stereoscopy is concerned with the visual perception of the three-dimensional space around us. Stereoscopy covers a wide range of phenomena including the physiological and psychological basis of visual perception, as well as the principles involved in the construction of equipment used to facilitate stereoscopic vision and perception.

Binocular vision is the ability to observe objects simultaneously with two eyes set at different positions. All people with two functioning eyes have the ability to observe with binocular vision. The differences in position of the two eyes means that the images they acquire are different. These differences are translated into differences in depth, or depth perception, by the observer. Most people, having binocular vision, are familiar with the perception of depth, and in fact take it for granted. Just as the perception of depth is important in day-to-day living, so it is important in many aspects of remote sensing.

The study of visual perception within psychology has not resolved all aspects of the visual perceptive process; some of these processes remain imperfectly understood. Since visual perception provides the theoretical and practical basis for stereoscopic perception, as our understanding of the processes involved increase, so our understanding of binocular vision will increase, and can be used to improve stereoscopic perception.

3.2.2 Monocular vision

The major components of the eye are shown in Diagram 3.1. Light from the object, incident on the lens (5) of the eye, is focused by that lens onto the retina (9) in much the same way as light is focused by the lens of a camera onto the focal plane of the camera. The retina contains light-sensitive elements that respond to the incident radiation by sending signals to the brain, in contrast to a camera where the film in the focal plane is exposed by this energy. The lens of a camera changes position relative to the film to focus objects at different distances from the camera onto the film. The lens of the eye cannot do this; instead it changes its shape and hence focal length when focusing objects at different distances onto the retina. This ability to change shape is called **accommodation**. Thus when looking at this page the lens will adopt one shape to focus objects at short distances onto the retina; when then looking out of the window the lens of the eye changes shape so as to focus objects at far distances onto the retina. As a person ages, the lens of the eye becomes more crystalline and tends to lose the ability to change shape.

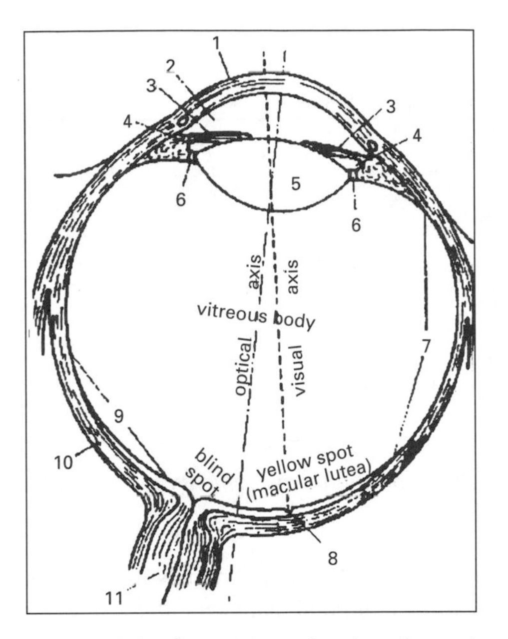

Diagram 3.1 Cross-section of the human eye showing the major components that affect vision.

For this reason, older people often have difficulty focusing on close objects (called short-sightedness).

The retina (9) consists of a matrix of light-sensitive filaments of two types, called **rods** and **cones** that are connected to nerve endings of the optic nerve (11) system. The rods predominate around the periphery of the retina, and the cones towards the centre. The rods are sensitive to light of all wavelengths in the visible region and so are like panchromatic sensors. There are three types of cones, sensitive to narrower wavebands that approximate the blue, green and red parts of the spectrum. The cones provide our perception of colour.

At the most sensitive part of the retina, the fovea centralis (8), there are no rods and each cone is connected to a single nerve ending of the optic nerve. Out from the fovea centralis groups of rods and cones are connected to single nerve endings. The greatest resolution is therefore achieved at the fovea centralis, but good illumination is required to use this part of the retina fully. The eye is usually oriented so that the object of interest is focused on the fovea centralis.

At lower light levels the cones more rapidly lose their effectiveness, being sensitive to narrower wavebands. With increasing distance away from the fovea centralis, more rods and cones are connected to each nerve ending, with only rods being connected at the periphery of the retina. The effect of connecting more rods and cones to a nerve ending is to increase the response of the nerve ending to low light levels. In consequence the periphery of the retina is much more sensitive to low light levels than is the centre of the retina. Objects tend to lose colour and become more black and white at night because of the domination of the rods. Indeed at very low light levels the fovea centralis is effectively blind. The rods and cones are connected to the optic nerve through simple pattern recognition processing centres called synapses. The optic nerve (11) leaves the eye at the blind spot where there are no rods or cones.

The iris (3) controls the amount of light that enters the eye in a given time, by changing the diameter of its aperture or pupil. The pupil is usually about 3–4 mm in

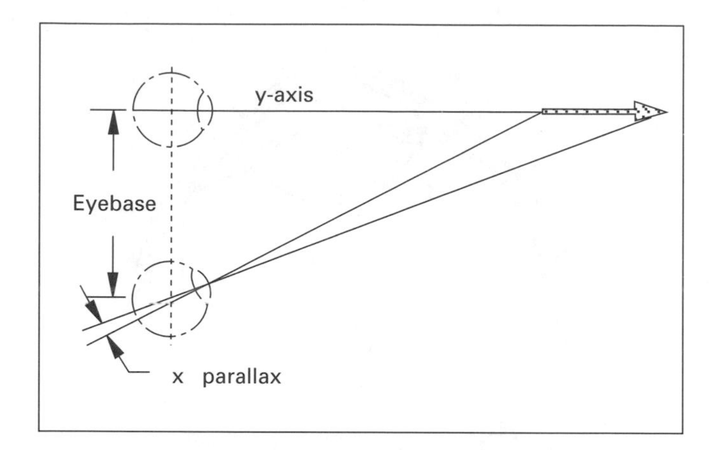

Diagram 3.2 Binocular vision and displacements in those images due to the eyebase during observation.

diameter, but it can vary between 2–8 mm. Animals that are nocturnal in habit, such as cats, can exhibit a greater range of pupil diameters, thereby having more control over the amount of light that enters the eye. The components of the eye are protected by a clear layer (1) and aqueous solution (2).

The image that is focused onto the retinal surface of a single eye does not contain information on the distance to the object from the observer, but it does contain clues to that distance that we have learnt to interpret as distances, with individual levels of reliability. There are several clues to depth in monocular vision.

- the relative sizes of objects implies depth, because we usually interpret identically shaped objects of different sizes to be identical objects but at different distances from the observer. Similarly, the size of familiar objects is usually used to estimate the distance to that object, and this characteristic can be used to trick an observer. For example, by showing a miniature version of an object, such as a toy car, next to a full scale model, the observer may 'see' two full sized cars but at different distances.
- shape or position can also provide clues. Parallel lines appear to become closer together the further away they are from the observer.
- interference, e.g. one object masking another, this will look as if the first is closer to the observer than the second.

3.2.3 Binocular vision

Binocular vision occurs when a scene is viewed simultaneously by two eyes at different locations. The images formed in either eye are flat images on the retina. Neither image, of itself, retains information on the distance to the object, but each image contains displacements due to differences in depth to different objects within the field of view, in a similar way as occurs in an aerial photograph.

As the two eyes are at different locations, so the displacements in either eye due to differences in distance or depth will be different. These displacements will be parallel to the eyebase in accordance with Diagram 3.2. The brain, on receiving these two different images, integrates them by interpreting the differences as depths or distances from the observer, and so creates one three-dimensional image of the scene observed by both eyes.

Conventionally the eyebase is taken as the *x*-axis, and so these displacements are called ***x* parallaxes**. The size or magnitude of the *x* parallaxes is a function of the eyebase,

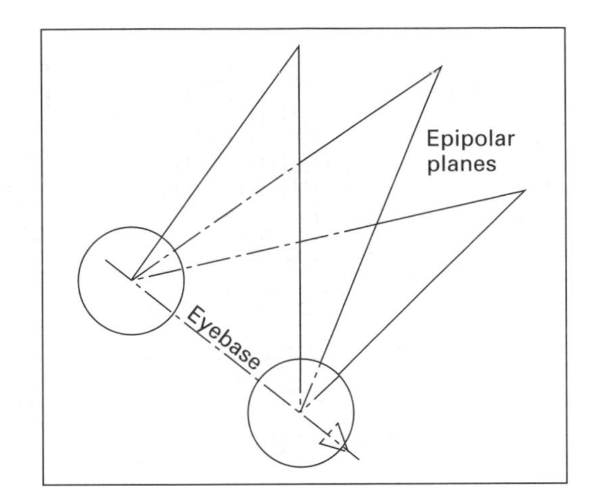

Diagram 3.3 Viewing in epipolar planes.

the distance from the observer to the object and the depth of the object. The existence of x parallaxes can be demonstrated by focusing on a distant object, such as a wall, and placing a finger in the line of sight. Two images of the finger will be seen. The displacement between these two images is a function of the x parallax difference due to the difference in distance between the observer and the fingers, and the observer and the wall. The displacement between the two fingers should be parallel to the eyebase, as can be demonstrated by tilting the head whilst viewing the wall.

When an object is viewed simultaneously with both eyes, the eyeballs are rotated so that the line of sight to the object passes through the fovea centralis of both eyes. This phenomenon is known as **convergence** of the optical axes. Focusing on a second object will cause the same phenomenon to occur, changing the convergence of the eye axes. The closer objects are to the observer, the greater will be the angle of convergence. Viewing a target means that both eye axes should bisect the target, so that the two eyes and the target define a plane known as the **epipolar plane**. There are clearly an infinite number of epipolar planes, with all of them having the eyebase as a common line, as shown in Diagram 3.3.

If the eyes do not intersect the target then the observer will not be observing in epipolar planes. In this case the shortest distance between the axes will be at right angles to the eye axes in the direction vertical to the eyebase. These vertical displacements are called y parallaxes, in accordance with the coordinate system shown in Diagram 3.4. The brain can accommodate small y parallaxes between features in an image. For larger y parallaxes the brain ceases to be able to match the detail recorded by either eye and hence cannot create a perceptual three-dimensional model of the object or surface. When this happens then competition occurs between the two images in the brain. This competition, called **retinal rivalry**, means that either one of the images dominates, or the two images alternatively dominate perception of the scene. In both cases three-dimensional or stereoscopic perception is destroyed.

3.2.4 The binocular perception of colour

If the images received by the left and right eyes contain small differences in colour or tone then the brain will allocate a colour or tone to the object that is some mixture of the two colours or tones. The finally perceived model will thus include colours or tones that

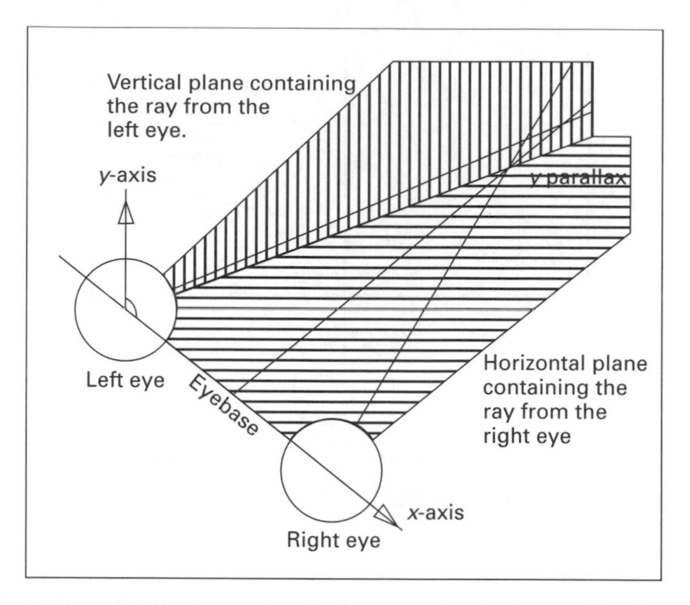

Diagram 3.4 The coordinate system used for viewing and associated x *and* y *parallaxes.*

may not occur in either of the originally perceived images. It is important to ensure that this is a negligible source of error in visual perception, by ensuring good rendition of colours in the viewed images, and ensuring that the analyst views images under good quality illumination.

The brain cannot accommodate large variations in tone or colour between the two images, setting up retinal rivalry when this occurs. A relatively common case of retinal rivalry occurs when water is recorded as being highly reflectant on one image and very dark on the other. In this case, observers will note that either one of the tones will dominate or the surface will appear to glitter. In either case the retinal rivalry is likely to cause stress to the observer, resulting in headaches and lethargy.

3.2.5 The general principles of stereoscopic vision

Binocular vision allows perception of a three-dimensional model from two flat images acquired at different positions so that they contain different displacements due to variations in the distance between the observer and the object or surface. We can also create three-dimensional models of an object using a pair of photographs of that object or surface, as long as both the photographs and the observer obey certain rules.

- The photographs must be taken of the same object detail, but from different positions where these positions are the ends of the camera base. The camera base is analogous to the eyebase of the observer.
- Viewing of the photographs must be oriented so as to reconstruct the geometry that applied during acquisition of the photographs. It is also important to ensure that both photographs are of similar quality, and that both are well illuminated.
- The two photographs must be viewed simultaneously, with the eyebase parallel to the camera base, where the eyebase is defined by a line passing through the two principal points, and their conjugate principal points (section 3.2.8). This requirement ensures that the photographs are being viewed in epipolar planes.

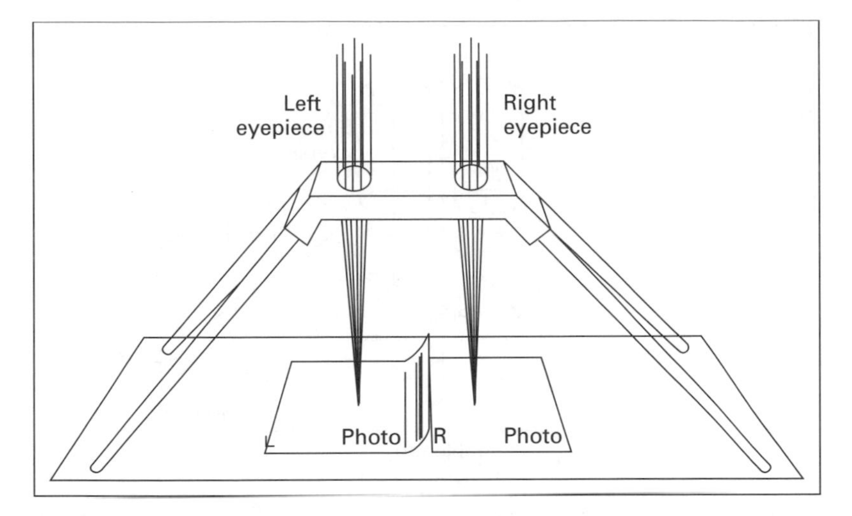

Diagram 3.5 The lens or pocket stereoscope.

3.2.6 Methods of stereoscopic viewing

There are a number of ways of satisfying the general principles of stereoscopic vision listed above. All methods require, in some way, the separation of the two images, so that each image is only seen by one eye.

It is possible to view simultaneously a pair of photographs with the unaided eyes and perceive a three-dimensional model. This is done by viewing the photographs either with parallel eye axes (left eye views the left photograph and right eye views the right photograph) or with crossed eye axes (left eye views the right photograph and vice versa). The point of focus with both of these techniques is different to that which the brain would expect from the convergence of the eye axes. Parallel eye axes are usually accompanied by focus at infinity but in this case the focus is at a shorter distance. Crossed eye axes means that the eye expects to focus at shorter distances than the actual distance to the photographs. For this reason the procedure takes some practice. It is, however, handy to use in the field and does not appear to damage the eyes at all.

Devices have been designed to eliminate this inconvenience in viewing stereoscopically, which achieve separation of the images either physically or optically.

3.2.7 Physical methods of separation using stereoscopes

A stereoscope is a device that achieves physical separation of the images by directing the left eye to receive either reflected or transmitted light from the left image and the right eye from the right image. Stereoscopes are convenient to use because their optical systems are designed to allow the observer to focus at the most comfortable viewing distance, about 300 mm for most observers, as well as providing magnification of the viewed photographs.

The two most common forms of stereoscope are the lens and mirror stereoscopes. The **lens stereoscope** (Diagram 3.5) is light and compact so that it is convenient for field use. However the limit of the eyebase on the separation of the photographs means that detail on the one photograph is often masked by the second photograph, because of the 60 per cent or more overlap between the photographs. This limitation can be overcome by bending the masking photograph so as to reveal the detail beneath.

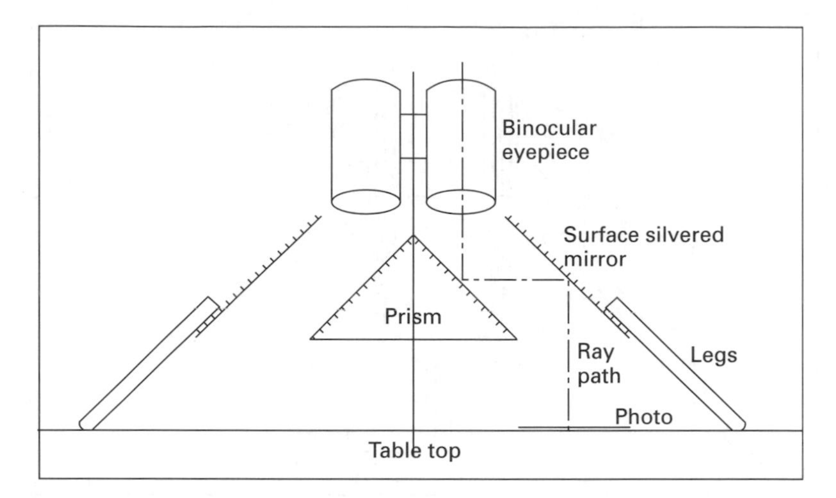

Diagram 3.6 The mirror stereoscope.

The **mirror stereoscope** (Diagram 3.6) overcomes this limitation by reflecting the light from the photographs off a mirror and a prism before it enters the binocular eyepiece. This design achieves wider separation of the photographs. The use of a binocular eyepiece allows a range of magnifications to be incorporated into the eyepiece in contrast to the lens stereoscope that has limited magnifications. The magnifications in most binocular eyepieces mean that only part of the stereo pair can be viewed at the one time. Most mirror stereoscopes overcome this disadvantage by allowing the observer to remove the binocular eyepiece so as to view the whole of the stereopair. The greater separation of the photographs means that they can be secured to a base during viewing and the stereoscope moved across the pair of photographs to view different parts of the stereopair in detail. Being able to fix the photographs to a base also allows the observer to use a parallax bar (Photo 3.1) to measure x-parallaxes and then calculate height differences as discussed in the next section.

The mirrors of the mirror stereoscope are surface silvered to eliminate double reflection. These mirrors are susceptible to etching by the acids of the skin; it is very important not to touch or otherwise handle the mirrors of the stereoscope with fingers. If the mirrors are inadvertently touched then they must be very carefully cleaned with alcohol on a soft cloth.

3.2.8 Viewing with a stereoscope

Proper stereoscopic viewing of an area common to a pair of photographs requires the observer to simultaneously view the pair of photographs in epipolar planes. These conditions are achieved as follows.

1. Mark the principal point (section 2.4) of each photograph. Using detail in the photographs, visually locate the principal points in the adjacent photograph. These two points are called the **conjugate principal points**.
2. Orient the photographs on the desk top or hard surface so that the parts of the photograph with common detail are closest together, and shadows are towards the observer. In this arrangement the conjugate principal points will be closer together than will the principal points. Secure one of the photographs to the base board.

3. Orient the second photograph to the fixed photograph so that the four principal and conjugate principal points all lie on the one line. Place a ruler along this line. Position the stereoscope above the photographs and orientate until the eyebase of the stereoscope is parallel to the photobase as represented by the ruler. The eyebase of the stereoscope is defined as a line through the axes of the binocular eyepiece. This arrangement sets the conditions necessary to view the photographs in epipolar planes.
4. Ensure that comfortable focus is achieved for each eye by viewing the respective images through the corresponding eyepiece of the stereoscope and adjusting the focus for that eyepiece as necessary. It is now necessary to set the convergence so that the viewer will see the same detail in both photographs. Select an obvious feature in the fixed photograph and move the stereoscope so that this feature is in the centre of the field of view of the binocular eyepiece, ensuring that the photobase remains parallel to the eyebase. Move the second photograph in or out along the line of the photobase until the same detail is in the centre of the field of view of the second binocular eyepiece.
5. Simultaneously view both photographs through the binocular eyepieces of the stereoscope. If the images do not fuse then the displacements between them will be in the x and y directions. Displacements in the y direction are caused when the observer is not viewing in epipolar planes. In this case, repeat step 3 to re-establish viewing in epipolar planes. If the displacements are in the x direction then the separation between the images as done in step 4 needs to be adjusted by moving the loose photograph either in or out along the photobase.
6. Once fusion has been achieved, find the most comfortable viewing position by carefully moving the second photograph in or out by small distances, to reach the best stereoscopic perception, and carefully rotate the stereoscope by small amounts to eliminate displacements in the y direction. Check that a comfortable viewing position has been achieved by leaving the stereoscope for five minutes and then coming back to it. If a comfortable viewing position has been achieved then fusion will be created immediately on viewing through the stereoscope. If this does not occur, but the images gradually fuse then further adjustment is required. Finally move the stereoscope across the whole of the pair of photographs to check on the stereoscopy across the whole area of overlap.
7. With fusion, the observer creates a perceptual model from fusion of the two separate images. The observer should become aware of differences in height in the model due to x-parallax differences created during acquisition of the photographs.

Most people can view a pair of photographs in this way and achieve stereoscopic vision in a reasonably short time. Some people, however, have a dominant and a lazy eye, the dominant eye doing most of the work. To achieve stereoscopy it is essential that both eyes do equal work; if this is not happening then the observer will see only one image, that which is being observed by the dominant eye. Initially the person with a dominant eye may only be able to maintain stereoscopic vision for a short period or have difficulty establishing stereoscopic vision in the first place. When this occurs perseverence is needed until the lazy eye becomes used to operating with the other eye and stereoscopy can be easily maintained for long periods.

3.2.9 Optical methods of separation

The two images are projected optically onto the one screen using an appropriate filter system. They are separated using the same filter system before entering the respective eyes of the observer. This optical separation can be achieved using either:

- coloured filters in two complementary colours, usually called the anaglyph method; or
- polarizing filters with the two filters set to accept light at polarizations that are at right angles to each other.

Coloured anaglyphs

Two black and white images are projected through complementary filters of blue–green and yellow–red onto the one screen. The images are then viewed by the observer who wears glasses containing similar filters. The filters used in the glasses will separate the viewed images so that the image projected in blue–green will be received by the eye behind the blue–green filtered glass and the yellow–red image will be received by the other eye.

The disadvantages of this method are that only black and white images can be used since colour is used to achieve separation; the levels of illumination can be quite low, and the complementary colours often create retinal rivalry.

Polarizing filters

The images are projected through polarizing filters set at right angles to each other onto a screen that reflects the images back towards the observer without a significant loss of polarization. The observer views the images through spectacles containing polarizing filters matching those used in projection.

Polarization can be used with colour or black and white photographs. The main disadvantage of this method is that the screens on which the images are projected often significantly degrade the polarization causing mixing of the images and degradation of the stereoscopic effect. Metallized paint or lenticular screens are suitable for reflecting polarized light.

3.2.10 Construction of a stereo-triplet

An interpreter who wishes to take photography into the field may wish to view those photographs stereoscopically. To do this a lens stereoscope will normally be chosen because of its compactness and lightness. A stereo-triplet is a way of creating two adjacent stereopairs using the three adjacent photographs, fixed so that they are suitable for viewing with a lens or pocket stereoscope. A stereo-triplet may be constructed as follows.

1. Select three overlapping photographs of the area of interest. On the centre photograph draw two lines, a and b at distance c on either side of the principal point (Diagram 3.7). The distance c is slightly less (about 5 mm) than the eye base of the observer, and is usually set to about 60 mm.
2. Visually transfer lines a and b to the two adjacent photographs at a' and b' and a'' and b''. Cut away the portions of the centre photograph outside lines a and b.
3. Secure the centre photograph to a piece of card along line a, the line joining the fiducial marks. Visually position the right-hand photograph, in the correct orientation, on the card with portion of the photograph left of line a, beneath the central photograph.
4. Using the stereoscope adjust the position of the right-hand photograph on the card so that the pair can be viewed comfortably during stereoscopic analysis. Secure the right-hand photograph to the card.
5. Repeat with the left-hand photograph.

This gives a stereo-triplet (see Photo 3.3). If only two photographs are used the result is a stereo-doublet.

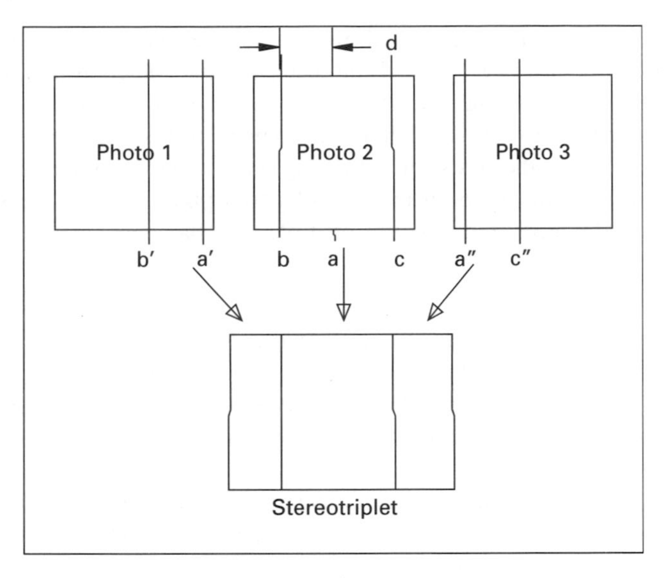

Diagram 3.7 Construction of a stereo-triplet.

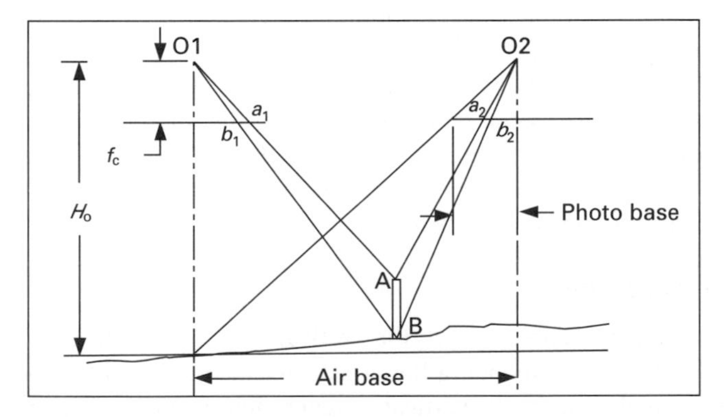

Diagram 3.8 The principle of the floating mark.

3.3 Measuring height differences in a stereoscopic pair of photographs

3.3.1 The principle of the floating mark

The height difference AB in Diagram 3.8 introduces displacements a_1b_1 and a_2b_2 into the left and right hand photographs respectively. If small marks are placed immediately over the photographs at points a_1 and a_2 then the brain is deceived into believing that these marks are in the left- and right-hand photographs respectively. As such the brain interprets the two marks as being images of the one mark, so that the observer perceives one mark at A in the fused model. If the marks are moved, say to b_1 and b_2 then the observer will perceive the fused mark moving from A to B in the model. Because of this characteristic the two points are called a floating mark.

The floating mark can be made to move up and out of, or down and into, the perceived three-dimensional model. At some stage the x-parallax differences between the mark and the model detail will become too great for the brain to accommodate, and the floating mark will break into two individual marks.

The principle of the floating mark allows us to place a point in the perceptive model as a fusion of two points, one imposed into each image. The coordinates of each point can be measured relative to the principal point of the respective image, and these distances can then be used to estimate height differences in the model. Because of this the floating mark is used as the basis for extracting quantitative information on height differences and positions of points in the pair of aerial photographs. One simple way to measure height differences is by the use of a parallax bar.

3.3.2 The parallax bar

A parallax bar consists of a rod supporting a mark etched onto a glass tablet at one end and with a micrometer, supporting a second glass tablet with etched mark at the other end (Photograph 3.1). The micrometer allows the distance between the floating marks to be read to the nearest 1 mm from the main scale and to the nearest 0.01 mm from the vernier scale of the micrometer.

The main scale of the micrometer is usually graduated with values that increase inward for reasons that will be discussed later. In operation the etched marks on the glass tablets are placed over corresponding detail in the two photographs. When the two marks exactly coincide with the separation of the photographic detail then the two marks will fuse into a floating mark at the same elevation as the perceived model. Slight changes in the x-parallax or separation of the floating marks, by rotating the micrometer screw, will cause the floating mark to either move up and out of, or down and into the model. The separation between the marks can be measured using the micrometer. Since differences in x-parallax between two points are a function of the difference in height between the two points, observing the two distances by means of the micrometer can be used to estimate the height difference as will be discussed in section 3.3.4.

3.3.3 Vertical exaggeration

When looking at a stereoscopic model, the vertical distances, or height differences, are exaggerated relative to corresponding horizontal distances. This difference between the vertical and horizontal scales is the vertical exaggeration, V. Vertical exaggeration occurs because the photographs are taken at one base length, the Air Base in Diagram 3.8 and are viewed at the second, shorter, photo base length. The x-parallaxes are greater than what would occur if the photography was taken at the shorter base length, causing the brain to interpret the larger x-parallaxes as being much greater height differences in the model than is actually the case. The existence of this vertical exaggeration means that it is very difficult to estimate height differences when viewing a stereo model without an aid such as a parallax bar.

3.3.4 Displacements due to height differences in an aerial photograph

Height differences introduce displacements in an aerial photograph that are radial from the nadir point as shown in Diagram 3.9. The magnitude of the displacement, $\delta p = p_a - p_b$ is a function of the distance of the point from the nadir point and the ratio of the height difference to the flying height of the aircraft above ground level.

Consider the case depicted in Photo 3.2, acquired at an altitude of 457 m above sea level using a camera with a focal length of 152.53 mm. The heights of the various buildings in the photograph can be determined using (3.1) below, particularly as the nadir point can be located.

In near vertical aerial photography, the nadir is usually assumed to coincide with the principal point. In Photo 3.2, the displacements of the buildings can be used to locate

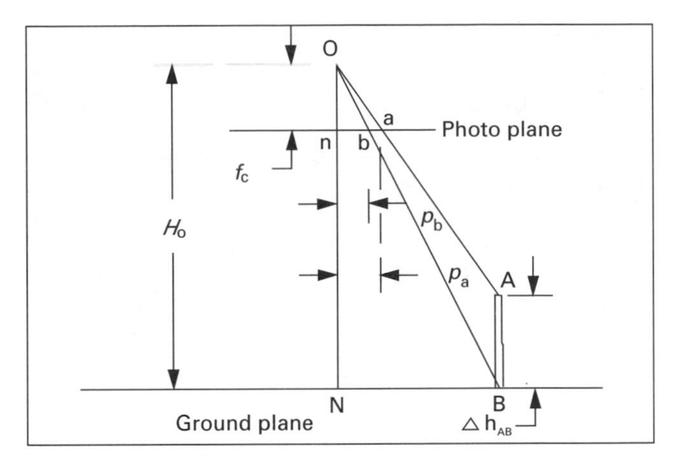

Diagram 3.9 Magnitude of displacements due to height differences.

the nadir point, since those displacements are radial from the nadir point. Extending the displacements from different buildings until they intersect locates the nadir point.

At the nadir, $NA = NB = 0$ so that (Diagram 3.9),

$$p_a = p_b = 0$$

$$\Rightarrow \qquad \delta p = 0$$

At distance $NA = NB$

$$\frac{p_a}{f_c} = \frac{NA}{H_o - \Delta h_{AB}} \quad \text{and} \quad \frac{p_b}{f_c} = \frac{NB}{H_o}$$

$$\Rightarrow \qquad p_a = \frac{NA \times f_c}{H_o - \Delta h_{AB}} \quad \text{and} \quad p_b = \frac{NB \times f_c}{H_o}$$

Therefore

$$\delta p_{AB} = p_a - p_b = f_c \left\{ \frac{NA \times H_o - NB(H_o - \Delta h_{AB})}{H_o(H_o - \Delta h_{AB})} \right\}$$

$$= \frac{f_c(NB \times \Delta h_{AB})}{H_o(H_o - \Delta h_{AB})}$$

But $p_b = \dfrac{NB \times f_c}{H_o}$ so

$$\delta p_{AB} = \frac{p_b \times \Delta h_{AB}}{H_o - \Delta h_{AB}} \tag{3.1}$$

which can then be rearranged to give

$$\Delta h_{AB} = \frac{H_o \delta p_{AB}}{p_a}$$

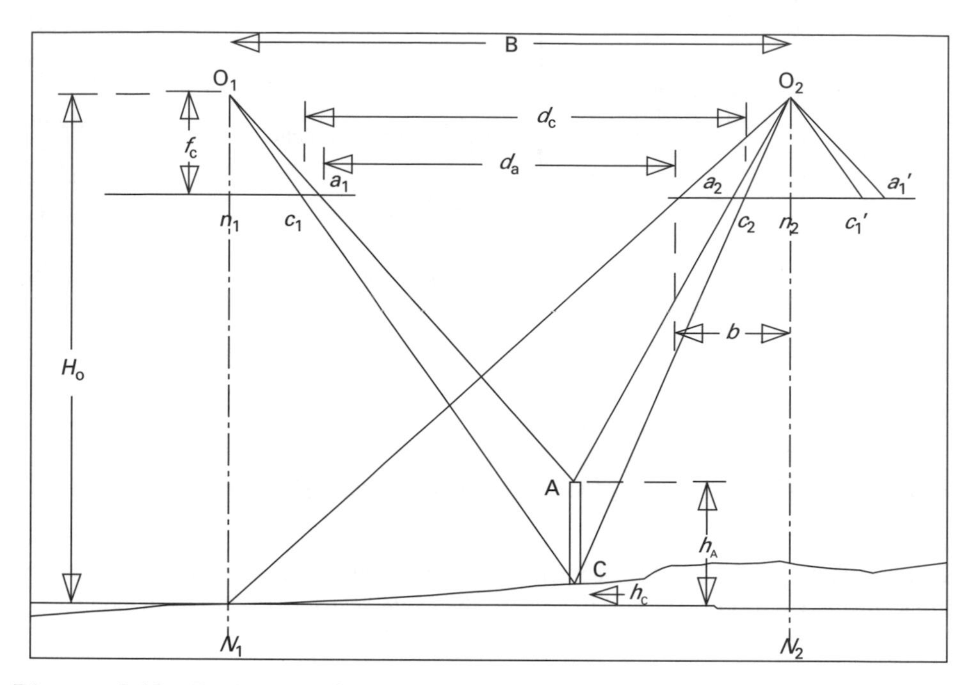

Diagram 3.10 Derivation of the parallax bar formula.

3.3.5 Derivation of the parallax bar formula

The displacements introduced by a difference in height in two adjacent aerial photographs will be in opposite directions when the height difference is located between the nadir points of the two photographs, as is normally the case (Diagram 3.10). Using the principle of the floating mark the distance between points of corresponding detail can be measured. The difference in these measurements, known as the parallax difference, is a function of the difference in height between the two points, as shown in Diagram 3.10.

The relationship between the parallax difference and the height difference is now developed, making the assumptions that:

- photographs are taken with vertical camera axes that are parallel; and
- both photographs are taken at the same altitude so that the air base, B in Diagram 3.10 is at right angles to the camera axes.

O_2a_1' is drawn parallel to O_1A, and O_2c_1' is drawn parallel to O_1C. The distances d_a and d_c are measured by means of the parallax bar. The parallax of A, p_a is given by

$$p_c = n_1a_1 + n_2a_2 = a_2a_1'$$

For C,

$$p_c = n_1c_1 + n_2c_2 = c_2c_1'$$

The parallax difference, δp_x is defined as

$$\delta p_x = p_a - p_c = (B - d_a) - (B - d_c) = d_c - d_a$$

The height difference, Δh_{AC}, is given by

$$\Delta h_{AC} = h_A - h_C$$

By similar triangles O_1O_2A and $O_2a_2a_1'$:

$$\frac{B}{H_o - h_A} = \frac{a_2a_1'}{f_c} = \frac{p_a}{f_c}$$

$$\Rightarrow \qquad H_o - h_A = \frac{Bf_c}{p_a}$$

By similar triangles O_1O_2C and $O_2c_2c_1'$:

$$\frac{B}{H_o - h_c} = \frac{c_2c_1'}{f_c} = \frac{p_c}{f_c}$$

$$\Rightarrow \qquad H_o - h_c = \frac{Bf_c}{p_c} \tag{3.2}$$

Now $\Delta h_{AC} = h_A - h_C = (H_o - h_C) - (H_o - h_A)$ so

$$\Delta h_{AC} = Bf_c\left(\frac{1}{p_c} - \frac{1}{p_a}\right)$$

$$= Bf_c\frac{(p_a - p_c)}{p_ap_c}$$

$$= \frac{Bf_c\delta p_x}{(p_c - \delta p_x)p_c} \tag{3.3}$$

Substituting (3.2) into (3.3) gives

$$\Delta h_{AC} = \frac{(H_o - h_C)\delta p_x}{p_c - \delta p_x}$$

where we either know, or can measure, H_o, Δp_x, but not h_C nor p_c. If C is placed on the datum surface then $h_C = 0$ and $p_c = b$ as can be proved using similar triangles. Then

$$\Delta h = \frac{H_o\delta p_x}{b + \delta p_x}$$

which is known as the **parallax bar formula**.

3.3.6 Characteristics of the parallax bar equation

To have δp_x positive for positive height differences, most parallax bars read increasing inward. The parallax difference, δp_x, is given by:

$$\delta p_x = p_a - p_c = d_c - d_a$$

If the parallax bar reads increasing values inward then reading $d_c > d_a$ when Δh is positive. If the parallax bar does not read increasing inward then the sign of δp_x will need to be changed in the parallax bar formula.

The important data are the parallax differences, not the actual parallax bar readings themselves.

By convention the datum is taken as the height of the left-hand principal point, that is the photobase, b, is measured in the right-hand photograph as the distance between the principal point and the conjugate principal point in that photograph. If the principal points are at different elevations then the bases measured in the two photographs will be different. The calculated height differences are relative to the elevation of the left-hand principal point.

When δp_x is small the equation can be approximated by

$$\Delta h = \frac{H_o \delta p_x}{b_o}$$

In this situation H_o/b_o is constant for a photo-pair so that all height differences are proportional to δp_x. When using the approximated equation then height differences can be calculated between two points without reference to the datum. In this case the parallax bar readings are taken at the two points of interest, and the parallax difference between the two is used to calculate the height difference.

Tilts in the photographs introduce warping into the perceived stereoscopic model created whilst observing a pair of stereoscopic photographs. Aerial photographs contain small tilts and slight differences in altitude from photograph to photograph. These tilts create significant height displacements in the perceptual model because the photographs, set up lying flat beneath the stereoscope, are not replicating the conditions that existed during acquisition.

3.4 Planimetric measurements on aerial photographs

3.4.1 Introduction

It is often necessary to make planimetric measurements from aerial photographs, either during the interpretation process or to transfer interpreted information from an aerial photograph to a map or grid of known scale. The planimetric measurements needed may include:

- determination of scale;
- measurement of distances;
- measurement of areas;
- transfer of detail from a photograph to update a map, add specific information to a map, or to construct a simple map if suitable ones do not exist.

The first three tasks are similar for both maps and photographs and will be dealt with as one topic in this section. The last task is specifically concerned with the transfer of information extracted from the interpretation of photographs.

All of the techniques discussed in this section deal with single photographs. They cannot measure, and therefore cannot take into account, displacements in the photograph due to height differences. In consequence, displacements due to height differences are normally converted into errors. Because of this source of error, these techniques are most accurate either when there are only small height differences in the area being mapped, or if the area is mapped in segments to meet this condition. If these restrictions are too harsh for a particular situation then sophisticated photogrammetric mapping techniques, routinely used in the production of topographic maps, must be used.

3.4.2 Determination of scale

Scale (1:x) is defined as the ratio of one unit of distance on the map or photograph to the equivalent x units of distance on the ground (Chapter 2). The larger the scale factor, x, the smaller the map scale and the larger the ground area covered within a given map area, and vice versa. Maps and photographs may be categorized as being small (1:50 000 and over), medium (1:50 000 to 1:10 000) or large (1:10 000 to 1:1000) scale.

3.4.3 Measurement of distance

Ground distance can be calculated as the product of the map or photograph distance and the scale factor. Similarly, distances on a map or photograph can be calculated by dividing the ground distance by the scale factor. Distance can be measured on a map or photograph using a graduated ruler or scale, a paper strip, a length of string and/or an odometer.

A **graduated rule** or scale can be used to measure a straight line distance on a map or photograph. Either the distance measured is compared with the scale bar on the map to give the equivalent ground distance, or it is multiplied by the scale factor to calculate the ground distance. The **paper strip** may be oriented along chords of a curve, marking both the position on the strip and on the map at the start and end of each chord. In this way the strip is gradually tracked around the total distance to be measured. The total length of the segments is used to calculate the ground distance. A length of **string** may be laid over a curve, with one end of the string placed at the start. The length of string used to cover the curved distance is compared with the scale bar to estimate the ground distance. Alternatively the length of string can be measured and the distance calculated knowing the scale of the photograph or map. The **odometer** is a device to measure distance on a map or photograph. It consists of a small wheel linked to a rotating dial against a graduated scale. When the odometer is tracked along a line, the wheel rotates, rotating the dial against the scale. The difference in scale readings at the start and end of the line is a function of the distance travelled.

3.4.4 Measurement of areas

There are three different types of areas to be measured on an aerial photograph: those with regular boundaries; those with irregular boundaries; and those with no clearly defined boundary as such. The first type of boundary is represented by paddock and fence lines, the second by creeks and watercourses and the third by scattered vegetation and soil associations. There are three ways of determining areas, all of which are suitable for measuring areas with regular and irregular boundaries: dot grid, planimeter and by use of a digitizer. Only the dot grid method is suitable for measuring areas with undefined boundaries.

A **dot grid** consists of a grid of dots at some set interval on a clear film base. The product of the distances in the x and y directions separating the dots is the area represented by

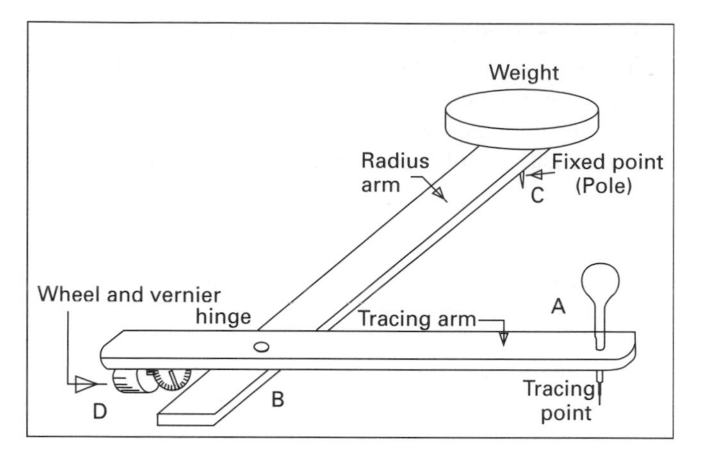

Diagram 3.11 Construction of the planimeter.

one dot. An area is determined by counting the number of dots within the boundaries of the area and multiplying this by the area represented by one dot. To reduce the work load a line grid may be drawn around set blocks of dots, say 10 by 10 or 100 dot blocks.

The accuracy of the technique can be increased by increasing the density of dots in the grid, but at the cost of increasing the workload. Dot grids are suitable for measuring areas with both well- and ill-defined boundaries. It could, for example, be used to measure the area of scattered trees across the image.

The **planimeter** is a mechanical device that can measure the area within a boundary on a map or photograph by integrating the areas of successively measured infinitesimally small arcs.

The planimeter (Diagram 3.11) consists of two arms AB and BC hinged at B. Point C is fixed to the base by means of a needle point. Point A traces out the area to be measured by sweeping out arcs with arm AB of known length. At the end D of the arm is located a wheel that touches the surface of the map or photograph, and rotates as the boundary is traced out by point A. The wheel at D is linked to a rotating, graduated drum. The drum is read at the start of the area, and again at the end. The difference between these two readings is a function of the area within the boundary traced out, and has to be multiplied by some factor to compute the area. This factor can be determined by measuring a known area, for example the area of a square or a circle, which also calibrates the planimeter.

The **digitizer**, is a computerized technique which is discussed in Chapter 7.

3.4.5 Transfer of planimetric detail

There are two theoretical bases from which methods have been developed to transfer planimetric detail from an aerial photograph to a map: the cross or anharmonic ratio; and the radial line principle.

The **anharmonic ratio** states that (see Diagram 3.12)

$$\frac{AC}{BC} \times \frac{BD}{AD} = \frac{ac}{bc} \times \frac{bd}{ad} = \text{constant}, k$$

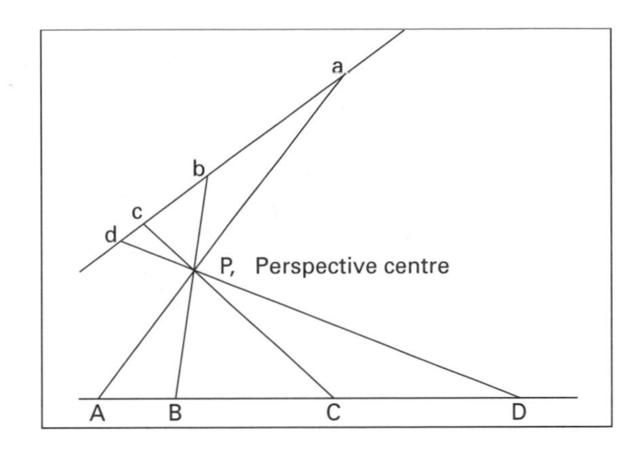

Diagram 3.12 The theoretical basis of the anharmonic ratio.

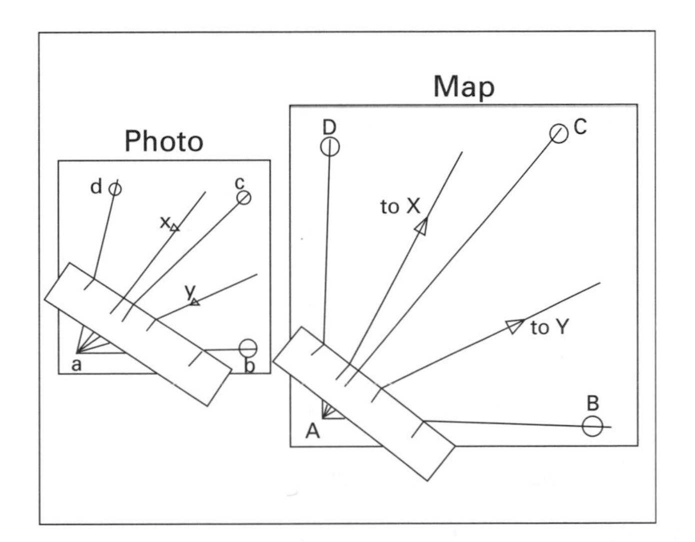

Diagram 3.13 The paper strip method of transferring detail.

There are three methods of transferring detail based on the anharmonic ratio: the paper strip method, using projective nets, and using the sketchmaster.

With the **paper strip method**, a set of four points in the photograph (a, b, c, d) and on the map (A, B, C, D) are selected so that they bound the area of interest (Diagram 3.13). To transfer a point x from the photograph to the map, place a paper strip on the photograph such that it crosses the lines radiating from one of the control points (point a in this case) to the other control points (b, c and d) and the line to *x*. Mark the paper strip where the four lines pass beneath the paper strip. Transfer the paper strip to an equivalent position on the map. Orient the strip so that the marks to the three control points coincide with the radial lines on the map to the other three control points. Mark on the map where the fourth line, to *X* will pass. Draw a line from the control point, A through this mark. Repeat the process from another control point to give a good intersection at *X*. Repeat this process for all new points being transferred.

The process of using a paper strip is time consuming and untidy if more than a few points are involved. The best results will be achieved if the points used give intersections

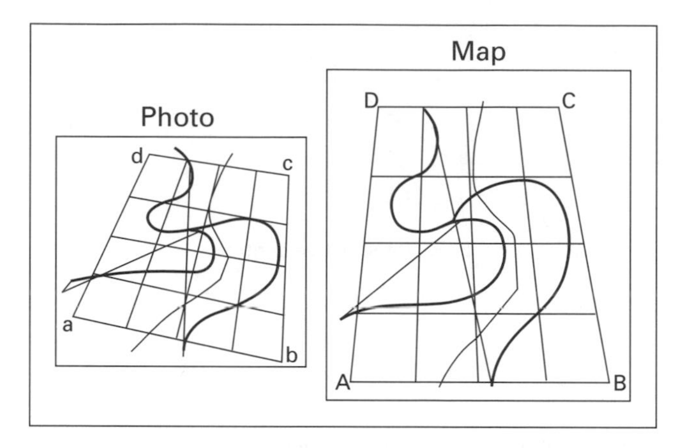

Diagram 3.14 Construction of projective nets between four corner points on an aerial photograph and a map.

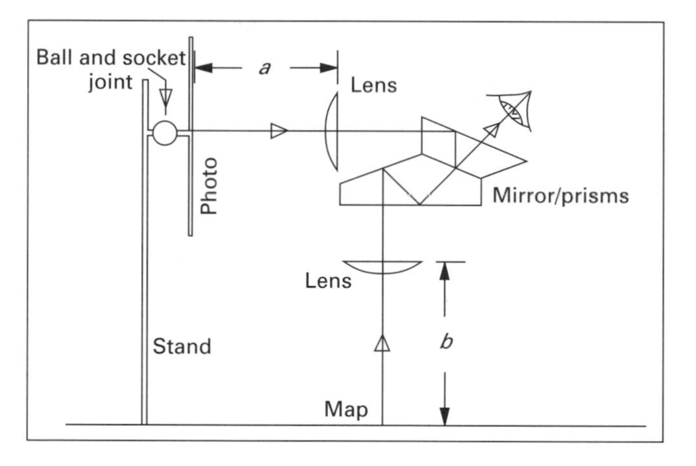

Diagram 3.15 Construction of the Sketchmaster.

at about right angles and if the four control points and the area enclosed contains only small height differences.

When using **projective nets**, again four corresponding points on the map and the aerial photograph are selected, ensuring that they bound the detail to be transferred. Join the points and create a grid by interpolating between the corner points, as shown in Diagram 3.14. The grid is created by linearly interpolating between the corner points along the sides of the grid, and then joining the equivalent points on opposite sides. Once the grids are drawn then detail is transferred by eye from individual grid squares on the photograph to the equivalent squares on the map. The net can be as dense as required. The method is less accurate than the others, but it is simple to construct and implement.

The **Sketchmaster** is an optical device that superimposes an image of the photograph onto the map so that detail can be transferred from the photograph. The photograph is fixed to the vertical plate which can be rotated about the three axes on a ball and socket joint. The lenses and prism provide focus for the photograph and the map, and combine the two images for viewing by the observer. The distances *a* and *b* can be adjusted to change the scale between the photograph and the map.

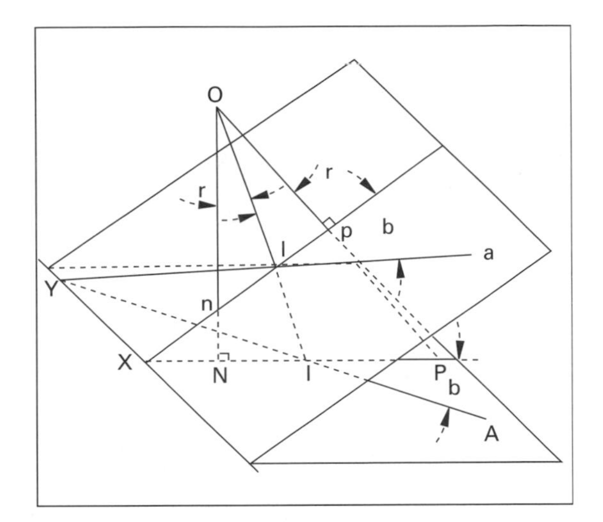

Diagram 3.16 Relationship between the principal point, nadir point and isocentre in a photograph.

The solution of the anharmonic ratio is achieved by superimposing the images of four points on the photograph and map. This superimposition is achieved by adjusting the relative scale, by changing *a* and *b*, and by adjusting the ball and socket joint to tilt the photograph. The technique is awkward due to the need to adjust the illumination of the photograph and the map.

The **radial line principle** states that angles with vertices at the principal point of a vertical photograph are the same as the equivalent horizontal angles on the ground surface. In a vertical photograph, the displacement of images due to relief and variations in flying height are radial about the principal point, thus leaving the subtended angles unaffected. In practice, aerial photographs are affected by tilts in the camera axis during exposure. Under these conditions horizontal angles in the surface are held true at the isocentre of the photograph rather than the principal point.

The **isocentre** is defined as the point of intersection of the line OI and the plane of the photograph where the line OI passes through the perspective centre, O (Diagram 3.16), and intersects the angle contained between the lines containing the principal point, OP, and the nadir point, ON, respectively. The isocentre cannot be readily located on a photograph as its position depends on the photographic tilt. For this reason techniques based on the radial line principle must use near vertical photography where the tilts are small and the isocentre is very close to the principal point. When this occurs, then sets of lines radial from the principal points of an adjacent pair of photographs and passing through common detail on each photograph can be used to locate the detail from the photographs onto a map.

At least two points (A and B in Diagram 3.17) must be located on the pair of photographs and on the map to set the scale of the plot. The principal points and conjugate principal points are located on each photograph. The known control points are located on both photographs, and then the additional points that are to be transferred are located on the photographs (points 1, 2, 3 and 4 in Diagram 3.17). A sheet of clear film is located over the first photograph and the principal point transferred to the film from the photograph. Radial lines from the principal point are then drawn on the film to pass through all of the other points including the conjugate principal point. The film is then

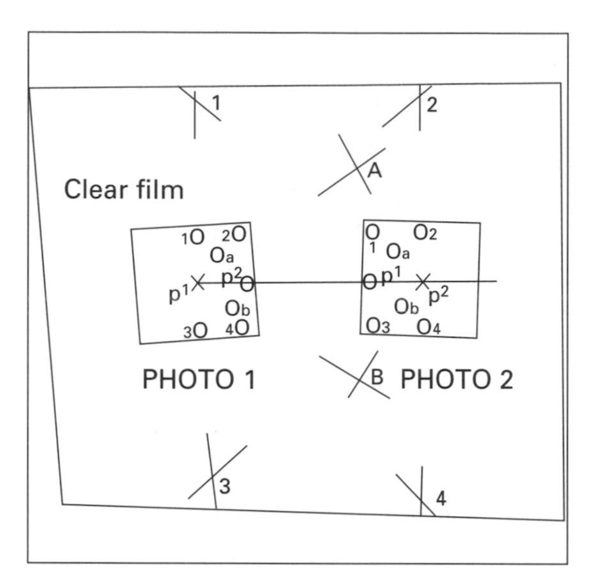

Diagram 3.17 The radial line principle and its use in transferring detail from a pair of photographs to a map.

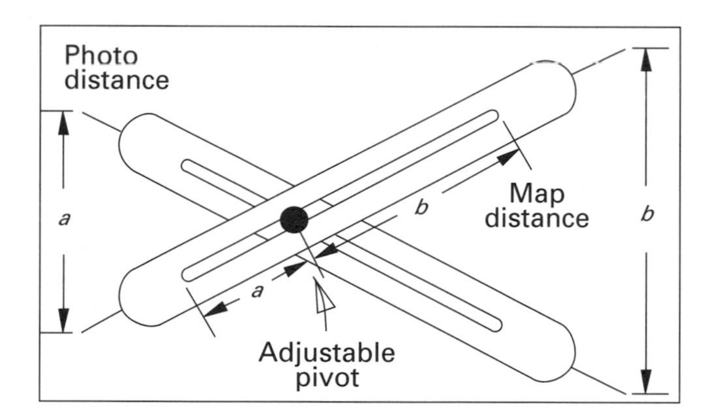

Diagram 3.18 Proportional dividers.

placed over the second photograph and oriented so that the line joining the principal point and the conjugate principal points from the first photograph overlays the same line on the second photograph. Radial lines are drawn from the principal point to intersect the radial lines drawn from the previous photograph. These intersections mark the positions of all points at a constant but unknown scale. The scale can be varied by changing the distance between the principal points before drawing the intersections from the second photograph. This plot, at a constant but unknown scale, is brought to the correct scale by enlargement or reduction using the control points.

The technique is valuable for locating a few points, or to fix a number of points to create a perspective net so as to plot detail from the photograph to the map.

3.4.6 Proportional dividers

Proportional dividers have an adjustable pivot point so that the distance from the pivot point to the pointers at either end can be changed. The ratio of the two distances gives

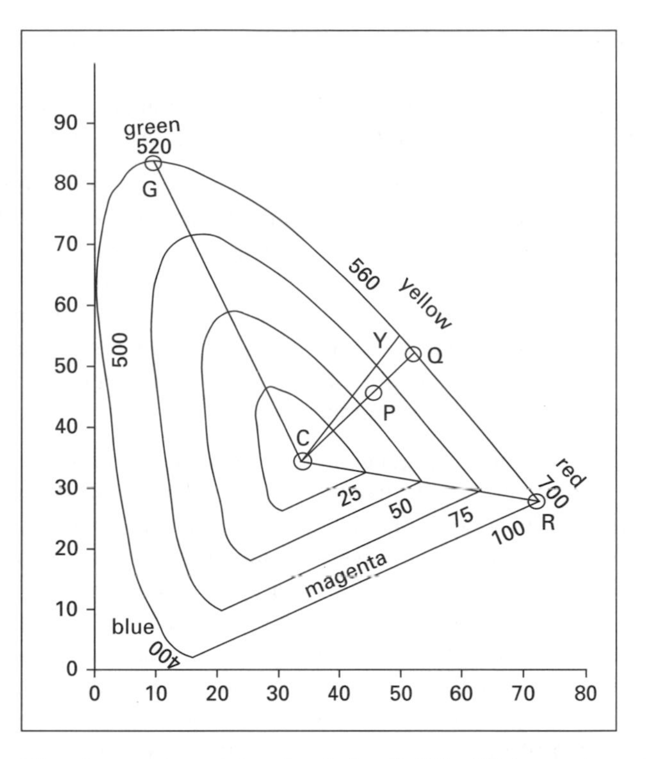

Diagram 3.19 The chromaticity triangle: Point P (46, 46) has a hue of 580 nm and saturation of 0.60; line CY bisects CR and CG, giving yellows as combinations of red and green.

the change in scale between the pairs of pointers. Once the scale of an aerial photograph and a map are known, the relative distances to the centre pivot on the proportional dividers can be calculated and set. Measuring a distance on the photograph or map will then allow the same distance to be marked on the other image.

3.5 The perception of colour

The generally accepted explanation for the perception of colour in the human eye was given in section 3.2 as being due to the sensitivity of three types of cones to blue, green and red light respectively. If blue, green and red light are combined then the resulting ray contains light of all three wavelengths. This process is called **additive colour mixing**, because it entails the addition of wavebands. The three colours, blue, green and red are called the three additive **primary colours**.

Additive mixing of the three primary colours will produce a range of spectral colours, as depicted within the chromaticity diagram (Diagram 3.19). The three primary colours are depicted on the outer edge of the colour domain at 400 (blue), 520 (green), and 720 (red) where the numbers refer to the mean radiation wavelength that is perceived as being that colour in the human eye. The centre of the diagram is white, created by mixing full intensities of all three primary colours.

The colours in the chromaticity triangle can be expressed in other coordinate systems than just the intensities of blue, green and red. The most common is in terms of hue, saturation and intensity.

Hue is the dominant wavelength in a colour, indicated as the number on the outside of the chromaticity triangle when drawing a line from the white point, C, through the location of the colour in the triangle, to the edge of the triangle. Consider the colour at P in Diagram 3.19. Drawing the line CPQ gives the hue as 580 nm.

Saturation is the degree of spectral purity of the colour, or the proportion of the colour due to the one wavelength. The closer the colour is to the outer edge of the triangle, the more pure it is, and the more saturated the colour. Saturation is measured as the ratio of the distances from the neutral point to the colour point and the neutral point to the edge of the triangle on the same line. In the case of colour at P, the saturation is the ratio CP/CQ, or 0.60. The maximum value of saturation is 1.0, and the minimum is zero at the neutral point. Pastel colours, containing a lot of white are desaturated, whereas intense, vibrant or strong colours are highly saturated.

Intensity is the total amount of illumination from the scene, or the scene brightness. Low levels of illumination affect the effectiveness of the cones and hence the perceived colours. It is important to have good illumination when viewing imagery.

Black and white (panchromatic) images acquired in any three wavelengths can be projected through red, green and blue filters to create a coloured image. The relative intensities of the three additive primaries at each point in the coloured image controls the resultant hue, saturation and intensity. If two of the primaries are full intensity, and the third is zero, then the resulting colour will have a hue value at the edge of the chromaticity triangle and at the point of bisection of the line joining the two contributing primaries. A combination of full intensities of green and red will lie at the bisection of the 520 and 720 points. The hues that result from the additive combination of the three primaries can be plotted on the chromaticity diagram. The blue, green and red primary colours lie on the edge of the triangle. As the intensities of the other primary colours are increased, the combined colour moves from the edge of the triangle towards the white point. Equal intensities of the three primaries will yield white, or grey, depending on the intensity.

Diagram 3.19 shows that combinations of pairs of saturated primary colours will yield the colours:

RED + GREEN = YELLOW
RED + BLUE = MAGENTA
GREEN + BLUE = CYAN

If panchromatic images are taken recording the radiation in the blue, green and red wavebands and these are combined additively using blue, green and red colours, then the result will be a true colour image, or an image that will depict the scene in the same colours as would be seen by the human eye. The coloured images produced will only be in true colour under the above conditions; under all other conditions they are called false colour images.

Consider Photo 5.3 where various combinations of images, recording the radiation reflected from the surface in the different thematic mapper wavebands, are additively combined. All of the resulting images are false colour images that are different because of the difference in reflectance in the TM wavebands.

The images in Photos 5.2 and 5.3 depict the differing information content of the different TM wavebands. In selecting imagery for interpretation the analyst should consider which sets of wavebands are most likely to depict the information in a way that allows the surfaces of interest to be identified and discriminated from other surfaces with the greatest accuracy and in the easiest manner. Selection of the most suitable wavebands

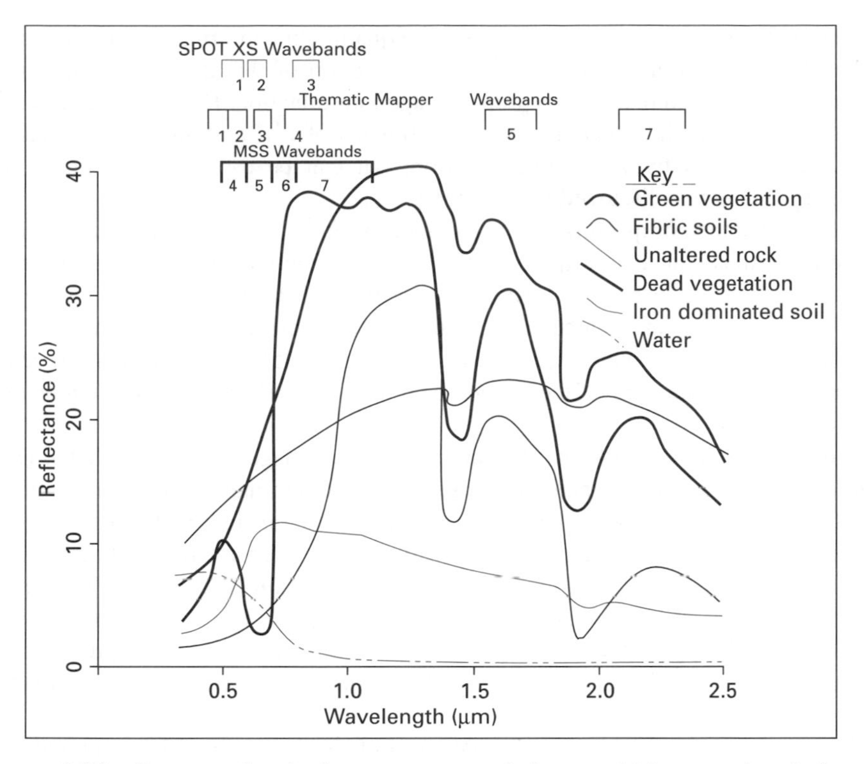

Diagram 3.20 Spectra of typical cover types and the acquisition wavebands for Landsat MSS, TM and the SPOT XS sensors.

should be based primarily on analysis of spectral reflectance curves as discussed in section 2.3, and shown in Diagram 3.20. In the ideal situation the actual spectral reflectance curves for the surfaces likely to be met would be known. Wavebands would then be selected so as to provide maximum discrimination between the surfaces.

Consider the surfaces depicted in Diagram 3.20. The most suitable wavebands to discriminate between these are where the separation between the curves is a maximum. The more separation there is between the wavebands, the better the separation of the classes in the image.

In practice the specific spectra for the different surfaces at the time of acquisition of the imagery are unlikely to be known in detail. In this case, generalized spectra should be used with the known wavebands to select those sets of wavebands that may be of most use. Once the waveband sets have been selected then the analyst needs to choose which additive primaries are to be projected through which of the images. The conventional selection when projecting green, red and NIR wavebands (MSS bands 4, 5 and 7, TM bands 2, 3 and 4 or SPOT Channels 1, 2 and 3) is to use blue, green and red respectively as this combination has been found to be the most useful for most purposes.

3.6 The principles of photographic interpretation

3.6.1 Introduction

Image interpretation is defined as the extraction of qualitative information on the nature, origins and role or purpose of the objects or surfaces viewed in the imagery. This qualitative information transcends the estimation of simple quantitative information of size, position, number and distribution of objects. Indeed, extraction of this quantitative information may be an essential prerequisite to proper image interpretation. However, there are some types of quantitative information, such as estimates of leaf area index or biomass that cannot be estimated by visual interpretation, but can be estimated by digital image processing techniques discussed in Chapter 5.

The importance of quantitative information in image interpretation can be assessed by example. An adequate description of the geology of an area may depend on measures of the dip and strike of bedding planes. Assessment of the importance of the different agricultural practices in an area may require measurement of the proportions of the different land uses. The interpreter must assess what quantitative information is required to develop an understanding of the phenomena influencing those objects of interest in the interpretation task.

But interpretation is not only dependent upon the imagery. The knowledge level of the interpreter influences the perceptions gained from the imagery and hence will affect both the quality and quantity of information gained. As an interpreter learns about an area, so perceptions change, causing modification of the interpretation and increasing the depth and range of information extracted. Knowledge of the sciences that influence the objects or surfaces of interest and hence better understanding relationships depicted in the imagery will also have a considerable impact on the quality of interpretation. Such knowledge can be supplemented by information from other sources, including maps and discussion with colleagues.

The interpretation process can be seen as a cyclical process of learning and deducing. The learning involves collecting information so as to formulate a hypothesis, (inductive reasoning), and then testing that hypothesis (deductive reasoning). Once the hypothesis has been accepted then it is used to extract information about the surface or object depicted in the image, before the process is repeated for another aspect of the image.

3.6.2 Levels of interpretation

There are three broad levels of analysis in the interpretation of imagery, differing in the degree of complexity in the interpretation process, and the depth of information extracted from the images: image reading, image analysis and image interpretation.

Image reading involves only a relatively superficial visual examination of an image, undertaken to identify quickly the main features on the image and determine their distribution and likely relationship. It is the level of analysis employed when, for example, attempting to locate the position and orientation of an image in an area on a map. It is the usual level of analysis of photographs in a magazine or book.

Image reading is an important preliminary level to analysis, since it allows the analyst to select appropriate imagery and to assess the relative merits of the different images available for more detailed analysis.

Image analysis is image reading with longer and more detailed visual inspection of the imagery to understand the more general and obvious features in the imagery, as well as the measurement of those quantitative features deemed necessary. In geology, it is likely to involve the marking of bedding planes, determination of dip and strike, delineation of drainage patterns and identification of discontinuities. In agriculture, it is likely to involve identification and measurement of the area of the different land uses, marking of drainage patterns, mapping slope classes, eroded areas and access roads.

Image analysis provides important information about the area, but it does not include the synthesis of that information. For some purposes image analysis is quite sufficient, but it may not be adequate if an understanding of causes of events is required.

Image interpretation is the inductive and deductive analysis of imagery, in conjunction with other data, to gain as full an understanding as possible of the features in the imagery, their functions and their inter-relationships. Image interpretation involves repeating the processes of inductive and deductive reasoning until such time as the interpreter is satisfied that the formulated hypothesis has been adequately tested, and then used to provide information on the surface or process.

For example an accumulation of facts about a building may allow the analyst to hypothesize that the building is a house rather than a shed or barn. This is the inductive reasoning process. However, a house would exhibit certain characteristics; if a majority of these exist then the hypothesis is likely to be accepted. Some of these typical characteristics are clothes lines, power and telephone lines, play area for children, barbecue areas, car parking areas, stove with chimney, washing and toilet facilities with waste outlet area, etc.

The context in which the building occurs can also be important. A property in an area could be expected to have a number of houses, where the number might depend on the size of the property and the land uses within the property. The number of houses that are occupied might also vary and may be important in the analysis. Thus information on all of the buildings may need to be collected before the analyst can finally test and either accept or reject the hypothesis.

Testing of the hypothesis that the building is a house is the deductive interpretation process. If the hypothesis is found to be wanting then the hypothesis needs to be either rejected or modified. The deductive process results in the collection of additional information that can then be used in a second inductive phase, to create a new hypothesis. The inductive phase thus uses all of the accumulated information to formulate a hypothesis.

Image interpretation is thus a continuous learning process, consisting of formulating and testing hypotheses, before extracting information. It requires high levels of concentration by the analyst, and as such must be done under conditions that are conducive to concentration and learning. It is also an extremely exhilarating and satisfying process of discovery of information about an area.

3.6.3 Principles of object recognition

Objects are identified on visual imagery primarily by use of the characteristics of those objects as recorded in the imagery, e.g. size, shape, shadow associated with the object, colour or tone, and pattern and texture.

These features are modified by the geometric, radiometric, spectral and temporal characteristics of the sensor system being used and by the processing done prior to interpretation. It is very important that the image interpreter be fully aware of, and understand, the characteristics of the imagery being used in the interpretation, since otherwise the wrong conclusions may be drawn as to why certain phenomena appear to exist in the imagery.

The **size** of objects is a function of the scale of the imagery, the perspective and geometric characteristics of the imagery, as well as the size of the actual object itself. Size can refer to the object's three-dimensional size. An interpreter should always measure the size of objects using the techniques discussed earlier and not depend on subjective judgement.

Photo 3.2 shows buildings of different sizes; the houses are easily discriminated from the high rise apartment buildings by the relative sizes of the two types of buildings. Again, roads are easily discriminated from pedestrian walkways by the widths of each feature.

The satellite images of Subic Bay (Chapter 5) include the city of Olongapo and the Subic Naval Base. Buildings in the city cannot be identified because they are too small relative to the resolution capabilities of the sensor system. Individual buildings can be identified in the naval base because they are quite separate from other buildings, and by the colour in the images. This demonstrates the dependence of size on photograph scale.

The **shape** of objects in visual imagery is a function of the perspective of that imagery. The shape of objects as seen from the ground is quite different to the shape as seen from the air, particularly when viewing vertically downwards. Oblique perspective images provide a transition between the horizontal and the vertical perspectives. Cultural features are designed with the horizontal perspective in mind. In consequence, the depiction of cultural features in the vertical perspective will often be quite different to their depiction in the horizontal perspective. There may, therefore, be a considerable dichotomy between these two perspectives, and the interpreter will often need to look for quite unfamiliar clues or characteristics when interpreting cultural features. Most other features are not designed with a particular perspective in mind, and indeed may favour the vertical perspective as in the case of agricultural features. In consequence there is usually a closer correspondence between the depiction of these other types of surfaces in both perspectives.

Shadows can indicate the nature of an object by displaying its profile, such as of trees and buildings. Imagery taken either early or late in the day can be particularly valuable for forestry and geology because the shadow effects emphasize features of interest.

Shadows mask detail contained within the shadow. Shadows in imagery are a function of the solar elevation and height of the object; the interpreter has to consider their effects in selecting the best time of day to acquire imagery.

Tone is the greyness in black and white photographs. The radiance at the sensor equation of Chapter 2 shows that the absolute values of the tones of an image are influenced by many factors other than the surface itself. Most of these external factors tend to influence the whole of the image in a similar way; they usually have little influence on relative tonal changes from location to location within the one image. Relative tonal changes are primarily due to changes in surface conditions and consequently are most important in the visual analysis of photographs.

In photography the relationship between tone and radiance is difficult to establish because:

1. the shutter speeds of cameras are set at fixed values, so that subtle variations in total radiance incident on the camera may not cause a change in exposure setting, but may still cause subtle changes in the resulting density on the film;
2. the light meter controlling exposure may not have the same response characteristics as the filtered film for the taking conditions; and
3. the temperature and duration of development and processing as well as the condition of the chemicals all affect the final tones in the photographs.

With electronic sensors the incident radiation levels are quantized in an analogue-to-digital converter (ADC) so that detector error is the major factor affecting the relationship between incident radiance and response. Enhancement of the digital data can be done in a number of ways (Chapter 5), many of which create complex relationships between radiance and the final tone. These tones are further modified in producing hard-copy images. In consequence the hue, saturation and intensity relationship within hard-copy images of satellite data can be quite complex and difficult to appreciate.

Changes in surface conditions, such as changes in water turbidity, the amount of green vegetation in the canopy, cultivation of the soil, surface soil moisture and the existence of fire scars, affect the response in the images. The analyst is concerned with information contained in these changes, but these changes can make it difficult to calibrate images one to another, and in consequence can make it difficult to quantify the changes that have occurred.

Tone and colour are important clues because of their relationship to energy received in specific wavebands. Where relative tone or colour in an image can be used, they can be very reliable indicators of variations in surface conditions. When interpretation has to be done between images then they may not be so reliable, unless the images have been radiometrically calibrated.

Because tones and colours are a function of the spectral and radiometric characteristics of the surface being imaged, it is important to be aware of the spectral bands depicted in the imagery, and how the different surface types will be recorded in that imagery. (Section 2.3 provides the basic reflectance information necessary.)

Texture is defined as the degree of roughness or smoothness of an object or surface in the image. **Pattern** is the regular repetition of a tonal arrangement across an image. Drainage lines form a pattern across the image since the arrangement is replicated from stream to stream, with different patterns indicating different important characteristics of the landform, soil and geology of the area. The paddocks of an agricultural area form a pattern; changes in that pattern may indicate changes in land-use activities and thus may indicate changes in physical, sociological or economic factors or processes at work, where differences in soil, climate, landform or costs of production are amongst the more usual factors that may be changing. Within a paddock the surface texture can be smooth (e.g. for good crops), rougher (for cultivated fields) or very rough for fields with variations in cover conditions. Sometimes the variations become sufficiently obvious as features to form patterns in their own right. Texture and pattern are related through scale; as the scale is reduced so patterns will degrade into textures. Scattered trees form a distinct pattern on moderate to large scale photography; as the scale is reduced the trees tend to merge and lose their pattern creating a rougher texture on the image.

Texture and pattern are very important identifiers on visual imagery. It is essential that when a task depends on texture and pattern the scale ensures that the features sensed are depicted as appropriate textures and patterns.

3.6.4 Interpretation strategies

The principles of object recognition provide us with the techniques necessary to identify individual objects or features on imagery, and understand many of their characteristics and functions. These principles focus on the object of concern, and its immediate environs and not the broader perspective in the analysis. This broader perspective is often essential to understand the linkages between objects and functions within an area, or to resolve ambiguities in the interpretation. These broader perspectives can be considered by the use of what are called interpretation strategies.

Location and association

The interdependence of objects or features frequently controls their relative locations. Analysis of the nature and function of the features in the vicinity of an object will often provide important clues as to the nature and function of the object itself. This appreciation of the interdependence of features starts at the regional scale and can proceed down to the microscale if required. Thus interpretation should start with a broad overview of the environs of the area of interest, so that the interpreter assesses features within the context of the environment. This overview is then followed by more detailed interpretation, that can be interrupted by a revision of the overview if the interpreter believes that subsequent analysis is raising issues that are in conflict with the currently held overview.

The location and association of buildings relative to transport corridors, such as roads or railway lines, can indicate the role or function of those buildings, just as the location and association of a set of buildings can be important in assessing the roles or functions of the different buildings. More subtle relationships can be even more important. The location of a stand of trees, and their association with particular slope values and soil classes, can be indicative of soil conditions; the soils may be either too thin or poor to justify clearing of the trees. Other information can strengthen or weaken such hypotheses; for example the mix of species within the stand of trees and the condition of pastures around the area, or on other areas that have similar physical attributes, can be valuable information in this type of assessment.

Temporal change

Imagery of an area can be quite different at different times of the day, and from season to season, (Photo 3.4) due to changes in surface conditions, in the sun–surface–sensor geometry, in shadowing and in the effect of the atmosphere. Many activities or phenomena have diurnal and seasonal patterns that can be used to improve the interpretation of these phenomena. The interpreter can often select, and must be aware of, the time of day and the date of the imagery so as to use these characteristics to improve the interpretation. Vegetation is dependent on meteorological history so that an interpreter must be aware of this history before conducting detailed interpretation.

Convergence of evidence

The interpreter collects a set of circumstantial evidence about the features of interest in the imagery. When sufficient evidence is deemed to have been collected then it is used to construct a hypothesis. For example, the circumstantial evidence that the analyst may collect about a feature could include the fact that the tone of the feature is dark in the image, the feature is linear in shape and creates a branching pattern, and that the feature is in the bottom of a valley. Clearly this evidence will suggest a watercourse as the nature of the feature. Similar processes occur with each feature in the image even though the process may occur quite automatically for many features.

There will always be cases where the interpreter does not believe that the evidence is strong enough to construct a hypothesis. In this case the interpreter has to gather further evidence from other sources. The most common sources of additional information are discussion with colleagues, archival information in the office and the collection of evidence in the field, or field data. Discussion with colleagues is a valued source of additional information because they, by approaching the problem from a different perspective, may identify evidence in the imagery that was missed by the first analyst. If this source of information, and other information available in the office, do not resolve the difficulty then a field visit may be necessary.

Field visits are best done either during or at the start of the interpretation. The analyst will also need to visit the field for other purposes such as accuracy assessment. If the ambiguities are not major then they may be able to wait until the accuracy assessment stage. If the ambiguities are major then they may influence the whole interpretation process. In this case it is usually best to conduct the field work as soon as possible.

3.6.5 Interpretation procedure

In approaching an interpretation task the analyst should proceed in accordance with a well established method in the conduct of the interpretation, adapted to suit the individual interpreter; the following provides the logical steps that can be followed.

1. Fully understand the information required, its type, levels of discrimination, spatial resolution and accuracy.

KNOW THY TARGET

2. Review the characteristics of the information sought and how it might be depicted on the imagery. Consider other classes and how they might also be depicted.

HOW DOES THY TARGET OPERATE

3. Evaluate the suitability of the imagery, facilities and staff skills to extract the information from the data.

ARE THY TOOLS SUFFICIENT TO THE TASK

4. Conduct an overview interpretation of the area using maps, photographs and other data or discussions. A preliminary visit to the area may be necessary.

KNOW THE ENVIRONMENT IN WHICH THY TARGET OPERATES

5. Conduct the interpretation.

HUNT THY TARGET

3.6.6 Visual interpretation of aerial photographs

The analyst's task is made easier by identifying first the clearly identifiable or relatively unambiguous objects or features, such as the drainage and transport corridors. The analyst then progressively addresses more complex features with the benefit of knowing more about the area from the interpretation.

During interpretation the analyst needs to record the results of the analysis either on the photograph, or on a clear film overlay to that image. Marking the interpretation results onto a map involves changing focus from the image to the map which tends to break concentration. Marking directly onto the photo also has the disadvantage that the annotation can mask detail in the images and can be difficult to erase. Marks on a clear film are easier to erase, using water soluble pens, and the film can be lifted off the photograph as required. However, the clear film can cause reflection that may interfere with visual analysis, scratching can occur on the surface of the film, and it is more difficult to take into the field.

3.6.7 Visual interpretation of scanner imagery

Scanner data have lower spatial resolution than aerial photographs, but usually enjoy better radiometric and spectral resolution. Spatial resolution affects the shape of features, the texture and patterns that will be depicted in the imagery. Shape, texture and pattern are usually more important interpretation clues in the visual analysis of photographs than of scanner image data, as will be discussed in Ch. 5. Interpretation tasks that depend very heavily on these clues can often use black and white or colour photography as successfully as scanner imagery.

Radiometric and spectral resolutions affect the colours and tones in image data. Because of the higher radiometric and spectral resolutions of scanner data, and the closer linkage between radiance from the surface and scanner data than is the case with aerial photography, colour and tone are more usually more important analysis tools in the visual analysis of scanner image data than of aerial photographs.

The loss of fine resolution detail in scanner data can be very disconcerting to an interpreter used to seeing fine detail both through normal viewing and through analysis of aerial photographs. Thus, a change in approach is required: the analyst can no longer depend on the fine detail to assist him in drawing conclusions as to the nature of the features imaged in the scene, but rather he uses shapes, texture and pattern to interpret broader, more macro-level features in the photo. The image, however, contains much more spectral and radiometric information, and so the analyst must learn to depend much more on the colours and tones in the image to identify smaller and more detailed features in the image. Proper exploitation of the radiometric and spectral information in the image requires the interpreter to learn more about the interaction of radiation and matter – more than is required with the visual interpretation of aerial photographs.

The interpretation of satellite images should therefore be easier in some ways than that of aerial photographs. The first interpretation should use all of the standard clues in the image to provide a good broad scale appreciation of the conditions in the area of interest. Once this phase is complete the analyst conducts detailed interpretation, depending primarily on the colours and tones in the image. Discussion in Ch. 5 will show that the colours and tones in satellite images can be changed to suit the convenience of the interpreter. The interpreter may choose to conduct this part of the interpretation using computer display facilities that allow the images to be readily changed in this way, rather than interpreting on photographic images.

3.6.8 The visual interpretation of thermal imagery

Thermal images are normally depicted as black and white images of the waveband sensed by the scanner or thermal camera as shown in Photo 5.2. The thermal image in this photograph is of the Subic Bay area of the island of Luzon in the Philippines, acquired by the Landsat TM sensor. Bands 2, 3, 4 of the same data set are also displayed in Photo 5.2.

Changes in tone in the thermal image are due to variations in the temperature, emissivity (Table 3.1) and roughness of the surface. The first two affect the level of radiation from each point on the object as discussed in Chapter 2. The roughness generally lowers the radiation from the surface due to shadowing, in much the same way as occurs in imagery in visible wavebands.

As the emissivity of the surface decreases, so the amount of energy radiated decreases, with the sensing instrument interpreting this decrease as a drop in temperature. Within an image, variations within the one cover type, such as within a water surface, can usually be assumed to be due to variations in the temperature of the surface, as long as that surface is consistent in its emissivity and texture.

The temperature of a surface depends on the incident radiation, the diffusivity and emissivity of the surface. The level of incident radiation affects the heating of the surface.

Table 3.1 The emissivities of common materials in the 8–14 micrometer waveband

Material	Emissivity
Granite	0.898
Sand, quartz	0.914
Basalt	0.934
Asphalt paving	0.959
Concrete paving	0.966
Water, pure	0.993
Water, with thin oil film	0.972
Polished metal	0.06

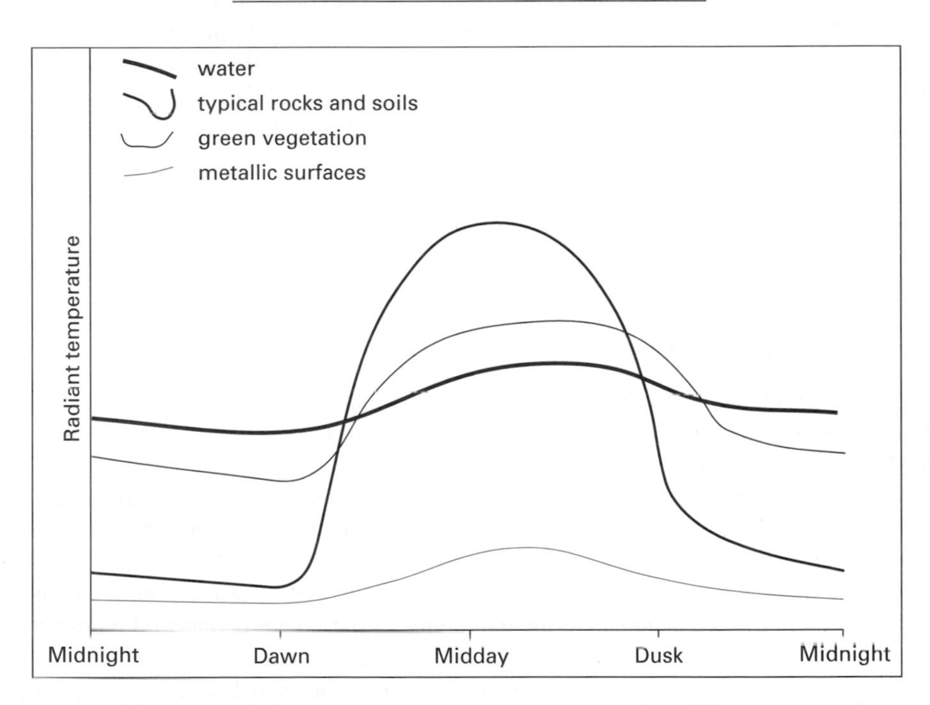

Diagram 3.21 Typical diurnal variations in temperature for different types of surface.

The diffusivity of a surface is its ability to store and transmit heat. It takes into account the specific heat and density of the different materials in the surface. Specific heat is a measure of the amount of energy that needs to be absorbed to raise the temperature of the surface by 1 degree, so surfaces, such as water, that need to absorb a lot of energy to raise their temperature by 1 degree have a high specific heat. The temperature of a material is also dependent on its ability to transmit heat to its neighbours. Good conductors can transmit heat in this way, whilst poor conductors (or good insulators such as air) cannot.

Since thermal imagery is affected by incident radiation, the time of day of the imagery is very important (Diagram 3.21). Water has a high specific heat and so it tends to stay cooler during the day than other surfaces, but it does not change its temperature at night as much as the surroundings. Surfaces that contain a lot of moisture thus generally appear cool during the day, and warm at night.

Photo 5.2 contains areas of water (Subic Bay), forest cover (Subic Bay Naval Base area), open grassland and soil, urban and commercial area (Olongapo City), industrial and airport areas (Subic Bay Naval Base). The image, taken at about 10.30 a.m. local time, shows the urban–commercial area of Olongapo city, and the industrial and airport areas of Subic Bay Naval base to be brighter than the surroundings. Some of the surfaces in these areas, such as concrete and galvanized iron roof sheeting, will have a lower emissivity than the vegetation of the forest areas, whilst other surface types, such as asphalt surfaces and rusted iron roofing, have a high emissivity. The surfaces in the urban and industrial areas of Olongapo and Subic naval base would generally have a lower emissivity than the vegetation, so they are certainly hotter than the forest areas. Because of this difference in emissivity their temperature, relative to the forest areas is probably greater than that indicated by the tones in the image. The forest areas are shown as being cool, despite their higher emissivity, indicating that they are significantly cooler.

The soil and grasslands areas would have similar emissivities to the Olongapo and Subic areas, yet they are darker in the image. This suggests that these areas are somewhat cooler than Olongapo and Subic. The water of Subic Bay is dark as would be expected.

The complexity introduced by emissivity means that thermal imagery of this type is often best used to compare different surfaces that should have similar emissivities. Variations in the tone of a water surface could be interpreted as differences in temperature since the water will have the same emissivity value. The same applies to clouds at different temperatures.

3.6.9 The visual interpretation of radar images

The SLAR image depicted in Photo 3.5 illustrates many of the characteristics of radar image data. An important factor affecting response from radar data is surface texture and roughness. The alluvial floodplain on the east side of the image is dark because the smoother surfaces of the floodplain are reflecting the majority of the signal away from the sensor. To the west of the image, the surface is generally dark but broken by some areas of higher response. These areas are usually trees along roads or watercourses, as seen in the north-west corner of the image, and of towns. The western part of the image is characterized by rice paddies, so the darkening of the image is due to the absorption and specular reflection of the radar energy by the water surface.

The response of towns is indicated by the return from San José city in the centre-right of the image, and a string of villages on the roads from San José to the north-west. Some town areas can also be seen in the south-west corner of the image. The strong response from towns is often due to corner reflector effects, and the strong returns that come from metal surfaces at a suitable orientation to provide specular return. The effects of metal features is also shown in the image by the response of power pylons across the southern part of the image, from west to east.

The western slopes of the mountains in the images are exhibiting a stronger return that the eastern slopes. This indicates that the radar was taken by an aircraft to the west of the area, with the radar pointing east. There is an area of radar shadow directly to the north of San José town, with an isolated peak erupting out of the shadow area. Some of the radar shadow areas indicate the lie-over effects on terrain as occurs in radar data. The main part of the Philippine Fault occurs in the north-east of the image. It is characterized by strong linear control on the topography. The river flows south-south-east along the fault and then debouches out of the valley and flows to the south-west.

4
Applications of visual interpretation

4.1 Introduction

Both visual interpretation and digital image processing play an important part in extracting information from remotely sensed images. The strengths of the two methods are complementary, so that an integration of the two will probably become the most usual method of extracting more complex information from image data. The use of visual interpretation will now be illustrated through four applications:

1. land assessment using stereoscopic aerial photography;
2. crop inventory using satellite imagery;
3. regional assessment using radar and satellite image data; and
4. visual interpretation of integrated data sets.

4.2 Land assessment using stereoscopic aerial photographs

4.2.1 User requirements

The town of Coffs Harbour (Map 4.1 and Photo 4.1), is expecting a period of significant growth, which will place stress on the existing town services, as well as creating pressures on the land use of the watershed. Harbour Creek is already subject to significant flooding; any developments in the watershed should not exacerbate this problem. Reduction in flood risk is a major policy of the Coffs Harbour council.

The town council also wish to retain a rural setting for the town, and plan to encourage the retention of good quality agricultural land for agriculture, to maintain aesthetically pleasing land for recreation and to retain steep lands as forested areas, whilst allowing urban development on other lands. Maintaining steep lands as forest areas will reduce runoff and soil erosion, mitigating flooding in the valley. Soil erosion and flooding can be further reduced by the adoption of appropriate agricultural practices.

This 'user specification' is stated, quite typically, in non-technical terms, but needs to be translated into a more precise definition of the information required by management. The information required to address user needs is rarely that extracted from the analysis of data; instead it is usually an integration of a number of different types of information derived from various sources. In general, information that is derived from data is the 'raw information' then used to derive the 'management information' necessary to support management decisions.

The steps that must be taken by the analyst will usually include:

- conversion of user specifications into management information needs by including technical precision in the definition of the information required;
- defining the Information Types that constitute the components of each type of management information – each component must be capable of extraction by some recognized process such as field observation, conversion of analogue map data, or the analysis of remotely sensed images;

- developing a plan to either purchase or acquire the required raw information, and then integrate these sets of information to provide the necessary management information;
- purchasing the raw information that is currently available;
- implementing a procedure to derive the raw information that is not currently available; and
- deriving the required management information by integrating the raw information data sets in a predefined manner.

4.2.2 Information requirements

Information on **land-using opportunities and constraints** is required to prepare Land-use Zoning plans, to promote the exploitation of the opportunities and to develop strategies to address the constraints, primarily through changes in land-using practices. The raw information required to prepare this management information includes: agricultural capability assessment; current land use; and land ownership.

An **assessment of erosion risk** is needed to identify constraints on land use so as to assist land-users to adopt or develop land-use practices that result in an acceptable level of erosion risk. The raw information required to prepare this management information is that necessary to drive models predicting erosion risk. If the universal soil loss equation is used as the basis for predicting erosion risk then the raw information required to drive that model includes: slope gradients and slope lengths; soil erodibility assessment; typical cover conditions; variations in rainfall intensity across the area; and land-ownership or the land administrative units that are of interest. In this particular case these units are the parcels of land-ownership.

An assessment of **flood risk, or changes in flood risk** is required to ensure that the risk of flooding is not increased, and preferably is reduced. At present the town of Coffs Harbour is subject to flooding from 1 in 50 year rainfall events preceded by significant rainfall, or 1 in 100 year rainfall events after periods of dry weather. The council desires that the probability of flooding be reduced if possible. The information required to meet this includes historical and current land use, proposed land-use zoning and land-ownership.

The raw information required can thus be summarized as:

- land-ownership, including maps and databases of the owner, lessee and their addresses;
- slope maps in appropriate class ranges;
- watersheds, sub-watersheds and drainage lines, with the drainage lines graded to assist in the construction of the sub-watersheds;
- agricultural land capability maps that categorize the lands of the catchment in terms of their suitability for agricultural production;
- soil erodibility maps suitable for assessment of the soil erodibility factors for the universal soil loss equation (USLE);
- historical land use matched to significant rainfall events that caused flooding, to be used as a basis for assessment of the potential impact of changes in land use on flood risk;
- current land use with classes that are suitable for identifying opportunities, comparison with historical landuse and assessing typical cover conditions for use in the USLE; and
- planned land-use zoning; each of which is now considered in detail.

4.2.3 Land-ownership

Spatial and other information on land-ownership is usually maintained by state agencies, often in digital form, as both spatial and attribute files. It is also held as cadastral maps

Table 4.1 Slope classes, class bound values and the contour separation at those bounds, for a 10 m contour interval

CLASS	BOUNDS	10 m CONTOUR SEP'N 1:25 000 scale (mm)	10 m CONTOUR SEP'N 1:50 000 scale (mm)
1	< 1°	> 23 mm	> 11 mm
2	1°–2°	23–11 mm	22–5.5 mm
3	2°–5°	11–5 mm	5.5–2.5 mm
4	5°–8°	5–3 mm	2.5–1.5 mm
5	8°–15°	3–1.5 mm	1.5–0.7 mm
6	15°–26°	1.5–0.8 mm	0.7–0.4 mm
7	26°–45°	0.8–0.4 mm	0.4–0.2 mm
8	> 45°	< 0.4 mm	< 0.2 mm

that depict the parcels of land and their ownership at some specified date (see Photo 4.2 for rural land-ownership of the watershed). Land-ownership is dynamic information so consideration needs to be given as to how up-to-date this information needs to be, to meet user needs.

4.2.4 Slope mapping

Slope mapping can be done in various ways: interpreting topographic maps; visual interpretation of aerial photographs; or processing of digital terrain models (DTM).

When **interpreting topographic maps**, the analyst determines the slope classes required for the task, where each class is defined as having lower and upper boundary slope or gradient values as given in Table 4.1.

The horizontal distance between adjacent contours is computed at each lower and upper gradient value. A class will be represented as areas that have contour separations between these lower and upper distance values. The lower and upper horizontal distance values of the classes used in this project are also given in the Table.

Mark these intervals on the edge of a piece of paper, and systematically map the slope classes across the map by determining which class represents the actual interval between the contours at each point on the map. The slope map of the watershed is shown in Photo 4.2.

For **visual interpretation of aerial photographs**, stereo aerial photographs are the best means of appreciating the shape of the landforms in an area, and hence provide the capacity to prepare slope and aspect maps that are most sensitive to subtle variations in landform shape. This characteristic can be valuable when mapping landforms that are irregular in shape. However the tilts in the photographs introduce a warping in the stereogram that is depicted as a tilt in the stereo model, which cause errors in the slopes. Another difficulty with stereo aerial photographs is that the vertical exaggeration in the stereo model means that the analyst cannot relate model slopes to ground slopes without the aid of a parallax bar or stereo wedges.

Stereo wedges are a connected pair of clear film sheets, each of which contains a set of vertical lines. The lines in one film are tilted relative to each other, so that when the line pairs are viewed through a stereoscope the resulting merged line appears to be sloping in the stereo model. Each pair of lines in the film set contains different displacements so that each merged line in the stereomodel appears to be at a different slope. The stereo wedges are placed on top of each photograph in a stereo pair so that the lines are vertical to the eyebase. The slope of the stereomodel is compared with the slopes depicted by the stereo wedge to determine which slope best approximates the slope of the stereo model.

Digital terrain models (DTMs) are usually stored as a mesh of grid cells across the area of the model, with an elevation value recorded in each cell. The height values in each cell can be used to compute slope and aspect. The accuracy of such techniques depends on the accuracy of the data in the DTM.

If the DTMs are constructed directly by stereo-photogrammetric techniques, using either aerial photographs or stereoscopic SPOT images, then each cell elevation value will have an accuracy value that can be specified by its standard deviation, $\pm \sigma_{\mathrm{h}}$. The gradient, g, is the ratio

$$g = \frac{\text{change in elevation}}{\text{horizontal distance}} = \frac{(h_2 - h_1)}{d} \tag{4.1}$$

where d is distance between the cells in the data, and h_1 and h_2 are the elevations of adjacent cells.

From the Law of the Propagation of Error, the variance of the calculated gradient can be estimated knowing the variances associated with the estimates of d, h_1 and h_2 as follows.

$$\begin{aligned}(\sigma_g)^2 &= \left\{\frac{(\sigma_{h1})^2}{d^2} + \frac{(\sigma_{h2})^2}{d^2} + (h_1 - h_2)^2 \times \frac{(\sigma_d)^2}{d^4}\right\} \\ &= \left\{\frac{(\sigma_{h1})^2}{d^2} + \frac{(\sigma_{h2})^2}{d^2} + (G)^2 \times \frac{(\sigma_d)^2}{d^2}\right\}\end{aligned}$$

Now $(\sigma_{h1})^2 = (\sigma_{h2})^2 = (\sigma_h)$ and $(g^2\sigma_d^2) \ll (2\sigma_h)^2$, so that

$$(\sigma_g)^2 \approx 2\frac{(\sigma_h)^2}{d^2}$$

or $\sigma_g \approx \pm 1.4142\sigma_h/d$.

SPOT stereoscopic data is claimed to be capable of producing 10 m contours. The normally accepted criteria for contouring is an accuracy of ± half the contour interval, 90 per cent of the time. Two standard deviations (section 5.2) from the mean value will include 95 per cent of the observations and 1.65 standard deviations will include 90 per cent of the observations. This means that 1.65 standard deviations are approximately equal to 5 m, or the standard deviation of each estimate of elevation is about 3 m. Most SPOT XS data is rectified to provide $d = 20$ m, so that

$$\begin{aligned}\sigma_g &\approx \pm\frac{1.4142 \times 3}{20} \\ &\approx \pm 0.2121\end{aligned}$$

which is a large range of gradient values. Clearly such data can only provide relatively coarse slope ranges. Better accuracies can be achieved using more expensive, larger scale aerial photography.

If the DTMs are constructed from contour data, then each cell value will be determined by the numerical interpolation used to estimate the cell values from the contour line data. The best accuracy that can be achieved will be that specified by the contouring criteria given above, i.e.

$$\sigma_g \approx \pm 1.4142 \times \frac{(\text{contour interval}/2 \times 1.65)}{d}$$

In practice the accuracy will be somewhat worse than this because of errors introduced due to the interpolation process.

4.2.5 Drainage interpretation

The drainage lines are identified on the stereo aerial photographs, and checked on the map, updating the detail on the map as appropriate (Photo 4.2). The interpretation should be started by following the main stream up from the mouth, through all the tributary streams in turn. Once this has been done then interpretation should start at the headwaters of the watershed, to see if any streams have been missed. Sometimes this sequence needs to be repeated a number of times until all streams have been identified and mapped.

This interpretation is not only important in itself, but it provides a good preliminary interpretation to the later more detailed analysis that will be conducted for both terrain classification and land-use assessment. The marked drainage lines are classified in accordance with the following rules: all initial streams are coded as Class 1; streams below the junction of two streams of the same class are graded to the next highest class. Thus the join of two streams of Class 1 will see the start of a Class 2 stream, the join of two Class 2 streams will start a Class 3 stream.

Once the drainage mapping is complete, the watersheds and sub-watersheds can be mapped on the topographic map, with supplementary interpretation from the stereo aerial photographs. A **watershed** is defined as the ground area about a drainage line, over which surface water flow is towards that drainage line. Watersheds typically match ridge lines that separate drainage lines, and they will close at the mouth of the stream. The watersheds and sub-watersheds are shown in Photo 4.2; analysis of the map shows that the sub-watersheds do not cover the whole of the watershed. The areas not contained in the sub-watersheds are areas that flow into the main drainage line itself. These areas are sub-watersheds that are usually far too small and difficult to define on the map, but they do contribute to the drainage of the main watershed.

4.2.6 Agricultural land capability and soil erodibility mapping

Agricultural land capability mapping is the assessment of the physical suitability of lands in the watershed for agriculture. The capability of the land for agriculture depends on:

1. soil fertility, both in the short and the long term (the depth and fertility of the soil are both important);
2. the erodibility of the soil, which depends on the cohesiveness of the soil, affected by the clay and humus content of the soil, the slope of the soil surfaces, the vegetative cover protecting that soil from the impact of the rainfall, and by the intensity of the rainfall itself;

Table 4.2 Classes typical of most currently adopted soil capability mapping schemes

Class	Soil type
A	Deep fertile loamy soils, capable of continuous cropping with little or no fertilizer.
B	Deep to moderately deep soils capable of continuous cropping over many seasons, but requiring fertilizer.
C	Shallow to moderately deep, fertile loamy soils that suffer moisture constraints, and for which care is required in tillage so as to maintain minimum soil loss.
D	Deep clayey soils that require particular practices or conditions for their use in cropping.
E	Deep sandy or erodible soils that require particular practices or conditions for their use in cropping.
F	Shallow to moderately deep clayey soil that can only be used for cropping on a rotation basis.
G	Shallow to moderately deep sandy or erodible soils that can only be used for cropping on a rotation basis.
H	Thin soils, or soils with a high rock content that are unsuited to cultivation and that are subject to high erosion.

3. the slope of the land, because of its impact on soil erosion, the ability to manage the land and the crops being considered;
4. the availability of adequate water; and
5. the climate of, and microclimate within, the area.

Classes that are typical of most currently adopted soil capability mapping schemes are shown in Table 4.2.

Terrain classification (section 4.2.7) will be used as the primary mapping technique, supplemented by soil coring in all of the mapped terrain units so as to categorize each mapped unit in terms of physical agricultural capability and soil erodibility.

4.2.7 Terrain classification

The surface and near surface characteristics of the terrain are important to the engineer, planner and environmentalist. To the engineer, this is because their shape and engineering characteristics will influence the construction costs of all engineering works. To the planner, the terrain is important because of its engineering characteristics as well as its social and aesthetic values. The environmentalist is interested because of the terrain's inter-relationship with the ecology of an area.

The foundation premise of terrain classification is that a geologic unit, when subjected to a particular climatic history, will create landforms that are typical of that geology and that history; variations in either or both of these factors will result in the formation of different landforms, and vegetation associations. Terrain classification utilizes this premise by mapping the different landforms and vegetation associations to discriminate units of the land surface that are likely to be different in terms of geology, climatic history and consequent soil and vegetation community development.

Landforms and vegetation associations can be readily identified in stereo aerial photographs. Interpretation of the images to identify the different landform units, integrated with the base geology of the area, provides a consistent classification of the area. By visiting each of the units of the classification in the field, the agricultural capability of each unit can be assessed thus mapping the agricultural capability of the area.

One difficulty with the technique occurs when the native vegetation associations in the area have been removed, so that the interpretation becomes entirely dependent on landform. The procedure is thus to:

- prepare appropriate aerial photographs and geologic maps of the area;
- interpret the images into distinct landform and native vegetation units, and integrate the interpretation with the geologic associations of the area so as to allocate each landform unit a distinct code; and
- transfer the integrated landform information to a topographic map of the area; use this as the basis for soil core sampling within each mapped unit, to estimate the soil capability of each unit; and prepare a final map of soil capability.

Areas of similar geologic history are known as **provinces**. Within each province are areas of near uniform influence by the other landscape forming influences, primarily climate and initial landforms. These areas, known as **land systems**, contain a pattern of a finite number of **land facets** occurring in some systematic manner across its surface. A **land facet** is defined as a landshape of simple geometric form, on a uniform rock or surficial deposit and having soil and water regimes that are constant or simply changing over the facet. The land facet is thus the basic mapping unit in terrain classification.

Classification of an area into provinces (Table 4.3), land systems (Table 4.4), land facets (Table 4.5) and determining the relevant characteristics of each facet provides information to the user for the whole area. Given a terrain classification of an area, the engineer, planner or environmentalist can evaluate the significance of the surface physiography of that area much faster and more reliably than by other methods.

Classification is by a numerical code of twelve digits where digits 1–4 represent the province, digits 5–7 represent the land system, and digits 8–12 represent the land facet as given in Tables 4.3–4.5. The study area has been mapped into land units using this scheme as shown in Photo 4.2. The terrain classification has identified the land units listed in Table 4.6.

4.2.8 Land-use and land-cover mapping

Land-use describes the use to which the land is being put, whilst land-cover describes the surface cover characteristics. A remotely sensed image, as a record of attributes of the land-cover and land-cover conditions at the time that the image was acquired, can be analysed to extract information on land-cover types, and their condition. The level of detail, or the resolution in the information, that can be mapped depends on the nature of the image data, in particular its spatial, spectral and radiometric resolutions. Most forms of satellite data have relatively low spatial resolution, but high radiometric resolution; the data can often provide higher informational resolution in areas where the covers are extensive, typical of agricultural and other rural areas, than it can in areas where the individual cover types are of small areas. Aerial photography has the opposite characteristic; it has high spatial resolution and low radiometric resolution so that it can often provide higher resolution land-cover information in areas of intensive cover variation, such as in urban areas.

Land-using activities cover a period of time or season. During this period the land-use is characterized by sub-periods of specific land-cover types or conditions. Imagery acquired during a sub-period will respond to the cover conditions that existed at that time; imagery acquired in different sub-periods can look quite different. There are also periods when some of the land uses have similar cover conditions so that images taken at these

Table 4.3 Coding for terrain classification at the province level

Age		Province	Coding
Archeozoic	1		
		General	10
Proterozoic	2		
		General	20
		Lower	21
		Middle	22
		Upper	23
Paleozoic	3		
		General	30
		Cambrian	31
		Ordovician	32
		Silurian	33
		Devonian	34
		Carboniferous	35
		Permian	36
Mesozoic	4		
		General	40
		Triassic	41
		Jurassic	42
		Cretaceous	43
Cenozoic	5		
		General	50
		Tertiary	51
		Quaternary	52

• Province: defined as an area of constant geology at the group level.
•• The first two digits allow grouping by age and name of structure. The last two are sequential numbering with that first digit pair.
••• Provinces within each group above are then numbered in numerical order as encountered, using a two-digit system which allows 99 provinces per group.

times will exhibit confusion between these land uses. It follows that it may not be possible to map all land uses from a single image, but rather a sequence of images may be necessary.

If the land use is agriculture then the period of time is at least a season (for cropping), and maybe longer (for pasture). Each agricultural land use undergoes a particular pattern of events during the season, and we can describe empirically with considerable accuracy and detail the sequence and nature of these events. The types of changes that can occur at the end of the season are also known, but the option that will be selected by the individual land-user is less predictable. Most of the options do involve significantly different practices; and these will be reflected in significantly different impacts on the image data of the area. The period of time is also affected by the land-use classes that are being employed; a 'winter cereal' land-use class has a period of a season whereas the 'cropping' or 'crop rotation' land-use classes have much longer time periods.

Land-use mapping is normally done by the use of a land-use classification scheme that categorizes all land uses into classes in the scheme. Classification schemes should strive to meet the following criteria.

Table 4.4 Coding for terrain classification at the land system level

	Land system numbering code									
Digit	0	1	2	3	4	5	6	7	8	9
Average relief (× 100 m)	<2	2–4	4–6	6–8	8–10	10–12	12–14	14–16	16–18	>18
Greatest local relief amplitude (× 10 m)	<1	1–2.5	2.5–5	5–7.5	7.5–10	10–15	15–20	20–30	30–50	>50
General physiography	Nearly flat	Gently rolling		Increasingly rugged terrain					Extremely rough terrain	

- • Land system: an area of constant geomorphic association, containing recurring physiographic, soil and vegetation associations.
- •• Land facet: landscape of simple geometric form, on a uniform rock or surficial deposit, and having soil and water regimes that are constant or simply changing over the facet.
- ••• The classification at the Land Facet level is partitioned into two parts: coding for the topographic component and coding for the soils component.
- •••• The three digits are for average elevation, greatest local relief amplitude, and general physiographic description of the area. The first two digits of this classification are for topographical shape, and the remaining three for soil characteristics identifiable on the imagery.

Table 4.5 Coding for terrain classification at the land facet level

	Topography
1.	Flat to undulating smooth surfaces usually with relatively deep soils
1.1	Flat surfaces
1.2	Gently undulating surfaces (to 2°)
1.3	Undulating surfaces (to 5°)
1.4	Strongly undulating surfaces (5°)
1.5	Sloping surfaces (to 2°)
2.	Irregular sub-horizontal to undulating eroded surfaces, usually with rock outcrop or shallow soils
2.1	Flat surfaces
2.2	Eroded surfaces
2.3	Benched surfaces
2.4	Undulating eroded surfaces
2.5	Strongly undulating eroded surfaces
2.6	Moderately dissected surfaces
2.7	Strongly dissected surfaces
3.	Slopes, including escarpments, between surfaces
3.1	Smooth steep slopes
3.2	Gentle slopes
3.3	Rough rocky steep slopes
3.4	Dissected slopes (to 5°)
3.5	Dissected slopes (5°)
3.6	Benched slopes
4.	Isolated hills
4.1	Conical hills including paps and buttes
4.2	Rounded hills
4.3	Complex hills
4.4	Razor-backed hills
4.5	Elongated rounded hills
5.	Isolated ridges, etc.
5.1	Low ridges (to 5° slope)
5.4	Rough irregular ridges
5.5	Linear strike ridges
5.9	Sand ridges
6–8.	Not allocated
9.	Drainage system
9.1	Major channels
9.2	Minor channels
9.3	Braided channels
9.4	Incised gullies, ravines, etc.
9.5	Lakes, lagoons, billabongs
9.6	Reticulated channel system (flood-outs)

Table 4.5 (Continued)

Soil Characteristics

Digit	Dominant soil colour	Dominant soil type	Significant surface cover
0	White	Rock outcrop, pockets shallow soil and gravel	Not present
1	Yellow	Clay soils	Silcrete, rounded
2	Orange/sienna		Silcrete, angular
3	Red		Ironstone, rounded
4	Red Brown	Silty soils	Ironstone, Platy
5	Chocolate	Sand over clay soils	Porcelainite
6	Organic fawn/brown		Quartzite
7	Grey	Sandy soils	Calcrete
8	Black	Stratified soils	Salt
9	Other organic soils	Rock outcrop and rubble	

Table 4.6 Land units identified in the Harbour Creek watershed area

Class	Name	Province	System	Facet
1	Coramba escarpment	3 501	169	33 400
2	Steep ridgelines	3 502	045	35 400
3	Low rounded hills	3 503	022	51 410
4	Beach and ridgeline	5 201	001	59 170
5	Alluvial flats	5 101	000	11 640
6	Gradual slopes	5 102	011	32 440
7	Alluvial floodplains	5 103	000	11 540
8	Swamp and marsh	5 202	000	11 540
9	Brooklana escarpment	3 504	169	33 400
10	Steep ridgelines	3 505	045	35 400
11	Low hills	3 506	022	51 410
12	Gradual slopes	5 104	011	32 440
13	Coramba plateaux	3 507	111	15 470

1. The scheme should recognize both land-cover resources and land-use activities within the classes of the scheme.
2. The minimum level of accuracy in mapping land-cover and land-use should be at least 85 per cent.
3. The accuracy of interpretation of all classes should be about equal.
4. The classification should be capable of replication using different interpreters or different data sets.
5. The classes adopted in the scheme should be suitable for use by the maximum number of resource managers.
6. The system should be suitable for mapping extensive areas.
7. The system should be hierarchical so that sub-categories can be mapped from more detailed remotely sensed data and ultimately (sub-sub-categories) from field surveys.
8. Aggregation of categories should be possible.
9. Comparison with future land use should be possible.

10. Multiple land uses should be acceptable to the classification.
11. Mapping should be done to the finest level possible with the data set, so that the derived information can be of use to the maximum numbers of resource managers.

The most commonly adopted classification scheme is that of the US Coast and Geodetic Survey (Table 4.7). This is a hierarchical scheme, in which the Level 1 classes should be capable of being mapped by the use of satellite data, the Level 2 classes either by high resolution satellite data or by high altitude aerial photography, the Level 3 classes from medium altitude aerial photography or scanner data and the Level 4 classes from large scale imagery or ground surveys. The classes are also designed for use over different areas, with the Level 1 classification being designed for national resource surveys, Level 2 classifications for regional and state wide surveys, and Level 3 classifications being suitable for Municipal or district level surveys.

The Coffs Harbour area was mapped to Level 2 using the 1:60 000 aerial photography depicted in Photo 4.1, with the resulting land-use map being depicted in Photo 4.2.

4.2.9 Mapping cover conditions

The cover conditions that could be expected in the watershed can be estimated from the land-use mapping. This estimate of cover conditions is useful for those covers that exhibit regular seasonal patterns that can be predicted from the land-use information, as is the case with most cropping. However there are a number of cover types that have cover conditions that cannot be predicted from the land-use data, particularly pasture, rangelands and forest cover. For these cover types an alternative approach would be to use satellite image data to estimate the cover, as will be discussed in Chapter 9. The difficulty with using satellite data for this purpose is the need either to select imagery that depicts representative cover conditions, or to adjust the derived cover values to those that are representative of the cover conditions for the area.

4.3 Crop inventory for district agronomy advisers

4.3.1 User needs

One important function of district agronomy advisory staff is to provide recommendations to farm managers on crop management practices including when to cultivate and sow, when to apply fertilizer, and crop rotations that should be practised. Advisers build up recommended practices in all of these activities, and prepare programmes to promote these recommended practices.

Agronomy advisers require information in support of these activities so as to:

- assess the effectiveness of advisory programmes by monitoring the rate and distribution of adoption of recommended practices; and
- identify the best from the various farming practices that are being used.

Whilst this information can be gained by other methods, they are expensive in time and travel when the districts are large. Even for smaller districts the expense of obtaining this information at the level of detail necessary to meet user needs suggests that it may be better to gather the information from remotely sensed data. Smaller field sizes can however create difficulties in deriving the information.

These information needs were first addressed for district agronomists in the marginal

Table 4.7 Land-use classification classes, from the US Coast and Geodetic Survey

Level 1	Level 2	Level 3
100 Urban or built-up	110 Residential	111 Single unit, low density
		112 Single unit, medium density
		113 Single unit, high density
		114 Mobile homes
		115 Multiple dwelling, low rise
		116 Multiple dwelling, high rise
		117 Mixed residential
	120 Commercial and services	121 Retail sales and services
		122 Wholesale, including trucking
		123 Offices and professional
		124 Hotel and motel
		125 Cultural and entertainment
		126 Mixed commercial and services
	130 Industrial	131 Light industry
		132 Heavy industry
		133 Extractive industry
		134 Industry under construction
	140 Transportation	141 Airports including terminals
		142 Railroads including terminals
		143 Bus and truck terminals
		144 Major roads and highways
		145 Port facilities
		146 Vehicular parking
	150 Communications and utilities	151 Energy facilities (gas and electrical)
		152 Water supply (including pumping stations)
		153 Sewage treatment
		154 Solid waste disposal sites
	160 Institutional	161 Educational institutions
		162 Religious centres
		163 Medical and health care
		164 Correctional centres
		165 Military facilities
		166 Government facilities
		167 Cemeteries
	170 Recreational	171 Golf courses
		172 Parks
		173 Marinas
		174 Stadiums and racetracks
		175 Zoos
	180 Mixed	
	190 Open land	191 Undeveloped land in urban areas
		192 Land being developed
200 Agriculture	210 Cropland and pasture	211 Row crops
		212 Field crops
		213 Pasture
	220 Horticulture	221 Evergreen tree and shrub crops
		222 Deciduous tree and shrub crops
		223 Nurseries and glasshouses
		224 Ornamental horticulture

Table 4.7 (Continued)

Level 1	Level 2	Level 3
		225 Vineyards
		226 Cane (sugar and bamboo)
		227 Annual vegetables and fruit
	230 Feedlots	231 Cattle
		232 Poultry
		233 Pigs
	240 Other agriculture	241 Inactive land
		242 Other
300 Rangelands	310 Grasslands	311 Improved pasture
		312 Native pasture
	320 Shrub and bushland	321 Sagebush prairie
		322 Coastal shrub
		323 Chaparral
		324 Second growth bushland
	330 Mixed rangeland	
400 Forest land	410 Evergreen	411 Pine
		412 Redwood
		413 Other
		414 Eucalypt and acacia
		415 Broadleaf
	420 Deciduous	421 Oak
		422 Other hardwood
	430 Mixed forest	431 Mixed forest
	440 Clearcut areas	
	450 Burnt areas	
500 Water	510 Streams and canals	
	520 Lakes and ponds	521 Lakes
		522 Fishponds
	530 Reservoirs	
	540 Bays and estuaries	
	550 Open sea	551 Open sea
		552 Coral reef
600 Wetlands	610 Forested wetlands	611 Evergreen
		612 Deciduous
		613 Mangrove
	620 Non-forested wetlands	621 Herbaceous vegetation
		622 Freshwater marsh
		623 Saltwater marsh
	630 Non-vegetated wetlands	631 Tidal flats
		632 Other
700 Barren land	710 Lake beds	711 Saline
		712 Claypans
	720 Beaches	
	730 Other sand and gravel	731 Sandplains and ridges
		732 Stony (gibber) areas
		733 Alluvial fans and alluvium
	740 Exposed rock	
800 Tundra		
900 Perennial snow	910 Perennial snowfields	
	920 Glaciers	
	930 Perennial icesheets	

cropping belt in Australia, where the typical practice is to cultivate after rainfall so as to prepare the seedbeds, control weeds and conserve moisture. Sowing is conducted after further rains.

Cultivation and land preparation can occur up to six months in advance of the normal sowing period in order to accumulate fallow period soil moisture. Rains usually occur from late autumn (April–May) until mid winter (July), with the area experiencing severe moisture deficiency by late spring (November). Generally the agronomists recommend early sowing practices so as to maximize growth before the onset of the hot dry weather that starts in mid to late spring (October–November), and continues until early autumn (March).

4.3.2 Technical requirements

The raw information that forms the basis for provision of this user information is:

- early and late cropping in each year;
- application of fertilizer and yield produced by paddock; and
- land-ownership.

The information on early and late crops and cropping can be derived from remotely sensed images. Since the information is only required to the paddock level, the information can be readily extracted by visual interpretation of photographic image products. This task requires the selection of images, their visual interpretation and the conduct of field work as required during the interpretation and for accuracy assessment as described in Chapter 6. This information can be integrated to provide the agricultural adviser with information on the crop rotations that are being practised, and how they are changing. Further integration with land-ownership information allows specific farmers to be identified for specific agronomy programmes.

The information on fertilizer application and yield could be sought directly from the farmers and compared with the cropping information to identify better practices.

The images required for interpretation of cropping need to be selected throughout the growing season so as to ensure that the covers of interest are reliably discriminated, in this case early from late crops. This is a monitoring function. The frequency of acquisition of images in monitoring may require one image every five years for some tasks, to images every few weeks for other tasks. The more images required the more expensive the monitoring programme; clearly there is a need to select the minimum set of images that can meet the needs of the programme. The minimum set of images will provide the best information if it is chosen when it provides the maximum discrimination between the covers of interest, and discriminates these covers from all other covers.

For vegetation studies, vegetation calendars are a valuable tool for indicating periods when the different covers may appear differently in image data. These calendars can then be considered, in conjunction with any constraints that may exist, to select windows, or periods when suitable imagery needs to be acquired.

4.3.3 Crop calendars

The reflectance characteristics of a vegetation community change throughout its phenological cycle as discussed in Chapter 2. Typical changes in reflectance throughout the phenological cycle of a crop have been described by Kauth and Thomas (1976) (Diagram 4.1).

The initial reflectance of a crop is due to the soil and litter on that field at cultivation.

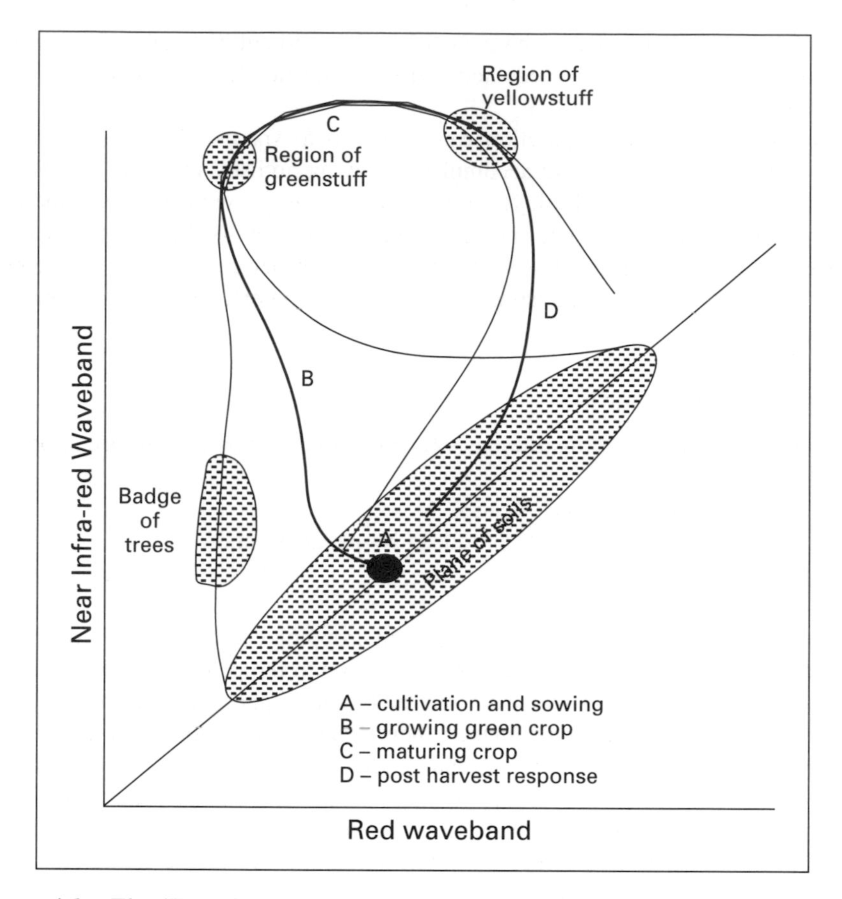

Diagram 4.1 The 'Tasselled Cap' description of the change in reflectance of agricultural crops with changes in phenological stage (Kauth and Thomas, 1976). The path ABCD represents the typical crop path during a cropping season.

This initial reflectance occurs within the vicinity of the plane of soils, as all soils and litter have reflectance values that lie within this area in these two wavebands. A typical reflectance value could be at A in Diagram 4.1. Cultivation dramatically darkens the response of the cultivated field primarily due to the greater shadowing component in the cultivated field, and the lack of vegetation reflectance. Surface soil moisture contributes to this change in the first few days after cultivation. Photograph 4.3 taken at cultivation shows cultivated fields as green due to the red soils that are predominant in the area. The more recently the field had been cultivated, the deeper the green colour. In Diagram 4.1, cultivated fields will have lower values in the plane of soils. As the cultivation becomes smoothed with the passage of time, so the reflectance increases, moving to the right in the plane of soils.

The reflectance at the seedling stage is very similar to the reflectance of the soil background since the seedlings will have a negligible contribution to reflectance. As the crop grows through tillering to stem extension, a number of factors affect canopy response.

1. The blue and red chlorophyll absorption bands, and high infra-red reflectance of the vegetation have an increasing effect on total reflectance. The contribution of the visible

bands is closely correlated with leaf area index (LAI) because the leaves are quite opaque at these wavelengths. By the time LAI = 2, there will be little contribution due to the soil background in the visible wavebands. The situation is quite different in the NIR because of the high level of transmission through green leaves. There will be some contribution by the soil background to NIR reflectance up to LAI values of 6–7 in the NIR, i.e. there will be changes in the NIR until near full canopy development.
2. As the canopy grows, so the shadow component increases, reducing the soil reflectance. This is most marked early in canopy development when the shadowing of each plant does not impinge on either the canopy or the shadow of other plants. As the canopy continues to develop overlaps occur and the effect of shadow decreases.
3. Surface smoothing occurs. As the cultivations are smoothed over time, the shadowing in the soil surface decreases and soil reflectance increases.

The response of the crop thus follows a trajectory that starts with the response of the soil surface within the plane of soils and ends in the vicinity to the response of green canopy or the region of greenstuff. Initially the trajectory moves strongly to the left due to the shadow effect, and the change in red reflectance. As the canopy develops these two effects become less important, and the others, of changing NIR reflectance and increasing soil reflectance, become more important. The direction of the trajectory changes so as to approach the region of greenstuff. If there is less green vegetation than about GLAI = 6 then there will be some contribution by the soil to reflectance and the response will not reach the region of greenstuff.

As the crop flowers, fruits and matures, the reflectance changes from that of a green to a yellow canopy. In the 'tasselled cap' this is indicated be a change in direction of the trajectory towards the region of yellowstuff. After harvest the cover consists of straw, in various stages of organic decay, and soil, so that the trajectory proceeds towards the plane of soils. A typical reflectance trajectory throughout this phenological cycle is shown by the line ABCD. Kauth and Thomas consider that the envelope of all possible crop trajectories looks like a tasselled cap, hence the name for their model.

The tasselled cap concept provides a very good description of the process, but it does not indicate the temporal sequence in a way that allows the analyst to select the most suitable images for a specific crop monitoring task in a specific district. A calendar of the phenological stages during the growth cycle of a crop, or crop calendar, can be used to indicate the reflectance characteristics that can be expected for the crop at the different stages throughout the growth cycle. A comparison of the crop calendars for the different crops that grow in an area will indicate periods when the reflectance characteristics of the different crops are both different and similar. Periods when the reflectance characteristics are different suggest windows, or periods, when it would be suitable to acquire imagery to discriminate between the crops under consideration.

Diagram 4.2 shows the crop calendars for wheat in three districts in eastern Australia. The three districts differ primarily in their latitude, with Moree having a warm subtropical climate and Barham a cool temperate climate. The calendars show that the growing period increases as the season becomes cooler, with earlier sowing and later harvesting.

Diagram 4.3 shows the crop calendars for wheat, oats and barley at Parkes in eastern Australia. Analysis of the crop calendars indicates that oats are sown and germinate earlier than the other two crops. Oats are often used as cattle fodder; if the crop is allowed to go to seed then it is harvested slightly earlier than other crops. If imagery can be acquired at the right time then it should be possible to discriminate oats from wheat and barley. Wheat and barley are planted at similar times, and harvested at similar times. It

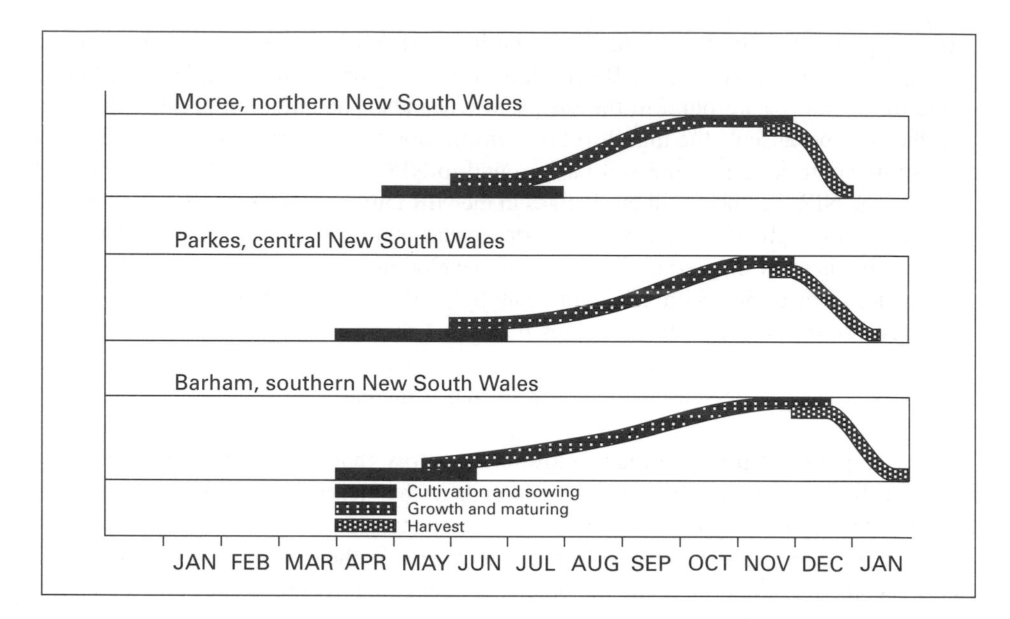

Diagram 4.2 Crop calendars for wheat for three towns in New South Wales – Moree (26°S), Parkes (29°S) and Barham (31°S) – where the main factor in the difference in the crop calendars of the three towns is the change in climate between the three districts.

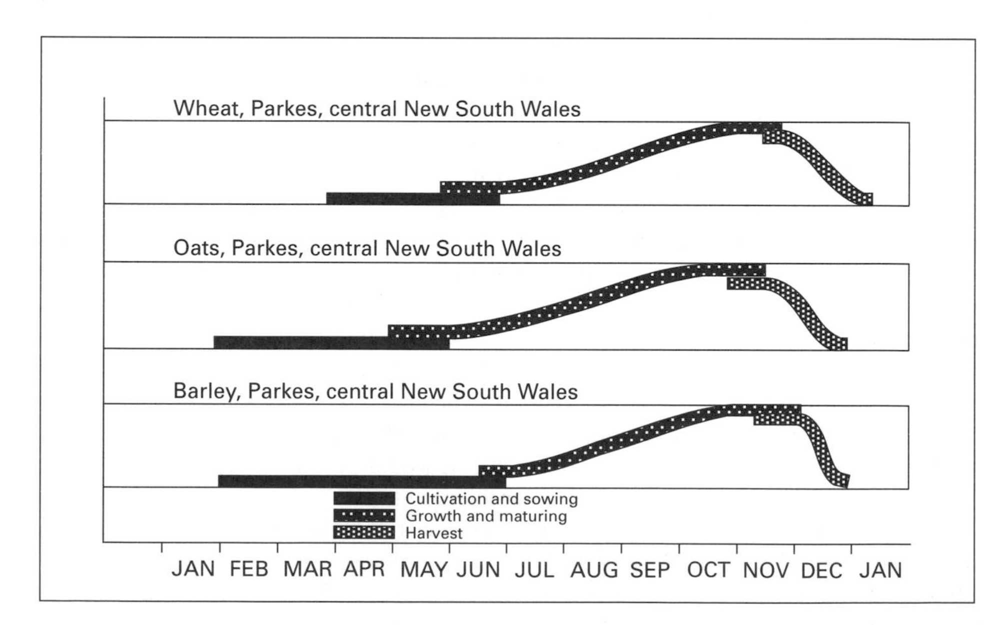

Diagram 4.3 Crop calendars for wheat, oats and barley for Parkes in central western New South Wales, Australia.

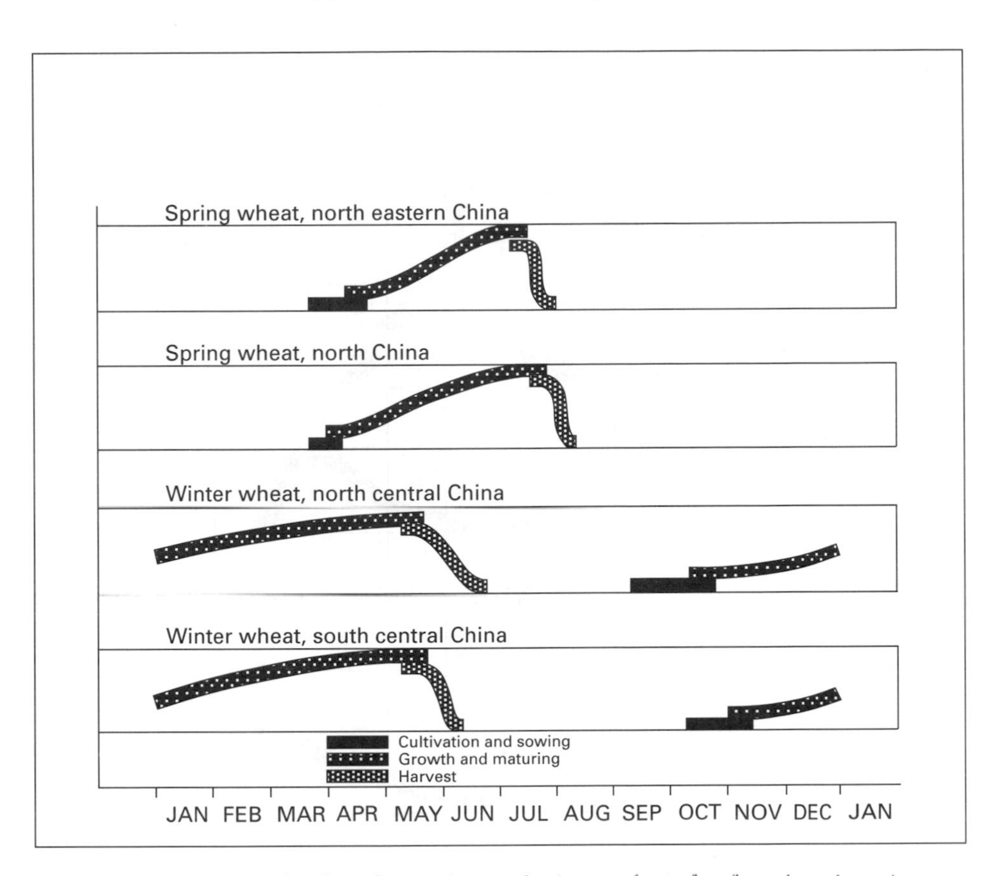

Diagram 4.4 Crop calendars for spring and winter wheat for four locations in China.

would seem difficult to discriminate these two crops on the basis of differences in phenological stage at Parkes in New South Wales.

The crop calendars in Diagram 4.4 for spring and winter wheat in China show that these calendars also change from region to region. The changes are often subtle. Sometimes imagery can be acquired at a suitable time to discriminate the earlier from the later crops. If this discrimination can be achieved on a routine basis then the derived information can be used routinely by managers who may be prepared to fund such a programme, but who will not support programmes that cannot provide reliable information.

It may be necessary to construct crop calendars for all of the crops grown in an area (Diagram 4.5) to allow the analyst to identify windows or periods during which acquired images would be able to discriminate between the various crops. More than one image may be required to discriminate between all of the crops.

Not only does the crop calendar assist in the selection of windows, but it also allows the analyst to weigh up the use of alternative windows since cloud cover may preclude data acquisition, or images at other times may be of more use to other resource managers concerned with related issues.

It will be noted that there is overlap between the phenological stages. This is because not all crops are sown on the same day. For a variety of reasons the crops are sown and harvested over a period of time. Generally the less reliable, or more variable, the rainfall, the longer the period of sowing. Clearly, the more precisely the boundaries

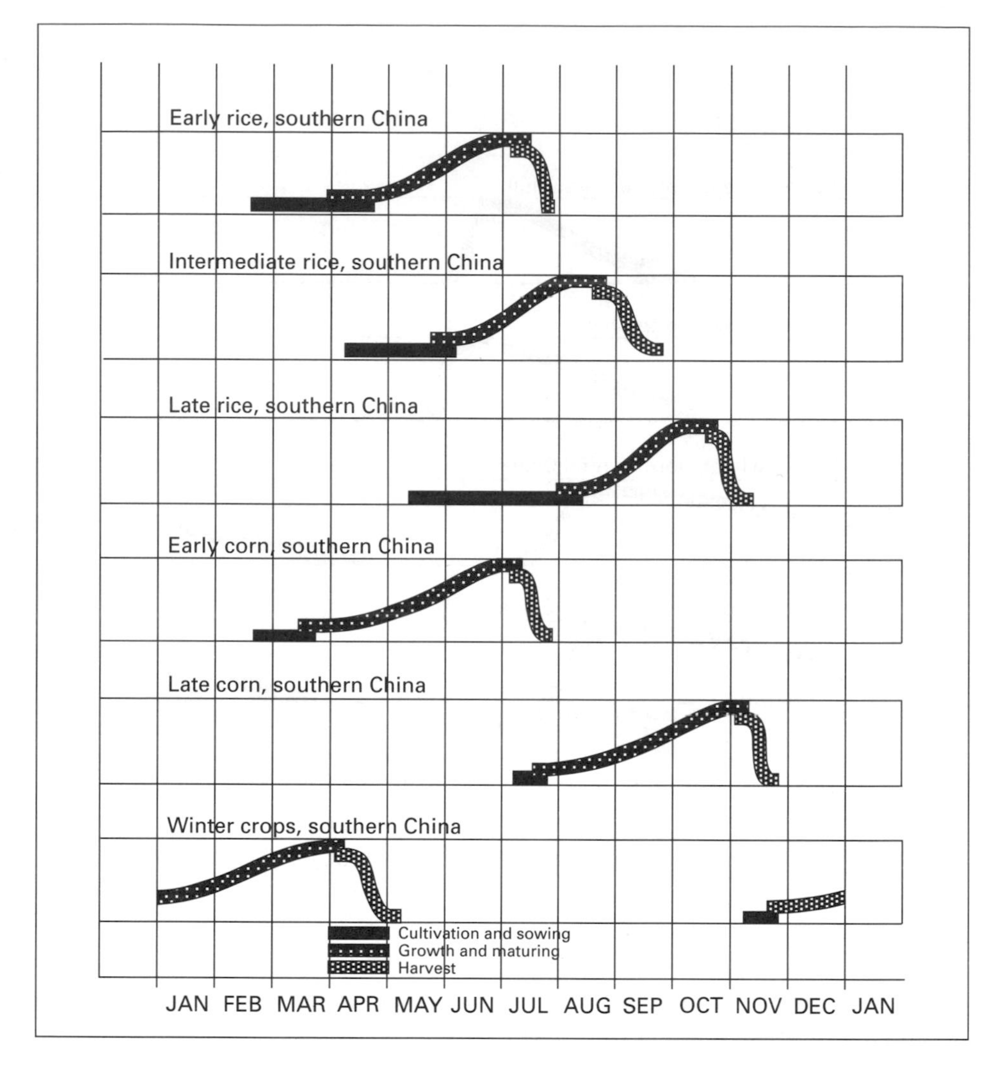

Diagram 4.5 Crop calendars for early, intermediate and late rice, early and late corn, and winter crops in the southern provinces of China.

between the phenological stages can be defined, the better the selection of windows within which to acquire images.

The location of the boundaries of the phenological stages are affected by various factors.

- The climate and weather with early, or late rains can affect planting dates and the range of planting dates in a district. The subsequent climatic history will affect the rates of growth of the crop, and the date at which it matures. The climate cannot be predicted so that systems that are designed to operate in a real-time mode must assume typical climatic effects, and allow latitude in the system to absorb the affects of significant variations in the weather on the crop.
- Farming communities may not have the resources to sow all of a crop within very short periods, say a few days, even if the weather conditions made this desirable. The limited availability of resources will cause a scatter in planting dates.

- Different management practices sometimes mean slightly different sowing dates, or slightly affect the onset of the different phenological stages. Thus aerially sown rice may be put under permanent flood before rice that is sown by direct drilling.
- Variations in soils, topographic slope and height influence the planting and subsequent growth history of a crop. If the differences in the phenological cycle due to one or more of these parameters are very marked then different crop calendars should be constructed reflecting the significant variations in the parameter(s).

For these reasons it is important to construct crop calendars for all crops that may be affected by one or more of the above factors to a significant degree. Crop calendars, to be most useful, should indicate:

- Phenological stages – the growth cycle of a plant can be partitioned into a number of stages:

cultivation	fallow
	cultivation
	sowing
growth	emergence
	growth
	flowering
maturation	fruiting
	maturing
	harvest

- transition periods – there is usually overlap between the phenological stages because not all crops are at exactly the same stage; and
- range of conditions – the range of conditions within the crop during each stage. The crop calendars depicted in Diagrams 4.2–4.5 depict the average height of the canopy throughout the season, but other indications of condition include average or range in LAI, biomass, response in a specific waveband or vegetation index values (Chapter 5).

The construction of a crop calendar should be done in a number of stages.

1. Identify the geographic region and crops to be represented by crop calendars. Analyse the climatic, soil and topographic variations in the region to determine whether one or more calendars may be necessary for each crop type.
2. Identify the phenological stages in the crops of interest and the time period of onset and completion of each stage. Determine whether more than one crop calendar is required for each crop type.
3. Collect data on the time of onset and completion of each phenological stage so as to construct the temporal sequence within the crop calendar(s).
4. Collect condition data throughout the phenological stages of the crop. Use this data to construct the shape of the calendars.

4.3.4 Designing monitoring systems

Designing resource information systems to monitor plant communities requires consideration of several factors.

Plant phenology The differences in plant and community phenology can be depicted as crop calendars as discussed. Comparison of these calendars will provide a good basis for:

- selecting windows for image data acquisition;
- assessing the likelihood of discriminating between the different cover types; and
- identifying potential confusions.

Potential interference with data acquisition The most obvious source of interference is cloud cover for systems dependent on solar radiation. Use cloud cover statistics to construct probabilities that the area will have less than 10 per cent cloud cover on any one day during each of the selected windows, $p_{c,\,w}$. Calculate the probability of acquiring an image during each selected window, $p_{a,\,w}$. If the satellite revisit cycle is every r days and a window is of d days duration then the probability of acquiring an image in one window is

$$p_{a,w} = p_{c,w}\frac{d}{r}$$

Determine the probability of acquiring images in all of the windows, p_{Σ}.

If there are m windows then p_{Σ} is the product of the probabilities in each window.

Consider a monitoring task with three windows where the probability of cloud cover being less than 10 per cent on one day ($P_{c,\,w}$) during the three possible windows is 0.10 (first two weeks of cultivation), 0.15 (second two weeks of cultivation to germination), 0.25 (late growth to fruiting) and 0.45 (maturation to post harvest).

The crop is to be monitored using a satellite with repeat coverage every 16 days. Use $p_{c,\,1} = 0.125$, $p_{c,\,2} = 0.25$ and $p_{c,\,3} = 0.45$ as the probability of less than 10 per cent cloud cover on one day in each of the three windows. The probability of acquiring an image in the first window is

$$p_{a,1} = p_{c,1}\left(\frac{28}{16}\right) = 0.22$$

The probabilities of acquiring images in the other windows are $p_{a,\,2} = 0.33$ and $p_{a,\,3} = 0.59$ respectively. The probability of acquiring images in each window is

$$p_{\Sigma} = 0.22 \times 0.33 \times 0.59 = 0.04$$

The probabilities of acquiring images in the windows can influence system design, for example by utilizing other satellites and by increasing the duration of the windows. If the monitoring programme depends primarily on images in one window then it may still provide valuable information even if images are not acquired in the other windows. In the above case information on cropping may be discriminated on the harvest image with an adequate reliability; if the information is to be used for purposes that do not require an in-season response then this may be acceptable.

Resolution required in the information The resolution required will significantly influence the methods of information extraction. If the information required is coarse, spatially and informationally relative to the raw data then visual interpretation is usually quite satisfactory and much cheaper than digital image processing (Chapter 5).

The area being considered in this study is of a marginal cropping area in eastern Australia. The agronomy advice for the area is to crop early so as to gain the maximum benefit from the winter rains before the onset of the hot and dry summer. Not all areas that are cultivated are sown, and of those sown, not all areas are harvested. Images acquired during the cultivation and harvest periods would indicate early and late cultivation, and areas that are harvested respectively. A mid-season image is required to identify early and late sowings.

4.3.5 Extraction of information

Rectified Landsat photographic imagery of the district is acquired at least three times, and preferably four times per season if it is available. The three images used for this study are shown in Photo 4.3. For this task Landsat MSS prints were chosen because the extensive crop fields are obvious on the images and because of their cost. As soon as the early season imagery is available it is distributed to district agronomy staff. It is also interpreted by skilled interpreters to identify prepared seedbeds from 1:100 000 photographic prints.

The image acquired soon after cultivation and sowing (June) is shown in Photo 4.3 where most of the cultivated fields appear as green because of the red soils in the area. The intensity of the green is an indication of how recently the field has been cultivated. There are some fields cultivated on the grey soils to the east of the image. Where possible the prepared fields are discriminated into early and late cultivations. The interpretation is finalized by visits to the districts and with discussion with the district officers. This interpretation becomes a cultivation overlay.

The mid-season image (Photo 4.3) acquired in September shows the cultivated paddocks as varying shades of red, with the best canopy development being on the most intensely red paddocks. This image is used to verify the cultivation interpretation and strengthen the discrimination between early and late sowings from the response of the crops. It is also used as an indicator of the better and the poorer crops, on the basis of their redness relative to the sowing date expected from the cultivation overlay. It can only be used as an indicator since the redness of the canopy in the image can be affected by a number of factors, including:

- the age of the crop since younger crops will have less canopy development;
- the vigour of the crop, affected primarily by moisture availability and soil fertility; and
- crop species which can affect the response.

Field visits are used to assess the accuracy of the mapping. This second overlay is effectively an updating of the cultivation overlay to create a mid-season crop overlay, and is shown in Photo 4.4. A number of fields were identified in the mid-season image that were not identified as crop in the early-season image. This can be the result of one of two situations.

- The field had not been prepared at the time of acquisition. The growing season in this part of Australia usually starts during the period March–May. The actual date of cultivation and sowing depends on the occurrence of rainfall, by time and amount, and on the farmer's preferred strategy of either early or late cropping.
- The field response was not easily discriminable from other non-cropping paddocks. This can occur for a number of reasons. If the farmers are using no or minimum tillage practices then there will be no cultivation of the soil, and hence no marked soil colour

Table 4.8 Summary of accuracy assessment for crop classification in the Bogan and Walgett Shires of New South Wales (Australia) in 1986

Shire	Image	No. properties in sample	No. paddocks in sample	Crop classific'n accuracy (%)	Errors of Omission (%)	commission (%)
Bogan	Cultivation	12	109	84	16	3.5
	Canopy	14	121	96.2	3.8	0.0
	Harvest	10	78	100	0.0	0.0
Walgett	Cultivation	22	166	65.3	34.7	4.3
	Canopy	19	138	66.1	33.9	0.0
	Harvest	10	96	93.8	6.2	0.0

since sowing occurs into the dead weed and herbage matter. Alternatively the crop may have been planted well before acquisition of the image so that the soil surface will have smoothed over, reducing the characteristic soil response used in the interpretation. Finally, the crop may have started to emerge, leading to a small contribution of green plant reflectance to the soil reflectance, further masking the tonal indicators used during interpretation of the cultivation image.

The third, or harvest, image is used to map the harvested and ripe crops. In general, the fields that have not been harvested appear greenish to straw in colour due to the red colouring of the ripe wheat. The recently harvested fields are highly reflective due to the straw lying on the surface, and they go bluish as the straw starts to break down. The image is interpreted to verify the interpretation from the mid- and early-season images, and field data are used to verify the accuracy of the interpretation. This information is the final estimate of crop area (Photo 4.4).

The results of the accuracy assessment are shown in Table 4.8. The image data is for Bogan Shire, whilst the table shows the results for both the Bogan and Walgett Shires. The table shows that difficulties were often experienced in discriminating cropping from pasture flushes in the Walgett area, but that the Harvest image provided a reliable discriminator. It meant that the data could not provide a reliable within season estimate of crop area in the Walgett Shire, not required in this particular application.

4.3.6 Summary

The accuracy with which cropping can be identified early, and during the season depends on the discriminability of that crop from other land covers. This discriminability depends on the variability within the crop class, and within other classes. Often these variabilities are much greater in marginal cropping country than in established cropping belts. The most successful indicator of cropping has been found to be the change in response that occurs at harvest. Not only is the change in response most dramatic but the period of harvest is much shorter than cultivation, and there is a much greater probability of acquiring images through the lower occurrence of cloud.

If the information is required for management purposes then it will be used not in the current season but in the future. The monitoring system design does not have to consider severe time-limit constraints. Time-limit constraints increase the difficulty of acquiring adequate images; it places pressure on the interpretation that increases the likelihood of error and reduces the time available for careful analysis, and reduces the options in accuracy assessment. Time constraints should not be included in a monitoring system unless they are absolutely essential in the provision of information to management.

If the user requires in-season information then the harvest image is too late. The most effective image for crop discrimination cannot be used and other approaches may have to be incorporated into the monitoring programme.

4.4 Estimation of crop production

4.4.1 User needs

A variety of agricultural industries depend on timely information on crop production. These industries include:

- the millers and processors who need information on both anticipated production and time of delivery so as to plan work schedules, employ casual staff, order packaging, arrange storage and distribution;
- grain sellers to negotiate sales;
- transportation industries to manage the allocation of transportation facilities; and
- agricultural advisers to identify the more, and less productive farmers, to identify better practices and to identify potential clients.

4.4.2 Technical needs

Crop production, p, is the product of crop area and crop yield, where the yield is the weight of grain produced per unit area:

$$P = \text{Area} \times \text{yield}$$

The law of propagation of error indicates that the variance of the estimate of production can be calculated from:

$$(\sigma_P)^2 = (\text{yield})^2 \times (\sigma_{\text{Area}})^2 + (\text{area})^2 \times (\sigma_{\text{yield}})^2$$

So, the higher the yield, the more critical that the area be estimated accurately, and the larger the area, the more critical is the measurement of the yield. The variance of the estimate of production can be reduced in the following ways.

1. Each component in the equation should contribute equally to the variance, i.e.

$$(\text{yield})^2 \times (\sigma_{\text{Area}})^2 \approx (\text{Area})^2 \times (\sigma_{\text{yield}})^2$$

   This requirement suggests the relative accuracies that need to be achieved in measuring $(\sigma_{\text{Area}})^2$ and $(\sigma_{\text{yield}})^2$ in each of the sub-areas within which the area and the yield are being estimated. In the production equation the yield parameter, will be a much smaller number than the area parameter. Analysis of the variance equation shows that under these circumstances, the variance associated with the estimate of the yield needs to be much smaller than the variance estimated for the area, if the two are to contribute similar amounts to the variance associated with production. The larger the area being considered, the more important it is to have accurate estimates of yield.
2. Each component in the equation should be minimized, commensurate with establishing a realistic cost of the sampling. The higher the accuracy and other constraints

imposed on a mapping task, the higher the cost. There must be a trade-off between the constraints, including accuracy, and the cost that must be finally decided from the value of the information to the client.

Crop yield is affected by a wide variety of factors, where the importance of the various factors can change from year to year, from crop to crop and from region to region. Thus the yield of rice in Australia is primarily dependent on the temperature at the time of flowering, since frosts and excessive cold destroy the ability of the flowers to pollinate. The yield of rice in most tropical countries is more dependent on the availability of water and the effect of disease.

Some factors that affect yield also affect the crop canopy and in consequence can be detected from remotely sensed images. Lack of water and disease affect the crop canopy and as such they can be detected on suitable imagery, although their impact on the imagery may be very similar. Other factors have little effect on the canopy.

Remotely sensed images are suitable for estimating those physical parameters that are highly correlated with canopy response, but are less suited for directly estimating parameters that are not highly correlated with canopy response. Thus remotely sensed images are suited for estimating the intercepted photosynthetically active radiation (IPAR), green leaf area index (GLAI) and usually biomass. Images may be suited to the estimation of yields under certain circumstances and for certain crops.

The technical specifications therefore are that the system must:

1. estimate crop area in the middle of the season to a satisfactory level of accuracy, whether images were acquired or not;
2. estimate yield in the mid- to latter part of the season, with a high level of accuracy; and
3. provide crop production information to clients for administrative districts or other boundaries as specified by various clients.

4.4.3 Inventory method

The system involves at least three layers of information:

1. crop area or density information;
2. crop yield information collected from mid- to late-season; and
3. administrative areas within which the information is required.

4.4.4 Measurement of crop area or density information

Crop area can be measured from either visual interpretation or by digital classification (Chapter 5), with examples given in Chapter 9. Such information will provide the specific area information required by (4.1), but it is relatively expensive to acquire a set of images throughout the cropping season, whether they be photographic or digital image data, unless the images are going to be used for more than one purpose. However, neither visual nor digital image analysis can provide adequate discrimination between crop types.

An alternative solution is to sample the crops, so as to estimate the density of cropping, for each crop type, within strata in the area of interest. The best strata to use are those related to cropping practices. Visual images are prepared at a suitable scale, and then the analyst delineates strata so that the areas within a strata are relatively homogeneous in terms of land-using patterns and intensities. Since these strata will not change very much from year to year, the same strata can be used for a number of seasons. This makes it easier to select the images for the analysis since the year of analysis is not that critical. An example of the results of this type of interpretation are shown in Photo 4.6. The interpretation of Forrest Shire reveals that there are seven land-use zones in the shire:

1. schist ridges, thin soils; little agricultural activity;
2. alluvial floodplains, used extensively for irrigated cropping and dairy production;
3. fertile slopes, used primarily for cereal cropping;
4. irrigation area used primarily for horticulture, with some vineyards and irrigated vegetables;
5. low hills, low-to-moderate fertility, used primarily for grazing of stock, with some cropping on an opportunity basis;
6. steep hills, of low fertility, used primarily for the grazing of stock; and
7. urban areas.

The strata are used as the basis for a stratified random sampling process. The sampling determines the density of cropping, for each variety/species, within each strata, in the season of interest. The stratified random sampling is thus conducted every season for which information on cropping is required.

It does this by defining a random set of samples within each strata. The samples must be of sufficient size to estimate the percentage of each crop type within the sample. The number of samples must be sufficient to ensure that they provide a representative sample of the strata. The percentage area of each crop type in each sample are calculated (Table 4.9). The data from the samples for a strata are averaged to provide an average proportion of each crop type for that strata. Similar sampling work yields average proportions for all crops in each strata. The sampling in Forrest Shire yielded the values given in Table 4.10.

4.4.5 Estimation of yield

Crop yield is commonly estimated visually in the field by those with experience in the technique. Yield can also be estimated from numerical models that start with information on the state of the crop at a given time, and soil conditions. Meteorological conditions for the remainder of the season are assumed to estimate yield.

Estimation of yield will be very imprecise in the first part of the growing season, but usually becomes more accurate as the season progresses. This is because many of the factors that affect yield are difficult or impossible to predict, and because the effects of these parameters are usually most marked at the beginning of the season.

If the yield is to be estimated visually, then this process can be made more rigorous by having the yield estimated for strata that are a function of plant productivity. Imagery, such as the daily NOAA AVHRR images acquired on a daily basis, can be used to derive a vegetation index that is indicative of the amount of green vegetative matter in the canopy. The effect of cloud cover can be reduced by compositing sets of the images together, so as to select, at each cell in the area imaged, that data least affected by cloud cover. The technique provides composite images that are relatively cloud free. The temporal range of these composite images is not significant in this task as long as negligible changes will have occurred in the cover conditions over that time. Such images are shown in Photo 4.5, representing images depicting conditions in late August, early October and mid December 1988.

Photograph 4.5 shows the effect of the compositing with sharp edges along the area in one strip of data that contained cloud cover, and its replacement by another strip with

Table 4.9 The percentage of each crop type in the fertile slopes strata

Sample	Wheat		OATS						
Variety	Cane	Albert	Dora	Dora2	Spain	Barley	Fallow	Pasture	Other covers
1	0.295	0.102	0.041	0.033	0.031	0.199	0.126	0.135	0.048
2	0.230	0.105	0.067	0.142	0.000	0.050	0.211	0.127	0.078
3	0.201	0.184	0.000	0.000	0.137	0.000	0.165	0.135	0.178
4	0.256	0.100	0.000	0.102	0.000	0.100	0.209	0.180	0.053
5	0.169	0.000	0.000	0.196	0.000	0.231	0.263	0.059	0.082
6	0.195	0.120	0.000	0.000	0.084	0.000	0.296	0.203	0.102
7	0.225	0.055	0.246	0.000	0.000	0.200	0.045	0.159	0.070
8	0.230	0.197	0.199	0.000	0.005	0.000	0.195	0.112	0.062
9	0.195	0.075	0.000	0.000	0.000	0.102	0.276	0.250	0.102
10	0.245	0.185	0.069	0.000	0.000	0.237	0.102	0.149	0.014
11	0.293	0.164	0.000	0.270	0.000	0.000	0.039	0.057	0.177
12	0.000	0.206	0.124	0.000	0.000	0.255	0.166	0.149	0.100
13	0.271	0.129	0.000	0.000	0.007	0.103	0.333	0.000	0.157
14	0.217	0.000	0.055	0.000	0.000	0.101	0.396	0.143	0.088
15	0.220	0.220	0.045	0.000	0.000	0.000	0.237	0.278	0.000
16	0.207	0.150	0.055	0.047	0.000	0.235	0.022	0.284	0.000
17	0.166	0.064	0.002	0.099	0.000	0.147	0.291	0.000	0.131
18	0.169	0.232	0.109	0.000	0.121	0.050	0.272	0.000	0.047
19	0.124	0.055	0.000	0.144	0.235	0.136	0.231	0.075	0.000
20	0.175	0.221	0.000	0.102	0.000	0.165	0.156	0.149	0.032
21	0.200	0.055	0.000	0.000	0.135	0.000	0.324	0.203	0.083
22	0.188	0.066	0.144	0.000	0.000	0.129	0.136	0.244	0.093
23	0.234	0.092	0.199	0.000	0.000	0.005	0.136	0.247	0.087
24	0.279	0.000	0.157	0.120	0.003	0.126	0.129	0.102	0.084
25	0.104	0.093	0.041	0.000	0.000	0.000	0.457	0.305	0.000
26	0.260	0.000	0.277	0.000	0.000	0.156	0.160	0.052	0.095
27	0.219	0.126	0.000	0.206	0.000	0.213	0.236	0.000	0.000
28	0.362	0.000	0.222	0.003	0.000	0.000	0.328	0.056	0.029
29	0.250	0.310	0.000	0.006	0.158	0.000	0.231	0.000	0.045
30	0.144	0.225	0.135	0.000	0.000	0.293	0.144	0.000	0.059
31	0.260	0.000	0.036	0.000	0.214	0.399	0.051	0.000	0.040
32	0.089	0.255	0.000	0.045	0.067	0.327	0.127	0.090	0.000
33	0.264	0.120	0.000	0.000	0.000	0.142	0.269	0.182	0.023
34	0.190	0.260	0.000	0.000	0.000	0.047	0.310	0.064	0.129
35	0.195	0.166	0.000	0.157	0.012	0.251	0.198	0.021	0.000
36	0.000	0.195	0.032	0.000	0.255	0.000	0.374	0.102	0.042
Mean	0.20	0.13	0.06	0.05	0.04	0.12	0.21	0.12	0.0

negligible cloud cover. All three images have the major roads superimposed on them, as an aid to navigation.

The images show the areas of maximum green vegetative material as lighter tones of green, going darker green, then yellow and finally to red when there is negligible green material in the canopy. By December the pasture and crop areas have matured and gone brown. This image depicts very well the forested areas of the south east corner of Australia. Comparison of this image with the previous (November) image, shows that

Table 4.10 Average proportion area of each crop type in each strata within Forrest Shire, from field sampling within the strata. The colours are related to the colour codes in Photograph 4.6

					Hills		
Strata	Alluvium (green)	Slopes (grey)	Irrigated (yellow)	Ridges (red)	low (magenta)	steep (cyan)	Urban (purple)
Wheat							
Cane	0.04	0.20	0.04	0.02	0.04	0.01	0.00
Albert	0.16	0.00	0.09	0.00	0.00	0.00	0.00
Dune	0.06	0.13	0.02	0.01	0.02	0.01	0.00
Oats							
Dora	0.02	0.06	0.00	0.00	0.00	0.00	0.00
Dora2	0.06	0.05	0.01	0.00	0.00	0.00	0.00
Spain	0.09	0.04	0.06	0.00	0.00	0.00	0.00
Barley							
Ibis	0.03	0.12	0.01	0.02	0.02	0.00	0.00
Ibis2	0.09	0.00	0.05	0.00	0.00	0.00	0.00
Triticale	0.03	0.00	0.00	0.00	0.00	0.00	0.00
Orange							
Valencia	0.00	0.00	0.11	0.00	0.00	0.00	0.01
Navel	0.00	0.00	0.15	0.00	0.00	0.00	0.03
Lemon	0.00	0.00	0.04	0.00	0.00	0.00	0.01
Vines							
Sultana	0.00	0.00	0.12	0.00	0.00	0.00	0.01
Vegetables							
Cabbage	0.03	0.00	0.01	0.00	0.00	0.00	0.01
Cauliflower	0.02	0.00	0.04	0.00	0.00	0.00	0.01
Carrots	0.04	0.00	0.06	0.00	0.00	0.00	0.02
Fallow	0.07	0.21	0.06	0.00	0.09	0.04	0.00
Pasture	0.18	0.12	0.05	0.93	0.80	0.90	0.03
Other	0.08	0.06	0.08	0.02	0.03	0.02	0.87

some of the pasture country in the southern tablelands are still very green in November, but that they have dried off by December. The October image, by contrast, shows that there is a considerable amount of green material around when that image was acquired. Comparison of the October and December images show that some areas of forest cover are depicted as yellow in the October image, but green in the December image. The forest cover has not become greener over this period, but rather the images are relative indicators of greenness, so that the absolute colour values in one image cannot be compared with the colour values in the other images. This indicates however that the pasture and crop areas are greener than the forest areas.

The October composite image was used as the basis for creating productivity strata, and these strata were sampled so as to derive estimates of yield, for each crop type, within each strata. The strata are based on the vegetation index values, smoothed to provide strata that can be readily sampled. The strata distribution across the Forrest Shire are shown in Photo 4.6. The yield samples are averaged for each strata, to give the values shown in Table 4.11.

Table 4.11 Average yield of each crop type in each yield strata within Forrest Shire, from field sampling within the strata (TONNES/HA). The colours are related to the colour codes given in Photograph 4.6

	Yield		
Strata	good (red)	moderate (green)	poor (grey)
Wheat			
Cane	2.6	1.9	1.3
Albert	2.8	2.1	1.4
Dune	2.6	2.0	1.4
Oats			
Dora	2.4	1.6	1.1
Dora2	2.4	1.6	1.0
Spain	2.5	1.8	1.2
Barley			
Ibis	2.5	1.7	1.1
Ibis2	2.6	1.8	1.2
Triticale	2.3	2.1	1.6

Table 4.12 Two-way count of land-use strata and yield strata showing the area in hectares in each combination of land-using and yield strata

	Land-using strata							
	1 Schist ridges	2 Flood plain	3 Fertile slopes	4 Irrigation dist.	5 Low hills	6 Steep hills	13 Urban areas	
1	7 334	3 211	26 478	13 942	10 223	13 754	3 635	Best yield
2	13 181	14 908	21 852	2 290	37 699	3 099	564	Mod yield
3	5 661	8 937	1 301	0	15 759	4 199	0	Low yield
	26 176	27 056	49 631	16 232	63 681	16 853	4 199	

4.4.6 Calculation of production

The production of each crop within the shire can now be estimated, by summing the production calculated for each polygon shown in Photo 4.6, where

$$\text{polygon production} = \text{polygon area} \times \text{proportion crop} \times \text{yield}$$

Manual calculation of production could be done for Forrest Shire. However, crop managers are interested in crop production over much larger areas than the individual shire, so that manual techniques are too slow and costly. A better way to conduct this computation is by means of a GIS (Chapter 7).

The areas of the polygons, shown in Photo 4.6, are given in Table 4.12. Each polygon is a combination of a particular land-use intensity and yield strata. The proportion of each crop in each land-use intensity strata is given in Table 4.10, and the yield for each yield strata for each crop is given in Table 4.11. The data in these tables can be used with the areas given in Table 4.12 to calculate total production for each crop in Forrest Shire.

4.5 Comparison of false colour composites derived from integrated image data sets for visual interpretation

4.5.1 Introduction

The analyst may have a range of data sets including remotely sensed data acquired in different wavebands and at different times, other types of spatial information such as gravity and magnetic data, elevation and slope data. The data sets may be acquired from different spatial, temporal or radiometric perspectives and again they will record attributes of the surface in quite different ways.

The analyst may have many more than three data sets as can be depicted in the one false colour composite, or four data sets that can be displayed in a stereoscopic pair of false colour composites where stereoscopic images can be created by transforming the data using the central perspective transformation. Radiance values in one channel of scanner data can be used as 'heights' to create stereoscopic images of other channels of the data.

Combining data sets requires that they be spatially registered. The differences between the geometries of different data sets need to be resolved before the data sets can be overlain. The process of spatially registering data sets is called rectification and will be discussed in Chapter 5.

Combinations of data sets will generally contain more information than individual bands of data, but some combinations contain similar information to other combinations. This occurs when there is a high level of correlation between the bands that are used to construct the different false colour composites. When two channels exhibit high levels of correlation then changes in one channel are matched by changes in the other channel. It is difficult to select and then use a number of combinations of data sets in visual interpretation; the use of one false colour composite is much more preferable. One technique for reducing the number of channels to be viewed is to transform the data so as to produce channels of data that exhibit low levels of correlation.

4.5.2 Description of the dataset used

This application illustrates the use of a combined data set of radar and SPOT XS data for regional analysis where Photo 4.7 depicts three bands from the original data:

depicted in red – Radar data
depicted in green – SPOT, Channel 3 (NIR)
depicted in blue – SPOT, Channel 2 (red)

The radar data set was registered to the SPOT image that had been rectified to match the topographic maps of the area. The registration was done on the flat terrain, achieving good registration on these areas. Because of the slant view of the radar data, significant displacements will occur in the radar data over the mountainous area shown in the image.

The four channels of data were transformed in two ways to reduce the number of channels containing usable information

1. the four channels were transformed to create four new, uncorrelated channels; and
2. the SPOT data were transformed to create a vegetation index channel that enhanced green vegetation.

The four new uncorrelated channels and the vegetation index channel were used to create Photo 4.8. At this stage it is less important to know what constitutes the three channels in Photo 4.8 than to compare the two images. Chapter 5 describes the techniques used to create this image, and so information on them will be included here. Photo 4.8 comprises two uncorrelated channels created by the principal components transformation, being principal components 2 and 4 and a normalized vegetation index channel:

Red – principal component 2
Green – normalized vegetation index
Blue – principal component 4

4.5.3 Visual interpretation of the false colour composites

The area depicted in the images contains portion of the Philippine Fault system in Luzon, in the north-eastern portion of the image. The sharp ridgelines evident in Photo 4.7 (image 1) are part of the uplifting associated with the fault. The western part of the image depicts portion of the central Luzon plain, a major rice-growing area in the country.

Image 1 shows the terrain in the area very well. The areas of vegetative cover are depicted as yellow green, due to the NIR reflectance of the green vegetation in the SPOT image and some reflectance in the radar image from the vegetation. The forested areas in Photo 4.8 (Image 2) are depicted as green–blue in contrast to the crop areas that are depicted as cyan. Image 2 has discriminated between these covers better than has image 1. Image 2 however does not show the terrain features very well at all. Bare areas of exposed rock and upland soil are dark reddish on Image 1, for exposed areas in the central plain (on the south-western part of the plain), on the mountains in the centre of the image, and on the plateau and low hilly area in the north-eastern corner of the image.

Image 2 shows some significant differences between these areas, particularly between the mountains in the centre and the plateau with low hills in the north-east. The area of exposed rock and upland soils in the central mountains are of quite different origin to the rest of the mountains in the image. Whilst the remainder of the mountains are formed, and continue to be formed, by upthrusting of crustal movement along the Philippine Fault, the hilly area (H1 and H2 in Image 2) has been formed by volcanic intrusions that have created quite different geomorphic shapes, and different cover conditions. All of the area below the line AA has been significantly affected by volcanic intrusions. In this area of more rounded geomorphic shape are a number of dome-shaped hills created by volcanic intrusion.

The area has been extensively dissected by crustal movement, creating a number of shear lines that are obvious in the images. The more obvious of these lines are shown on Image 2. The Philippine Fault passes from where the river just touches the edge of the image, on the eastern edge of the image, in a north-westerly direction (line GG). It contains the San José river for part of its length, until the river breaks to the east, and then rounds an isolated shear ridge to return to the line of the fault. The Philippine Fault branches in the image to create two shear lines that are obvious in the image. There are a number of shear lines that are parallel to the main fault, and others that are approximately at right angles (lines BB, CC and DD). Most of these lines are more obvious on Image 2 than Image 1.

4.5.4 Conclusion

All visual interpretation depends on a knowledge of remote sensing, the environmental processes that influence the area of interest and the characteristics of the area itself. Lack of knowledge about some of these factors limits the amount of analysis that can be done by the reader. However, the images show that transforming the data can enhance the information content of the derived images in a way that facilitates visual interpretation.

5

Image processing

5.1 Overview

The purpose of image processing is to derive raw information from remotely sensed data. It is defined as the computer processing of digital image data to either extract, or facilitate the extraction of, information from that data. In extracting information it is often advantageous to utilize both visual interpretation and image processing because of their relative strengths and limitations. Chapter 3 considered visual interpretation but here we consider image processing. The relative strengths of image process (compared with visual interpretation) are as follows.

1. Faster and cheaper in processing large amounts of quantitative data.
2. Can conduct routine tasks in a fast and consistent way.
3. Produces quantifiable information.
4. Produces comparable results.
5. Makes it easy to integrate quantitative information sets.

The relative weaknesses should also be noted.

1. Cannot resolve unexpected situations; 'uninspired'.
2. Cannot learn; 'must be told'.
3. Derives less information.
4. Cannot easily handle either unstructured data sets or qualitative data.

Image processing is the better of the two methods for conducting more routine tasks; visual interpretation is better where the intuitive and creative skills of the analyst are required. It is important, therefore, to see visual interpretation and image processing as complementary techniques in the extraction of information. The designer of resource management systems needs to look continuously for opportunities to optimize the extraction of information from remotely sensed data, and often these opportunities involve integration of both techniques into the information extraction process.

There are many different facets of image processing that are not, as yet, particularly useful in the management of physical resources including artificial intelligence (AI) which concentrates on topics like robotic vision and identification of writing. These facets of image processing are not covered in this text. However the relevant facets of image processing include preprocessing, enhancement, classification and estimation. These are now considered in detail.

5.1.1 Preprocessing

Remotely sensed data is uncalibrated in some or all dimensions of the data (spatial, radiometric, spectral and temporal) although there will be an increasing trend, partially or completely, to calibrate data in all of these dimensions. If this lack of calibration introduces unacceptably large errors during the process of extracting information from the data then the data will need to be corrected prior to information extraction. The most important calibrations that may be needed will correct the data for radiometric and spatial errors.

Radiometric errors affect the digital values in the data for each picture element or pixel. They are caused by sensor errors as well as atmospheric scattering and absorption. By changing the response values in the data they affect the ability of the analyst to derive estimates of parameters such as leaf area index (LAI) or biomass from the data since variations in response values are interpreted as variations in these parameters. Radiometric errors introduced by the sensor are corrected by calibration. Errors introduced due to atmospheric effects are corrected by the application of atmospheric corrections as discussed in section 5.3.

Spatial errors affect the ability of the analyst to relate field data to the imagery, to relate the imagery to other data, and to provide spatially accurate information or advice. Spatial errors are corrected by rectification of the data (section 5.4).

Spectral errors arise when the actual spectral transmission characteristics of the sensor filters are different to those expected, causing differences in the response values. The spectral transmission characteristics of the filters are known at instrument construction, and they rarely change or deteriorate. This is rarely a source of error in image processing.

Temporal errors arise due to differences between actual and expected acquisition times. These differences are usually small relative to the rate of change of events in the environment, so temporal errors are usually insignificant.

Both spatial and radiometric errors involve systems or processes outside of the sensor. Spatial errors are affected by the platform altitude and orientation as well as by the effects of topographic relief on the image data. Radiometric errors are affected by the state of atmospheric transmission during acquisition. In consequence neither of these sources of error can be fully corrected without external sources of information.

Preprocessing is defined as the calibration of the data so as to create **ratio data**, i.e. the processed data is in commonly used, standard, physical units of measurement, both spatial and radiometric.

5.1.2 Enhancement

Enhancement techniques are processes that enhance subsets of the data so as to improve the accuracy, reliability, comfort and speed of visual interpretation. Most digital image data has a much greater dynamic range and sensitivity than the human eye. Consequently, there are levels of information in the data that cannot be seen visually when all of the data is being depicted. For example, the human eye can detect about 20 grey tone levels in black and white images, but there are often many more than this in the scanner data, up to the full radiometric range in the data. Most modern sensors such as TM and SPOT have 256 radiometric levels with the actual data contained in a significant portion of this range. To see all of the information embedded in the data requires enhancement of subset ranges of the image data before display either on a screen or in a photograph.

Enhancement is therefore defined as the visual emphasis of subranges of the data so as to facilitate visual interpretation. The emphasis of a subrange of the data is normally at the expense of the enhancement of the residual subranges of the data.

5.1.3 Classification

If remotely sensed images are considered to be an *m*-dimensional data set, where there are *m* wavebands in the image data, then classification is the process of partitioning the

m-dimensional response domain into a discrete number of class subdomains. These subdomains may match the range of response values that would be expected from the different land-cover and land-use classes.

Classification can provide either a final information layer, say of land-covers, or it can provide an indicator or pointer to specific conditions to assist visual interpretation. Use of classification as a pointer allows the interpreter to identify areas of similar radiometric conditions quickly. This technique exploits the advantages of image processing – the fast processing of large sets of remotely sensed data – and the advantages of visual interpretation in using all clues in the data to make decisions.

In collecting information on renewable resources, the most important first matter to be resolved is, 'what is the landcover or land use being analysed?'. Addressing this question is the province of visual interpretation and classification. Once resolved then the analyst can address the second question of, 'what is the land-use/land-cover condition?'. Estimation is concerned with this second question.

5.1.4 Estimation

Estimation is transformation of the data, sometimes with other data, into an estimate of one or more physical parameters using numerical models. Usually these models will depend on the type of land-cover being considered, for example models to estimate LAI from remotely sensed data are dependent on the species mix being sensed. It will, therefore, usually be necessary to map land-covers and land uses before estimation of canopy condition parameters.

The radiometric data used to drive estimation models are affected by radiometric errors and atmospheric attenuation. Errors due to these sources will introduce errors into the derived estimates so that radiometric calibration and atmospheric correction need to be applied before estimation models can be accurately used.

The statistical nature of remotely sensed data means that all of these techniques must be based on sound statistical theory. Because statistics is critical to good image processing, students must have an appreciation of some aspects of statistical theory including probability distributions, hypothesis testing, sampling and regression. The more essential statistical theory, including some specific to remote sensing, are included in section 5.2, as a prelude to image processing.

In practice, image calibration may be done by the supplier before the analyst receives the data. Correcting for atmospheric effects can be treated as an enhancement and so will be treated in section 5.3. Spatial calibration or rectification uses some of the techniques discussed in image enhancement and so it is treated after that section. The remainder of this chapter then considers image classification, clustering, classification strategies and estimation.

5.2 Statistical considerations

Digital image data is like a huge grid of cells covering the area of interest, whether the data are acquired by a remote sensor, or by sampling of an aerial photograph. Each cell is of a finite size, with the relative position of this cell to all other cells dependent upon certain factors.

1. Consider first the geometric relationship between this cell and other cells. Cells are sampled sequentially along scanlines, with a number of scanlines constituting an image.

Some aspects of the geometry are internal to the sensor and are usually known to a high degree of accuracy. Others are external to the sensor and are known less accurately.

2. Random positioning errors for all cells in an image scanline may occur. There is a random component in the positioning of the start of the image and all pixels in the image that affects the data values for all of the pixels in the image.
3. Errors are also possible in the position of a cell relative to the adjacent cells, due to displacements introduced due to height differences.

In consequence the actual position of a cell during acquisition has a random component associated with it. A second image would have cells in different positions even if all of the known and measured parameters were identical for both images. The response within a grid-cell or pixel is dependent upon the reflectance within the field of view and its immediate environs due to the point spread function at the time of sampling the pixel in that waveband. The point spread function states that light emanating in a direction from a source is scattered in accordance with the sin(x)/(x) function (Diagram 5.18). This curve shows that energy from a source is emanated in such a way that it will affect the response of adjacent pixels.

An image is thus a sampled record of the radiance from the surface being sensed. There are many implications of this, but the first and most important is that all methods of processing or analysing digital image data must be statistically valid i.e. all data processing methods must abide by appropriate statistical rules as apply to the analysis of statistical data, and the assumptions on which statistical procedures are based must be met within statistically acceptable tolerances. If a procedure assumes that the data being analysed is normally distributed then the data actually used should have a distribution that is a statistically valid approximation of the normal distribution, otherwise erroneous results may be achieved.

Remotely sensed data is acquired in accordance with a specific geometric model, in specific wavebands, at specific times and the response is quantized into a finite number of levels. In consequence, remotely sensed data is sampled in four ways: spatially, radiometrically, spectrally, and temporally.

Slight differences in canopy geometry or surface conditions introduce differences in reflectance which cause differences in the data values recorded in remotely sensed data. Large differences can cause a change in the pattern of response in the image data that can be explained in terms of changes in the physical conditions of the surface. However, subtle changes in canopy geometry or surface conditions introduce an apparent or quasi-random scattering in the data.

Scattering of the data is also caused by atmospheric effects. The net result is that the data for a surface that may be considered as being homogeneous or uniform by an observer will usually have data values that are scattered in some way.

A **frequency diagram** records the number of occurrences of each data value, i.e. the count of each data value, in a set of data as a graph with the horizontal axis being the range of values that the data can have, and the vertical axis being the count for each value or group of values. If the *area* of the column (range of values × count) is proportional to the frequency then the diagram is called a **histogram**. Typical histograms are shown in Photo 5.6. A **cumulative frequency diagram** has similar axes, but the count recorded at a value is the sum of all of the counts up to and including the count for that value. A cumulative frequency diagram must therefore have values that either increase or remain constant across its full extent.

A 2–D frequency diagram records the number of occurrences of each combination of data values in two separate parameters, and can be displayed by printing the count values at each location in the two-dimensional space with axes as the range of values in the two parameters, or as **scattergrams** which have the same axes but the count data have been contoured and are shown as contour or density slice images. Typical scattergrams are shown in Photo 5.7.

All statistical methods assume that data has a distribution or spread function of some form, called a probability density function. **Parametric methods** assume that the characteristics of the distribution and the parameters that describe its shape are either known or closely approximate a known distribution. To use parametric methods, it is necessary to derive estimates of these parameters and ensure that the actual data fits this theoretical distribution within acceptable tolerances. **Non-parametric methods** of statistical analysis do not assume a knowledge of the shape of the distribution and as such they are more flexible, but they require a lot more data to define the nature of the distribution during processing, and as such they can be more complex and costly to use.

5.2.1 Probability density functions

A frequency diagram (or histogram) graphically depicts the probability distribution of the data in the sample. In remotely sensed data the data observations for a class are usually distributed over a range of values as shown in Diagram 5.7. The probability that a particular value will occur is the count of that value divided by the total count. The probability distribution is thus the histogram normalized for the sample population used to create the histogram.

Probability density functions (PDFs) that have been taken from observed data can have many shapes. If the data are of a homogeneous surface type then its PDF would usually be **unimodal** and have data values that are distributed equally about the mean value as shown in Diagram 5.1(b). Data sets that contain the data collected for many different types of surface will often be **multi-modal**, as shown in Diagram 5.1(e). Multi-modal histograms can also be found in Photo 5.6.

All PDFs can be defined by the probability at each point in the distribution. But this method of specification is untidy and difficult to use. If there is some symmetry in the PDF then a mathematical function might adequately describe it by means of a few parameters. This approach is much more usable and generally applicable in a processing sense than PDFs based on specific data sets. The use of standard models in processing results in lower processing costs and increases the flexibility of the processing system. If the model is a good fit to the data sets, and the sample is representative of the whole population then the model can be assumed to be a good fit to the data for the whole population.

The use of models to create PDFs for a population requires training data to estimate the model parameters with sufficient reliability to represent the whole population and to test that the sample population is a good fit to the data. Some models of probability density functions that are commonly used include the binomial distribution and the normal distribution.

Binomial distribution

Consider the situation where there is a probability, p, of an event occurring (i.e. a success), and a probability, $(1 - p)$ that the event will not occur i.e. a failure. The probability of x occurrences (successes) from n events (trials), assuming the binomial distribution is given by

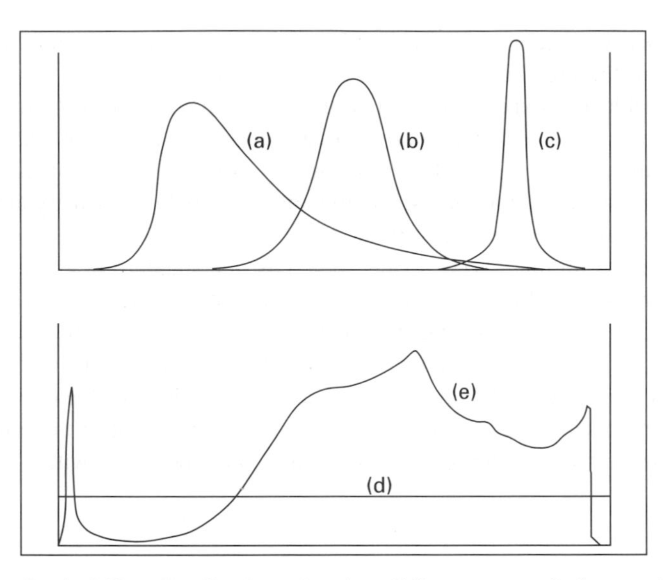

Diagram 5.1 Probability distributions for five different sets of observational data: (a) skewed unimodal; (b) normal unimodal; (c) leptokurtic unimodal; (d) uniform; and (e) multi-modal.

$$\Pr(x) = \frac{n!}{x!(n-x)!}\, p^x (1-p)^{n-x}$$

where $n!$ is factorial n.

A binomial distribution concerned with the probabilities associated with simultaneous occurrences of two or more events is called the **multinomial distribution**. A use of the multinomial distribution is in the analysis of the accuracy of visual interpretation or classification as will be discussed in Chapter 6. In comparing a map of land use with field data on land use, each class in the map has probabilities of being wrongly classified into the various classes. These probabilities should be a good fit to the multinomial distribution.

Normal distribution

The normal distribution is unimodal and symmetrical about the central value. The value of the central peak is determined by the mean, and the spread is measured by the standard deviation. The importance of these parameters is indicated by the equation for the normal distribution:

$$\Pr(x) = \frac{1}{\sigma\sqrt{2\pi}} \exp \frac{-(x-\mu)^2}{2\sigma^2}$$

where μ is the mean and σ is the standard deviation (σ^2 is the variance).

The normal distribution has two parameters, the mean and standard deviation. The normal curve is expressed in the form:

$$\Pr(x) \sim N(\mu, \sigma^2)$$

The area under the curve and above the x-axis between the values a and b represents the probability that x lies within the range a to b, written in the form, $\Pr\{a < x < b\}$. Since the total area between the normal curve and the x-axis represents the total probability for that distribution, the area beneath the curve equals one.

If the variable x is expressed in units of $z = (x - \mu)/\sigma$ then the normal distribution associated with z is the standard normal distribution with a mean value of 0 and a variance of 1, that is $\Pr(z) \sim N(0, 1)$. Values in any normal distribution can be converted into values in the standard normal distribution and so probability density tables need only be kept for the standard normal distribution.

In the standard normal curve, a value $z_a = 1$ is one standard deviation either side of the mean. There is a 68.26 per cent probability that an observation will fall within one standard deviation of the mean, i.e. 68.26 per cent of the population falls within one standard deviation of the mean. This distance out from the mean also indicates the point of inflection of the curve, that is the rate of change of the slope of the curve changes sign at this point. In the same way, it can be shown that 95.5 per cent of the observations fall within two standard deviations of the mean, and 99.7 per cent within three standard deviations.

5.2.2 Correlation

Correlation is an indication of the predictability of the value of one parameter relative to the values of one or more other parameters. If field data are collected on two or more parameters (reflectance in wavebands, leaf area index or biomass) then correlated parameters will usually see increases in one parameter accompanied by changes (increases for positive correlation or decreases for negative correlation) in the other parameter(s) proportional to the increases in the first parameter. Similarly, decreases in the first parameter will be accompanied by consistent changes (this time decreases or increases respectively) in the other parameters. With uncorrelated data, any increase or decrease in one parameter has equal probability of being matched by changes in the other parameters, with the magnitudes of these changes being quite random. Photo 5.7 shows the three 2–D scattergrams for the Landsat TM image depicted in Photo 5.3. In these scattergrams the data in (a) are highly correlated whilst the data in other scattergrams exhibit less correlation. These scattergrams are also depicted in Diagram 5.4 to show the typical response values for water (w), urban (u), forest (f), green pasture (p) and brown herbage (h) in these scattergrams.

Highly correlated data between two or more parameters does *not* establish a physical cause for the correlation; if, however, there is a physical link between parameters then observations of those parameters would be correlated.

The correlation between observations of different parameters is determined by first calculating the variance and covariance. For two parameters (x, y), pairs of observations $(x_1, y_1), (x_2, y_2), \ldots, (x_n, y_n)$ are used to compute the mean and covariance arrays:

$$\text{mean array} = U = |\mu_x \quad \mu_y|$$

$$\text{Covariance array} = \underset{=}{\Sigma} = \begin{vmatrix} \sigma_{xx} & \sigma_{xy} \\ \sigma_{xy} & \sigma_{yy} \end{vmatrix}$$

where

$$\sigma_{xx} = \Sigma\,(x_i - \mu_x)(x_i - \mu_x)/(n - 1) = \text{variance of } x$$
$$\sigma_{xy} = \Sigma\,(x_i - \mu_x)(y_i - \mu_y)/(n - 1) = \text{covariance between } x \text{ and } y$$
$$\sigma_{yy} = \Sigma\,(y_i - \mu_y)(y_i - \mu_y)/(n - 1) = \text{variance of } y$$

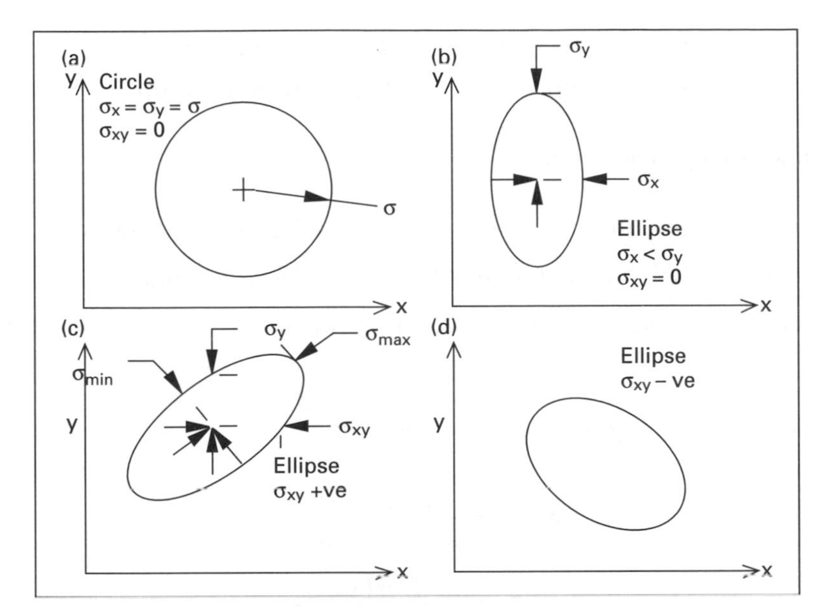

Diagram 5.2 The shapes of the normal distribution at one standard deviation for populations in two variates, with different correlations between the parameters.

If the values in each column and each row in turn are divided by the standard deviation in the column and row, then the covariance matrix becomes a correlation matrix:

$$\text{Correlation matrix} = \mathbf{C} = \begin{vmatrix} \dfrac{\sigma_{xx}}{\sigma_x \sigma_x} & \dfrac{\sigma_{xy}}{\sigma_x \sigma_y} \\ \dfrac{\sigma_{xy}}{\sigma_x \sigma_y} & \dfrac{\sigma_{yy}}{\sigma_y \sigma_y} \end{vmatrix}$$

$$= \begin{vmatrix} 1 \dfrac{\sigma_{xy}}{\sigma_x \sigma_y} \end{vmatrix}$$

The correlation value will vary between +1 and –1 where 0 indicates zero correlation. Positive values mean that the semi-major axis is oriented so that positive differences $(x_i - \mu_x)$ in one parameter are accompanied by positive differences $(y_i - \mu_y)$ in the second parameter. Negative values occur when positive differences in one parameter have negative differences in the other parameter. The data distributions shown in Diagram 5.2 are depicted as plots at one standard deviation from the mean as indicated by the covariance arrays. The plot is circular in Diagram 5.2(a) because the standard deviations are equal and there is negligible correlation. If the standard deviations are different and there is negligible correlation then the distribution will be elliptical, with the semi-major and semi-minor axes being parallel to the coordinate axes (Diagram 5.2(b)). In Diagrams 5.2(c) and (d) the variances are different and the covariances are positive and negative respectively.

Covariance and correlation matrices can be extended to n-dimensional space. They are a valuable means of analysing the correlation between sets of parameters as occurs with multi-spectral remotely sensed data.

The relationship between data values in old dimensions (x, y) and new dimensions (u, v) of rotation θ between them such that the new axes coincide with the semi-major and semi-minor axes of the ellipse are

$$u_i = x_i\cos(\theta) + y_i\sin(\theta) - [\mu_x\cos(\theta) + \mu_y\sin(\theta)]$$

$$v_i = -x_i\sin(\theta) + y_i\cos(\theta) - [-\mu_x\sin(\theta) + \mu_y\cos(\theta)]$$

By the law of propagation of error

$$\mathrm{var}\{u_i\} = \cos^2(\theta)\mathrm{var}\{x_i\} + \sin^2(\theta)\mathrm{var}\{y_i\} + 2\sin(\theta)\cos(\theta)\mathrm{cov}\{x_i, y_i\}$$

$$\mathrm{var}\{v_i\} = \sin^2(\theta)\mathrm{var}\{x_i\} + \cos^2(\theta)\mathrm{var}\{y_i\} - 2\sin(\theta)\cos(\theta)\mathrm{cov}\{x_i, y_i\}$$

$$\mathrm{cov}\{u_i, v_i\} = \sin(\theta)\cos(\theta) \times [-\mathrm{var}\{x_i\} + \mathrm{var}\{y_i\}] + [\cos^2(\theta) - \sin^2(\theta)] \times \mathrm{cov}\{x_i, y_i\}$$

Now θ is such that $\mathrm{cov}\{u_i, v_i\} = 0$, so by definition

$$\tan(2\theta) = \frac{2\sin(\theta)\cos(\theta)}{[\cos^2(\theta) - \sin^2(\theta)]} = \frac{2\mathrm{cov}\{x_i, y_i\}}{\mathrm{var}\{x_i\} - \mathrm{var}\{y_i\}} \tag{5.1}$$

Equation (5.1) can be used to calculate θ and hence solve for var $\{u_i\}$ and var $\{v_i\}$.

5.2.3 Statistical characteristics of satellite scanner data

The same areas of four different cover types are identified on images acquired by three different sensors. The data are analysed to derive mean and covariance arrays as shown in Table 5.1. The values given in Table 5.1 are not exactly comparable since the images were acquired in 1987 (XS) and 1991 (TM and MOS–1), so that differences in cover, atmospheric and solar zenith conditions will introduce variations in the data values. However the data do provide an insight into a number of characteristics of the data as acquired by these three sensors.

1. Satellite response data follows the general shape of the spectral reflectance curves for the different surfaces. This is particularly noticeable in the case of the water and forest cover types for the TM and XS data.
2. The variances of the TM data are larger, in almost all cases, than the variances associated with either the XS or MOS–1 data. Busy surfaces, such as urban cover could be expected to have very large variances in the SPOT XS data because the smaller pixels have less opportunity to average across the range of surface conditions. This occurs for SPOT XS band 1 which has about twice the variance of TM Channel 2, but does not occur for the other bands. The variance in the TM data is about 1.5 to 2.5 times the variance in the XS data, suggesting that TM data is more sensitive than SPOT or MOS–1.

   The XS data are next most sensitive, with variances that are usually more than twice that of the MOS–1 data, although the averaging effect of the larger MOS–1 pixels 50m by 50m will be contributing to the difference in variance.
3. The correlations between bands for one sensor are generally matched by similar correlations in the other sensors. Thus high correlations occur between the visible bands for all three sensors for the urban and rural cover types. This suggests consistency between the sensors.

Table 5.1 Mean, covariance and correlation arrays for four cover types as acquired using the three scanners, thematic mapper, SPOT XS and MOS–1.

Cover sensor	Ch	Mean array	Covariance and correlation arrays							
Water										
MOS–1	1	20.19	0.27	0.01	0.03	0.00	1.000	0.068	0.118	0.000
	2	15.08		0.08	0.00	0.00		1.000	0.000	0.000
	3	8.58			0.24	0.00			1.000	0.000
	4	7.00				0.00				1.000
TM	1	65.42	3.44	0.37	0.01	0.11	0.03	0.12	0.02	
	2	19.39		1.13	0.11	–0.02	0.04	–0.04	–0.02	
	3	13.85			0.68	0.05	0.02	–0.04	0.00	
	4	7.51				0.34	0.00	0.02	0.03	
	5	4.40					0.57	–0.01	0.05	
	6	135.62						0.24	–0.01	
	7	3.14							1.25	
			1.00	0.188	0.007	0.102	0.021	0.132	0.010	
				1.000	0.125	–0.032	0.050	–0.077	–0.017	
					1.000	0.104	0.032	–0.099	0.000	
						1.000	0.000	0.070	0.046	
							1.000	–0.027	0.059	
								1.000	–0.018	
									1.000	
XS	1	32.71	0.58	0.13	0.06		1.000	0.267	0.158	
	2	20.69		0.41	0.14			1.000	0.437	
	3	13.71			0.25				1.000	
Urban										
MOS–1	1	31.57	5.02	3.88	1.36	1.02	1.000	0.817	0.532	0.529
	2	32.31		4.49	1.48	1.06		1.000	0.613	0.582
	3	20.21			1.30	0.79			1.000	0.805
	4	15.60				0.74				1.000
TM	1	95.10	74.13	31.57	46.37	21.04	33.66	1.44	27.47	
	2	37.23		16.96	22.85	13.43	19.52	0.32	14.36	
	3	46.85			40.21	20.13	34.10	1.46	26.03	
	4	43.49				40.76	33.37	–2.90	15.26	
	5	61.02					68.01	0.07	40.65	
	6	156.93						1.97	1.37	
	7	36.38							31.37	
			1.000	0.890	0.849	0.383	0.474	0.119	0.570	
				1.000	0.875	0.511	0.575	0.055	0.623	
					1.000	0.497	0.652	0.164	0.733	
						1.000	0.632	–0.324	0.427	
							1.000	0.006	0.886	
								1.000	0.174	
									1.000	
XS	1	48.16	38.76	32.94	16.24		1.000	0.926	0.543	
	2	42.28		32.63	15.09			1.000	0.550	
	3	40.60			23.06				1.000	

Table 5.1 (Continued)

Cover sensor	Ch	Mean array	Covariance and correlation arrays							
Forest										
MOS–1	1	20.64	0.71	0.07	0.36	0.39	1.000	0.163	0.314	0.283
	2	15.45		0.26	0.11	0.11		1.000	0.159	0.132
	3	24.39			1.85	1.91			1.000	0.858
	4	21.76				2.68				1.000
TM	1	64.70	3.10	0.87	1.20	6.78	3.57	0.24	1.27	
	2	23.11		1.07	0.89	6.34	3.61	0.15	1.21	
	3	19.92			1.96	6.47	4.00	0.04	1.54	
	4	73.45				86.49	38.34	1.70	10.98	
	5	44.21					33.08	0.85	9.01	
	6	137.20						0.49	0.19	
	7	13.48							4.16	
			1.000	0.477	0.487	0.414	0.353	0.195	0.354	
				1.000	0.615	0.659	0.607	0.207	0.574	
					1.000	0.497	0.497	0.041	0.539	
						1.000	0.726	0.261	0.579	
							1.000	0.211	0.768	
								1.000	0.133	
									1.000	
XS	1	30.17	0.98	0.46	4.08		1.000	0.615	0.703	
	2	19.61		0.57	2.03			1.000	0.458	
	3	61.77			34.41				1.000	
Rural										
MOS–1	1	28.77	4.05	5.37	0.23	–0.16	1.000	0.911	0.093	–0.078
	2	26.69		8.58	0.42	–0.32		1.000	0.117	–0.107
	3	28.29			1.50	1.00			1.000	0.797
	4	24.30				1.05				1.000
TM	1	84.63	26.23	19.55	41.81	34.75	50.62	5.08	28.47	
	2	37.71		18.22	35.89	40.88	50.01	2.86	22.65	
	3	52.14			77.76	76.65	100.6	6.97	49.47	
	4	68.63				199.8	179.1	–7.36	37.63	
	5	91.88					281.7	8.29	152.4	
	6	149.74						7.37	13.28	
	7	41.92							197.39	
			1.000	0.894	0.926	0.480	0.589	0.365	0.396	
				1.000	0.954	0.678	0.698	0.247	0.378	
					1.000	0.615	0.679	0.291	0.399	
						1.000	0.754	–0.192	0.189	
							1.000	0.182	0.646	
								1.000	0.348	
									1.000	
XS	1	40.92	13.81	18.09	23.60		1.000	0.934	0.563	
	2	36.27		27.19	19.23			1.000	0.327	
	3	55.20			127.2				1.000	

5.2.4 Shannon's Sampling Theorem

A series of photographs taken as a person walks across a field, timed such that they are taken at intervals of about 100 m, may give a good indication of the general direction of travel and average speed of the walker. But they will not tell the observer anything about the fluctuations in direction or velocity that occurred between the photographs. In the same way, annual images may indicate general land use and land quality, but will tell the analyst very little about what actually happened in between.

In renewable resources there are certain frequencies about the events being monitored: imagery must be acquired at a frequency that allows this cyclical pattern of physical events to be adequately recorded. Shannon's Sampling Theorem is concerned with the development of a sampling frequency sufficient to represent a population that exhibits a cyclic fluctuating pattern.

Shannon's Sampling Theorem states that a **bandlimited** function at frequency W can be reconstructed exactly from samples taken at a frequency interval of $W/2$ or higher using a sinc function, sinc = $\sin(x)/x$ as the shape of the sample function. An image is bandlimited if the variability in the data does not contain variations that have frequencies higher than the band limit, i.e. higher than W. A strict definition is that a picture function, $g(i, j)$ is bandlimited if its Fourier transform, $G(i, j)$, is zero whenever either $|f_i|$, or $|f_j|$ is greater than W.

In a spatial sense the importance of Shannon's Sampling Theorem lies in its ability to assist us in selecting suitable pixel sizes to sample the surface, or to assess the likelihood that data of a known pixel size can adequately record the radiance from the surface. In a temporal sense the theorem's role is to assess whether the sampled data is representative of cyclic data, either to reconstruct that data or to derive information from the data, at a suitable level of accuracy.

If seasonal events are to be monitored then at least two images per season are required, assuming that higher frequency cycles, for example due to weather events, do not introduce significant variations from the seasonal cycle. If in-season variations are significant then more frequent images are required.

5.3 The enhancement of image data

Image enhancement is the process of transforming digital image data so as to create an image within which either a selected density range, or selected detail in the image, is enhanced or emphasized relative to the remainder of the data or detail. Enhancement selectively enhances some information in the data and may suppress other information. An analyst may well decide to use a number of complementary enhancements in the conduct of the one interpretation task, where each enhancement is developed to assist with the interpretation of specific aspects of the project.

It is important that the interpreter be familiar with the reflectance characteristics of the surfaces that have to be imaged, as well as the effects of the enhancement used, to ensure that he does not either misinterpret the imagery or suppress information that is important to him. One tactic that the interpreter may employ would be to conduct a preliminary interpretation on imagery that has a broadly based enhancement that suppresses little information, but does not reveal detailed subtle information, before focusing on detailed interpretation using imagery that has been optimized for this purpose.

Image enhancement, by suppressing irrelevant data and information, and emphasizing that which is relevant, assists the interpreter to extract the required information in the most economical, reliable and accurate manner. Visual interpretation involves both physiological and psychological processes; different interpreters will often prefer different enhancements for their interpretation. Selection of the enhancement type, and more particularly the parameter values in that transformation, can only be done empirically by the individual interpreter. For this reason, facilities that are used to create enhanced images for visual interpretation must include a high quality colour video monitor, a rapid and cheap display for the enhanced image data.

There are an infinite number of ways that imagery can be enhanced, but experience has shown that there are a number that are used for the majority of visual interpretation work. In addition, each time a new enhancement is used the interpreter has to learn the significance of the displayed colours, and if other interpreters are to be involved, then the same will apply to them. For this reason, it is best to use a small armoury of enhancements wherever possible, even though this might cause some small loss in information content in the display.

5.3.1 The pseudo-colour density slice

A single channel of image data, as depicted in Photo 5.2 for the seven Landsat TM wavebands contains up to 256 levels of information, far more than can be discriminated by the human interpreter. Individual, or ranges of data levels can be colour-coded to different colours so as to convey more information to the analyst. This process is called pseudo-colour density slice.

5.3.2 Linear enhancement

The simplest way to enhance a selected subset of the data values is to use a linear enhancement of the form:

$$\begin{aligned} \text{NEW} &= A + B \times \text{OLD} && \text{when } 0 \leq (A + B \times \text{OLD}) \leq M \\ &= 0 && \text{when } (A + B \times \text{OLD}) < 0 \\ &= M && \text{when } (A + B \times \text{OLD}) > M \end{aligned}$$

where M is the maximum value that the transformed data values can take. Such a transformation is shown in Diagram 5.3 for a data set with histogram shown in (a), the cumulative frequency diagram in (b), the enhancement in (c) and the histogram of the enhanced data in (d). The linear enhancement shown has been selected to enhance the central peak in the dataset; in consequence all of the low data values have accumulated at zero value (or the zero bin) in the enhanced data, or black in the displayed image. A large part of this image will thus appear as black. There will also be a significant component of the data that will be white, from the large count in the highest bin value. The central range of values will exhibit good tonal contrast.

If the histograms in Photo 5.6 arc converted into cumulative frequency diagrams, then the optimum linear stretch for all of the data would be a line that closely coincides with the curve of the cumulative frequency diagrams. A simple linear enhancement provides a good fit when the data are distributed in a unimodal manner as is the case with the data from the first frequency diagram in Photo 5.6, but does not provide a good enhancement for data with a multimodal distribution, such as the data in the fourth frequency diagram. Some histograms can be even more markedly multimodal than that depicted for the data in Photo 5.6, and in these cases the linear enhancement provides a poor enhancement of the data.

The advantages of the linear transformation are that its effects are easy to visualize

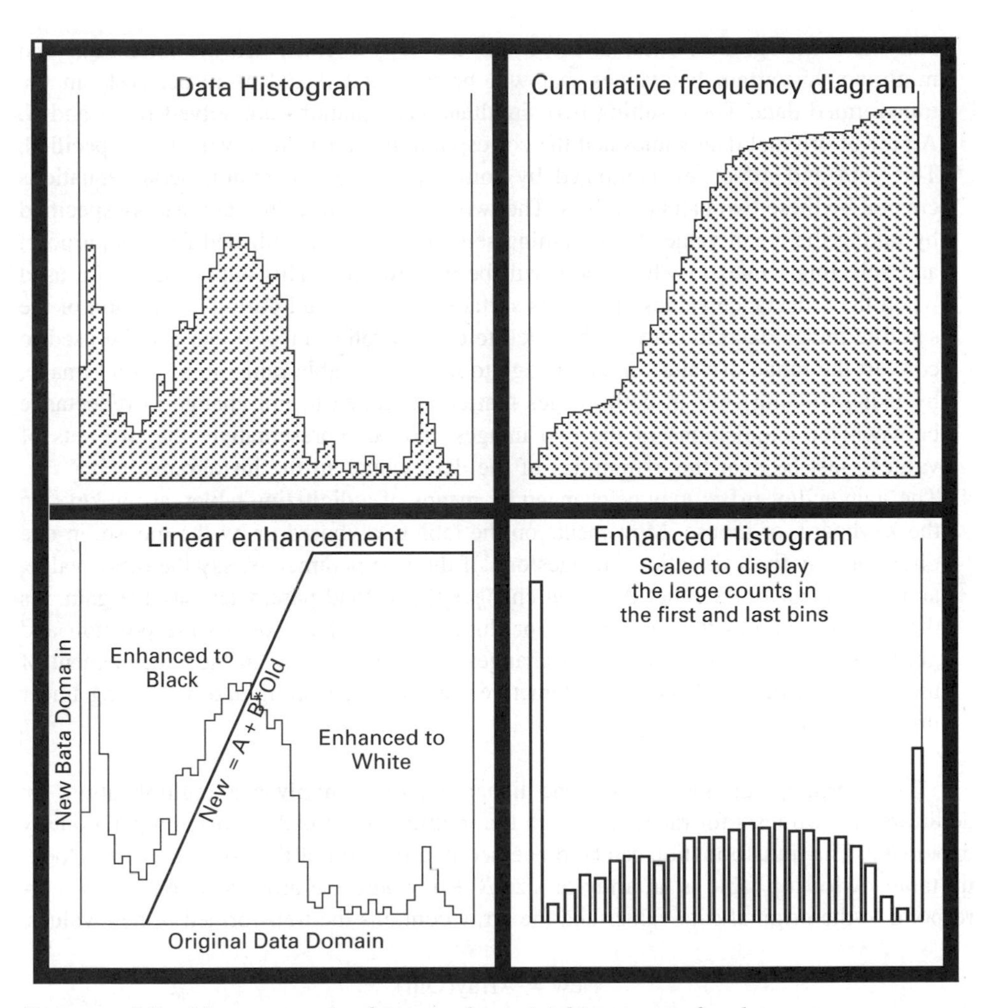

Diagram 5.3 Linear stretch of image data: (a) histogram of a data set; (b) cumulative histogram of the data; (c) the linear stretch enhancement and (d) the histogram of the enhanced image data.

by the interpreter, and the transformed values are linearly correlated with the original response values. These advantages make the linear transformation one of the more regularly and widely used enhancements.

The parameters of the linear transformation, A and B, are known as the **offset** and **gain** respectively. They can be determined in a number of ways.

1. The gain and offset are specified by the user. The difficulty in this approach is the time involved in determining the two values by the user. It is not a commonly used approach.
2. The percentage of the data in the tails of the enhancement is specified, i.e. the percentage of the data that will have values of 0 or M as a result of the linear transformation. Often the cut-off percentages are set as default values of 1 or 2 per cent, but can be overridden by the operator if required. The total pixel count in the histogram is used to calculate the number of pixels that should be in the lower and upper tails. The histogram values are summed upwards, and downwards until this value is exceeded,

and these histogram values, at the lower and upper ends, are the maximum and minimum histogram bin values that will be assigned 0 and M respectively in the transformed data. The resulting two simultaneous equations are solved for A and B.

3. A pair of original data values and the corresponding transformed values are specified. The gain and offset are computed by solution of the two simultaneous equations created by these two pairs of values. The two original data values are usually specified by taking the mean values from training sets of the data established for this purpose, and selecting values to which these will be transformed. This approach can be used when the reflectance values of various surfaces are observed during acquisition of the satellite image data, to convert the satellite data to reflectance. It can also be used to convert the range of values in one image to the comparable range in a second image, by using training sites in both images that are assumed to not change in reflectance between the acquisitions of the two images. To be representative, the two sets of values must be near the extremities of the data values in the histograms.
4. The gain and/or offset may be changed by means of a digitizing tablet, arrow keys on the keyboard, or mouse. Movements on the tablet, keyboard or by the mouse, in one direction, say the x direction, changes one of the two parameters, say the offset value, and movement in the other direction changes the second parameter, say the gain. As the operator moves the mouse, key or digitizing tablet cursor, so the position and gradient on the linear enhancement changes causing a change in the enhancement of the displayed image. This is an interactive way of establishing and modifying linear image enhancements.

Transforming an image using the linear equation involves a multiplication, an addition and two tests for range, of which the multiplication is the most computationally expensive. The transformation can be processed in a fraction of the time by use of a look-up table. A lookup table is an array of size $N + 1$ where the array locations, 0–N, correspond to the original data values and the array contains the transformed or new values.

$$\text{New} = \text{Array(Old)}$$

The thematic mapper data for the Subic Bay image is displayed as histograms in Photo 5.6. The data does not cover the full range, 0–255, and so it has been linearly enhanced before being displayed in Photo 5.2. The data shown in Photo 5.2 is the same enhancement as has been used to create the colour composites in Photo 5.3.

5.3.3 Non-linear enhancements

Criticisms of linear enhancements are that they:

- do not match the logarithmic response of the human eye;
- create sharp cut-offs at the extremities of the enhancement; and
- do not provide a good enhancement with multimodal data, particularly if it has peaks at both ends of the histogram.

These limitations of the linear enhancement can be corrected by the use of non-linear enhancements that are usually a function of the form 'New = f(Old)' where 'f(Old)' is a non-linear function of the old parameter, e.g.

$$\text{New} = A \times \log(\text{Old}) + B \qquad \text{or} \qquad \text{New} = A \times \sin(B \times \text{Old}) + C$$

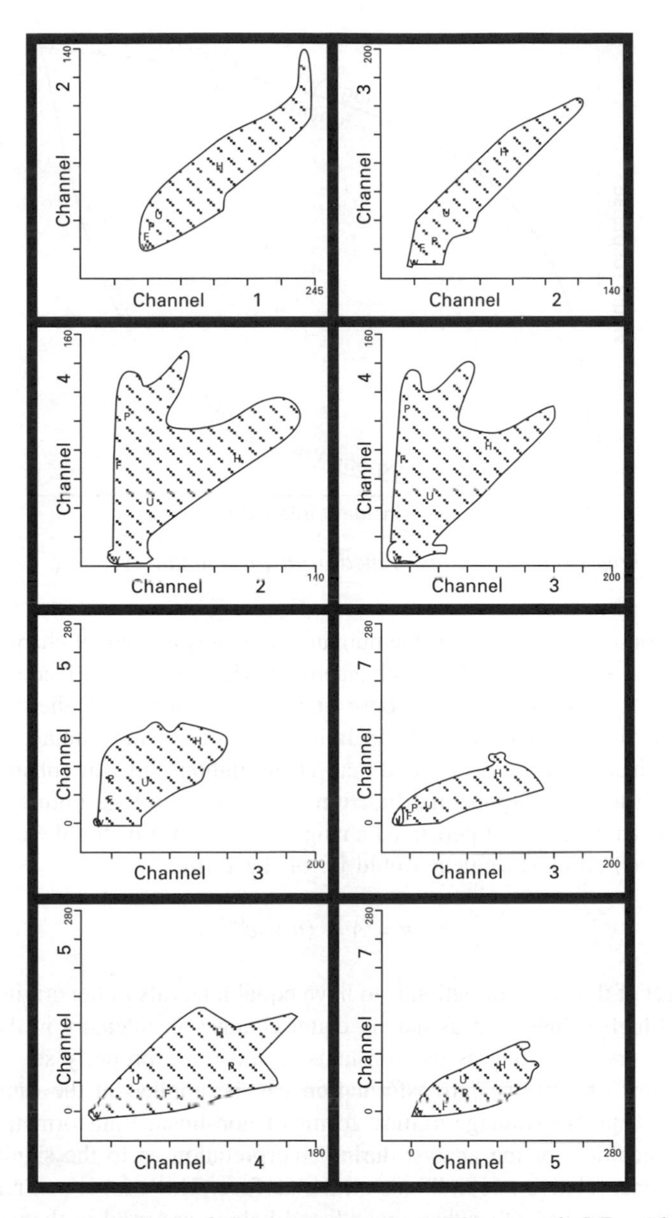

Diagram 5.4 Representation of the scattergrams in Photograph 5.7, with mean values for training data for selected landcovers in the image.

Another approach is to use a series of linear functions as segments across the range of the data instead of the one non-linear function.

The response of the human eye is a logarithmic function to base e of the radiance incident on the eye, of the form:

$$\text{eye response} = \ln(\text{radiance})$$

which means that the eye is much more sensitive to changes in response at low light levels than at high light levels as shown in Diagram 5.5.

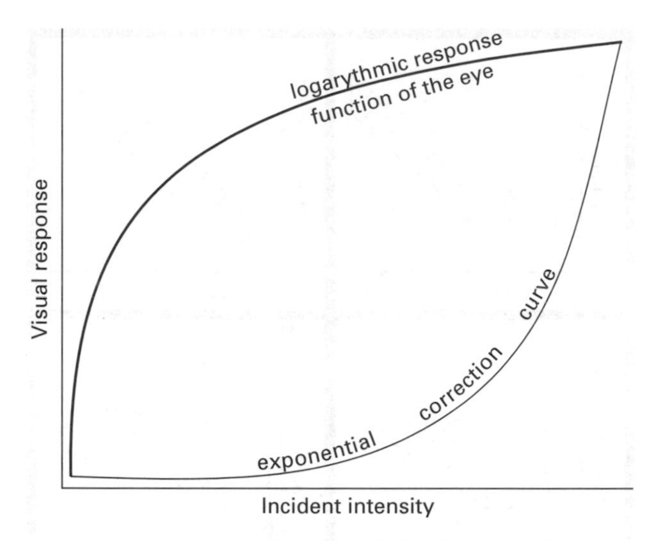

Diagram 5.5 Logarithmic response function of the human eye.

This response characteristic of the human eye means that equal changes in response at the darker and lighter ends of the enhancement will not be perceived as having equal tonal changes. A unit change in response at the darker end will elicit a larger visual response than will a unit change at the lighter end, since the human eye will detect smaller tonal changes at the darker end than at the lighter end. An enhancement can be introduced, by use of an exponential function, to reverse the logarithmic response function of the human eye, so that perceived changes in response match the original changes in response. Such an enhancement would be of the form

$$\text{New} = A + (b \times e^{\text{Old}})$$

The effect of this transformation is to have equal intervals in the original data at low, moderate and high values seen as having equal changes in intensity in the transformed data by the observer, as long as the constants A and b are properly set.

Other forms of non-linear transformation can be applied in the same way as described above. One disadvantage to most forms of non-linear transformation is that they can create difficulties for the analyst during interpretation as to the significance of the variations in tone, relative to conditions on the surface. This cost is one reason why few non-linear enhancements, other than those listed below, are used in the enhancement of image data. Specific non-linear enhancements that are routinely used include piecewise linear stretch, histogram equalization and logarithmic response enhancement.

5.3.4 Piecewise linear stretch

The piecewise linear enhancement or stretch is a variation on the linear stretch that goes some of the way towards reducing the disadvantages of the linear enhancement. In this enhancement there are a number of linear enhancements, each with different gain and offset values, that are used to transform parts of the range of response values in the original data. The transformed data can be made to be a good fit to the cumulative frequency diagram of the actual data, so as to produce good quality enhanced images.

The usual way to compute the piecewise linear enhancement is to construct the cumulative frequency diagram, identify the break points in the gradients of the histogram,

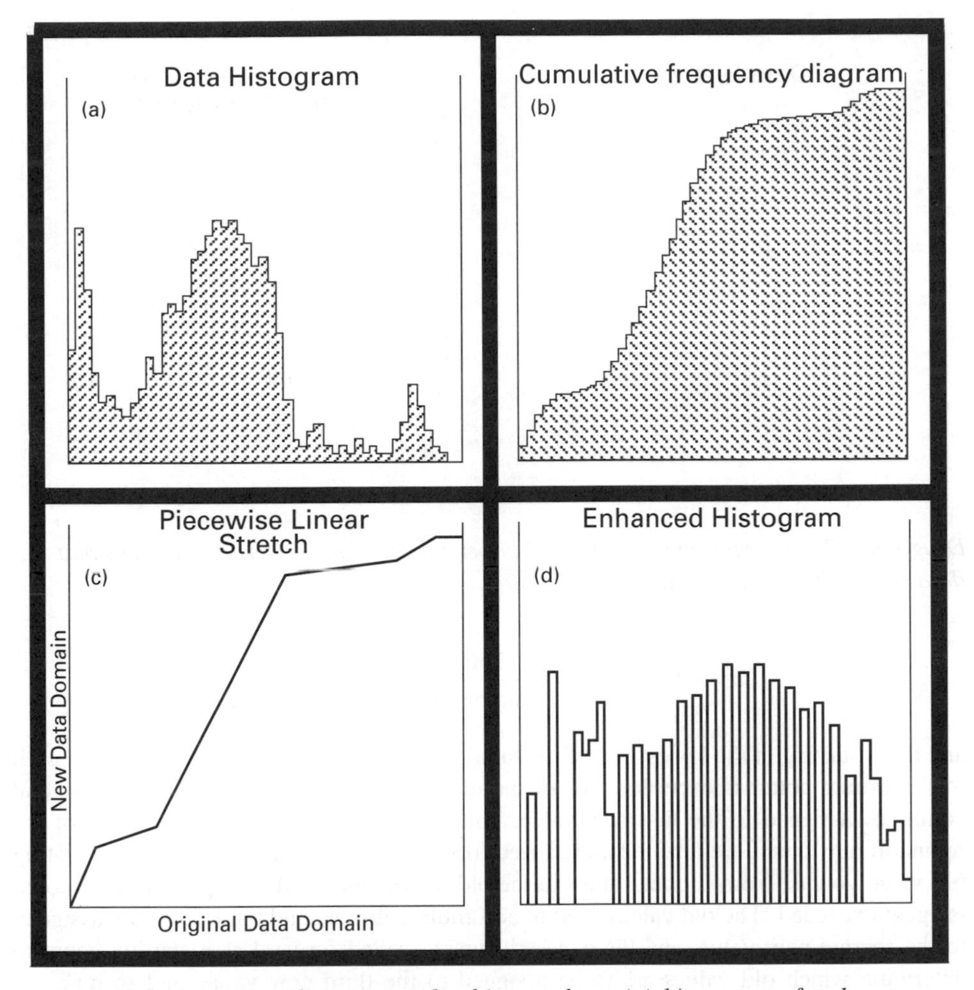

Diagram 5.6 Piecewise linear stretch of image data: (a) histogram of a data set; (b) cumulative frequency diagram of the data; (c) the piecewise linear stretch enhancement and (d) the histogram of the enhanced image data.

and fit individual linear enhancements to the segments of the histogram between these breakpoints. Consider the histogram, and corresponding cumulative frequency diagram, for a data set as shown in Diagrams 5.6(a) and (b). A piecewise linear stretch of this data could have an enhancement as depicted in Diagram 5.6(c), which would produce an enhanced image with a histogram like that shown in Diagram 5.6(d).

The piecewise linear enhancement gives a much better result than the linear enhancement when the data exhibit a multimodal distribution across the histogram of the data values. The cost, as with all non-linear enhancements, is that the simple linear correspondence between the image data and the original data is lost in the transformation.

5.3.5 Histogram equalization

The objective of histogram equalization is to create images which have approximately equal numbers of pixels, or pixel counts at each transformed density value, i.e. the transformed data will have an approximately uniform distribution. The enhancement does this by determining the optimum number of pixels to be assigned each value, from

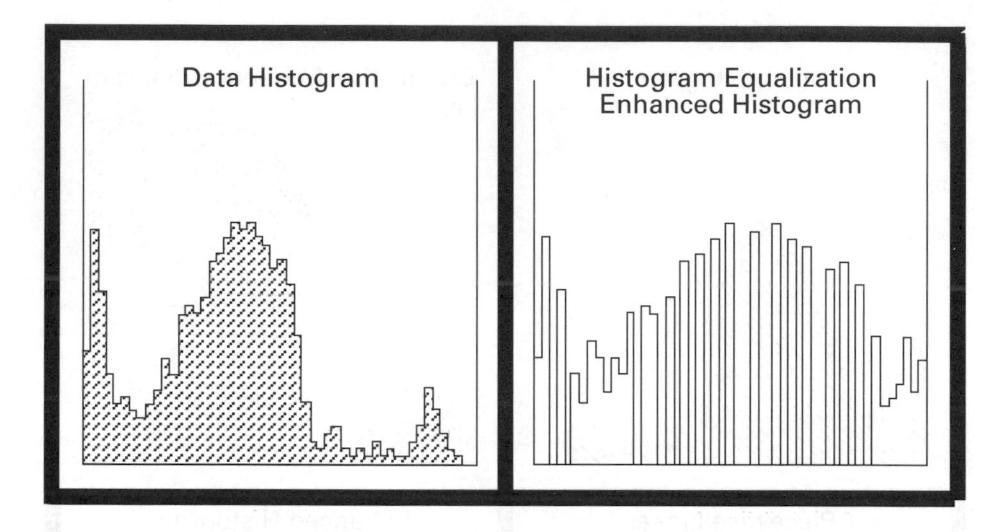

Diagram 5.7 A histogram of data and the enhanced data set produced from that data by the histogram equalization transform.

$$\text{Optimum count} = \frac{\text{Total number of pixels}}{\text{Number of values in the data}}$$

and then, starting at the lowest old value, sums the value counts in the old histogram until this optimum count is exceeded. All of the old values used to accumulate to this total count are assigned to the first new value. The count of the old values in excess of the optimum new count are retained and carried forward as a starting count for accumulating old value counts, from further values in the old histogram, until the optimum new count is again exceeded. The old values used to accumulate this second total count are assigned to the second new value, and the excess is again carried forward as a starting count to determine which old values are to be assigned to the third new value, and so on.

If the histogram count for an old value (and the total count that includes this histogram count) is larger than the optimum new count then that one old value will be assigned to the new value. If the total count exceeds n times the optimum count where $n > 1$, then all of the old values will be assigned to the one new value, but the next $(n - 1)$ new values will have zero counts. In this way the average count in each bin or value is kept the same.

This process continues for the remainder of the data. Once the mapping from old values to new values is known through this process, then a look-up table can be constructed and the image transformed by the use of this look-up table.

Diagram 5.7 shows the enhancement produced by the histogram equalization transformation for the same data set as used to show the effects of the linear and piecewise linear enhancements.

5.3.6 Ratioing

Ratioing is the procedure of forming a ratio of one function of a subset of the N bands in the data, to another function of another subset of the N bands in the data.

$$\text{Ratio} = \frac{\sum(\text{Const}_1 \times \text{Band } i)}{\sum(\text{Const}_2 \times \text{Band } j)}$$

The parameters $Const_1$ and $Const_2$ are real. A further constant may be added to the denominator to ensure that division by zero does not occur. In practice the functions are normally quite simple ones, consisting of only one band in each of the numerator and denominator to give ratios of the form:

$$\text{Ratio} = \frac{\text{Band } i}{\text{Band } j}$$

For data in the range 0–255 a ratio of Band *i*/Band *j* will give values in the range (1/256) to 1 when Band $i <$ Band j and in the range 1 to 256 when Band $i >$ Band j, assuming a constant of unity is added to the values in both bands so that they are in the range 1 to 256. A simple linear enhancement of the ratio will severely compress the values less than one relative to the values greater than one, so that ratios are normally enhanced using a non-linear transformation. Suitable enhancements could be a piecewise linear or logarithmic stretch as described earlier.

Assessment of the usefulness of ratioing requires an understanding of the reflectance characteristics of the different surface types as described in Chapter 2. Most discrimination between green vegetation, soil and water occurs in the red and infra-red regions, so that a ratio of Band 7/Band 5 for Landsat MSS should yield high values for green vegetation, values slightly greater than unity for soils and less than one for water.

A form of ratioing that enhances the relative intensities in the different bands, called **normalization**, is of the form:

$$\text{New}_i = \frac{(n \times \text{Old}_i)}{\sum_{j=1}^{n} \text{Old}_j}$$

for n wavebands. This transformation retains the relative intensities in the bands but eliminates variations in total intensity.

Another special group of ratios are the **vegetation indices**. The vegetation indices are used both to enhance the response of green vegetation, and to estimate green plant matter, either by estimating above-ground biomass or leaf area index (LAI). The vegetation indices do this by quantifying the distance from the line of soils in the direction of the point of green herbage as indicated in Diagram 5.8. As the amount of green herbage cover increases within a pixel from no herbage, the response of the pixel moves from the line of soils towards the point of green herbage as discussed in Chapter 2. The distance from the soil line, towards the point of green herbage is thus correlated with the amount of green herbage in the top of the canopy.

The two commonly used indices are

$$\begin{aligned} \text{RVI} &= \text{Ratio vegetation index} \\ &= (\text{NIR})/\text{Red} \\ &= \text{Band 4/Band 3} \quad \text{in Landsat TM} \\ &= \text{Ch3/Ch2} \qquad\quad \text{in SPOT XS data} \end{aligned}$$

and

$$\begin{aligned} \text{NDVI} &= \text{Normalized difference vegetation index} \\ &= \frac{(\text{NIR} - \text{Red})}{(\text{NIR} + \text{Red})} \end{aligned}$$

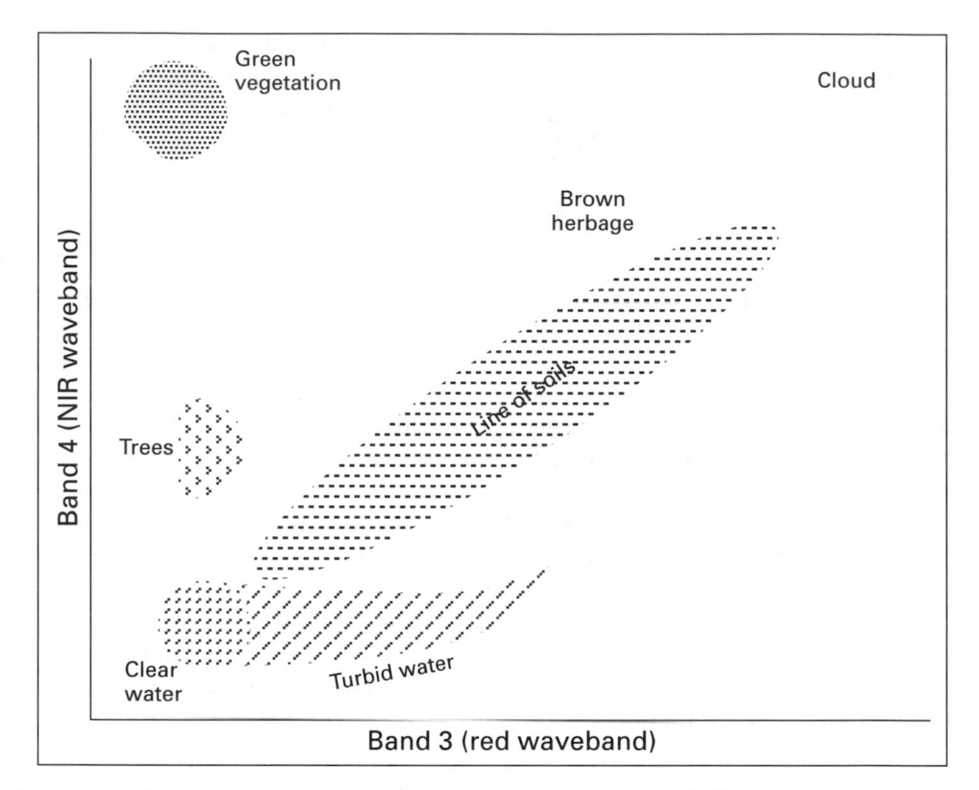

Diagram 5.8 Typical response values in Landsat MSS band 5/Band 7 space for the common cover types.

$$= \frac{(\text{Band 4} - \text{Band 3})}{(\text{Band 4} + \text{Band 3})} \quad \text{in Landsat TM}$$

$$= \frac{(\text{Ch3} - \text{Ch2})}{(\text{Ch3} + \text{Ch2})} \quad \text{in SPOT XS data}$$

These ratios have lines of constant value or contours that are radial from the origin as shown in Diagram 5.9. In practice, the NDVI has been found by various authors to give slightly better estimates of biomass or LAI than RVI under most conditions, so that it is the more commonly used index. An example of the normalized difference vegetation index is shown in Photo 5.4(f). An example of different vegetation indices is shown in Photo 5.8.

5.3.7 Orthogonal transformations

Remotely sensed data, in N bands, can be transformed by a rotation, to give values relative to a new set of axes that are at a specified orientation to the original axes. The most common orthogonal transformation, the principal components transformation, is designed to transform the data to create uncorrelated data sets.

Consider the data in the 2–D scattergrams in Photo 5.7 for the image in Photo 5.2. The data are highly correlated between some bands. A principal components transformation of the data would use a rotational transformation to create new data that have axes that coincide with the direction of maximum data scatter, then at right angles to this line,

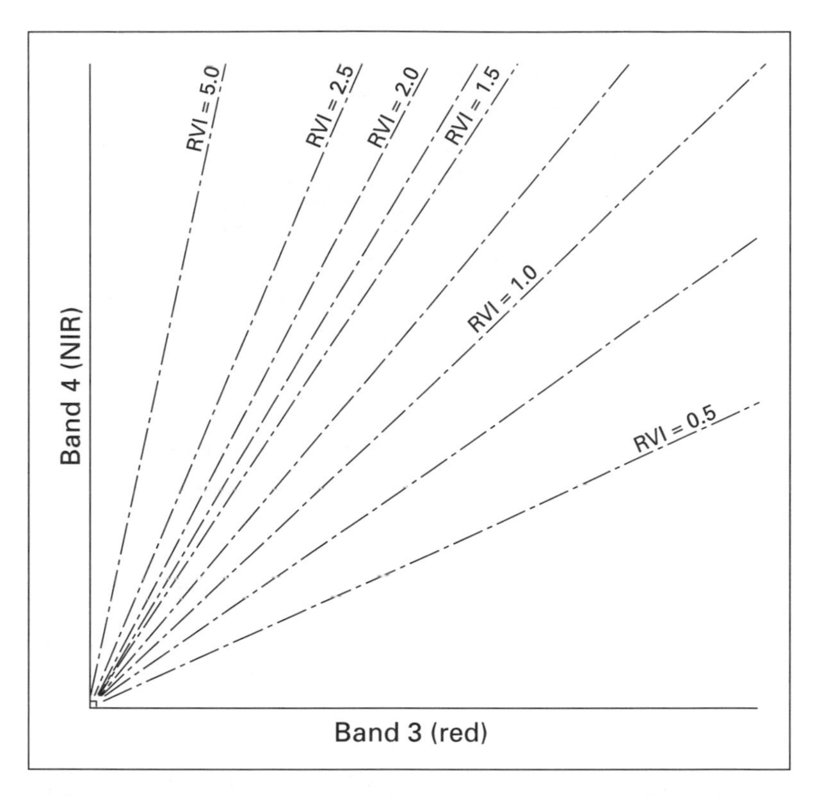

Diagram 5.9 Relationship between the line of soils and lines of equal value for the ratio vegetation index (RVI).

in the direction of maximum remaining scatter, then at right angles to both of these axes, in the direction of the maximum remaining scatter, until all of the dimensions in the original data are covered.

A 2–D frequency diagram or scattergram represents the distribution of the data in two dimensions. Consider the data in Diagram 5.10 that represents two bands with the ellipsoid at one standard deviation distance from the mean. If these two bands are to be transformed on their own by the principal components transformation then the transformed data will lie along the semi-major and semi-minor axes of the data. The transformed data are in the same number of components as the original data, with the maximum scatter in Component 1, the next maximum scatter in Component 2.

The same principles apply to imagery in *N*-dimensional space. The first primary axis, coinciding with the first semi-major axis of the *N*-dimensional ellipsoid defined by the covariance array, will contain the maximum amount of variation in the sample data used to derive the covariance array. With image data, the greatest variation is usually due to illumination so that Component 1 is often termed the brightness component. The next highest primary axis will contain the second highest amount of variation in the transformed data. With imagery in the visible and NIR, this can be due to the contribution of green matter in the data and so the second component is sometimes called the greenness component. For highly correlated data, the first two or three components of the transformed data will usually contain most of the information in the original data, the remaining components being primarily noise.

For the Landsat TM data in Photo 5.2, the seven bands have been transformed by means of the principal components transformation using the correlation array in Table 5.2

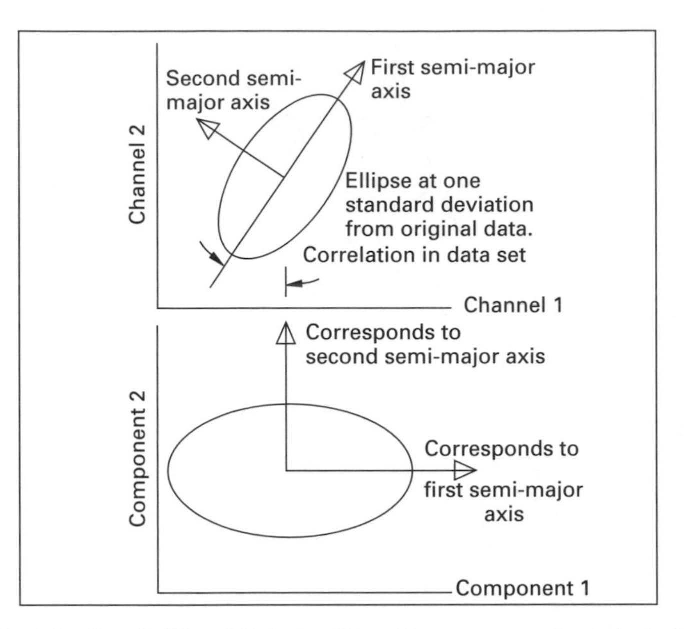

Diagram 5.10 The Band 5/Band 7 2–D ellipsoid at one standard deviation, with semi-major and semi-minor axes as specified by the covariance array.

and the first four components displayed in Photo 5.4(a) to (d). A colour composite of the first three components is shown in Photo 5.4(e). There is high correlation between some of the bands in the original data Table 5.2.

Preparing the orthogonal transformation, given in Table 5.3 results in principal components images in which the variance in the components, and the percentage of the variance in each component are shown in Table 5.3. This data shows that 76.3 per cent of the variance, i.e. of the information in the data, is contained in the first component, 91.7 per cent in the first two components and 96.7 per cent in the first three components. Only the first four components contain a significant amount of information, the variance in the last three components being essentially noise. This observation is illustrated in the images of the first four components in Photo 5.4(a) to (d) inclusive. Whilst these components have all been linearly enhanced, the components exhibit increasing levels of noise that are particularly noticeable in the third and fourth components.

The covariance array for image data changes from image to image, primarily due to changes in surface and atmospheric conditions. In consequence each principal component image will be different – the components will mean somewhat different things in each image. The interpreter has to relearn, with each image, what is the significance of each component in relation to the interpretation task at hand. This learning component with each transformation can form a significant cost overhead.

One attempt to resolve this limitation has been to develop standard transformations that are a close approximation of the principal components transformation, so that application of this transformation will produce more consistent component images. One version of this approach, developed by Kauth and Thomas (1976) for Landsat MSS data, creates the index images:

$$\begin{aligned} \text{SBI} &= \text{Soil brightness index} \\ &= 0.433 \times \text{Band4} + 0.632 \times \text{Band5} + 0.586 \times \text{Band6} + 0.491 \times \text{Band7} \end{aligned}$$

Table 5.2 Summary of image statistics used to create a principal components transformation of the image data

	Mean array	Covariance array Ch1	Ch2	Ch3	Ch4	Ch5	Ch6	Ch7
Ch1	74.47	156.35	89.53	157.07	56.47	208.61	50.02	123.50
Ch2	28.38		59.93	104.98	89.56	176.52	32.80	88.59
Ch3	29.60			194.62	151.61	326.88	61.12	165.13
Ch4	61.80				743.63	603.72	44.54	185.75
Ch5	54.61					854.22	113.90	349.26
Ch6	142.39						30.84	59.48
Ch7	21.10							168.35

Correlation array Ch1	Ch2	Ch3	Ch4	Ch5	Ch6	Ch7
1.0000	0.9249	0.9004	0.1656	0.5708	0.7203	0.7612
	1.0000	0.9721	0.4242	0.7802	0.7630	0.8820
		1.0000	0.3989	0.8017	0.7889	0.9123
			1.0000	0.7575	0.2941	0.5280
				1.0000	0.7018	0.9210
					1.0000	0.8255
						1.0000

The principal components transformation used the seven wavebands in the TM image displayed in Photo 5.2. In practice the analyst may choose to use a subset of the channels, particularly the six reflectance channels

Table 5.3 The principal components transformation parameters and the contribution of the variance to each of the components

Component	Principal components transformation matrix Ch1	Ch2	Ch3	Ch4	Ch5	Ch6	Ch7
1	0.3693	0.4146	0.4175	0.2362	0.3893	0.3682	0.4177
2	0.4199	0.1538	0.1583	−0.7776	−0.3697	0.1756	−0.0527
3	0.3847	0.3355	0.1990	0.2132	−0.1283	−0.7760	−0.1890
4	−0.2903	−0.1023	0.1580	−0.4835	0.4170	−0.4713	0.5005
5	0.5956	−0.3168	−0.6152	0.0555	0.0665	−0.0898	0.3885
6	−0.0769	−0.3206	0.3727	0.2392	−0.6535	−0.0331	0.5167
7	−0.3072	0.6916	−0.4677	−0.0073	−0.2997	−0.0095	0.3445

Component	Contribution of components to the variance in the data PCA variance	Percentage of variance
1	706.3	76.3
2	142.1	15.4
3	45.8	5.0
4	19.7	2.1
5	6.0	0.6
6	3.0	0.3
7	2.4	0.3

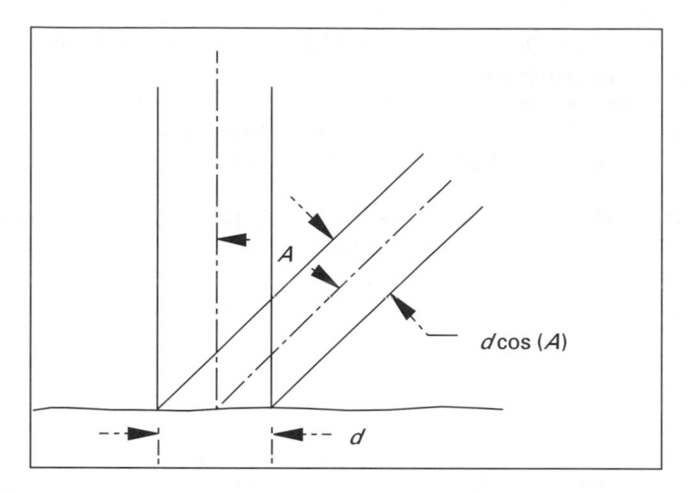

Diagram 5.11 Relationship between solar elevation and energy incident on the target area.

GVI = Greenness vegetation index
$= -0.29 \times \text{Band4} - 0.562 \times \text{Band5} + 0.60 \times \text{Band6} + 0.491 \times \text{Band7}$

In this transformation the SBI and GVI correspond to the first two components of the principal components transformation for conditions that the authors believed were typical for Landsat MSS data. These two axes coincide with the line of soils, or brightness and towards the response of green vegetation respectively (Photo 5.8).

5.3.8 Atmospheric calibration

Two primary sources of noise are seasonal variations in the intensity of solar radiation and atmospheric attenuation of the optical signal during passage from the surface to the sensor. The amount of energy incident on a defined area, such as a pixel, is a function of the angle of elevation of the Sun as shown in Diagram 5.11. As the angle of elevation, *A*, increases, the cross-sectional area projected in the direction of the Sun decreases relative to the ground area and, in consequence, the lower the amount of energy available to illuminate the pixel area. The relative dimensions of these two areas are a function of the cosine of the solar elevation, so that satellite data can be corrected for this effect by dividing the radiance data incident on the sensor by cos(solar elevation). Whilst this transformation will correct for variations in incident solar radiation due to changes in elevation, it does not correct for variations in:

- radiance due to changes in atmospheric path length;
- absorption and scattering in that path length; or
- reflectance due to the effect of the surface on reflectance, as a function of the solar incident angle.

The effects of variations in atmospheric conditions on satellite data, and the potential impact on image processing, particularly of temporal data sets, can be great. An important issue in monitoring renewable resources will be the satisfactory correction of satellite data for atmospheric effects. The energy incident on a scanner is given by

$$E_{s,\lambda} = [\, r_\lambda s_\lambda t_\lambda E_{i,\lambda} + s_\lambda E_{a,\lambda} \,]/\pi$$

which can be rearranged to give, as in equation (2.8)

$$r_\lambda = \frac{\pi(E_{d,\lambda} - E_{A,\Pi})}{(t'_\lambda E_{i,\lambda})}$$

Now the scanner response values are calibrated to be linearly related to radiance, i.e.

$$E_{d,\lambda} = k_1 R_{d,\lambda} + k_2$$

If e_λ, t'_λ, $E_{i,\lambda}$, $E_{A,\lambda}$, k_1 and k_2 are assumed to be constant then

$$\begin{aligned} r_\lambda &= \frac{\pi k_1 R_{d,\lambda}}{t'_\lambda E_{i,\lambda}} + \frac{\pi(k_2 - E_{A,\lambda})}{t'_\lambda E_{A,\lambda}} \\ &= c_1 R_{d,\lambda} + c_2 \end{aligned} \tag{5.2}$$

where r_λ is the surface bi-directional reflectance, $R_{d,\lambda}$ is sensor response value, and k_1, k_2 are constants related to radiance from the surface and the atmosphere respectively.

This linear model is a good approximation at a point where the assumptions about constant path transmissivity and atmospheric radiance are nearly correct. However, across an image these assumptions are not always correct and can introduce significant errors. There are two recognized ways of calibrating remotely sensed data for atmospheric effects: image-to-image correction; and calibration using field spectrometer observations.

Image-to-image correction

A minimum of two points, with low and high response values in the image data, and considered to be of constant radiance in both the control image and the image to be corrected, are selected. Data values for these two points are determined in both images, and used to compute the gain and offset values in a linear enhancement. The linear enhancement is applied to the second image so that the data for both images occupy the same data space. The control image either can be a standard image that is nominated as the control, or may be calibrated to reflectance using the techniques described below.

This method of correction is relatively simple to use. However, it assumes that sufficient points can be found and that the radiance of these points has not changed significantly between the images. It also assumes that the simple linear model can be applied across the whole of the image.

Calibration using field spectrometer observations

A number of fields are chosen and the reflectance of these fields determined at the time of overpass using field spectrometers as discussed in Chapter 6. The average response values for the fields are determined from the satellite data, and the two data sets are used to derive the gain and offset values in a linear enhancement. The sites chosen will need to be of fairly constant reflectance across an area much larger than a pixel (as discussed in Chapter 6), to ensure that the observed reflectances are representative of the pixel, and both low and high reflectance values should be measured in close proximity, to calculate the linear model parameters, effectively at a point. It would be an advantage to observe sites that do not exhibit significant diurnal variation in reflectance and are a close fit to the lambertian reflectance model so that the data can be collected over a longer time period.

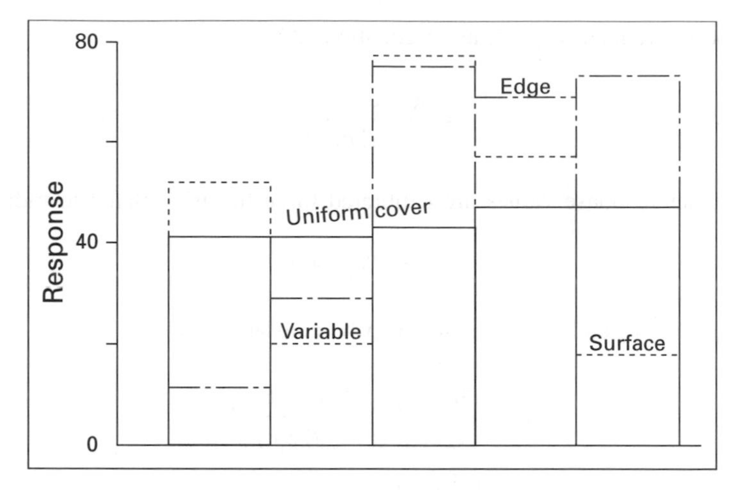

Diagram 5.12 Response profiles across three differently textured surfaces: (a) low texture; (b) an edge; and (c) a highly textured surface.

If the gain and offset values are calculated using field sites in close proximity to each other then they can be compared with values calculated using field data taken at other locations to see if there is significant variation in atmospheric conditions across the image. If there is a significant variation then the one linear model cannot be used to calibrate the image data. The technique is expensive in terms of field data collection, but can allow the analyst to determine the impact of variations in atmospheric conditions before conducting the calibration.

Another approach is to measure or calculate the parameters that constitute the constants in (5.2). k_1 and k_2 can be taken from calibration tables for the scanner of interest, $E_{i,\lambda}$ measured using a spectrometer and t'_λ and $E_{A,\lambda}$ either estimated by making assumptions about atmospheric conditions or determined from spectrometer observations taken for that purpose.

Ratios of bands are less affected by atmospheric absorption and scattering than are the actual band values because the atmospheric absorption and scattering usually affects both bands in the same direction, but at slightly different magnitudes.

5.3.9 Measurement of texture

Texture is a measure of the roughness or smoothness of image data. Texture can be of interest of itself when different land-covers exhibit different amounts of texture (Diagram 5.12). For example turbid water, commercial centres and some soil and rock conditions can be displayed as cyan colour in standard false colour composites, but the three will exhibit quite different textures. A measure of texture should measure the variability of the data. The texture in the vicinity of a pixel is calculated using a window or set of pixels centred on the pixel of interest.

Consider a 3 × 3 window of pixels (Diagram 5.13) across four different types of surfaces: (a) a smooth surface; (b) a rough surface; (c) an edge parallel to the lines in the data; and (d) an edge at about 45° to the lines in the data.

$$\text{Smooth} = \begin{pmatrix} 43 & 48 & 41 \\ 44 & 42 & 42 \\ 43 & 41 & 45 \end{pmatrix}$$

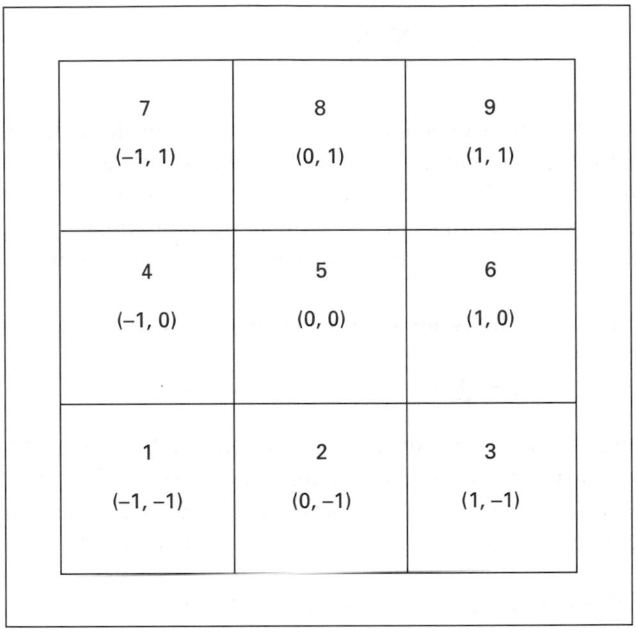

7 (−1, 1)	8 (0, 1)	9 (1, 1)
4 (−1, 0)	5 (0, 0)	6 (1, 0)
1 (−1, −1)	2 (0, −1)	3 (1, −1)

Diagram 5.13 Layout of the pixels in a 3 × 3 window, and their local coordinate system.

$$\text{Rough} = \begin{pmatrix} 26 & 48 & 48 \\ 58 & 77 & 83 \\ 72 & 92 & 10 \end{pmatrix}$$

$$\text{Vertical edge} = \begin{pmatrix} 29 & 59 & 55 \\ 24 & 50 & 50 \\ 20 & 57 & 57 \end{pmatrix}$$

$$\text{Oblique edge} = \begin{pmatrix} 21 & 23 & 52 \\ 23 & 58 & 52 \\ 50 & 56 & 53 \end{pmatrix}$$

One measure of texture would be the **sum of absolute differences** from the central pixel across an $n \times n$ window.

$$T_a = \sum |(R_i - R_c)| \quad i = 1, 2, \ldots, n \times n$$

This measure of texture gives values of 15, 221, 105 and 134 for the four sample surfaces given above. While it provides a measure of texture, it cannot discriminate between rough surfaces and edges.

A second measure of texture is the **sum of absolute adjacent differences** across the rows of the data, and down the lines. This actually gives two measures of texture, one in the x direction and the other in the y direction.

$$T_x = \sum_{j=1}^{n} \sum_{i=1}^{n-1} |(R_{i,j} - R_{(i+1),j})|$$

$$T_y = \sum_{j=1}^{n-1} \sum_{i=1}^{n} |(R_{i,j} - R_{i,(j+1)})|$$

where i is the count in the x direction and j is the count in the y direction.

The values for the four surfaces are $T_x = 20$, 198, 97, 81, and $T_y = 13$, 134, 37, 67, with averages of 17, 166, 67 and 74 respectively. This measure will discriminate rough from smooth textures, and will identify edges that are approximately aligned with the x and y axes. However, it will not separate edges that are not parallel with these axes from rough texture.

The third measure is the **sums of squares about the centre pixel**.

$$T_s = \sum (R_i - R_c)^2 \quad i = 1, 2, \ldots, n \times n$$

This measure gives values of 53 for (a), 9419 for (b), 2221 for (c) and, 3984 for (d).

The sums of squares can also be a good measure of texture, but very large individual differences have a larger impact on this measure than on T_a. Edges can thus be more difficult to discriminate from highly textured surfaces, particularly when using smaller windows.

Specific applications of texture analysis are to identify edges, i.e. edge detection, and image smoothing.

Edge detection

Edge detection can be a valuable aid in visual interpretation where the analyst is looking for boundaries, edges and other linear features. Edges are features in the image characterized by very narrow but very high gradients in one direction and very low gradients at right angles, with this pattern extending over a number of pixels in the data. Large gradients occur where the differential of the surface equation is a maximum. If a picture function, $g(i, j)$ consists of an ideal picture function, $s(i, j)$ and a noise function $n(i, j)$ such that

$$g(i, j) = s(i, j) + n(i, j)$$

then an ideal edge detector will determine the maximum gradient of $s(i, j)$ and be unaffected by variations in response due to $n(i, j)$. This can be done in a number of ways: fitting a surface; use of differences of averages; image smoothing; and by Fourier transform.

Fitting a surface

If a surface is taken to be of the form

$$A_o + A_1x + A_2y + A_3xy + A_4x^2 + A_5y^2 = R \tag{5.3}$$

then the gradient, G, at any point in this surface is the differential of this equation, i.e. in the x direction

$$G_x = A_1 + A_3y + 2A_4x \tag{5.4}$$

and in the y direction

$$G_y = A_2 + A_3x + 2A_5y$$

while the maximum gradient will be

$$G_{max} = [G_x^2 + G_y^2]^{0.5}$$

If the nine of these equations, for the pixels in the 3 × 3 window about the central pixel, are used to solve for the six unknowns, then the easiest solution is to assume unit coordinates in the x and y directions with the origin at the central pixel as shown in Diagram 5.13. The equations are then of the form:

Point	
1, 1	$A_o + A_1 + A_2 + A_3 + A_4 + A_5 = R_{1,1}$
1, 0	$A_o + A_1 + A_4 = R_{1,0}$
1, −1	$A_o + A_1 - A_2 - A_3 + A_4 + A_5 = R_{1,-1}$
0, 1	$A_o + A_2 + A_5 = R_{0,1}$
0, 0	$A_o = R_{0,0}$
0, −1	$A_o - A_2 + A_5 = R_{0,-1}$
−1, 1	$A_o - A_1 + A_2 - A_3 + A_4 + A_5 = R_{-1,1}$
−1, 0	$A_o - A_1 + A_4 = R_{-1,0}$
−1, −1	$A_o - A_1 - A_2 + A_3 + A_4 + A_5 = R_{-1,-1}$

which can be solved to find A_1 and A_2, the gradients in the x and y directions at the centre pixel of coordinates (0, 0):

$$A_1 = \frac{1}{6}(R_{1,1} + R_{1,0} + R_{1,-1} - R_{-1,1} - R_{-1,0} - R_{-1,-1})$$

$$A_2 = \frac{1}{6}(R_{1,1} + R_{0,1} + R_{-1,1} - R_{1,-1} - R_{0,-1} - R_{-1,-1})$$

hence $G_x = A_1$, $G_y = A_2$ at point (0, 0) and $G_{max} = (A_1^2 + A_2^2)^{0.5}$ or use $G = |A_1| + |A_2|$ as a computationally simpler approximation of the gradient parameter. If the 3 by 3 example data set given previously is used then the calculated gradients of the four surfaces are:

	SMOOTH	ROUGH	VERT. EDGE	TILTED EDGE
G_x	1.167	−2.667	22.5	18.167
G_y	0.0	11.833	11.833	20.5
$G = \lvert A_1\rvert + \lvert A_2\rvert$	1.2	14.5	34.333	38.667
$G_{max} = \sqrt{A_1 + A_2}$	1.167	12.130	25.422	27.391
$G = \lvert A_1\rvert + \lvert A_2\rvert$	1.167	14.500	34.333	38.667

Differences of averages

Equation (5.4) gives the gradient in the x and y directions at the pixel centre, assuming a second-order polynomial surface (5.3), but does not give the maximum gradient in the range ±0.5 coordinate units in the x or y directions, nor does it allow for variations from the second-order polynomial. Because of these limitations many other forms of edge detecting algorithms have been developed, many of which use differences from the central pixel, such as

$$G = (2R_{0,0} - R_{1,0} - R_{-1,0}) + (2R_{0,0} - R_{0,1} - R_{0,-1})$$

which is the same as developing the matrix product

$$G = \begin{vmatrix} 0 & -1 & 0 \\ -1 & 4 & -1 \\ 0 & -1 & 0 \end{vmatrix} \times \begin{vmatrix} R_{-1,1} & R_{0,1} & R_{1,1} \\ R_{-1,0} & R_{0,0} & R_{1,0} \\ R_{-1,-1} & R_{0,-1} & R_{1,-1} \end{vmatrix}$$

$$= S \times \begin{vmatrix} R_{-1,1} & R_{0,1} & R_{1,1} \\ R_{-1,0} & R_{0,0} & R_{1,0} \\ R_{-1,-1} & R_{0,-1} & R_{1,-1} \end{vmatrix}$$

Where S could be considered as a weighting array for the contribution of the response values for the pixels surrounding the central pixel in calculating the gradient value, G.

Image smoothing

Image smoothing filters out high-frequency variations in response. It is used to reduce the impact of noise on image data, and to reduce the effect of sharp edges. Image smoothing can be done by means of various averaging techniques where the influence of the pixel and its surrounding pixels are weighted for the calculation of the transformed pixel value. Smoothing can also be done from a Fourier transform of the data, before using this modified Fourier transform to reconstruct the image data.

The matrix multiplication

$$\begin{vmatrix} 0 & 0 & 0 \\ 0 & 1 & 0 \\ 0 & 0 & 0 \end{vmatrix} \times \begin{vmatrix} R_7 & R_8 & R_9 \\ R_4 & R_5 & R_6 \\ R_1 & R_2 & R_3 \end{vmatrix}$$

returns R_5, the original value for the central pixel, whereas

$$\frac{1}{mn} \begin{vmatrix} 1 & 1 & 1 \\ 1 & 1 & 1 \\ 1 & 1 & 1 \end{vmatrix} \times \begin{vmatrix} R_7 & R_8 & R_9 \\ R_4 & R_5 & R_6 \\ R_1 & R_2 & R_3 \end{vmatrix}$$

gives the average of all $m \times n$ response values. Changes in the values in the first matrix, S, will change the weighting of the different pixels on the derived value, and hence affect the derived image.

Fourier transform

The Fourier transform assumes that a data set can be reconstructed as an additive combination of different frequencies, each of a specific amplitude. The contributing frequencies and their contributions or amplitudes, are called the frequency distribution of the curve. Diagram 5.14 shows four contributing frequencies and their contributions, both as frequency curves and as a frequency distribution. The resulting composite curve is also shown. The Fourier transform does the reverse in that it determines the frequency distribution for an original curve. Once the frequency distribution is known, then it can be modified before reconstructing a new composite curve.

In image data there are frequency distributions in both the x and the y directions. The frequencies in each direction have a sine wave profile in that direction in the image space, and a uniform profile at right angles.

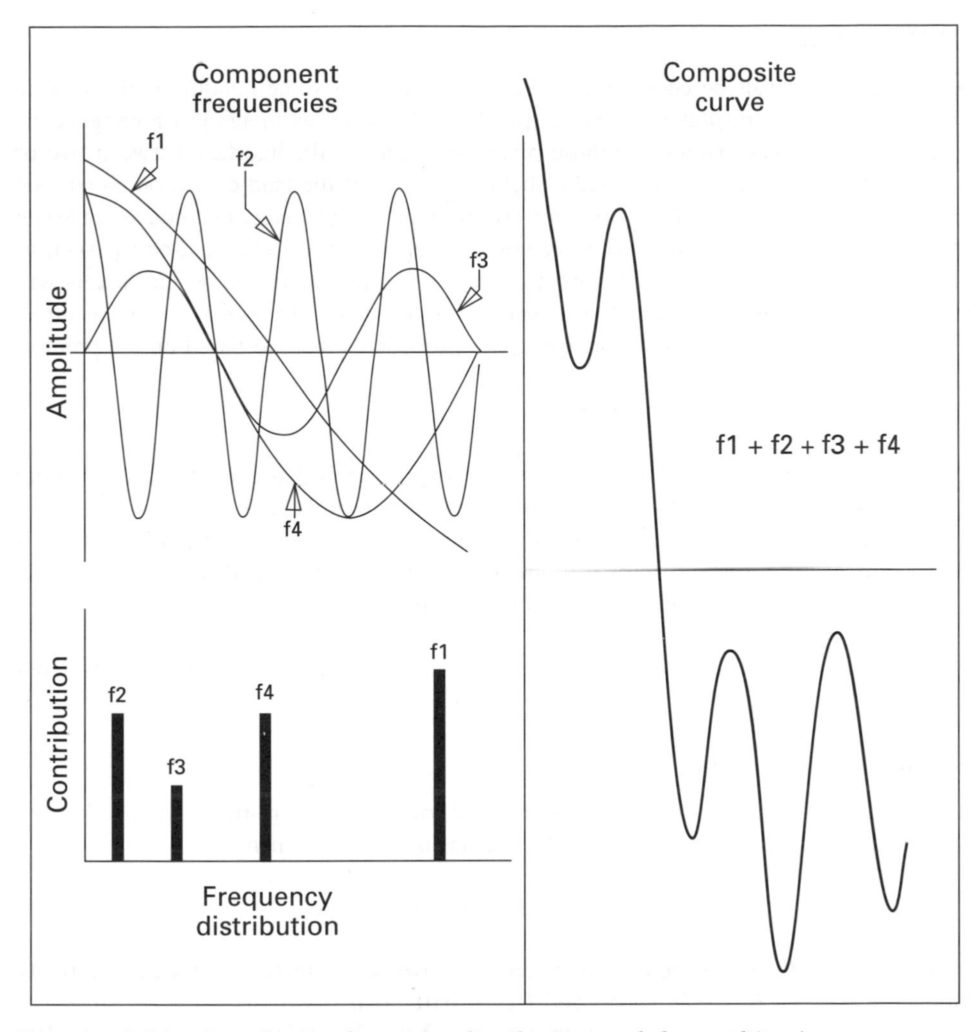

Diagram 5.14 Contributing frequency distributions and the resulting image distribution as seen in a profile through the data.

The Fourier transform takes the picture function, $g(x, y)$ and determines which frequencies are contributing to the variation in response in the x and y directions. It states these contributions as a frequency distribution function, $G(x, y)$ in which the ordinates, x and y, are the frequencies and the abscissae is the contribution (weight or amplitude) of that frequency. It follows that a picture function, $g(x, y)$ can be analysed to get its Fourier transform, $G(x, y)$, and then the original image can be reconstructed by integrating the frequency distributions by their weight or contribution, as specified by $G(x, y)$. The analyst can also choose to reconstruct the image but delete some of the frequency distributions from the reconstruction. Thus deleting the high frequency components will smooth the image, and deleting the low frequency components will sharpen the image or enhance edges.

An example of both image smoothing and edge detection are shown in Photo 5.4(g) and (h).

5.3.10 Mixtures

All image pixels can be considered to be mixtures. With some conditions the mixing occurs at very fine resolutions, and the pixel can be taken as imaging a homogeneous surface. At the other extreme are those pixels that straddle the interface between two or more land-covers; these are accepted as being mixtures of the land-covers on either side of the interface. Between these extremes are the majority of pixels that consist of some proportion of vegetative, soil and water cover. Even within the one land-cover, such as pasture or grassland, a pixel will consist of areas of different covers; pasture pixels will include areas of soil, green and brown vegetation and hence be a mixture of the three. For a variety of tasks it is an advantage to use procedures that are based on the premise that all pixels are mixtures.

There are two situations that occur:

1. mixtures of finite areas of different cover types, or pixels that consist of two or more land-covers that exist on either side of a definite interface; and
2. variations in surface response within the one cover type, where the different components of the surface, such as soil, green and brown herbage and water, have different densities and distributions in the different parts of the surface.

The first case will be considered here under mixtures; the second will be considered under use of the vector classifier in section 5.5.5.

Mixtures of discrete cover types

If a pixel is a mixture of two land-covers, each of which has response values that are distributed in close approximation to the normal distribution, then:

$$R_\lambda = p_1 R_{1,\lambda} + p_2 R_{2,\lambda} \tag{5.5}$$

where p_1, p_2 are the proportions of the pixel area covered by surfaces 1 and 2 respectively (and $p_1 + p_2 = 1$), $R_{1,\lambda} \sim N(\mu_1, \sigma_1)$ and $R_{2,\lambda} \sim N(\mu_2, \sigma_2)$.

If the mean band values for surfaces 1 and 2 are determined from the image data and substituted into (5.5) then the equation, for any pixel, R_λ, contains two unknowns with the condition that the proportions sum to unity. This means that the equation can be solved using one or more wavebands. To derive the equation, substitute the constraint $p_2 = 1.0 - p_1$ into (5.5):

$$R_\lambda = p_1 R_{1,\lambda} + (1 - p_1) R_{2,\lambda}$$

so that

$$p_1 (R_{1,\lambda} - R_{2,\lambda}) = R_\lambda - R_{2,\lambda}$$

and extend this for n wavebands which can be compacted into the one equation to solve for p:

$$p_1 \sum (R_{1,\lambda} - R_{2,\lambda}) = \sum (R_\lambda - R_{2,\lambda})$$

$$\Rightarrow \qquad p_1 = \frac{\sum (R_\lambda - R_{2,\lambda})}{\sum (R_{1,\lambda} - R_{2,\lambda})}$$

and p_2 (= $1 - p_1$) can also be found. The resulting p_1 and p_2 are the proportions of the pixel covered by the two surface types.

In practice the user is normally concerned with the boundary between two covers, the boundary around a specific cover type, or estimating the area of isolated pockets of one surface type (such as water) within other surface types (such as vegetation). In this situation, the cover response values $R_{1,\lambda}$ and $R_{2,\lambda}$ are either taken as the class means as discussed above or are derived from pixel values adjacent to the boundary pixels to be analysed, where the class and the boundary pixels may have been identified previously in classification of the image. This derivation assumes that the covariance arrays for the response distributions for the two land-covers either side of the interface are the same. This assumption will introduce errors into the estimates when the covariance arrays for the two covers are significantly different, as can occur when looking at estimates of (say) water and (say) forest cover. In most other situations the errors introduced by this assumption will usually be small.

If a pixel is a mixture of three or more covers then the proportions p_1, p_2, etc. have to be calculated using least squares as discussed in section 5.5.5.

5.4 Rectification

5.4.1 Theoretical basis for rectification

Rectification is the process of transforming image data so as to create an image oriented to map coordinates in a specific map projection. The original image pixels will contain distortions characteristic of the geometry of the scanner and due to relief in the terrain. Because of these errors scanner image data will not be a good fit to map data, and indeed one image may not be a good fit to another image. However, it is essential to be able to integrate data sets, either by registering an image to another image or to a specific map projection. It is most usual to adopt the projection used for the topographic maps of the area.

Rectification involves three processes:

1. correction for systematic errors (section 5.4.2) to establish the correct internal geometry in the scanner data;
2. absolute orientation to transform the data to a specified map projection by fitting the image data to ground control (section 5.4.3) and then resampling (section 5.4.4) the original scanner data to create rectified image data; and
3. windowing and mosaicing (section 5.4.5) the rectified data to create images that correspond to the map areas, called geocoding the data.

5.4.2 Correction for systematic errors

Whilst the systematic errors vary by type and magnitude between the various sensors, they all normally include:

- earth rotation effects during acquisition of the data;
- mirror velocity errors due to variations in the mirror velocity as it sweeps to acquire the scanlines of data;
- platform velocity error causing displacements along the scanline as it is acquired due to the velocity of the satellite relative to the surface during acquisition;
- distortions in the objective lens can incur small systematic errors in the data;

- variations in the resampling rate, due to variations in the electronics or in the spacing of the detectors in the sensor – these errors are usually very small; and
- detector lag which means that the response signal from the detector lags behind the incident radiation signal. Whilst the effect of this is to introduce radiometric errors in the data, it can affect the perceived positions of features in the image data thereby affecting the positions chosen for those features in the image data.

External systematic sources of error are those that are independent of the internal characteristics of the sensor. The external sources of error include:

- orientation of the sensor relative to the Earth's surface;
- position of the sensor relative to features on the earth's surface;
- rates of change in the orientation and position of the sensor relative to the surface; and
- height displacements created in the imagery due to either relief on, or Earth curvature of, the surface. With some sensors, height differences can introduce significant errors whilst with others Earth curvature can be significant.

The systematic errors should be expressed as a mathematical correction to the image pixel coordinates. When the correction can be calculated then it should be applied to create corrected pixel values. With many errors the shape but not the magnitude is known. When this occurs the correction equation must be included in the adjustment conducted in the next stage. In practice, this is usually not difficult because most of the systematic errors have shapes the same as the correction surface adopted in the next stage.

Theoretically, once the data are corrected for the internal and external errors, the image data should be a perfect fit to ground detail. In practice this will not be the case because:

- the actual nature of all the errors are not known precisely, particularly as some may change with time;
- the cumulative effect of many small errors will not be known where these errors can accumulate or cancel, with possibly both occurring in different parts of the image; and
- the effect of unknown errors which generally will be very small and so can be treated as in the previous case.

5.4.3 Fitting image data to ground control

Once the internal systematic errors have been corrected it is necessary to correct for the external sources of error by fitting the image data to ground control. This involves identification of a sufficient set of control points on both the image and the ground, or on a map and then developing a transformation to estimate ground coordinates for any pixel from the image pixel coordinates. This transformation is usually a first or second order polynomial equation.

If a first order polynomial equation then it is of the form

$$P_x = a_0 + b_0E + c_0N$$
$$P_y = a_1 + b_1E + c_1N \qquad (5.6)$$

where P_x, P_y are the pixel coordinates that have been corrected for internal systematic errors.

A first-order polynomial contains six unknowns. A minimum of three points need to be identified in both the image and on the ground to solve the equations. In practice, a lot more points should be used and the unknowns determined by use of least squares.

If a second-order polynomial equation of the form

$$P_x = a_0 + b_0E + c_0N + d_0EN + e_0E^2 + f_0N^2$$
$$P_y = a_1 + b_1E + c_1N + d_1EN + e_1E^2 + f_1N^2 \quad (5.7)$$

is used, this requires a minimum of six control points to solve the 12 unknowns. A second-order polynomial is the most common form used in rectification. A third- or higher-order polynomial can be used, having more unknowns and requiring more control points for their solution. Generally the higher the order of the polynomial the smaller the residuals, as the curve can provide a better fit to the data. However the trade-off is that the polynomial can also fit distortions in the data due to errors in the control point identification or coordinates.

Many of the known systematic errors can be described by a second-order polynomial function. In consequence, a second-order polynomial will provide a good fit between image and ground control, once the other systematic errors, i.e. those that have a shape which is not a second-order polynomial function, have been corrected. There is only one major geometric error in Landsat data that is not a reasonable fit to a second-order polynomial: the errors introduced due to variable mirror velocity. The only major error in the SPOT data that is not a reasonable fit to a second-order polynomial are those due to height differences in the terrain being imaged. A second-order polynomial is the recommended correction surface for most types of image data.

The parameters in (5.6) or (5.7) are derived by using ground control points to solve the polynomial equations by means of least squares. Whilst the minimum number of ground control points can be selected, this does not protect the result from errors in identification of individual control points, nor does it allow for assessment of the accuracy of the transformation. It is essential that more control points be used than the minimum required, and the equations solved by least squares. Least squares uses all of the equations to derive estimates of the unknowns that are a best fit to all of the equations. The criteria used to decide on best fit is to minimize the sums of squares of the residuals at all of the control points, i.e. to minimize the variance.

Least squares is based on the assumption that the residuals are distributed in accordance with an uncorrelated normal distribution (Diagram 5.15). Systematic errors introduce correlation and so least squares should not be used for data that has not been corrected for systematic errors. This problem is overcome by either correcting for the systematic errors before conducting, or including the systematic errors in, the adjustment as discussed above.

Control points are identified in the image and on the ground or on a map, with the points being distributed evenly across the area of the image, or that part that is to be used. If the transformation contains N unknowns then a minimum of $N/2$ points need to be identified. However, for least squares, a minimum of $5N/2$ control points should be used. For a second-order polynomial of 12 unknowns, at least 30 points should be identified and used to estimate the parameters in the transformation.

Control points that are clearly identifiable on both the image and the map must be used. One difficulty in control identification is the influence of the pixel structure on the images of ground features, which often makes it difficult to accurately identify point features on the image. The best solution is to use intersections of linear features if they

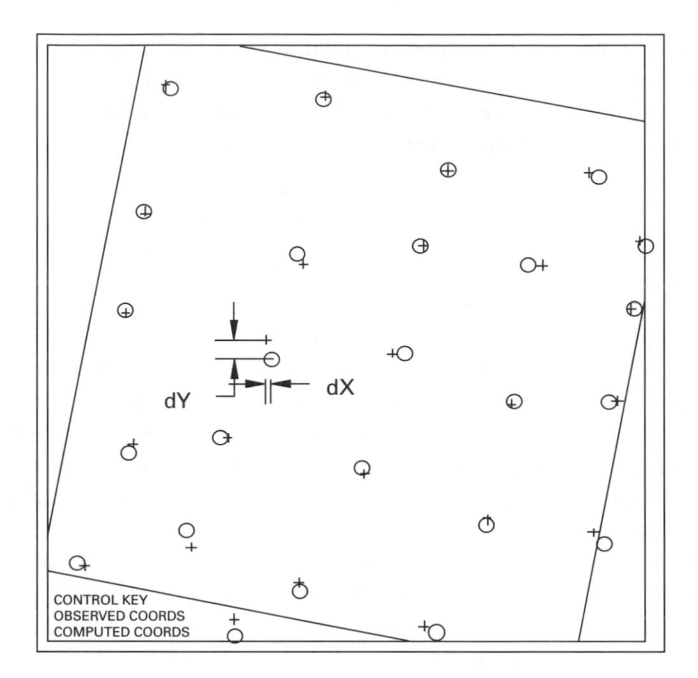

Diagram 5.15 Plot of residuals after least squares adjustment. A pattern in the magnitude and direction of these residuals would be of concern to the analyst. The plot scale for the residuals is much larger than the plot scale for the control points.

exist, or points on a linear feature where they do not. Thus road intersections are good control points, particularly if the analyst can interpolate along the linear features and use the intersections of these interpolations as the control points. The intersections will not usually coincide with the centres of pixels in the image data, so that the coordinates of the control points need to be determined to the sub-pixel level, usually to one decimal place, and to the equivalent resolution in the map data.

The least squares adjustment will provide estimates of the transformation parameters that minimizes the variances of the residuals. These residuals can be printed, and plotted, as shown in Table 5.4. The analyst should inspect these residuals to identify large individual residual values as these will usually indicate errors in identification of a control point or calculation of the coordinates in either the map or the image. The residuals, as plotted in Diagram 5.15, are depicted at two scales. The plot scale of the residuals is much larger (in this case 1:5000) than the scale of the diagram (in this case 1:1 000 000) so as to clearly show the residuals. A plot of this type quickly indicates patterns in the residuals that may be due to incorrect identification of control, or the existence of systematic errors that have not been adequately taken into account. Once the analyst is satisfied that the magnitude of both the sums of squares of, and the individual, residuals are sufficiently small then resampling can be done.

5.4.4 Resampling the image data

The pixels in the original image will be in a different position, orientation, and be of a different size to the pixels to be derived for the rectified image as shown in Diagram 5.16. Resampling is the process of calculating the response values for the pixels in the rectified image by use of the response values in the original image data. The new pixels can overlap a number of the pixels in the original data, the number depending on the size and

Table 5.4 Distribution of residuals from least squares adjustment between control points in the image and the map

Point	Pixel	Line	Est(P)	Est(L)	R(E)	R(N)
1	657.1	275.4	656.5	275.8	−0.6	0.4
2	1394.9	320.3	1395.1	320.6	0.2	0.3
3	1981.3	669.4	1980.2	668.0	−1.1	−1.4
4	2700.6	703.4	2699.7	703.9	−0.9	0.5
5	524.6	880.9	524.7	880.9	0.1	0.0
6	1268.4	1074.0	1269.0	1072.9	0.6	−1.1
7	1847.9	1037.5	1848.3	1037.5	0.4	0.0
8	2365.1	1123.2	2366.5	1123.0	1.4	−0.2
9	2925.7	1030.3	2925.2	1030.8	−0.5	0.5
10	2880.4	1340.7	2880.1	1340.8	−0.3	0.1
11	425.6	1351.9	424.6	1350.3	−1.0	−1.6
12	1142.0	1583.8	1141.3	1585.6	−0.7	1.8
13	1776.2	1555.0	1775.1	1555.0	−1.1	0.0
14	2298.5	1782.1	2298.3	1781.8	−0.2	−0.3
15	2749.7	1775.6	2749.5	1775.6	−0.2	0.0
16	451.0	2030.5	451.3	2031.4	0.3	0.9
17	891.6	1956.0	892.2	1955.9	0.6	−0.1
18	1570.7	2108.2	1570.8	2107.6	0.1	−0.6
19	2165.3	2389.7	2165.5	2390.1	0.2	0.4
20	2530.5	2476.0	2533.2	2477.1	2.7	1.1
21	191.4	2580.2	192.1	2580.0	0.7	−0.2
22	727.9	2413.0	728.3	2411.3	0.4	−1.7
23	1270.0	2704.3	1269.9	2705.0	−0.1	0.7
24	1921.8	2908.9	1920.8	2909.4	−1.0	0.5

Sums of squares = 33.32

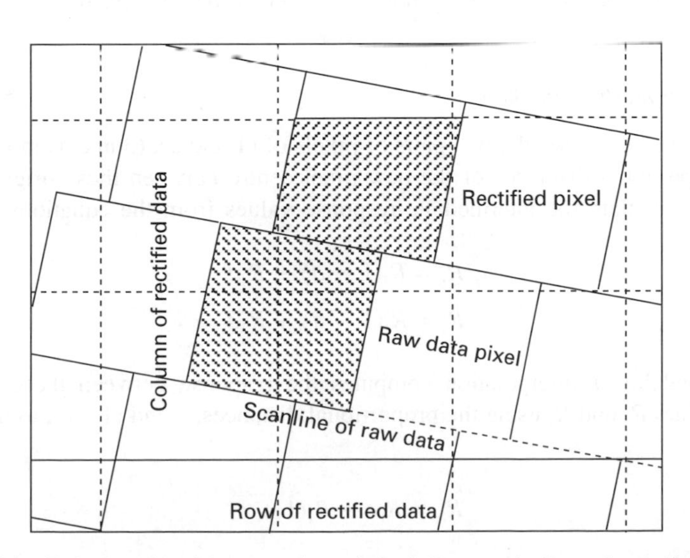

Diagram 5.16 Relationship between the original image pixels and rectified pixels.

Table 5.5 Typical rectified pixel sizes for different sensors

Sensor	*Original pixel size* (m × m)	*Resampled pixel size* (m × m)
AVHRR	1100 × 1100	1000 × 1000
MSS	80 × 80 (at 60 m centres)	50 × 50 or 100 × 100
TM	30 × 30	25 × 25
MOS–1	50 × 50	50 × 50
SPOT MSS	20 × 20	20 × 20 or 25 × 25

position of the new pixels relative to the original pixels. Usually the rectified pixels will be of the sizes as given in Table 5.5. Each size is generally an integer multiple of each smaller size, making the merging of data sets with different sized pixels much easier and faster for data analysis.

Resampling involves selection of pixels in the original image that will influence the resampling process and then calculation of the new pixel response using the response values of these influencing pixels and the location of the new pixel relative to the location of the old pixels. The image-to-control transformation is used to calculate the coordinates of each new pixel in turn in the original data, deriving fractional image coordinates for each pixel. Knowing the original coordinates for the new pixel allows the resampling process to identify those pixels whose response values will be used to calculate the response of the new pixel. Once the response has been calculated, it is entered as the response value for the new pixel and processing proceeds to the next new pixel. There are three commonly used methods of calculating the response of the new pixel from the original image data values: nearest neighbour, bi-linear interpolation and cubic convolution.

Nearest neighbour method

The value of the nearest original pixel to the required pixel location is adopted as the response value for the new pixel. This is fast and cheap, appropriate for theme or classified image data, but it is generally not considered to be suitable for computing image data.

Bi-linear interpolation method

A linear interpolation uses the two pairs of pixels of (1 and 2), (3 and 4) in Diagram 5.17 and the proportional distances of the new pixel centre between these original pixel coordinates to calculate the intermediate response values from the equations

$$R_r = R_3(1 - d) + R_4 d$$

$$R_s = R_1(1 - d) + R_2 d$$

A second linear interpolation computes the response between these intermediate response values R_r and R_s using the proportional distances, c and $(1 - c)$, to the new pixel location.

$$R_N = R_r c + R_s(1 - c)$$

These equations can then be combined to give

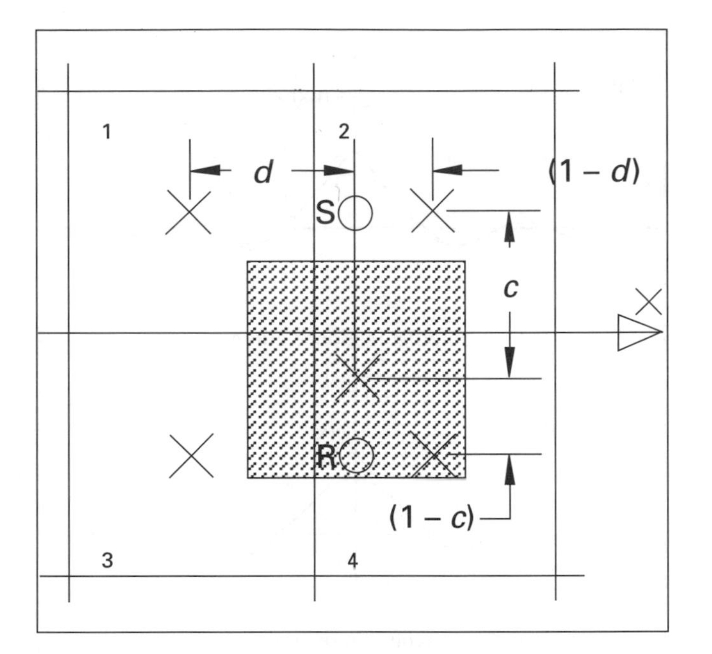

Diagram 5.17 Relationship between old and new pixels in bi-linear interpolation.

$$R_N = R_1 + cd(R_1 + R_4 - R_2 - R_3) + c(R_3 - R_1) + d(R_2 - R_1)$$

Cubic convolution method

Sampling theory states that a continuous signal can be sampled at discrete intervals and the $\sin(\Pi x)/\Pi x$, or sinc, function can be applied to the sampled data to reconstruct the continuous signal, as long as the sampling frequency obeys Shannon's Sampling Theorem (section 5.2.4). The cubic convolution filter is an approximation of the sinc function in which the main positive and the first negatives lobes are retained as shown in Diagram 5.18.

Cubic convolution is accomplished using a 4×4 box of pixels surrounding the location of the new pixel in the original data. A vertical axis through the new pixel location intersects lines along the four scanlines of data (Diagram 5.19). Interpolation values are computed for these four intersections using the data along each scanline in turn in the cubic convolution algorithm. Finally, the response for the new pixel is calculated using the cubic convolution algorithm with the four interpolated values as a transect at right angles to the scanlines. The interpolation formula is

$$\begin{aligned} R_i = R_1\{(2 + 5d + 8d^2 + d^3) \times R_1\} + \\ R_2\{1 - 2d^2 + d^3\} + R_3\{d + d^2 - d^3\} + \\ R_4\{d^3 - d^2\} \end{aligned}$$

where d is a measure of distance between pixel centres as shown in Diagram 5.20.

In practice the assumptions on which the cubic convolution are based are rarely met because the radiances from the surface frequently contain higher frequency components than will be recorded in the image data. This means that the requirements of Shannon's Sampling Theorem are usually not met and so the resampling cannot properly reconstruct the data. Despite this limitation, and the higher processing cost associated with use of the cubic convolution, it is the most commonly used method of resampling image data.

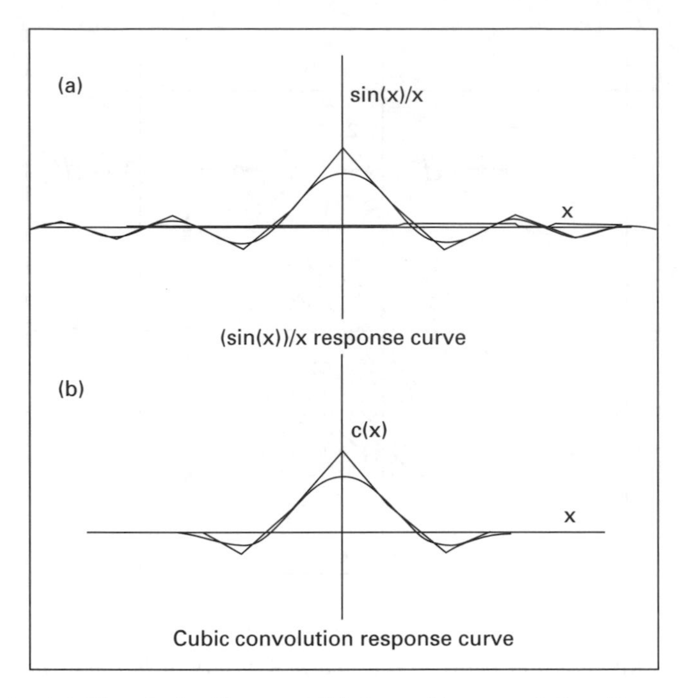

Diagram 5.18 (a) The sin (πx)/πx and (b) the cubic convolution filters.

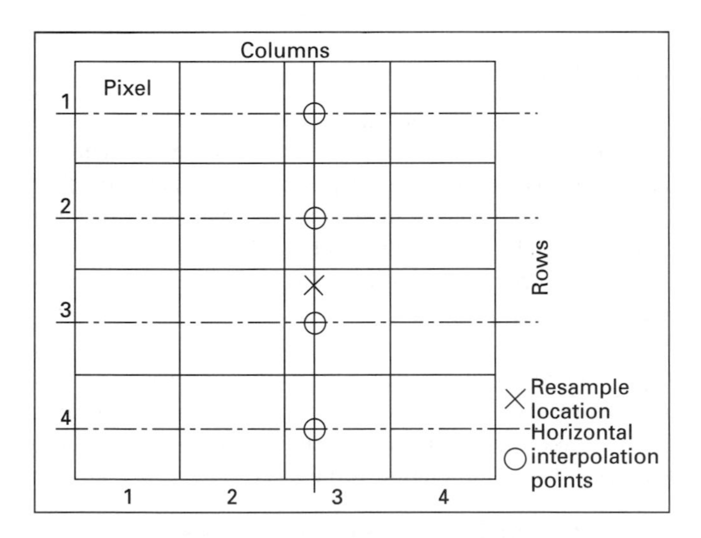

Diagram 5.19 The interpolation axes for cubic convolution resampling.

5.4.5 Windowing and mosaicing

The resampled image data will have pixels of a specified size and interval oriented along the coordinate axes of the map coordinate system, i.e. the lines of data will be parallel to the x axis and the paths will be parallel with the y axis. The rectangular area coverage created during resampling will be partially or wholly covered by image data; some areas within the rectangle may not be covered by image data and will be assigned 0 or 255, i.e. black or white, in the rectified image data. The rectangle may not, however, coincide with the coverage of the topographic maps of the area. For operational purposes, it is

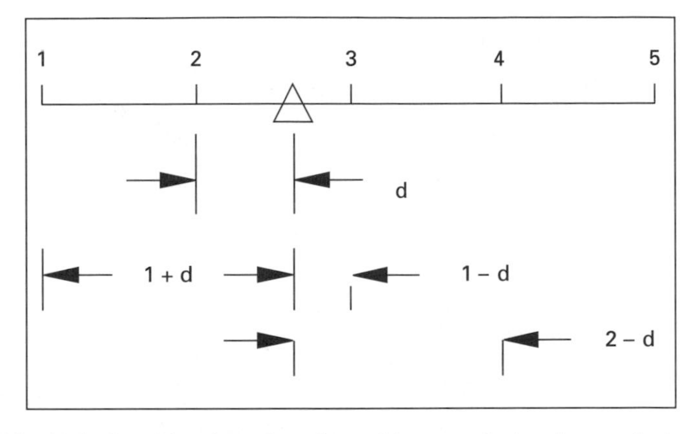

Diagram 5.20 Relationships in using the cubic convolution interpolation equation.

better to create image areas that coincide with map areas. This can be done by mosaicing together image files to cover more than the map area and then defining a window subset of the data that coincides with the map area.

Map areas have boundaries that are constant lines of longitude and parallels of latitude. These boundaries do not provide a straight edge parallel to the map coordinates. The solution is to create a mask of the map boundaries by digitizing those boundaries, and using this mask in a GIS (Chapter 7) to create an image file of the data within the mask. As the mask will often cover two or more images the procedure is to:

1. mosaic the relevant files to cover the whole area of interest; and
2. window this new file using a mask so as to create image files that correspond to map coverages.

5.5 Image classification

5.5.1 Principles of classification

Classification is the process of partitioning an image data set into a discrete number of classes in accordance with specific criteria that are based, in part, on the individual image point data values. In image processing, the image data set are the set of data values for all of the pixels in the image and the image pixel data values are the set of data values associated with an individual pixel. Classification is the process of using the pixel data values to allocate that pixel to one of a discrete number of classes. The pixel data set can contain other data, including data on the texture that uses the data for the surrounding pixels, and added to the data set for the pixels as additional dimensions. The data values used in image processing are usually numerical values, but this is not essential in classification in the more general sense where coded and logical data can form the data set for classification.

It is often the case that pixel response values are such that the pixel could belong to a number of classes. Since that pixel is to be assigned to one class out of the alternatives, some criteria has to be established on which the assignment criteria are to be based. These assignment criteria set decision surfaces that partition the data domain into class sub-domains. A pixel with data values on one side of a decision surface will be assigned to one class; pixels on the other side will be assigned to another class.

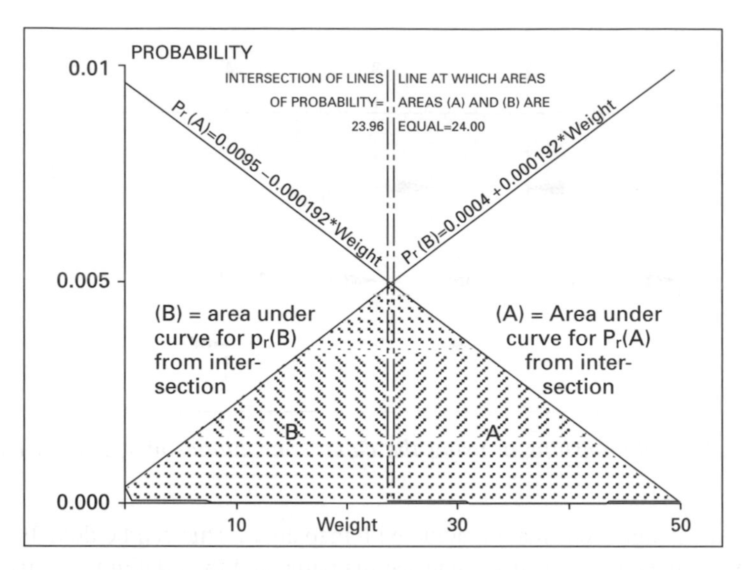

Diagram 5.21 Probability distributions that an animal is either species A or B from its measured weight, with errors of classification (a) and (b) for the decision surface set at w = 23.96.

Consider the case of the weights of 100 animals of two species, A and B, with all of the animals weighing 0–50 units of weight. From this sample of animals, it has been determined that the probability that an animal is of species A can be determined from the weight of the animal provided $0 \leq \text{weight} \leq 50$ as follows.

$$\Pr(A) = 0.0096 - 0.000192 \times \text{weight}$$

and the probability distribution for A is shown in Diagram 5.21.

All of the animals are either species A or B, so the probability distribution of B is given by

$$\Pr(B) = 1 - \Pr(A) = 0.0004 + 0.000192 \times \text{weight}$$

If Pr(A) = Pr(B) then there is equal probability of the animal being species A or B. Below this weight, found to be 23.96 by simultaneous solution of the above equations, then Pr(A) > Pr(B), so the decision is made to assign the animal to species A. Above this weight the probability of the species being B is greater than the probability that it is A and so all animals will be assigned to species B. This value, where the decision changes, is called the **decision surface**, and is defined as the surface in the data domain where the decision assignments change from one class to another.

If a population of animals of species A and B are classified using this probability distribution then some animals will be assigned to the wrong species class, since some species A animals are likely to weigh more than 23.96, and some animals of species B will weigh less than 23.96. This gives rise to errors in classification. These errors of misclassification, shown in Diagram 5.21 as the shaded areas, are of two types:

- **errors of omission** when an area of class A is assigned to class B, or omitted from its correct class A by the classification process; and
- **errors of commission** when areas other than class A are classified as being in class A.

In the case of the animals being classified into species A or B, the categories of errors of omission and commission depend on your perspective. From the perspective of species A, the errors of omission are all animals that are species A but that weigh more than 23.96 and hence are classified as being species B. These errors are depicted as area (A) in Diagram 5.21. The errors of commission are all of those animals that are species B but weigh less than 23.96, or shaded area (B) in Diagram 5.21. With more complex situations, involving many classes, the errors of omission and commission can be constructed as an error matrix as is discussed in Chapter 6.

Misclassification incurs costs and the costs of misclassifying one class may be different to the costs of misclassifying another class. Consider in the case discussed above, that it may be more expensive to misclassify animals of species A than animals of species B. If the costs of misclassification are to be considered then the decision surface needs to be changed so that the costs of misclassification are minimized. Consider the situation where it is more expensive, by a factor c, to misclassify A than B. Consider Diagram 5.21 in which the errors of misclassification of species A and B are the shaded areas A and B respectively. If the costs of misclassification are identical then equal areas (A) and (B) will represent equal numbers misclassified and hence equal costs. If the costs of misclassification are not equal then the areas (A) and (B) need to be adjusted so that

$$\text{misclassification count of A} \times \text{cost}_A = \text{misclassification count of B} \times \text{cost}_B$$

Since the counts are proportional to the area under the curves then:

$$\text{Area}_A \times c = \text{Area}_B$$

where c, the ratio of the costs of misclassification, is equal to $[\text{Cost}_A/\text{Cost}_B]$. If the areas (A) and (B) are adjusted so that

$$\frac{\text{Area}_A}{\text{Area}_B} = \frac{1}{c}$$

then the costs of misclassification will be minimized. To achieve this in the above example requires the decision surface to be shifted to make area (A) smaller and area (B) larger so that the ratio of these areas reflects the ratio of the costs of misclassification. Consider the example discussed above.

$$\text{Area}_A = \frac{1}{2}(50 - w)(0.0096 - 0.000192w)$$

$$\text{Area}_B = \frac{1}{2}w(0.0004 + 0.000192w)$$

so that

$$c(50 - w)(0.0096 - 0.000192w) = w(0.0004 + 0.000192w)$$

If $c = 1$ then this equation can be solved to give $w = 24.00$, different to the point of intersection at 23.96 since the two lines, of Pr(A) and Pr(B) are not symmetrical and hence the two areas (A) and (B) as shown in Diagram 5.21 are not the same. If $c \neq 1$ then the equation can be put in the form of a quadratic equation in w for solution:

$$0.000192[(1/c) - 1]w^2 - [0.0096(1/c) + 0.0004]w + 0.48(1/c) = 0$$

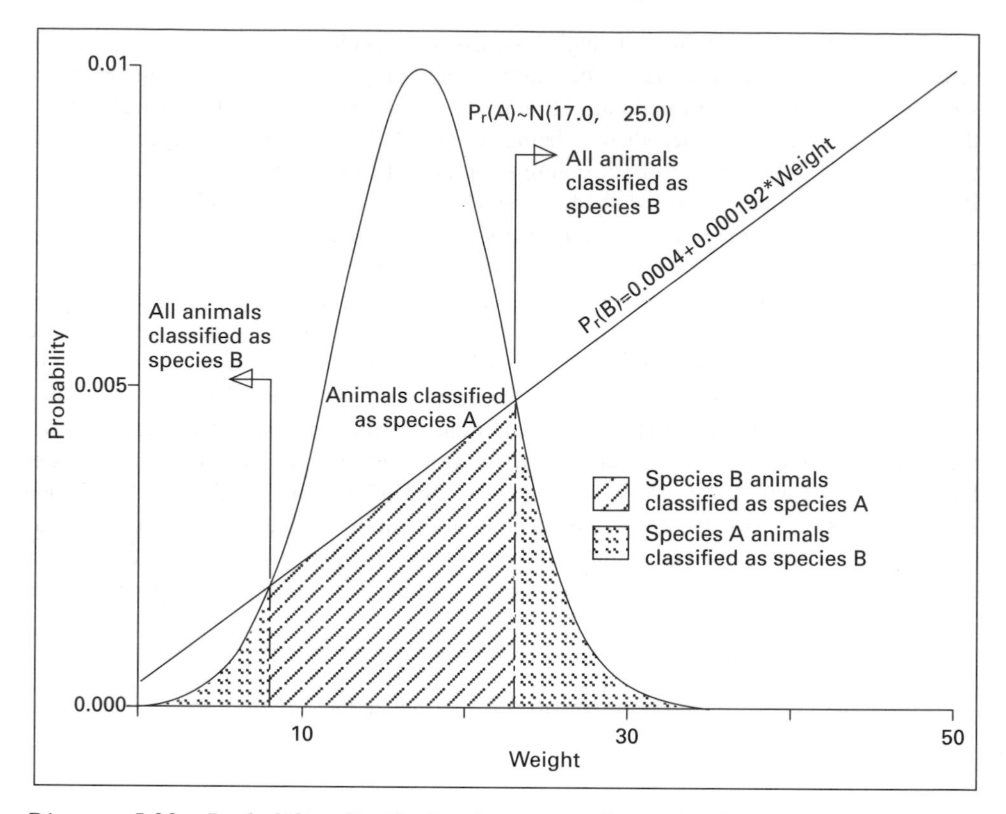

Diagram 5.22 Probability distribution for a second sample of animals from the population sampled to construct the probability distributions in Diagram 5.1.

In the case where $c = 2$, the equation can be solved to give $w = 29.8$ as the decision surface.

Classification involves a number of important concepts, the underlying assumptions and philosophies of which need to be understood to achieve reliable and consistent results.

The use of numerical models to establish decision surfaces

Partitioning the data sets into class domains is usually done by means of a classifier algorithm. The shape and position of the decision surfaces depend on the characteristics of the algorithm. If the algorithm makes assumptions about the distribution of class data that do not accord with how the data are actually distributed in the data set then the decision surfaces will be wrongly positioned and errors will occur in the classification. It is therefore essential to understand the theoretical model on which the classifier algorithm is based, and to be sure that the actual data is a satisfactory fit to this model.

In the earlier example of the weights of 100 animals of two species, the probability distribution was shown for that sample in Diagram 5.21. However a second sample of animals from the population may give a probability distribution of the weights of animals of species A as being normally distributed as shown in Diagram 5.22. This diagram shows the probability distribution of species A as being narrow, with a mean value of 17 and standard deviation of 5. If the probability that an animal is species B is the same as before, i.e. $\Pr(B) = 0.004 + 0.000192w$, then animals weighing less than 8 and more than

23 will be classified as being in species B, assuming that the decision surface is set at Pr(A) = Pr(B).

If the analyst is in doubt as to the suitability of the theoretical model on which the classifier algorithm is based then the goodness of fit of the actual data to the model that forms the basis of the classifier should be determined using appropriate statistical procedures such as the chi-square goodness of fit test.

Use of parameters that are usually derived from field data

The values used for the algorithm parameters significantly affect the shape and position of the decision surfaces. In the example above, both of the algorithms discussed had two parameters: in the first the parameters were a gain and offset; in the second the parameters were a mean and variance. Changing any of these values affects the probability distribution, the position of the decision surfaces and hence the accuracy of the classification.

Estimation of parameter values is usually done by use of a sample of data selected for that purpose, or training data. The sample of data must be representative of the class, i.e. it needs to represent the range of conditions that apply to the class, and it must be of an adequate size to give representative values for the class. For example, in the case of the two samples of animals from the one population as discussed above, there is clearly something wrong if two samples can result in such different probability distributions. The samples may be too small, or they may not truly represent the class conditions. Discussion on methods of collecting field data to establish class parameters is included in Chapter 6.

An important factor to consider in selecting the size of the sample is the sensitivity of the algorithm to slight variations in the derived parameter values. For example the mean of a set of data is relatively insensitive to the inclusion of a few wild or erroneous pixels within a large sample data set, whereas the variance is very sensitive to such data values. If the algorithm is sensitive to the inclusion of such data then care needs to be taken to ensure that erroneous data are filtered out of the data set before determining the class parameters.

Decisions based on 'spectral' data

Classifier algorithms may allocate a pixel to a class on the data set belonging to the pixel, or they can utilize additional data about the pixel, or data derived from the surrounding pixels. Most classifiers limit classification to the use of the spectral data for the pixel and as such are called **spectral classifiers**. However, there are many situations where other data sets can significantly improve classification accuracy. For example, certain classes of tree species may have similar spectral values in the image data, but they may grow in distinct climatic regimes reflected in differences in elevation, gradient and aspect or temperature and rainfall, or a combination of these. In a situation like this, inclusion of additional data sets, effectively as additional bands of data, can improve the classification accuracy.

Another type of information that may improve classification accuracy under some conditions is to include texture as a source of discrimination. This can be done by processing the image data to create a data file of texture which can then be used as an additional band or dimension of data in the classification process.

Partitioning the spectral domain into discrete classes

Once the classification parameters are defined, they create decision surfaces within the data domain such that all pixel values within the one sub-domain, as specified by the decision surfaces, will be assigned to the one class. The resulting classification is uniform within the class sub-domains.

In some circumstances this partitioning is a valid reflection of actual conditions, such as the mapping of land-cover into major classes of cultural, herbage, woody cover, water, snow and ice, and soil when these land-covers are quite discrete. In other situations where the different conditions are not so discrete then partitioning the image in this way introduces artificialities that can introduce errors in the classification. Consider the case of pastures, that can consist of varying amounts of soil, green and brown herbage, woody vegetation and water. The one class of pasture can consist of a wide variation in conditions and contributions by the different covers where these variations can occur in quite gradual and subtle ways. The variations, rather than being discrete steps from one cover type to another, are a gradual change from one to another as part of a continuum of conditions within the one pasture class. Most of the spectral classifiers cannot handle these within-class variations in a satisfactory way for a number of reasons; inadequate training date; a poor fit of the data to the normal distribution; and/or inadequacy of spectral classes.

Inadequate training data occurs because the response values are changing from pixel to pixel due to changes in surface conditions, and so it is often very difficult to define areas that are physically homogeneous and can therefore act as a training area of adequate size to define class parameters.

Whilst the whole of the class area can be defined to establish training areas, the data within this area may be a **poor fit to the normal distribution** due to variations in response caused by variations in physical conditions. One solution is to cluster the data (section 5.6), or create a number of spectral classes within the separate physical classes, in the hope that the clustering process will create classes that are a reasonable fit to the normal distribution. However, the **inadequacy of spectral classes** for estimating condition parameters may arise in a situation where because average class conditions do not match actual conditions across the class. Introducing errors into images of conditions, and into statistical estimates derived from the image data.

One algorithm that has been designed to address these conditions is the vector classifier (section 5.5.5). The most commonly used types of spectral classifiers are: lookup table, parallelepiped, maximum likelihood, and vector, each of which are now considered in turn.

5.5.2 Look-up table classifiers

An array, each dimension of which represents one band of the data to be used in the classification, is constructed with the range of values in each dimension matching the range of values in the corresponding band of data. If the resulting array is too large then groups of values in the original data may be assigned to single values in the array. The simplest way to achieve this is by dividing the values in each band by a constant to give values in the array. Each location in the array is assigned a class assignment symbol. Pixels with a combination of band values coinciding with an array location will be assigned to the class assignment at that location. Classification occurs by reading the pixel data values, converting them to array dimension values, reading the class symbol at the relevant array location and giving this assignment as the classification for the pixel.

A three-dimensional array for data values that can be in the range 1–256 would require an array of capacity 256^3 (= 16 777 216) values. One solution is to reduce the size of the array by dividing the data values by a constant, the resulting values being the new

```
0000000000099999999999999998888888888877777777777777777776666666
0000000000099999999999999988888888888777777777777777777766666666
0000000000099999999999999988888888887777777777777777776666666666
0000000009999999999999998888888888777777777777777777666666666666
0000000009999999999999988888888887777777777777777766666666666666
0000000009999999999999988888888877777777777777775666666666666666
0000000009999999999999888888888777777777777777755666666666666666
0000000009999999999998888888887777777777777775555666666666666666
0000000009999999999998888888887777777777777555556666666666666666
0000000099999999999988888888777777777777775555555666666666666666
0000000999999999999888888887777777777775555555555666666666666666
0000000999999999998888888777777777777555555555555666666666666666
0000000999999999888888887777777777555555555555555666666666666666
0000000999999998888888877777777775555555555555555666666666666000
0000009999999988888887777777777555555555555555555666666666000000
0000009999999988888877777777774455555555555555555666666000000000
0000009999999888887777777777444455555555555555555666000000000000
0000099999998888887777777774444455555555555555555000000000000000
0000099999988888777777774444444455555555555550000000000000000000
0000099999888887777777444444444455555555500000000000000000000000
0000999998888777777744444444444555555555000000000000000000000000
0000999998888777777444444444444455500000000000000000000000000000
0000999988877777744444444444444400000000000000000000000000000000
0009999888777773344444444444442200000000000000000000000000000000
0009998887777333344444444422222200000000000000000000000000000000
0099988877733333344442222222222200000000000000000000000000000000
0099888777133322222222222222222200000000000000000000000000000000
0998887711133322222222222222222200000000000000000000000000000000
0988771111122222222222222222222200000000000000000000000000000000
0987711112222222222222222222222200000000000000000000000000000000
0871122222222222222222222222222200000000000000000000000000000000
0722222222222222222222222222222200000000000000000000000000000000
```

Diagram 5.23 The look-up table classifier after Weigand et al. *(1977) with band 5 along the horizontal axis and band 7 along the vertical axis. 0 – not classified, 1 – dark water; 2 – turbid water; 3 – dark soils; 4 – moderately dark soils; 5 – light soils; 6 – reflective sand and cloud; 7 – dead vegetation; 8 – mixed vegetation; and 9 – green vegetation.*

array locations. If the data values in the above case are divided by 4 then the range of values in each dimension are 1–64 requiring a capacity of 64^3 (= 262 144) locations.

One of the difficulties with look-up table classifiers is in the construction of the table itself. The most common way is to use one of the other algorithms to assign classes to the array locations. However, an advantage of the look-up table classifier is that the decision surfaces in the array can be arranged in ways that defy easy numerical definition. A look-up table constructed using one of the other algorithms can be modified by hand. Alternatively, field data for each class can be digitized and inserted into the array to modify the initial assignment of classes to array locations. As the array is used, and errors are identified in the classifications then these errors can be used to update the array assignment.

Look-up table classifiers are also subject to errors if the image data is affected by atmospheric and sensor effects so that the decision surfaces in the array no longer coincide with the decision surfaces that should apply to the data. This problem can be overcome by calibration of the data.

The most commonly used look-up table classifier with Landsat MSS data is that due to Weigand *et al.* (1977) as shown in Diagram 5.23. It consists of a two-dimensional array, representing Landsat MSS bands 5 and 7, usually of size 64 by 32, so that the

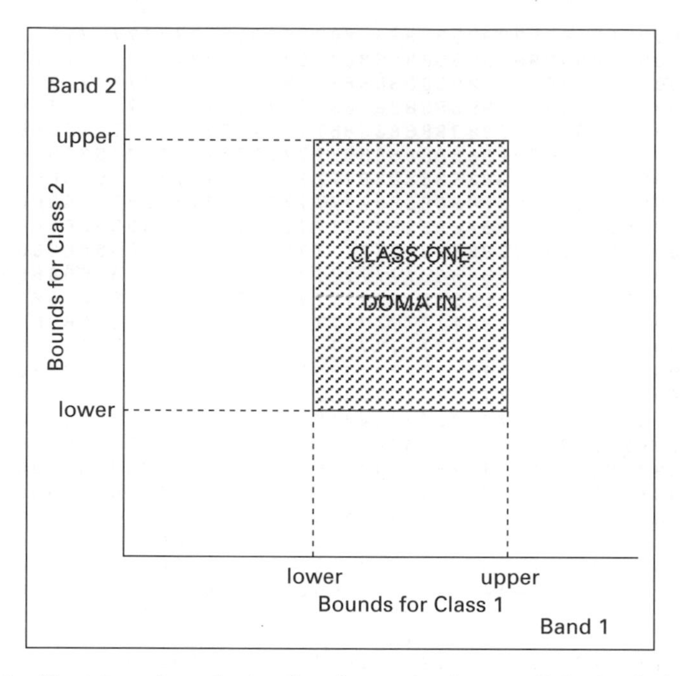

Diagram 5.24 Decisions boundaries for classes in the parallelepiped classifier in two dimensions.

original data is divided by 4 and 2 respectively before interrogation of the array to determine the class assignment.

Classification is done by dividing the Landsat MSS data in bands 5 and 7 by 4 and 2 respectively to give the corresponding array location, and then assigning the pixel to the class that occurs at that location. The classification process is very fast, training is required to create the look-up table, but little training is required for the operator who conducts the classification process. This particular look-up table classifier is suitable for the broad classification of land-covers. Its reliability in conducting this classification would be significantly improved if the image data is radiometrically calibrated before classification.

5.5.3 Parallelepiped classifier

Parallelepipeds are class sub-domains in the image data that are specified by upper and lower bounds in each band in the data, as shown in Diagram 5.24 for two bands of data. In two bands of data the parallelepiped is actually a rectangle, in three dimensions it is a true parallelepiped and in four or more dimensions it defines an area with similar shape characteristics as a parallelepiped in three dimensions. The actual class area depends on the bound values in each band for each class, as well as the decision criteria during classification in relation to areas that are common to two or more parallelepipeds. Consider the two classes in Diagram 5.25 that have overlap in the data domain depicted by the shaded area in the diagram and the values for the pixel located at P. Many parallelepiped algorithms assign pixels with values that are common to two or more classes to the first class to which it is assigned as this reduces processing time. In the case of the pixel at P it will be assigned to Class 1 or to Class 2 whichever class is first in the interrogation. Clearly the sequence of interrogation of the classes will affect the assignment

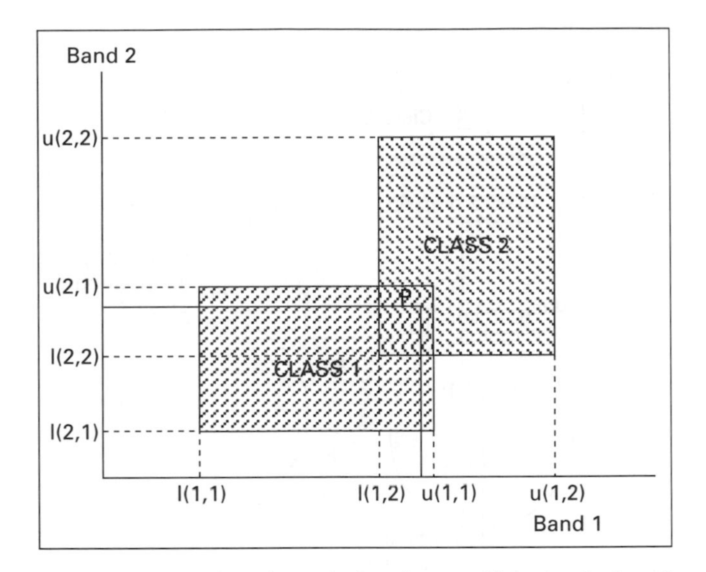

Diagram 5.25 Overlapping class bounds for the parallelepiped classifier.

of the pixels and hence the accuracy of the classification result, a disadvantage of this approach. An alternative is for the classification process to determine how many class domains overlap the specific pixel values, and then use a second criteria to decide between those classes for assignment of the pixel. The usual second criteria is to either allocate the pixel to the class with the nearest mean value, or to allocate P to the class in which values of the class edge are furthest from the values of P.

The normal way to define the upper and lower boundaries of the classes is to use the mean and variance values in each band in turn. The class boundaries are then specified as the mean value ± the standard deviation times k where k is a factor, usually 2 or 3. An alternative is to use the range of data values in the training data, setting upper and lower bounds when a percentage of the data, say 2 or 5 per cent, will be outside of that bound. Classification is conducted by testing, for each class in turn, whether the pixel values fit within the class bounds for that band. If the pixel values fall outside of the class bounds in a band then interrogation for that class ceases and commences for the next class. Interrogation is done for all classes, and the second criteria are invoked if the pixel is found to fall within the parallelepipeds for two or more classes.

The parallelepiped classifier is fast, but it can incur a number of significant errors. The actual classification result depends on the decision process in the algorithm as discussed. The class domains are all parallelepipeds, and if the classes have probability distributions that are quite different to this then large errors can occur. Particularly large errors can occur if the bands are highly correlated.

5.5.4 Discriminant function classifiers

Consider an image containing a number of identifiable classes, from which the analyst has determined the mean and covariance arrays for each class. If a pixel has values that coincide with the mean values for a particular class then the pixel could reasonably be expected to belong to that class – it has a high probability of belonging to the class. As the pixel values move away from the mean value for a class, so the probability that the pixel belongs to that class will decrease. If the pixel is simultaneously moving towards

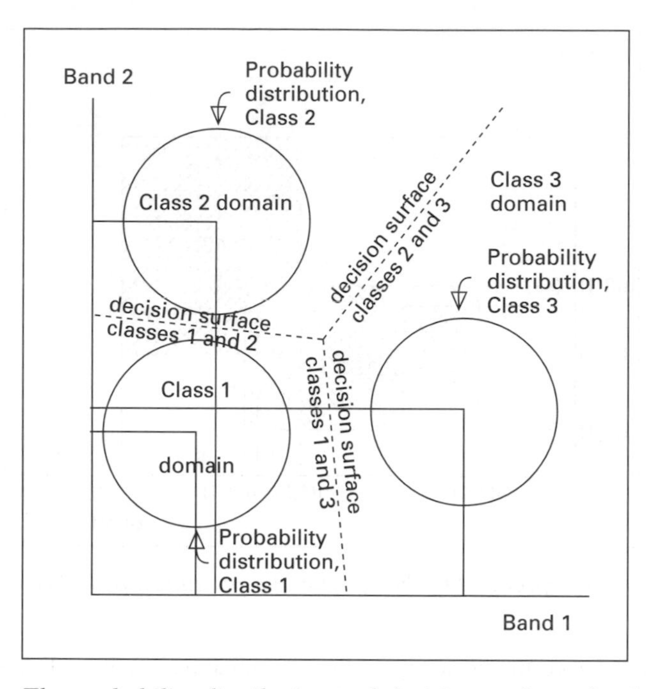

Diagram 5.26 The probability distribution and decision surfaces for three classes in two dimensions for the minimum distance classifier.

another class mean then the probability that the pixel belongs to the second class is increasing. At some stage these probabilities, weighted by the cost of misclassification, will become equal. The decision surface between those two classes is set at this point. On one side of the decision surface the pixel will be assigned to one class and to the other class on the other side of the decision surface.

The simplest decision surfaces are when the euclidean distances from the pixel values to the class means are calculated, and the pixel assigned to the class with the shortest euclidean distance. This type of algorithm is called the **minimum distance classifier** (Diagram 5.26). It is a special case of the **maximum likelihood classifier** where all of the classes are assigned the same variance in all of the bands, and the covariance is assumed to be zero. The decision surfaces for the minimum distance classifier are planar surfaces in the *n*-dimensional space.

In practice, the covariance arrays for the different classes are usually quite different, indicating different variances in the different wavebands in the different classes, and different amounts of correlation. Furthermore, if one class occurs more frequently in an image then there is a higher probability that a pixel is of that class rather than other classes that occur less frequently. Finally, there may be costs associated with misclassifying a pixel as one class relative to misclassifying the pixel as another class. Incorporation of these concepts into classification has led to the development of the maximum likelihood classifier.

Development of the maximum likelihood classifier

Consider that there are k classes, $W_1, W_2, \ldots, W_k$ and a pixel with values in array X_p is to be assigned to one of these classes. The probability of the different classes being in the area sensed by the pixel is $\Pr(W_1), \Pr(W_2), \ldots, \Pr(W_k)$ for each class in turn where

$$\sum_i \Pr(W_i) = 1$$

If the area covered by all classes is equal, or assumed to be so, then there is equal probability of each class coinciding with the area of the pixel, i.e.

$$\Pr(W_i) = \frac{1}{k} \quad \text{for } i = 1, 2, \ldots, k$$

In practice, it is usual to assume equal probabilities for all classes even though this assumption is rarely true. It would be much better to estimate the area of each class within the total area and use these values to calculate the probabilities of the occurrence of the classes for use in the subsequent classification.

The probability density function for a class, $\Pr\{X_p|W_i\}$ is the probability that a point, p with values in array X_p is in the class W_i. The function assumes a distribution associated with the data for the class, and the parameters for that distribution have to be determined before the classification can take place. Usually the normal distribution is accepted as the probability distribution of the data and the parameters for this distribution are the mean vector, μ_i and the covariance array, σ_i^2.

The parameters of the distribution are usually determined either using training data or by the use of clustering algorithms (section 5.6). The first results in supervised classification, the second in unsupervised classification with the relative merits and applications of both being discussed in section 5.7.

Use of the normal distribution means that the probability density function for $\Pr\{X_p|W_i\}$ is multivariate gaussian of the form

$$\Pr\{X_p|W_i\} = \frac{1}{\sigma\sqrt{2\pi}}\exp\left(\frac{-(x-\mu)^2}{2\sigma^2}\right)$$

Bayes' rule states that the probability of class i occurring for a given value of X_p is given by

$$\Pr\{W_i|X_p\} = \frac{\Pr\{X_p|W_i\} \times \Pr\{W_i\}}{\sum[\Pr\{X_p|W_j\} \times \Pr[W_j\}]}$$

Bayes also states that there are costs incurred in misclassification, and that these costs might not be the same in misclassifying the different classes. Construct these costs as a loss function, $L\{W_i|W_j\}$ which is the loss or cost incurred when a pixel is assigned to class i when it is in class j. The loss function is a two-dimensional array of size $k \times k$ when there are k classes. If the loss function specifies unit loss when $i \neq j$ and zero loss when $i = j$, then

$$\begin{aligned} L\{W_i|W_j\} &= 1 \quad i \neq j \\ &= 0 \quad i = j \end{aligned}$$

so that when $k = 3$

$$L\{W_i|W_j\} = \begin{pmatrix} 0 & 1 & 1 \\ 1 & 0 & 1 \\ 1 & 1 & 0 \end{pmatrix} \tag{5.8}$$

Consider an example where a pixel is to be mapped into one of three classes. The probabilities that the pixel is of class i, given the actual pixel values, X_p are given by $\Pr\{W_i|X_p\}$, and calculated to be 0.3, 0.5 and 0.2 for Classes 1, 2 and 3 respectively. Now the weighted probability, with the weighting given by the loss function (5.8) above, is

$$R\{W_i|X_p\} = \sum_{\substack{j=1 \\ j \neq i}}^{k} L\{W_i|W_j\} \times \Pr\{W_j|X_p\}$$

and in the above example

$$R\{W_1|X_p\} = 0.00 \times 0.3 + 1.00 \times 0.5 + 1.00 \times 0.2 = 0.7$$

$$R\{W_2|X_p\} = 1.00 \times 0.3 + 0.00 \times 0.5 + 1.00 \times 0.2 = 0.5$$

$$R\{W_3|X_p\} = 1.00 \times 0.3 + 1.00 \times 0.5 + 0.00 \times 0.2 = 0.8$$

As the intention is to minimize cost then the class with minimum $R\{W_i|X_p\}$ will be chosen, in this case Class 2. It will be noted that for the pixel values used in this case, stored in array X_p, the $\Pr\{W_2|X_p\}$ has the highest value. The zero loss function may not always be the most suitable to use. Consider, for the case above, the use of a loss function of the form:

$$\mathbf{L} = \begin{pmatrix} 0.00 & 0.33 & 0.67 \\ 1.00 & 0.00 & 1.50 \\ 0.33 & 0.50 & 0.00 \end{pmatrix}$$

then $R\{W_1|X_p\} = 0.00 \times 0.3 + 0.33 \times 0.5 + 0.67 \times 0.2 = 0.30$

$$R\{W_2|X_p\} = 1.00 \times 0.3 + 0.00 \times 0.5 + 1.50 \times 0.2 = 0.60$$

$$R\{W_3|X_p\} = 0.33 \times 0.3 + 0.50 \times 0.5 + 0.00 \times 0.2 = 0.35$$

so that Class 1 would be chosen. The procedure then is to compute $[R\{W_i|X_p\}, i = 1, 2, \ldots, k]$ and select the class with the minimum value of $R\{W_i|X_p\}$. Because the selection is done by comparison or relativities, components of the equation that have an equal effect on all values of R can be ignored thereby eliminating unnecessary computation. This is done by defining a discriminant function, $G_i(X)$ for Class i as

$$G_i\{X\} = -R\{W_i|X_p\}$$

then the pixel will be assigned to the class with the largest value in the discriminant function.

$$G_i\{X\} = -\sum_{j \neq i} L\{W_i|W_j\} \times \Pr\{W_j|X_p\}$$

$$= \sum_{j \neq i} L\{W_i|W_j\} \times \frac{\Pr\{W_j\} \times \Pr\{X_p|W_j\}}{\sum_j [\Pr\{W_j\} \times \Pr\{X_p|W_j\}]}$$

and if the normal distribution is adopted as the probability density function for $\Pr\{X_p|W_j\}$ then

$$G_i\{X\} = -\sum_{j \neq i} \frac{L\{W_i|W_j\} \times P\{W_j\} \times e^{-0.5\times(X - U_i)\times(X - U_i)/(\Sigma)^2}}{2\pi \times |\sum_i| \times \sum[P\{W_j\} \times P\{X_p|W_j\}]}$$

Where U_i and Σ_i^2 are unbiased estimations of μ_i and σ_i^2 respectively. The constant components of the denominator can be ignored without affecting the relative discriminant function values between the classes. The generalized form of the discriminant function is:

$$G_i\{X\} = -\frac{\sum_i^2 [L\{W_i|W_j\} \times P\{W_j\} \times e^{-0.5\times(X - U_i)\times\{1/\Sigma_i^2\}\times(X - U_i)}]}{|\sum_i|}$$

It can be seen that there are k products $\Pr\{W_j\} \times \Pr\{X_p|W_j\}$ to be computed and then k multiplications of $L\{W_i|W_j\} \times \Pr\{W_j\} \times \Pr\{X_p|W_j\}$ for each discriminant function so that there are $k \times k$ computations for each pixel. This is a very considerable processing cost that means that the zero loss function is usually adopted to reduce the processing load to k computations per pixel. When the zero loss function is adopted the discriminant function becomes:

$$G_i\{X\} = -1 \times \sum_i^2 \frac{P\{W_i\} \times e^{-0.5\times(X - U_i)\times\{1/\Sigma_i^2\}\times(X - U_i)}}{|\sum_i|}$$

The -1 is common to all of the discriminant functions so that it can be ignored without affecting the relative values of the functions. The discriminant functions can be further simplified by taking the logarithm of both sides:

$$g_i\{X\} = \ln(\Pr\{W_i\}) - \frac{(X - U_i) \times (X - U_i)}{2 \times \sum_i^2} - \ln\left(|\sum_i|\right) \tag{5.9}$$

The form of the discriminant function in (5.9) is inherently quadratic because of the $(X - U_i) \times (X - U_i)$ term. The decision surface can be circular, elliptical, parabolic, hyperbolic or linear depending on the parameter values for the different classes.

Further assumptions can be made which will result in simpler forms of discriminant functions. Consider the case where all variances are assumed to be equal to S, and there are zero covariances.

$$\sum_i = SI$$

where **I** is the identity matrix. Then:

$$g_i\{X\} = \ln(\Pr\{W_i\}) - \frac{(X - U_i) \times (X - U_i)}{2 \times S} - \ln(S)$$

Since S is the same for all classes, the value ln(S) will be the same for all classes and so will not affect the relative values of the discriminant functions. It can thus be discarded. If the remainder of $g_i\{X\}$ is expanded then the X^2 term will be the same for all discriminant functions and can be discarded. Removal of these two components gives discriminant functions of the form:

$$g_i\{X\} = \ln(P\{W_i\}) + \frac{X \times U_i}{S} - \frac{U_i^2}{2S}$$

In this equation the first and last components are independent of the values in the individual pixels and so they can be computed once at the start of the classification to give a constant, K_i for each class so that the discriminant functions simplify to the form:

$$g_i\{X\} = K_i + \frac{X \times \mathrm{U}_i}{\mathrm{S}}$$

This form of the discriminant function is a linear function in X resulting in linear decision surfaces. Since $(X - \mathrm{U}_i) \times (X - \mathrm{U}_i)$ is the square of the euclidean distance between the pixel values and the class means, this algorithm is called the minimum distance classifier, mentioned at the start of the section.

A second simplification of the main form of the discriminant functions as given in (5.9) is to assume that all classes have the same covariance array, Σ_i^2. From

$$g_i\{X\} = \ln(\Pr\{W_i\}) - \frac{X^2}{2 \times \sum^2} + \frac{X \times \mathrm{U}_i}{\sum^2} - \frac{\mathrm{U}_i^2}{2 \times \sum^2} - \ln\left(|\sum{}^2|\right)$$

in which $X^2/(2 \times \Sigma^2)$ and $\ln(|\Sigma^2|)$ are the same for all classes and can be deleted as they will not affect the relative values in the discriminant functions. The discriminant functions simplify to

$$g_i\{X\} = \ln(\Pr\{W_i\}) + \frac{\mathrm{U}_i \times X}{\sum^2} - \frac{\mathrm{U}_i^2}{2 \times \sum^2}$$

This form of the discriminant function is also a linear equation in X so that it also gives linear decision surfaces.

If the probabilities of the classes occurring at a pixel are assumed to be equal, that is $\Pr\{W_i\}$ is the same for all classes, then this can also be eliminated. This assumption is usually made in classification without good justification. Often the reverse of applying different probabilities $\Pr\{W_i\}$ to the different classes is more valid, either as constant values across an area, or varying as a function of conditions. If a pixel is surrounded by pixels of a specific class then there is a higher probability that the pixel belongs to the same class. In the simplest case, the probabilities $\Pr\{W_i\}$, can be taken as the same as the proportions of the classes within the area being mapped even if these proportions are estimated using simple classification methods.

Summary

The four versions of the maximum likelihood discriminant function classifier are:

1. Full maximum likelihood classifier

$$G_i\{X\} = \frac{-L\{W_i|W_j\} \times \Pr\{W_j\} \times e^{-0.5 \times (X - \mathrm{U}_i) \times (1|\Sigma_i^2) \times (X - \mathrm{U}_i)}}{|\sigma^2|}$$

   This form of the algorithm is rarely used because of the high processing costs that would be incurred.

2. Maximum likelihood classifier

$$g_i\{X\} = \ln(\Pr\{W_i\}) - \frac{(X - \mathrm{U}_i) \times (X - \mathrm{U}_i)}{2 \times \sum_i^2} - \ln\left(|\sum{}_i|\right)$$

   This adopts the zero loss function, but still provides a very powerful classification algorithm. It is the commonly used version of a 'full' maximum likelihood classifier.

3. Minimum distance classifier

$$g_i\{X\} = \ln(P\{W_i\}) + \frac{X \times U_i}{S} - \frac{U_i^2}{2S}$$

4. Common covariance classifier

$$g_i\{X\} = \ln(Pr\{W_i\}) + \frac{U_i \times X}{\sum^2} - \frac{U_i^2}{2 \times \sum^2}$$

Characteristics of the maximum likelihood family of classifiers

The characteristics of the maximum likelihood family of classifiers are as follows.

1. The classes are weighted by the probability of occurrence, $Pr\{W_i\}$, of the class. Usually all classes are assumed to have the same probability of occurrence; using better estimates of this probability will have a significant effect when the probabilities of all classes are dissimilar.
2. The classes are weighted by the cost of misclassification.
3. The classifier assumes that the remotely sensed data for all classes are normally distributed, so that deviations from normality in the data will incur errors in the classification. If the classes are of surfaces that are fairly homogeneous in their land-cover characteristics, such as agricultural crops, then the data are usually a good fit to the normal distribution. If the classes contain significant variation in land-cover condition, as can occur in other land-covers including forested areas, water surfaces and rangelands then the data may not be a good fit to the normal distribution.

   The goodness of fit of the data to any specified distribution can be tested statistically. The data should be so tested if the analyst is in doubt about the goodness of fit of the data to the statistical model being used. This analysis can only be done after the collection of field data, and not during the planning stage. To make reasonable judgements during the planning stage requires the analyst to consider the reflectance characteristics of the surfaces in question, and how changes in surface conditions might affect reflectance.

   A method of handling the problem of the data not being a good fit to the assumed distribution is to have the domain covered by a land-cover class represented by more than one spectral class in the classification. Partitioning the training data into a number of spectral classes is usually done by means of clustering algorithms (section 5.6).
4. The decision surfaces of the maximum likelihood classifier are curvilinear whereas simplifications of the algorithm have linear decision surfaces. The magnitude of the covariance arrays can significantly affect the classification when using the maximum likelihood classifier. For example, a class with a small covariance array near a class with a large covariance array may have decision surfaces that define an island within the domain of the second class.

Implementation of the maximum likelihood classifier

Implementation of the linear simplifications of the maximum likelihood classifier can be done by simplifying some of the steps in the implementation of the full classifier. The steps in implementing the full classifier are set out below.

1. Set class parameters, $N\{\mu_i, \Sigma_i\}$, $i = 1, 2, \ldots, n$ for the n classes. Normally this is done by defining training areas for all information classes using the image data as a base map. If the spectral classes for the individual information classes are not a good fit to

the normal distribution, or if the analyst considers that they may not be, then clustering techniques are used to create a number of spectral classes within the information classes of concern.

2. Decide on class weights, $Pr\{W_i\}$. Often these are made to be equal for all of the n classes, i.e. $Pr\{(W_i\} = 10/n, i = 1, 2, \ldots, n$.
3. Set classification threshold. The classification process uses (5.8) to derive values for each $g_i\{X\}$ at each pixel. The pixel is then assigned to the class for which $g_i\{X\}$ is a maximum. Adoption of this practice will assign each pixel to a class since each pixel will give values for each $g_i\{X\}$. However the response of some pixels may be such that their response is a long way from all of the classes, i.e. $g_i\{X\}$ will be large and negative for all classes. In this case the probability that the pixel belongs to any of the classes is low. The operator may decide to set a threshold value as a minimum value of $g_i(X)$ for classification of the pixel.
4. Implement the classification. The classification may be done for the whole area of the image, or for a portion of the image, for example for all areas that are held freehold, or all areas on specific soil types. If the classification is to be constrained to certain conditions in this way then the classification will be conducted within a mask of the areas to be classified. The source of the mask will depend on its purpose, but would normally be derived either from maps or from previous image processing.
5. Conduct accuracy assessment. The conduct of accuracy assessment is an essential component in any classification task, and must be conducted in accordance with strict statistical rules, as discussed in Chapter 6.

Examples of classification are shown in Photo 5.5, in which photo images of the same area but taken at different dates over a two year period using the MOS–1, TM and SPOT sensors have been classified. The classification was based on a number of established training areas used for each image, which were meant to provide representative samples of all the major cover types in the area. However the complexity of the area meant that they might not represent all covers, nor would they necessarily have distributions that are a good fit to the normal distribution, the theoretical basis of the discriminant function classifiers. In consequence, the next step of the analyst was to use these training areas as seed areas for clustering the data (section 5.6) to create a number of spectral classes that were then used to classify the data using the minimum distance classifier. The results of this classification are shown in Photo 5.5 (a), (c) and (e). These spectral classes are grouped together, using the class statistical data, visual inspection of the images and field visits, to create informational classes as shown in Photo 5.5 (b), (d) and (f).

5.5.5 The vector classifier

There are many surfaces that have response characteristics that do not readily fit the assumptions of the previous classifiers. Some of these surfaces have response characteristics that closely approximate a linear surface in the response domain. The vector classifier has been developed to handle situations where the response within the land-cover class varies as a continuous linear distribution due to variations in land-cover condition within the total land-cover class domain.

Derivation of the vector classifier

The power per unit wavelength received by a scanner from an instantaneous field of view (IFOV) covered by a 'uniform' surface of apparent spectral radiance, N_λ at height h is given by

$$P_\lambda = \int_o^\Omega \int_o^{A_c} N_\lambda \eta_o \, da \, d\oint$$

where

N_λ $= N_{\lambda,0}\, t_\alpha + N_{\lambda,a}$
$N_{\lambda,0}$ = spectral radiance of 'uniform surface' at zero altitude in watts per m^2 per steradian
$N_{\lambda,a}$ = spectral radiation scattered into IFOV and emitted by the intervening atmosphere
t_α = atmospheric transmission coefficient at height, h
η_0 = optical efficiency of the scanner
A_c = area of collecting optics of scanner in m^2
Ω = solid angle of field of view scanner in steradians

A uniform surface should be considered to be one where those factors affecting response in the nominated waveband do not vary sufficiently to affect the response recorded by thc sensor. In consequence, the response of a uniform surface will not vary with changes in surface condition, but will change with sensor and localized atmospheric effects. Generally these factors will influence response by less than 1–2 count values in the data of all current sensors. If the IFOV contains two or more uniform surfaces of radiance values $N_{1,\lambda}$, $N_{2,\lambda}$, . . . , etc. then the power per unit wavelength received by the sensor is the sum of the radiances from the areas covered by the individual uniform surface types:

$$P_\lambda = \int_o^{\Omega_1} \int_o^{A_c} N_{1,\lambda} \eta_o da \, d\oint + \int_o^{\Omega_2} \int_o^{A_c} N_{2,\lambda} \eta_o da \, d\oint \ldots \quad (5.10)$$

Integrating for n uniform surfaces gives

$$P_\lambda = (N_{1,\lambda}\Omega_1 + N_{2,\lambda}\Omega_2 + \ldots)\, \eta_o A_c \quad (5.11)$$

where Ω_1, Ω_2, . . . are the solid angles within the IFOV of the scanner, covered by surfaces 1, 2, . . . respectively. If the surface area of the IFOV is S, and the sub-IFOV areas covered by uniform surfaces 1, 2, . . . , n are $S_1, S_2, \ldots, S_n$ then

$$\frac{\Omega_i}{\Omega} = \frac{S_i}{S} \quad i = 1, 2, \ldots, n$$

Substitution in (5.11) gives

$$P_\lambda = [(S_1 N_{1,\lambda}) + (S_2 N_{2,\lambda}) + \ldots] \frac{\Omega \eta_o A_c}{S} \quad (5.12)$$

so that the power received by the IFOV is linearly proportional to the ratios of the sub-IFOV areas of the different uniform surfaces. The satellite digital data have been calibrated so that the pixel response values, R_λ, are linearly related to radiance and hence power received, i.e. for some constant, c,

$$R_\lambda = c N_\lambda \quad (5.13)$$

$$\Rightarrow \quad N_\lambda = \frac{1}{c} R_\lambda$$

Integration of (5.10) gives

$$P_\lambda = N_\lambda \Omega \eta_\circ A_c \tag{5.14}$$

Substitution of (5.14) into (5.12) gives

$$N_\lambda = \frac{1}{s}(S_1 N_{1,\lambda} + S_2 N_{2,\lambda} + \ldots + S_n N_{n,\lambda})$$

Replacing N_λ, $N_{1,\lambda}$, . . . with R_λ, $R_{1,\lambda}$, . . . from (5.13) gives

$$R_\lambda = \frac{1}{s}(S_1 R_{1,\lambda} + S_2 R_{2,\lambda} + \ldots + S_n R_{n,\lambda}) \tag{5.15}$$

The equation shows that the response of a pixel is the sum of the products of the response of each uniform surface by the proportion of the IFOV covered by that surface. This equation is of the same form as the mixed pixel algorithm discussed in section 5.3, in which mixtures of n surface types, each with a normal distribution were considered as a mixture within a pixel. This equation is more important, however, than for just considering mixtures because it provides an explanation for the distribution of data values of certain land-cover classes in Landsat and other scanner response domains. Analysis of various land-cover classes, including the response of pastures and rangelands, show that the response from these covers occupies a linear sub-domain within both Landsat and SPOT response domains. If an area of grasslands is used to create 2–D histograms then analysis of those histograms will show that the data are distributed across a triangular area that is bounded by the response of green herbage, brown herbage and the background soil response. Analysis of rangelands data will show that the response from rangelands has similar characteristics. In both cases the distribution of response for that class is not a good fit to the normal distribution, and variation in response within the class contains information on class condition.

The objective is to develop (5.15) into an algorithm that creates a class domain consistent with that expected for spatially variable surfaces, yet which retains information within that domain which can be used to estimate selected physical attributes of the cover type. The pastures domain of a planar triangle contains significant information on the condition of the pastures. If the pastures are to be classified by the use of a parallelepiped or maximum likelihood classifier then the resultant partitioning of the response domain will not retain information on pasture conditions contained in the data. A classifier that creates a domain that contains all variations in response within the land-cover, yet also contains information on cover condition will thus act both as a classifier of the total class whilst providing information on the condition of the class at each pixel. Let

$$\frac{S_i}{S} = p_i \quad i = 1, 2, \ldots, n$$

so that

$$R_\lambda = \sum_{i=1}^{n} p_i R_i \tag{5.16}$$

with the condition that

$$\sum_{i=1}^{n} p_i = 1$$

Then

$$p_n = 1 - (p_1 + p_2 + \ldots + p_{n-1})$$

which is substituted into (5.16) to reduce the number of unknown proportions, p_i, $i = 1, 2, \ldots, (n - 1)$ to be solved

$$R_\eta = p_1(R_{1,\lambda} - R_{n,\lambda}) + p_2(R_{2,\lambda} - R_{n,\lambda}) + \ldots + p_n(R_{n-1,\lambda} - R_{n,\lambda}) + R_{n,\lambda} \quad (5.17)$$

If we consider m discrete wavebands then there is one equation (5.17) for each waveband:

$$R_1 = p_1(R_{1,1} - R_{n,1}) + p_2(R_{2,1} - R_{n,1}) + \ldots + p_{n-1}(R_{n-1,1} - R_{n,1}) + R_{n,1}$$

$$R_2 = p_1(R_{1,2} - R_{n,2}) + p_2(R_{2,2} - R_{n,2}) + \ldots + p_{n-1}(R_{n-1,2} - R_{n,2}) + R_{n,2}$$

$$\vdots$$

$$R_m = p_1(R_{1,m} - R_{n,m}) + p_2(R_{2,m} - R_{n,m}) + \ldots + p_{n-1}(R_{n-1,m} - R_{n,m}) + R_{n,m} \quad (5.18)$$

where $(R_{i,1}, R_{i,2}, \ldots, R_{i,m})$ is the signature for uniform surface i in wavebands $1, 2, \ldots, m$ and designated by the column vector $\mathbf{A}_i$. $(R_1, R_2, \ldots, R_m)$ is the actual signature in the sensor, or the actual pixel response, designated by the column vector $\mathbf{A}_p$.

Equations (5.18), when put in matrix form, are

$$\mathbf{A}_p - \mathbf{A}_n = \sum_{i=1}^{n-1} \mathbf{p}_i(\mathbf{A}_i - \mathbf{A}_n)$$

Rearranging this gives

$$\sum_{i=1}^{n-1} \mathbf{p}_i(\mathbf{A}_i - \mathbf{A}_n) = \mathbf{A}_p - \mathbf{A}_n$$

that can be stated in the form:

$$\mathbf{AU} = \mathbf{R} \quad (5.19)$$

where $\mathbf{A}$ is the $(n - 1) \times m$ rectangular matrix derived from $(\mathbf{A}_i - \mathbf{A}_n)$ $\mathbf{U}$ is the $(n - 1)$-column matrix of unknown proportions, p_i $\mathbf{R}$ is the m-column matrix of $(\mathbf{A}_p - \mathbf{A}_n)$.

Equation (5.19) is a matrix of m simultaneous equations containing n–1 unknowns that can be solved if $n - 1 \leq m$. These equations can be solved to give a unique solution when $n - 1 = m$. The least squares method of solving these equations, when $n - 1 \leq m$, finds a 'best fit' solution of all equations that satisfies the criteria that the sums of squares of the residuals will be a minimum. This could be visualized as choosing the solution that minimizes the euclidean distance between the actual pixel response in m-dimensional space, and the surface defined by (5.19).

To solve these equations by least squares, the equations need to be rearranged so as to create $(n - 1)$ simultaneous equations. Let $\mathbf{B} = \mathbf{A}^T \times \mathbf{A}$ and multiply both sides of (5.19) by $\mathbf{A}^T$ to give:

$$\mathbf{A}^T \times \mathbf{A} \times \mathbf{U} = \mathbf{A}^T \times \mathbf{R}$$

Substitute B in this equation and transfer to the right hand side to give

$$\mathbf{U} = \mathbf{B}^{-1} \times \mathbf{A}^{T} \times \mathbf{R}$$

where $\mathbf{A}^{T}$ is the transpose of matrix **A**, and **B** is a $(n - 1) \times (n - 1)$ square matrix. Then

$$\mathbf{U} = \mathbf{B}^{-1}\mathbf{A}^{T} \times \mathbf{R}$$

where $\mathbf{B}^{-1}$ is the inverse of **B**. Let

$$\mathbf{C} = \mathbf{B}^{-1}\mathbf{A}^{T}$$

then

$$\mathbf{U} = \mathbf{CR} \tag{5.20}$$

where **C** is a $(n - 1) \times$ m rectangular matrix.

Array **C** will be constant if the values in array $\mathbf{A}_i$, $(i = 1, 2, \ldots, n)$ remain constant. Computation of the proportions in **U** for a pixel of response values $\mathbf{A}_p$ requires determination of **R**, where $\mathbf{R} = \mathbf{A}_p - \mathbf{A}_n$ and the matrix multiplication **CR**.

The shape of the figure defined by (5.20) is linear in m-dimensional space. The vertices of that figure are n node points with response values as given in $\mathbf{A}_i$, $(i = 1, 2, \ldots, n)$. The position of any point on, or in, the figure is precisely specified relative to the nodes by the proportion vector **P**, where **P** is the proportions in matrix **U** plus $\mathbf{p}_n$. The proportions can be described as coordinates of a point relative to the nodes. In a conceptual sense (5.20) is transforming the data from one coordinate system to another defined by proportion coordinates relative to the node positions.

From this description it can be seen that the proportions of pixels with response values identical to a node point will all be zero, with the exception of the proportion of that actual node, which will be one. If a pixel has response values that lie on or in the figure defined by the nodes then (5.20) will derive proportions that will solve all equations in (5.18) without residuals. Often pixels will lie outside of the figure defined by (5.20) in which case the solution will derive proportions for a point on the surface of the figure that is the shortest euclidean distance from the pixel point. This point, called the **footpoint**, F, is the foot of the normal from the pixel onto the surface. The distance from it to the pixelpoint is called the footpoint to pixelpoint distance, or FP distance. The FP distance can be used as a measure of the likelihood that the pixel belongs to the class by setting a threshold value. Below this value the pixel is considered to belong to the class and above this value it is considered to not belong to the class.

If the pixelpoint is on the figure defined by the nodes then all of the derived proportions will be positive and will sum to one. If, however, the footpoint is on an extension of planar surfaces defined by subsets of the nodes as shown in Diagram 5.27, beyond the nodes, then some of the proportions will be negative. Since they will still sum to unity, some of the proportions may be larger than 1. When one or more of the proportions is negative then the least squares process has identified the shortest distance to the nearest planar surface defined by subsets of the nodes, beyond the locations of the nodes that define the planar surface. These extensions are not part of the figure defined by the nodes, and so it is necessary to find a footpoint on the figure rather than on surfaces that are an extension of the figure. This is done by the approximation:

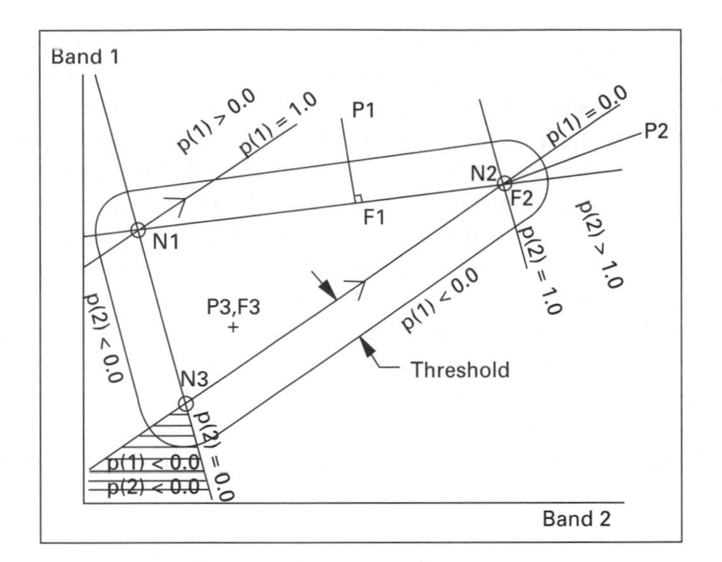

Diagram 5.27 The domain of the vector classifier for three nodes in two-dimensional space and the relationship of pixels to that node in various positions in the two-dimensional space.

1. assign zero to all negative proportions and sum the remaining proportions; and
2. compute new proportions that will sum to unity by dividing the individual proportions by the sum.

The effect of this process is shown in Diagram 5.27, in which there are three nodes that define a planar surface in 2–D space. A pixel located at P_1 will have three derived proportions that are all positive and sum to unity, with a FP distance shown by P_1F_1. A pixel at P_2 will have one negative proportion (p_3) and two positive proportions for the footpoint at F_2. Derivation of the footpoint at F_2' is shown on the diagram.

Characteristics of the vector classifier

The **intrinsic characteristics** of the vector classifier are that it:

1. depends on the uniform surface criteria being met, with the uniform surfaces being distribution independent except that they need to be peaked unimodal so that one value provides a good estimate of the response of the node;
2. defines an $(n - 1)$-dimensional linear surface connecting n operator-defined limiting conditions or nodes;
3. transforms response coordinates into proportion coordinates relative to the n nodes;
4. tests for inclusion in the class by use of the shortest euclidean distance between the pixel values and the surface defined by the nodes; and
5. cannot be solved if the equations are numerically unstable, as occurs when the n nodes define, or approximately define an $(n - i)$-dimensional surface where $i > 1$.

The **extrinsic characteristics** of the vector classifier are set out below.

1. Determination of the node response values can be done either from statistically adequate training sets or by extrapolation from the distribution of the response data

in the response domain. For mixtures of monocultures (section 5.3) there are usually satisfactory opportunities to collect adequate training data to determine mean and covariance array values for each node. In conditions of environmental change, there are rarely the opportunities to identify statistically valid training data sets. Frequently there are no pixels representing the desired uniform surfaces and indeed the class data distribution may be bounded by mixed pixels. In consequence, it is often impossible to define the full nodal domain from training data in situations of environmental change. The alternative is to estimate the nodes by extrapolation from analysis of the data response distribution.

2. The nodes identified by analysis of the spectral distribution of data may not be uniform surfaces, but mixtures themselves, of other uniform surfaces. These nodes can be used to process the data by means of the vector classifier, but the resulting proportions are proportions of proportions. The mixtures involved in the nodes would need to be determined by use of field data. The response of a pixel in relation to node response is given by (5.16) as;

$$R_\lambda = p_1R_1 + p_2R_2 + p_3R_3 + \ldots + p_nR_n$$

If $R_1, R_2, \ldots, R_n$, are themselves mixtures of different surfaces, then

$$R_1 = p_{a,1}R_a + p_{b,1}R_b + \ldots$$

i.e. surface 1 is a mixture of surfaces a, b, . . . , and

$$R_2 = p_{a,2}R_a + p_{b,2}R_b + \ldots$$

and so on, then

$$\begin{aligned} R_\lambda = {} & R_a\{(p_1p_{a,1}) + (p_2p_{a,2}) + (p_3p_{a,3}) + \ldots + (p_np_{a,n})\} \\ & + R_b\{(p_1p_{b,1}) + (p_2p_{b,2}) + (p_3p_{b,3}) + \ldots + (p_np_{b,n})\} \\ & + R_c\{(p_1p_{c,1}) + (p_2p_{c,2}) + (p_3p_{c,3}) + \ldots + (p_np_{c,n})\} \\ & + \ldots \end{aligned}$$

Thus the vector classifier can be used to calculate proportions 1, 2, 3, . . . , and by knowing proportions a, b, c, . . . , from field data, the contribution of each uniform surface to the pixel response can be calculated.

3. The reliability of the classification depends upon the data used to define the nodes being representative of the nodes. The derived nodes will apply to one area, at one time, and in a specific image response domain, unless the data have been calibrated for atmospheric and sensor effects.
4. The surface defined by the algorithm is a linear surface, and variations from this linearity in the class data distribution may incur errors. Unless the reasons for response changes between the nodes are well known and can be accepted as approximating a linear domain then the requirement of linearity needs to be verified for each data set.

Diagram 5.28 shows a high level of correlation between the Landsat MSS bands for ten lakes, where these saline or brackish lakes vary in response due to variations in turbidity. The vector classifier, with suitable nodes, would classify all of the lakes

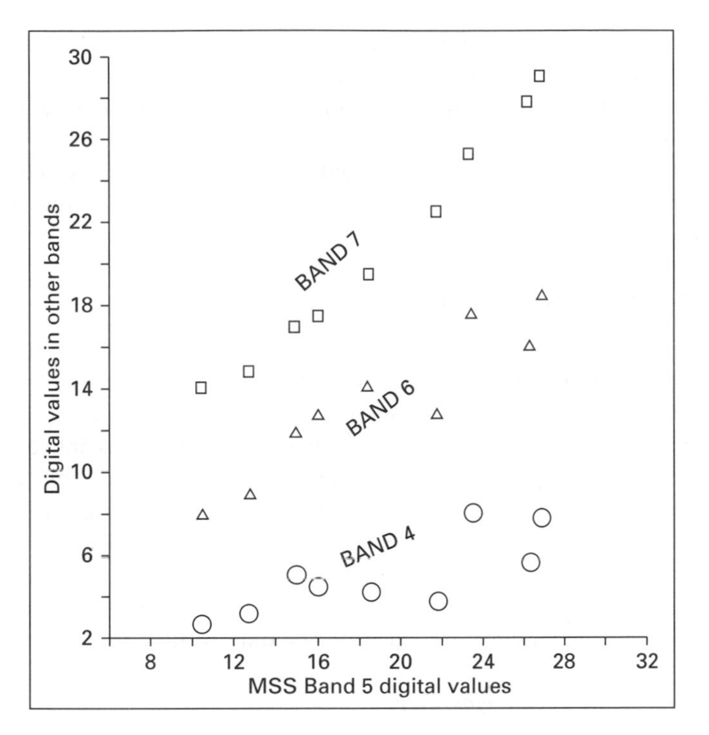

Diagram 5.28 The response characteristics of ten lakes in Western Victoria (Australia), illustrating the linear change in response that can occur across certain land-covers.

as being in the one class, and derive proportion coordinates relative to these nodes. Since variations in the response of these lakes is due to variations in turbidity, it is possible that these proportion coordinates would be significantly correlated with the turbidity levels measured in the lakes.

5. All pixels falling within the class are assumed to be proportions of the specified node surfaces. Without further information the classifier can do nothing else. Situations will occur where either the response from a quite distinct surface type or a mixture of other surface types coincides with a location in the class, causing classification errors.
6. The nodes define a sub-domain of the scanner data domain so a change of one response value in the image data may result in large changes in the estimated proportions. The size of the sub-domain controls the meaningful resolution that can be achieved in transforming the data.
7. The more nodes being processed then, in general, the shorter the distance between the nodes and the more sensitive the processing to changes in response value. With five nodes in Landsat MSS data the inherent noise level in the data often affects the derived proportions. This sensitivity to small changes in response values also applies to node locations; slight changes in node position can significantly affect the derived proportions.
8. If a pixel response is affected by more uniform surfaces than are used in the processing then the resulting proportions will be erroneous. However, if more nodes are used than is necessary then the derived proportions for the unrepresented nodes should be close to zero.

9. The proportions of a cover type that do not change significantly, in area, over a period of time, can be determined using one image and then applied as a known proportion in subsequent images so as to improve the analysis of the proportions of the other surface types.
10. One or more node values can be changed from pixel to pixel, at the cost of extra processing time, as long as the node values are stored in the appropriate form. Soil response data could be used in this way.

Identification of nodes for the vector classifier

Nodes can be identified either by use of training data, or by analysis of the response distribution of the image data. It is very difficult to ensure that training set data are in the vicinity of the extremities of the response distribution of the data of interest to the analyst. Whilst the analyst must strive to have them as close as possible to the optimum position, by accepting that the selected nodes are themselves proportions, the analyst can use field data to convert the calculated proportions to actual proportions. If the image data is in four or less bands (or contain four or less significant components from principal components analysis), then displaying all band pair combinations as 2–D histograms, and using training data mean values, can be used to select optimum node positions. The use of band pair histograms requires a good understanding of how to read the histograms.

Implementation of the vector classifier

The vector classifier can be implemented in three ways.

1. with **standard form** the response values of all of the nodes are known, and the proportions are to be determined.
2. with **known proportion form** the proportion of one type of surface is assumed to be known at each pixel, having been derived using earlier data. Thus the area of water surface, or woody cover may be assumed to be constant over a period of time. In this case then both the proportion and the response of the surface are known and can be subtracted from the response, reducing the number of unknowns.
3. with **variable response form** the response values for a node vary from location to location in the image. An example of this situation is the response of soils that vary across an image. In this case the response of this surface type has to be determined separately and used as input to the processing.

The steps in implementation are:

Step 1. **Create training signatures** which are used to support the analysis in Step 3, or provide the ‘mixed’ nodes for the classification, in which case Steps 3 and 4 are bypassed.

Step 2. **Analyse class data distribution** to assess the distribution of response of the cover to ensure that it is a good fit to the linear model.

Step 3. **Create 2–D scattergrams** of the image data and use the training areas (Step 1) of the dominant cover conditions within the land uses of interest to locate the response of those conditions in the scattergrams. With pastures, this means creating training areas of green and brown herbage, bare soil or soils and of trees if they occur. Typical 2–D scattergrams for Landsat TM data are shown in Photo 5.7.

Step 4. **Define the node response**. If possible simultaneously display all, or up to six scattergrams on the screen, with the option of displaying them either in black and white or density sliced. Superimpose the mean values for the training classes, and other node values on the scattergrams as appropriate. Then decide where the node response values should be. Often the response determined in the training areas is not at the extremity of the domain but is part way into the domain because there are usually few pixels of just one surface type. Either move the node to the selected position or create a new node at the chosen position. Do this in one scattergram at a time, the band values for the node being specified, or modified, as the position is defined in the scattergrams. Amend the position as necessary. Once all of the nodes have been specified, save the node file and check for singularity in the array, that is ensure that the n nodes do not lie on a linear figure that can be defined by $(n - 1)$ or less points. For example, if three nodes lie on a line then the array containing node response values would be singular. If the nodes are singular then either one or more nodes need to be dropped, or redefined.

Step 5. **Specify the threshold for classification** that will define the envelope of the class domain around the linear figure defined by the nodes. All pixels with footpoint-to-pixelpoint distances less than the threshold will be classified as being in the class, whilst those with larger distances will not be included.

Step 6. **Specify output requirements**. The normal output is images of the proportions of the nodes, but additional images that may be chosen.

- **Classification mask** shows those pixels classified or not classified.
- **Distance image**, or image of the distances from the pixelpoint to the footpoint, that form the basis of inclusion or exclusion from the class, can be very helpful in assessing whether the classification is capturing all of the pixels that it should, and what would be appropriate threshold levels. It also gives a good indication of the distances of different conditions from the figure and can be used to assess the goodness of fit of the data to the model.
- **Footpoint response**, or image of the response of the footpoint on the linear figure defined by the nodes, can also be of use.

Step 7. **Specify input controls**. The classification may be constrained to occur within specific classes or mask areas that are controlled by the means of mask images derived from other sources.

Photos 9.3–9.8 show two examples of the use of the vector classifier. In the first example, once it was determined that the response of rice during the growing season follows a linear trajectory from the response of water to the response of green canopy, then the vector classifier was used with two nodes, water and green canopy to define the rice class domain, as discussed in Chapter 9. Changes in position in this domain were recorded by the classifier as variations in the proportion of water and cover response, being depicted in the photograph as variations in greenness–redness.

The second example considered the response of rangelands when the grasses are dry, so that only the woody canopy is exhibiting the typical response of green vegetation. It was determined that the response domain in the rangelands is a close approximation of a linear domain with three nodes, one of which was the response of the woody cover. The vector classifier was used to estimate the proportion of woody cover within each pixel in the image data. Field data were used to develop a regression model to use this estimate to derive estimates of actual woody canopy cover. This model was used to transform the proportion of woody cover image into estimates of actual cover as depicted in

Photos 9.6 and 9.7 as variations in greenness. This process was repeated in 1975 and 1985. Once the percentage cover has been quantified in this way for two dates, the estimates can be compared to derive estimates of change, as discussed in Chapter 9.

5.6 Clustering

Remotely sensed images contain large amounts of data. In order to analyse that data it would be advantageous to know of natural groupings or clusters in the data. Because of the number of dimensions in remotely sensed data, it is impossible to visually identify clusters. Clustering techniques have been developed to partition multi-dimensional data sets into discrete clusters or groupings of the data, in accordance with some criteria.

If the data can be partitioned into a number of clusters, then the statistical parameters describing these clusters can be determined and used as class parameters in classification. In addition to this application, if field data have identified information classes for which the data are not a good fit to the normal distribution, then clustering may be able to sub-partition the data into a number of spectral classes before classification.

Clustering can therefore be used in a number of ways in the classification process: to define spectral classes within informational classes; to identify spectral clusters for classification; and to refine training statistics.

1. An informational class is a class that contains unique information in relation to user needs. An urban planner may wish to discriminate urban from non-urban areas; these represent information classes to that resource manager, so information classes are thus related to user needs. There is no inherent reason why these information classes will fit the assumed distributions of the various classifiers. Clustering can be used to look for clusters in the data for an informational class, that partition the data into spectral classes that may be an acceptable fit to the normal distribution. If this is achieved then those classifiers based on the normal distribution can be used to classify the informational class into a number of spectral classes which can then be grouped to map the informational class. This process is an application of clustering within supervised classification, since the classification process is supervised through definition of training areas to control the development of class parameters.
2. The objective is to develop class parameters for the classification without interference by the analyst, i.e. unsupervised classification. The optimum situation would be to identify natural groupings in the data, but this is unlikely because natural groupings of information to the user need not create natural groupings in the response data. The default position is to search for spectral groupings that allow classification of the image data into classes that can then be grouped into useful informational classes.
3. Training areas can include the occasional 'wild' pixel, and clustering can isolate these pixels from the training data. Clustering processes to refine training statistics are a specific application of clustering.

5.6.1 Clustering criteria

The notion of clusters implies the notion of groups of data such that the average distance between values in a group is much smaller than the distance between groups. The usual criteria in clustering is thus to maximize the difference between the between-class and within-class distances. Most clustering algorithms also assume that the required clusters need to be a reasonable fit to the normal distribution. The usual methods of implementing these criteria are by minimizing the Variance or by maximizing the Divergence.

Minimizing the variance

Let n_i be the number of data values in cluster i of k clusters, $x_{j,i}$, $j = 1, 2, \ldots, n_i$ being the set of n_i data values in the cluster, and m_i be the cluster mean value. Hence

$$m_i = \frac{1}{n_i} \sum_{j=1}^{n_i} x_{j,i} \tag{5.21}$$

and the sums of squares is

$$S = \sum_{i=1}^{k} \sum_{j=1}^{n_i} (x_{j,i} - m_i)^2$$

where m_i are values that minimize the sums of squares within the clusters. The optimal assignment of a new data value, or pixel value, is to that class which minimizes S. If the new data value is shifted from one class to another and reduces S in the process, the pixel is assigned to the second class. This process is repeated until the pixel is assigned to that class which minimizes S. Once a pixel is assigned to a class its value is used to recompute the mean for that class, to ensure that the mean will minimize the sums of squares in the class. In consequence the means for the classes will gradually shift as new data values are assigned to the different classes. It is for this reason that this approach to clustering is called the **migrating mean** clustering procedure.

Since the means migrate in a pass through the data, some of the early data might well be better assigned to other classes than to that class to which they were initially assigned. This problem is solved by passing through the data a number of times, until the changes in the class statistics become negligible or tend to oscillate from one pass to the next.

The minimum variance criteria work well if the data form compact clouds which are well separated from each other. This criteria can, however create artificial clusters when a cluster has a large scatter and hence the sums of squares may give lower values of S by splitting the cluster into a number of smaller and tighter sub-clusters.

When considering multi-dimensional data, as is typical with remotely sensed data where each data value $x_{j,i} = (v_1, v_2, \ldots v_l)$ in 1 wavebands, then the above approach has to be expanded. If m_i is now defined as the mean array for cluster i and m is the mean array for all data values then

$$\mathbf{M} = \frac{\sum_{i=1}^{k} n_i m_i}{\sum_{i=1}^{k} n_i}$$

and

$$\mathbf{S}_t = \sum_{i=1}^{k} \sum_{j=1}^{n_i} (x_{j,i} - \mathbf{M})(x_{j-1} - \mathbf{M})^T$$

is the total sum of squares for all data points about $\mathbf{M}$. Expanding this gives

$$\mathbf{S}_t = \sum_{i=1}^{k} \sum_{j=1}^{n_i} \{(x_{j,i} - m_i) - (m_i - \mathbf{M})\} \{x_{j,i} - m_i) - (m_i - \mathbf{M})\}^T$$

$$= \sum_{i=1}^{k} \sum_{j=1}^{n_i} (x_{j,i} - m_i)(x_{j,i} - m_i)^T + (m_i - \mathbf{M})(m_i - \mathbf{M})^T$$

since . . . the cross-products in each class will sum to zero. Hence

$$\mathbf{S}_t = \mathbf{S}_w + \mathbf{S}_b$$

where $\mathbf{S}_w$ is the within-class scatter and $\mathbf{S}_b$ is the between-class scatter. Now the total scatter matrix does not depend on the allocation of the individual pixels to the classes, but both $\mathbf{S}_w$ and $\mathbf{S}_b$ do depend on this allocation. In consequence if one of the matrices is reduced then the other matrix will be increased. The normal criteria is thus to reduce the within-class scatter, $\mathbf{S}_w$, which automatically increases the between class scatter, $\mathbf{S}_b$. This is done by assigning each pixel to each class in turn, and allocating the pixel to that class that minimizes $\mathbf{S}_w$.

The size of the matrices, $\mathbf{S}_w$, are measured using either the trace or the determinant, where the trace is the sum of the diagonal elements of the covariance array.

Maximizing the divergence

Consider maximizing the ratio $(\mathbf{S}_w)^{-1}\mathbf{S}_b$ of the between- to within-class distances. The eigenvalues of an array are the variances along the semi-major and semi-minor axes of the data distribution depicted in the covariance array. The eigenvalues thus measure the maximum value of this ratio and so the criterion is to measure the trace of the eigenvalues derived from the ratio matrix $(\mathbf{S}_w)^{-1}\mathbf{S}_b$.

5.6.2 Clustering of training data

Training areas, whether they are regular or irregular in shape, can include pixels that do not properly belong to the class. Inclusion of these pixels will usually have little effect on the class means as they will represent only a small proportion of the pixels in the training data set. However they will have a large impact on the covariance array for the class and hence on the location and shape of the class decisions surfaces.

One way to minimize this problem is to cluster the data in the training area using clustering processes. Another way is to assume that the class in the training area is normally distributed, and reject pixel values that diverge from the class mean by more than some specified limit. This approach will reject all pixels, both legitimate and otherwise that are further than this limit from the mean value for the class. To ensure that the number of legitimate pixels being rejected is small, the threshold has to be set at quite a distance from the mean, at least two standard deviations and preferably three from the mean.

A better alternative is to use the deviations, $d_{i,j}$, defined as

$$d_{i,j} = \sum_{p=1}^{l}(x_{j,p} - x_{j,p})^2 = (x_i - x_j)(x_i - x_j)^T$$

where l is the number of bands in the data, x_i, x_j are data arrays for pixels i and j in a training set of n pixels such that $j > i$, i.e. deviations are computed for each pixel from all subsequent pixels in the data set so as to not duplicate computation. There are therefore $n' = n(n - 1)/2$ deviations to be computed in each training set.

Now x_i and x_j are normally distributed so that $z = x_i - x_j$ is normally distributed and $d = z.\, z^T$ is a gamma distribution with $\alpha = 0.5$ and $\beta = 4\sigma^2$ where σ^2 is the average variance for the data in the training set.

The clustering process is thus to compute the mean and covariance array for the data within the training area. For each pixel in turn the deviations, $d_{i,j}$ is calculated and the probability of this value occurring in the gamma distribution, $\Gamma\{0.5, 4S_i\}$ is tested. If the probability is low, typically less than 2 per cent, then the pixel is rejected from the class. After a pass through the data the retained pixels are used to recompute the class statistics.

5.7 Classification strategies

The analyst must not only know the characteristics of the tools that are available to him, but must develop a plan to extract the required information from the data. The objectives of this plan are to extract the information required in the most cost effective manner. This section analyses the factors that must be considered in the development of such a plan.

5.7.1 Types of classes

There are two important types of classes to consider in remote sensing: information and spectral classes. Information classes are those that convey information on the physical world. The criteria used to separate one informational class from another are usually quite different from the criteria that will separate statistical or spectral classes in image data. In consequence, information classes will not always have spectral data that are a good fit to the normal distribution.

Consider land-cover or land-use mapping. Land-covers are the physical covers at a point in time, such as, herbage, soil, woody cover, water, and cultural features. These classes are based on surficial physical attributes that are one of the dominating causes of changes in response or reflectance in remotely sensed data at most stages throughout the year. In consequence, individual remotely sensed images can usually provide information on the required land-covers at some level of resolution and accuracy.

Land uses are the use to which land is put by man and include cropping, pasture, forests, reservoirs, lakes, streams, urban commercial activities and waste land. Land uses involve land-covers and how they are used by man over time. Land-use information cannot automatically be gleaned from single remotely sensed images, since the one land-cover, such as herbage, can occur on different land uses, such as cropping, pasture, waste land and urban/commercial. Land use is thus a seasonal or longer-term activity and mapping of land use must reflect this longer-term interval. Land uses have characteristic land-cover patterns over time, and in consequence the usual way of expressing land uses is through these changes.

Spectral classes are spectral sub-domains within which the data are distributed in close accordance with a probability distribution appropriate to the classifier being used.

If the physical land-cover of an area is reasonably homogeneous then the data could be expected to be a reasonable fit to the normal distribution. If, however, there are systematic influences, such as major changes in surface conditions, within the land-cover then the data might not be a good fit to the normal distribution. Under some circumstances the data may be distributed in a manner suitable for the vector classifier. There will be circumstances where it is not even a good fit to this algorithm. If the land-covers are not normally distributed then it may be necessary to partition the data into a number of spectral classes.

Consider the complication of using imagery taken on a number of occasions for the purpose of using seasonal variations to improve the classification accuracy. Even if areas are identified as being reasonably homogeneous and hence a good fit to the normal distribution in an image, different areas may change in response in different ways from one image to the next, and hence the land use may not be homogeneous over time.

Classes are less likely to be a good fit to the normal distribution in multi-temporal data than in an individual image. Yet in this and in other situations it is not easy to revert to a look-up table classifier because of the difficulty of constructing the table and the limitations of such classifiers. Consequently most classifications are done using parametric classifiers and so it is necessary to arrange the data so that the assumptions of the selected parametric algorithm are met within acceptable limits.

5.7.2 Selecting classes and classifiers

We have seen that there will be optimum decision boundaries in the spectral data for the information classes, but that it is often difficult to define them adequately. This problem can be made more tractable if we choose imagery in wavebands and at a time such that these boundaries are most clearly discernible. It is important to choose imagery when there is maximum spectral discrimination between the information classes. It is not important to ensure good discrimination between the spectral classes within the one information class because this does not affect the information derived.

The imagery must be chosen in anticipation, from assessment of physical conditions in the field. This can be done if the analyst has a good appreciation of the distribution of data for the different information classes in the spectral data. To summarize then, analysts need to be familiar with the spectral characteristics of the information classes to choose the most suitable imagery, and then they need to choose imagery in appropriate wavebands and at the best time to maximize the discriminability of the information classes.

The information classes in this imagery may not be a good fit to the various parametric classifier algorithms, in which case it is necessary to partition the information class domain into spectral class sub-domains that are a satisfactory fit to one of the classifiers. The usual way of doing this is to create information class training areas that are both large enough and distributed across the area so as to be representative of the class. It is poor statistical practice to take one training area as a sample to represent a class. It is much better practice to select training areas for each class distributed across the whole area to be classified.

Once this data is accumulated it should be analysed for goodness of fit to the distributions of the various classifiers that are available. If the data are a good fit to one of the classifier distributions then that classifier should be used for the classification. If the data for an information class is not a close fit to any of the distributions then the class data should be partitioned into spectral classes using clustering techniques.

5.7.3 Improving interclass discriminability

The more discrete the information classes in the spectral data, the more reliable and accurate will be the classification. If the discrimination is particularly good then simpler, and cheaper algorithms may be used, yet achieve the required accuracy standards. Discrimination is increased by using imagery acquired in appropriate wavebands and taken at the right time as discussed in section 5.7.2. It is also increased if the confusion between different land-covers is reduced. Confusion can be reduced in various ways.

Use of masks It is possible to specify areas that are not to be classified, or are to be processed in different ways, by identifying the boundaries of these areas on a map or image and digitizing this information to create an image mask. For example, in mapping irrigation, it would be logical to mask out dams, rivers and other water surfaces that are clearly not irrigation areas, but which will respond very similarly to irrigation areas in remotely sensed imagery in some seasons. A mask can be used to specify areas that are *not* to be classified, to reduce confusion and/or the cost of the classification. A mask can also be used to define areas that *are* to be classified using different techniques, for example a different algorithm, different spectral classes or even different data sets. For example, localized rainfall may have caused quite different surface conditions, and management response, in some parts of the image than in others. It may be appropriate to partition the images on the basis of differences in condition and conduct quite different classifications in both areas. Different land-covers or land uses may be controlled by topographic effects so that inclusion of appropriate topographic data (slope, height, aspect) may significantly improve classification accuracy. Cloud cover may also convince the analyst of the advantage of using different data sets for different subimage areas.

The use of masks is very much an integration of local and environmental knowledge in the classification process. By use of the masks the analyst is achieving greater control over the classification process.

Use of temporal change Different land uses are represented by different land-covers and land-cover conditions, at different stages in the season. Careful selection of temporal imagery might assist in discriminating between the various land uses, such as crops and pasture. The temporal changes that occur in spectral data for the one land use will often vary over the area being classified, creating different classes for the different sub-areas. Consequently, the more temporal images that are used in the classification, and the larger the area, the more spectral classes are likely to be required, significantly increasing processing time and costs.

Incorporation of additional data sets into the classification Classification is generally taken to use spectral data, controlled by masks as appropriate. However other types of data, useful in increasing inter-class discriminability, can be included in the classification process. Some examples of other types of data are:

- a measure of texture – with some land-covers exhibiting higher textural variability, the use of which may improve discriminability between two classes;
- information on topographic height, slope, aspect, temperature, rainfall and other parameters that influence land use may improve classification accuracy; and
- classification of adjacent pixels. If a majority of adjacent pixels are classified into one class, then that may suggest the classification of the central pixel. Whilst this approach needs to be used with care, as there are instances of stands of trees in the centre of cropping fields and dams in pasture fields, the probability of classifying these adjacent pixels should often be a factor in considering the classification of the central pixel.

5.7.4 Cost minimization

A number of the techniques discussed above will improve classification accuracy, and will also reduce costs. There are other techniques that reduce costs, but their use needs to be assessed, particularly as they may affect the accuracy of the derived information. These other techniques are set out below.

1. Costs can be reduced by using only the **essential data set**, and not data values that do not contribute significant information on the classes. Contribution is usually seen as a function of correlation, so if the data values in two bands are highly correlated, as indicated in the correlation matrix, then one of the bands will contribute nearly as much information about the surface as will the two bands. However computation of

the correlation matrix using large data sets can mask divergences between classes with small populations. The best solution is therefore to compute correlation matrices using sample populations for each information class, that are of similar size for each class, and reject a band or bands on the basis of analysis of the correlation in these matrices. The final decision on which bands to use in the classification should be based on evaluation of repetitive classifications of an area of test field data, with highly correlated bands being eliminated at each repetition. Comparison of the classification accuracies and processing costs for each classification of the test data will indicate the cost/benefit trade-off of using various band combinations.

2. Another approach is to use the **simplest algorithm** that will achieve the desired accuracy. The accuracy required for any type of management information has to be considered in relation to the accuracies of the other types of information available, as well as the costs incurred by improving the accuracy. There is usually no benefit in having information that is more than an order of magnitude more accurate than other information being used, because errors in the information are going to swamped by the errors in the other information sources. The important aspects are the influence of the information on the decision-making process, and propagation of error through that process. In general, the more accurately the information is required, the more costly it is to acquire that information.
3. **Sampling techniques** can sometimes be used to achieve the information required, but may not always be applicable, e.g. where it is required to locate every area of specific land-covers. Sampling can be used when the information required is at resolutions that are coarse relative to the resolution of the data, e.g. if the information required is the area under crop in the catchment area for a grain silo. Sampling can significantly reduce processing costs for little loss in accuracy. There are a number of approaches to sampling.

   (a) **Pixel skipping** involves classifying only one pixel every n pixels, and classifying along one line every m lines. This approach is effectively sampling the data by taking a regular sample of $1/(mn)$ part of the total data set. Different sampling fractions could be adopted for different parts of the classification if required. For cropping a sampling fraction of 1–5 per cent has been found to give almost identical results to classification of the whole of the data set.
   (b) A **property sample** requires the analyst to classify a sample of the properties within the area to be classified, with the classification controlled by a mask that contains the required properties. The sample of properties could be chosen either as a random or a stratified random sample. This approach usually takes longer than pixel skipping because each pixel and line of the data has to be interrogated, even if only briefly.

4. The analyst could choose to **stratify the area** and use either different data sets, different algorithms or different spectral classes in each strata. This technique is justified if there are variations in either the class distributions, the within-class quality, or the management practices across the area, that suggest different strategies for classifying the different strata to obtain the desired information.

5.7.5 Field data in classification

The need for field data is considered in detail in Chapter 6. Collection of field data is likely to be one of the most expensive, and logistically difficult, aspects of any mapping or monitoring task. There are significant benefits in using procedures that reduce the need for field data. Field data are required for two purposes: to train the classifier algorithm; and to evaluate the accuracy of classification into the information classes.

1. The more parameters in the classifier algorithm, and the more spectral classes, then the more field data that are likely to be required to train the classifier algorithm. In addition, the poorer the fit of the spectral data to the classifier probability distribution, the more field data that may be required to ensure that all class conditions are adequately represented. Design of classifier algorithms that have decision surfaces in close accord with those of the classes of interest can significantly reduce the amount of field data required for training purposes. This is particularly so if the algorithm exploits our understanding of the reflectance, phenology and the other physical properties, of the surfaces of interest in making a decision.
2. The field data required to evaluate the accuracy of classification is dependent on the number of information classes, and the reliability required from the accuracy assessment. This field data must be quite separate and independent of that used for training the classifier.

With any mapping it is essential to analyse the accuracies achieved using the field data collected for that purpose, and display this assessment on the map, or with the statistics produced. Only in this way can users of the information make a judgement on the weight that they will give to this information, relative to other information, knowledge or intuition in making a decision.

5.7.6 The use of different classification strategies

Classification of data to produce land-cover or land use information can be done using either parallel or sequential classification strategies.

In a **parallel strategy** all of the data are used simultaneously to arrive at a decision. Multi-temporal image classification, in which the data for a number of images are used in the one classification process is the classic illustration of the approach. Parallel classification strategies suffer the disadvantage that they are expensive for each classification and they duplicate processing costs when updating the classification, as will routinely occur in the monitoring of vegetation, water and other resource conditions. With the maximum likelihood classifier the cost of classification is a function of the square of the number of dimensions, or wavebands, in the data. Multi-temporal classification of five Landsat MSS images (twenty dimensions) would have costs that are a function of 20^2 for parallel classification instead of being a function of 5×4 in the case of sequential classification. If new data are to be added to the classification, either within the existing data set range of dates to improve classification accuracy, or after the existing data to update the classification, then parallel classification will incur much greater costs than will sequential classification.

Sequential classification strategies process sets of data relevant to a date, to produce interim information, such as land-cover information, for that date. These interim sets of information are then used to update or revise earlier provisional information, for example of land use, to map the land use as at the date of the imagery. Sequential classification can be done by the use of the following strategies, either individually or together.

1. Use **phenological rules** to control the updating process. Classify each image in turn to derive land-cover information. Use these images of land-cover within the framework of environmental and phenological rules to derive land-use information. This

will often be done by starting with a land use mask, however simple it might be, which is updated using the land-cover information in accordance with the environmental and phenological rules, to create a revised land use mask. In some situations the updating rules can be simple logic rules. Consider mapping flooded rice in an area that contains water surfaces and other agricultural crops that are regularly flood-irrigated throughout their growing season, coinciding with the growing season of rice. Only rice, and swamp, of all of these land uses, is put under permanent flood and may develop a green canopy as the season progresses, although all of these land-covers can look like rice on individual images. The following phenological rules will separate the various land uses.

(a) Only rice, swamp and water bodies have the same land-cover of water to green canopy on two or more sequential images in the growing season. Other land uses will look like rice on individual images, often more than one, but not sequential sets of images. Therefore if a pixel looks like rice on two or more sequential images then it can only be rice cultivation, swamp or water land uses.
(b) Of these three options, only rice and swamp may develop a green canopy during the growing season. Therefore a pixel must exhibit an increase in green canopy cover during the growing season for it to be rice cultivation, but it may still be swamp.
(c) Swamps are normally flooded before the season starts, and may be known beforehand. Swamps can be deleted either by exclusion from the classification, or by mapping them prior to flooding of the rice bays.

Similar sets of rules are known for other conditions. The process can be built as an expert system designed to monitor the physical parameters of interest.

2. Use **hierarchical classification** in which the data are partitioned into broad classes using one combination of data and processes, and then each of these broad classes is analysed in more detail using a different set of data and processes. Consider the above example of monitoring rice in which Rule (b) uses increasing green canopy to discriminate rice and swamp areas from other water bodies. The first level of image classification is to discriminate the 'looks like rice' from the 'does not look like rice' classes which is adequate for Rule (a). For Rule (b), however, more information is required, by analysing only those pixels that fit into this category to determine whether they are exhibiting increasing green canopy. This second level of information satisfies Rule (b).

Sequential classification strategies provide the basis for a more powerful and flexible approach to the establishment of a classification procedure, as well as providing interim information, than do parallel strategies.

5.8 Estimation

5.8.1 Introduction

Estimation is the transformation of remotely sensed and other data into estimates of specific resource parameters by the use of numerical mathematical algorithms or models. It provides information on the condition of the sensed surfaces, either at the time of the acquisition of the remotely sensed data or at some time in the future. Classification provides information on what is the nature of the surface land-cover or land use and estimation provides information on the condition of that surface.

Estimation models can be of two types: current condition models that provide information on the conditions of the land-cover or land use at the time of data acquisition; and future condition models that provide predictive information on future conditions relative to the date of acquisition.

Current condition models derive estimates of selected physical parameters, such as leaf area index (LAI) or biomass of the different covers or land uses. These models may vary from cover to cover so it is essential to determine cover or use before attempting to estimate condition.

Future condition models provide predictive information designed to assist resource managers in choosing between alternate management strategies. These models use quantitative information on the conditions at a time and project this information into the future by use of models that estimate changes assuming certain specified inputs, particularly of weather conditions and economic activity. For example models of plant growth may depend on measures of LAI as the starting point and then use plant growth models to estimate LAI at some point in the future, where these plant growth models may use as inputs soil types, slope and aspect as well as anticipated weather conditions and plant consumption by herbivores.

5.8.2 Development of current condition models

Resource managers are interested in a range of condition information on the different surface land-covers including:

- Cropping
  Crop Height
  Per cent cover
  Leaf area index
  Biomass
  Plant water status
  Evapo-transpiration from the crop
  Yield
- Pastures
  Per cent cover
  Leaf area index
  Per cent green material
  Biomass
  Plant water status
  Evapo-transpiration from the pasture
- Forest and Woody Covers
  Per cent cover
  Species composition
  Tree height
  Woody biomass
  Evapo-transpiration from the cover
  Water status
- Water Resources
  Turbidity
  Chlorophyll content
  Water temperature

The common approach to the implementation of models to estimate these types of parameters is:

- selection and calibration of the appropriate image data;
- classification of land-cover and land use;
- conversion of the image data into indices that are correlated with the physical parameters of interest; and
- implementation of a model to convert the indices into estimates of the physical parameters.

Selection of the appropriate imagery

The imagery chosen must depict, either directly or through surrogate parameters, information on the physical parameters of interest. The stronger the link between the imagery and the physical parameters and the fewer sources of confusion or conflict then the more likely that the analyst will be able to develop or use models that provide good estimates of the physical parameters of interest. For example, intercepted photosynthetically active radiation (IPAR) is a measure of the radiation absorbed by the green vegetation relative to radiation in other wavebands. It is thus highly correlated with measures of red radiation relative to radiation not absorbed, such as NIR radiation. Imagery in these two wavebands should be able to provide good estimates of IPAR at the time of data acquisition.

Green leaf area index (GLAI) is often highly correlated with IPAR and hence image data using the red and NIR wavebands to estimate IPAR should be able to estimate GLAI when they are highly correlated. Imagery may not be able to provide a good estimate of GLAI when a proportion of the green material is not significantly contributing to IPAR at the time of overpass. This occurs when some of the green material is positioned in the canopy such that it does not contribute to IPAR at that time. For example, the green canopy material may be growing from the base of the plant canopy, and at acquisition this material may not be intercepting a significant amount of radiation. In this situation the amount of green material estimated from the remotely sensed data is likely to be less than the amount that actually exists in the canopy. Another example concerns the canopy of *Eucalyptus* Spp. plants where the leaves are of similar reflectance to that from the leaves of other species, but the canopy reflectance is dark in comparison with that of most tree species. The eucalyptus trees hang their leaves vertically to minimize exposure to the sun during the heat of the day, reducing IPAR for midday imagery but not the GLAI of the plant.

The correlation between IPAR and biomass is usually not as high as with GLAI so that imagery may not be able to estimate biomass as well as it can estimate GLAI. The closer the linkage between the physical parameters of interest and the image data, the better the image data will be in estimating the physical parameters.

This linkage is dependent upon the wavebands being sensed, the time of day of data acquisition, and the phenological stage of the crop at the time of acquisition. Changes in response data due to atmospheric and sensor effects will be interpreted as changes in the physical parameters, unless the processing corrects for these effects through calibration.

Classification of land-cover and land use

The interdependence of models on the land-cover or land use being measured has been noted, necessitating the mapping of land-cover and/or land use before estimation.

Selection of indices

The image data within the class of interest are usually converted into indices that are highly correlated with the parameters of interest. There are two general approaches:

1. conversion of the image data directly into indices; and
2. conversion of the image data into proportion indices using the vector classifier

Conversion into indices

For herbage, a number of vegetation indices have been developed. These vegetation indices transform the image data into one independent index, which can then be regressed against observations of the physical parameters of interest to derive regression models to estimate those physical parameters from the index data, as discussed in Chapter 6. Disadvantages of this approach are that the indices do not allow for variations in atmospheric conditions in the data, and that one index can only estimate one independent parameter. Various authors have developed estimates of biomass and LAI for monocultures and pastures using vegetation indices. These are more successful in estimating the physical parameters of monocultures than of pastures because of the influence of brown material in the canopy on the indices.

Conversion into proportion indices

The vector classifier (section 5.5), converts image data into coordinates within a sub-domain covered by the land-cover or land use of interest. With pasture or herbage this land-cover will include at least green and brown herbage, and soil. With swamps and water surfaces this land-cover will include at least water and green vegetation. For other conditions the node or parent cover conditions can usually be readily identified. Processing by the vector classifier derives proportion coordinates relative to the node response values, where the node response is determined by the analyst. Since the nodes are defined in the image data, atmospheric effects on this data are taken into account in defining the nodes, in a general or average sense across the image. For pastures and herbage the resulting proportion coordinates will be at least p_g (proportion green vegetation), p_b (proportion green vegetation) and p_s (proportion soil cover), but may well also include p_t (proportion tree cover). The amount of each of these cover conditions varies between maximally to negligibly within the domain defined by the nodes. Position in this domain, as defined by these proportion coordinates, are regressed against the physical parameters of interest so as to develop regression models to estimate the physical parameters from the proportion index data.

With the vector classifier the number of independent parameters is $(m - 1)$ where m nodes are used. Thus with pastures, using the three nodes of green and brown herbage and soils, there are two independent proportions determined by the vector classifier for each pixel. This means that up to $(m - 1)$ independent physical parameters can be estimated using these $(m - 1)$ proportion parameters. Work using this approach has been done in Australia as shown in Photo 5.9, in which the image data was converted into proportion of green and brown herbage response that were then converted into estimates of the percentage green herbage by the use of a regression model developed with the use of field data.

In pastures the vector classifier has been shown to provide better estimates of pasture status than the vegetation indices as shown in Diagram 5.29. The vector classifier can similarly be used for rice, water surfaces and any other land-covers that have a linear

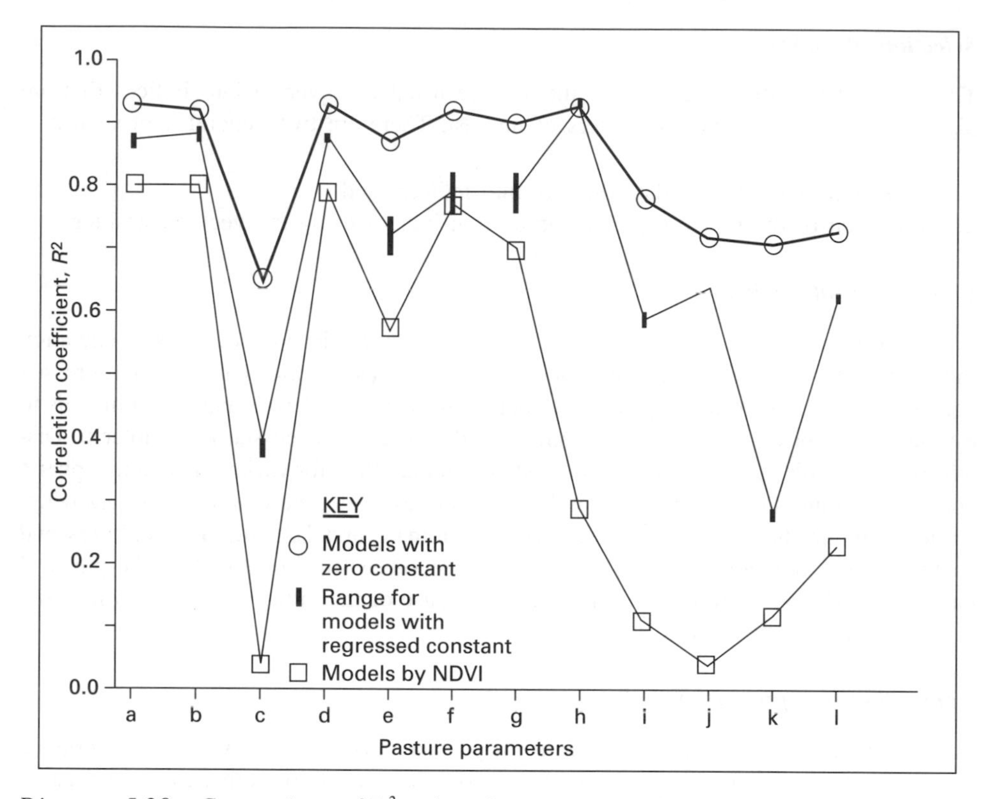

Diagram 5.29 Comparison of R^2 *values for estimating the different pasture parameters of: (a) total leaf area index; (b) GLAI; (c) BLAI; (d) wet biomass; (e) dry biomass; (f) green wet biomass; (g) green dry biomass; (h) brown wet biomass; (i) brown dry biomass; (j) per cent green wet biomass; (k) per cent green dry biomass; (l) per cent GLAI by use of* (p_g, p_b) *model with zero constant, all three models as range and NDVI, from McCloy* et al. *(1992).*

response sub-domain within the Landsat response domain. In these cases the proportion coordinates are used to develop a regression model to estimate selected physical parameters.

5.8.3 Development of future condition models

Models to predict changes in some physical environmental parameters with changes in some input parameters, including time, have been, and continue to be developed in the various disciplines. Many of these models need to be integrated with remotely sensed data because they provide:

1. specific starting condition information on some parameters in the model; and
2. a means of monitoring and adjusting the model over time and/or space.

As an example, many models that estimate GLAI within growth models, are of the form:

$$GLAI_{time} = f[GLAI_{start}, \text{rainfall, temperature, soil type}]$$

For this model, remotely sensed data can provide information on $GLAI_{start}$, and can monitor $GLAI_{time}$. If $GLAI_{time}$ is too far from the actual GLAI at that time then the measured values can be used as the basis for further modelling. The discrepancy between the estimates by modelling and from remotely sensed data provides a key to investigate either model performance or to determine what other factors are at work.

Most applications of future condition models will require the integration of a number of data sets within the model. This processing is one of the major strengths of GIS and will be discussed in more detail in Chapter 7.

6
The use of field data

6.1 The purpose of field data

6.1.1 Definition and description of field data

Field data are defined as verified, or verifiable, data collected using proven, relatively close-range techniques such that the resultant data are much more accurate, consistent, and at a higher resolution than the information derived from remotely sensed data. Remotely sensed data are used to provide area extensive information, while field data are used to improve this information extraction, to calibrate either the data or the information, and to assess the accuracy of the derived information.

Field data can be of many types: data collected in the field, derived from statistical data in the office, from aerial or ground photographs, or from aerial observation. It must be capable of independent verification, duplication or replication using similar or different techniques. This is in contrast to the information derived from the remotely sensed data that can be duplicated using the same data sources but which can only be verified using field data.

The techniques used in acquiring field data must be well established, of known characteristics, accuracy and reliability. These techniques can include:

- physical site visits to collect specimens of the physical characteristics of interest, e.g. observations on geomorphology and samples of vegetation, rocks and soils;
- measurement of parameters in the field, e.g. the measurement of radiometric, gravimetric and magnetic data using remote, but close sensors, and the measurement of temperature and rainfall using direct contact measurement techniques;
- measurement of parameters using established remote sensing techniques such as the measurement of field areas on aerial photographs and the use of contours derived by photogrammetry, as long as the measurements meet the criteria set above; and
- sampling surveys to extend field data to cover larger areas, e.g. to estimate the biomass for a pixel area.

Information derived from both field data and image processing contains errors, the magnitude of both being of concern to the analyst. In this context, the accuracy of information derived from remotely sensed images is defined as the closeness of the estimated values to the values determined by the use of field data. As field techniques are not perfect, so the estimate of accuracy is not an estimate of the closeness of the extracted information to 'true' values.

The accuracy of image classification is usually determined using a sample of sites for which the physical characteristics are measured using field techniques and then compared with the characteristics as estimated from the image data, as described in section 6.6. The accuracy of a class is defined as the probability of being classified in that class which the area to belongs, i.e.

$$A_{class} = Pr(\text{Classified in Class/Actually in Class}) \tag{6.1}$$

If F_{class} is the set of field samples in the class and O_{class} is the set of image based samples that correspond to the field samples and are in the class, then

$$A_{class} = \frac{\text{Count of } O_{class}}{\text{Count of } F_{class}} \tag{6.2}$$

Resource managers are concerned not only with the accuracy of the information that they are using but also with its consistency. Information is consistent if its accuracy, from location to location, or from image to image, are similar, i.e. the differences in the accuracies are not statistically different at the 95 per cent confidence level. Clearly consistency can be assessed in various terms: are the accuracies of mapping different land uses consistent, one to each other; are they consistent over the image; and are they consistent from image to image.

The consistency within a set of classes in a classification or of a class across a set of classifications is defined as

$$C^2 = \sum_{i=1}^{n}(A_i - \overline{A})^2/(n-1) \tag{6.3}$$

Information derived from different images can all have low levels of accuracy, but they, as a set, will be consistent if the accuracy is similar in all of the information. The information may be accurate – it will only be consistent if the information derived from each image is of similar accuracy.

The spatial resolution of the field data must also be significantly better than that of the information derived from the remotely sensed data so as to ensure that errors are not introduced due to this cause. An associated problem is the need to be able to register the field data to the image data so as to avoid the development of 'apparent' errors caused by misregistration of the data sets. It is easy to register the location of field data as grid coordinates in the local mapping system, but difficult to rectify the image data without creating spatial errors in the comparison.

6.1.2 Role and types of field data

Field data have a role to play in all facets of information extraction from remotely sensed data, including:

- correction for atmospheric effects;
- support for both visual interpretation and image processing in information extraction;
- development of estimation models and in estimation processing; and
- accuracy assessment in both classification and estimation.

The techniques used in collecting and analysing field data are often relevant to more than one of these facets. Among the techniques available are the following.

- Spectrometer observations are used for atmospheric corrections, in the development of estimation models and improving the conceptual models used by the interpreter to relate physical conditions to remotely sensed image data.
- There is the collection of physical field data on surface conditions. This type of field data uses all of the standard techniques for the conduct of surveys and the collection of data including botanical surveys, collection of botanical, geologic and other samples,

and the observation of geomorphic conditions. These techniques, being well covered in texts on the relevant disciplines, will not be covered here.
- Observation of close range measurement of parameters includes gravity, magnetics, electromagnetic induction as well as surficial features from aerial photographs. In general, these techniques are also well covered in the relevant texts and will not be covered here.
- Sampling surveys in both estimation processing and accuracy assessment are valuable.
- The analysis of survey data for accuracy assessment is also important.

Whatever method, or methods, are used in collecting field data, the analyst must start with:

1. a clear definition of the purpose of the field data; and
2. a specification of the criteria that the field data must meet, including the types, resolution, timeliness and accuracy of the data to be collected.

The purposes of and criteria to be met by, field data can be best understood by analysing first the nature of the information extraction process. A remotely sensed image, as a record of the energy reflected or radiated from the surface in selected wavebands at a moment in time, is not the information on characteristics of the surface that are of interest to resource managers, but are indicative of these characteristics to some degree. They may also be indicative of other characteristics to a greater or lesser degree. In consequence there is considerable potential for confusion and conflict, resulting in the drawing of wrong inferences or conclusions, during extraction of information from the data. Generally these inferences are drawn by a process of extrapolation rather than interpolation.

Interpolation is defined as the filling of intermediate values between two or more known boundary conditions often by using a linear interpolation model. During interpolation the boundary or outer limits are known so that errors arise due to differences between the assumed model and the actual conditions. If linear interpolation is used then errors will occur when the shape of the surface deviates from this model. However, the resulting errors can be small, particularly if the locations of the spot heights are carefully chosen. **Extrapolation** is defined as projecting from known information to estimate unknown information *outside* of the known conditions. In the analysis of remotely sensed and other data, the extrapolation is supported by the availability of the data for both conditions, and its relationship to information only available for the known conditions. The process assumes that extrapolation conditions found valid elsewhere can also be applied under these conditions.

During information extraction the known attributes about the features of interest are the image data itself. Other data (such as field and training data), information (such as soil and topographic map data) and knowledge (including heuristic, numerical and other models) form the basis for the development of specific relationships between the image attributes and the information being sought under particular conditions or locations. The information extraction process, by identifying the known attributes and using the available accumulated data, information and knowledge, selects the most valid of these relationships to extrapolate from the known image attributes to estimate the causal physical characteristics. Often the assumptions linking physical characteristics to causal attributes (on which this process is based) are valid and the extrapolation will lead to correct results. However, extrapolation is fraught with risk because it is often very difficult to know whether the assumptions used in the extrapolation are valid in the specific conditions pertaining to the particular image and surface. Auxiliary information and knowledge are invaluable in determining whether the assumptions are being, or are likely to be, met. Despite the use of such information and knowledge there are always likely to be circumstances where a different conjunction of environmental conditions than those considered by the analyst will create the specific data characteristics as depicted in the imagery. When this occurs the imagery will often be depicting different conditions than those expected by the analyst. The analyst, by using his or her constructed relationships or models may draw the wrong conclusions from the data, leading to errors in the information extraction process.

With extrapolation, the errors can be very large if the assumptions create a model which is a long way from actual conditions. Consider the case of estimating LAI from a vegetation index image using a numerical model. This numerical model would have been developed to operate under a range of conditions, and if all of these conditions are replicated in the mapping then the estimates of LAI could be expected to be accurate. If they are not replicated exactly then the estimates of LAI may be good or poor depending on the differences between the original and these actual conditions, and how these differences affect both the LAI and the vegetation index. The only way that an analyst can verify the quality of his or her work is by the conduct of field evaluation using field data and through a knowledge of the likely impact of the difference between the two canopies on LAI and the vegetation index. This means that the analyst needs a good understanding of the conditions pertaining in the development of the model, and must have a good conceptual understanding of the interaction of radiation with matter.

Important components of any interpretation task are for the analyst to:

- be aware of the assumptions being made during each extrapolation task;
- identify other conflicting, or potentially conflicting sets of assumptions, so as to determine which are the more likely to apply under the actual image conditions; and
- ensure the validity of the assumptions, as practically as possible during the analysis. If there is a discrepancy between the physical conditions interpreted from imagery and other evidence available to the analyst, then the analyst has to be very cautious in accepting that set of assumptions as an appropriate extrapolation model.

Field data have an important role to play in meeting these requirements during image analysis, since the field data are assumed to be significantly more accurate than the information being derived from remotely sensed data. For this reason, the collection of field data should not contain the same processes of inference making by extrapolation as occurs in image analysis. If similar processes are used, then the observations must be taken under tightly controlled conditions so that all of the significant boundary conditions are known. In the example given above, of estimating LAI for a pixel, sampling can be used as long as a sufficiently large and representative sample is taken to keep the errors well below those that might occur in the image processing. However, it is important to remember that field data are not error free. (The unfortunate term 'ground truth' is both misleading and incorrect; it should never be used.)

All users of information derived from remotely sensed data should be aware of the errors that exist in the information. This realization is likely to prompt three responses from users:

1. a demand that the information derived from remotely sensed imagery undergo rigorous accuracy and consistency assessment in accordance with established or agreed criteria;

2. access to the accuracy assessment during their use of the information; and
3. the need for back-up systems with verification processes designed to protect users from legal actions that may arise out of actions taken on the basis of information provided. Thus for some purposes the information provided is used as a 'pointer', and action is only taken once the actual conditions have been verified in the field.

6.2 Collection of field spectral data

6.2.1 Purpose of collecting field spectral data

Spectral data are collected in the field using spectrometers or scatterometers.

Spectrometers record the electromagnetic radiation entering the optics of the instrument in set waveband ranges between about 300 μm (ultraviolet) and 15 μm (beyond the thermal infra-red). They may be designed to collect the energy in many, but very narrow wavebands, or they may sense a much smaller number of broader wavebands that will often match the wavebands of a specific scanner, such as the Landsat TM.

Scatterometers generate microwave radiation in one or more wavebands, and receive the return signal where both the generated and recorded returned signals may be of like or of opposite polarization. The wavebands used in scatterometers would normally correspond to one or more of the standard radar wavebands.

Spectral data are usually collected for the following reasons.

1. They optimize bands used in data acquisition. The spectral data will be used to identify the spectral bands, phenological stages and seasonal conditions that optimize discrimination between the different physical features or conditions of interest to the analyst. Once the optimum wavebands and conditions have been determined then imagery acquired in accordance with these criteria should provide the best opportunity to discriminate the features in the imagery of interest to the analyst, at minimum cost and with maximum accuracy. The ability of the imagery to discriminate the covers of interest should make the information extraction process faster, easier, more accurate and consistent – in this way reducing processing costs and increasing usability.
2. They understand better the interaction mechanisms, or the relationship between surface conditions, environmental processes and the reflectance characteristics of the surface. This will facilitate the development of better sensors and sensing systems, lead to improved analysis techniques and develop more robust and consistent strategies to extract information from the data.
3. In the development of Estimation Models, they estimate the values of physical parameters or attributes of the surface from spectral data acquired of that surface.
4. They facilitate the incorporation of environmental rules based on our environmental knowledge into the information extraction processes so as to improve their accuracy and consistency or reduce their costs.

6.2.2 Measurement of field spectral data

The radiance emanating from a surface depends primarily on the amount of energy incident on the surface, and the reflectance characteristics of that surface. The radiance emanating from a surface is not constant for a given surface condition because of atmospheric and path geometry effects. Reflectance is much more consistent for given surface conditions. Reflectance, r_λ is defined, to be

$$\text{reflectance} = \frac{\text{reflected spectral radiance}}{\text{incident spectral irradiance}}$$

$$r_\lambda = \pi L_\lambda / E_\lambda$$

The reflectance is usually determined in either of two ways – by the direct measurement of reflectance which measures incident and reflected radiation to calculate the reflectance directly from the pair of observations, or by the measurement of reflectance factors, in which the reflected radiation from the target and from a control surface are used to derive the reflectance factor of the target, or target reflectance relative to the control surface. If the reflectance of the control surface is known then the target reflectance can be calculated. Both methods are now considered in detail.

The direct measurement of reflectance

Either one, or a set of instruments are used to measure both the incident irradiance on, and the reflected radiance from, the target surface. Both observations are calibrated and the reflectance calculated as the ratio of the reflected to incident radiation. The general observational procedure is as follows.

1. Measure the incident irradiance by means of a spectrometer with opaque diffuse objectives set horizontal so as to receive all incident (hemispheric) radiation.
2. Measure the reflected or target radiance using objective lens with a defined field of view.
3. Under cloud-free atmospheric conditions, groups of observations of various targets can be observed between observations of incident irradiance, as long as irradiance observations are taken before and after observing the targets, and no more than one hour separates each observation of irradiance.
4. Repeat the procedure (Steps 1–3) until all of the required observations have been taken.
5. Calibrate the irradiance and radiance observations using calibration curves for the individual instruments – it is important that the instruments be regularly calibrated to maintain accuracy.
6. Compute the reflectance for the different surfaces. How the reflectance is calculated depends on the manner in which the observations were taken as discussed below.

If two or more instruments are available then the optimum situation of taking simultaneous observations of irradiance and radiance can be implemented. This is essential if there is any cloud; reflectance off the cloud can cause local fluctuations in irradiance that will be unnoticed by the observer. In this case the irradiance and radiance observations can be calibrated and the reflectance calculated as the ratio of the two observations.

If only one instrument is used, or if one instrument is set up to record the irradiance at a regular interval then the irradiance observations will not correspond, in time, to the radiance observations. The solution is to use the calibrated irradiance observations to fit a surface, such as a sine curve, between the observations knowing that the irradiance rises to a peak at solar midday and then decreases (Diagram 6.1). Use the fitted curve to interpolate a value of irradiance at the time of observation of radiance of the target and use this value with the target radiance to calculate target reflectance. To use this technique the irradiance observations must be taken:

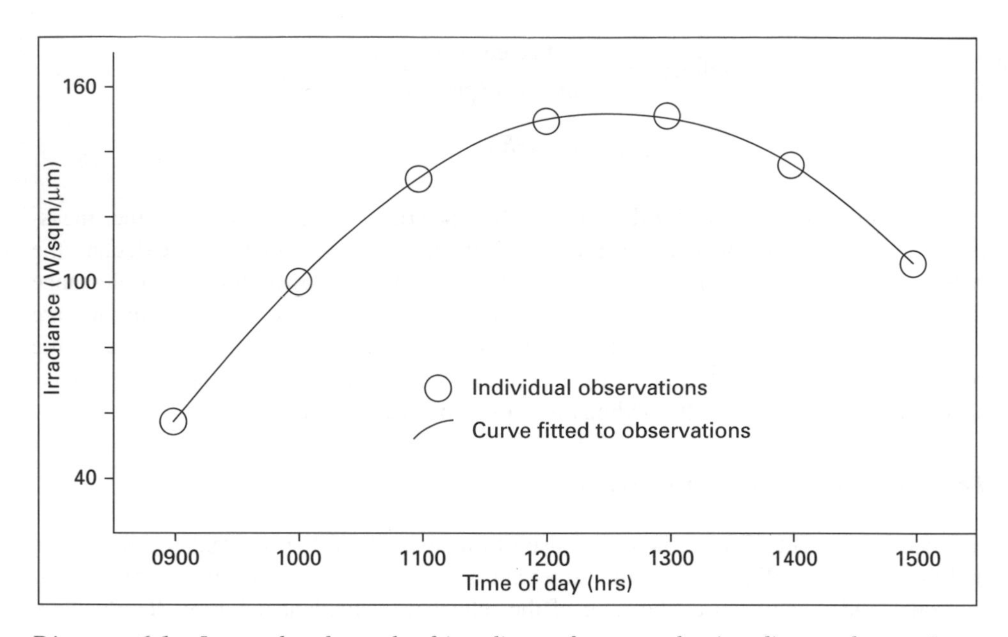

Diagram 6.1 Interpolated graph of irradiance from regular irradiance observations taken throughout one day.

- at a regular interval, of no more than hourly separations, from before until after the radiance observations;
- at the same location as the radiance observations; and
- under cloud-free conditions.

The irradiance observations can be fitted to either a linear, sinusoidal or second-order polynomial curve by regression on a computer. Generally linear interpolation will only be used when only two observations of irradiance have been taken, before and after the radiance observations of the target surfaces. Since the irradiance does not change linearly, a linear model should only be used over short time periods, e.g. when taking radiance observations of the one target. The derived regression curve is then used to estimate irradiance at the time of observation of target radiance, to calculate target reflectance.

$$\text{Irradiance} = A_o \sin\left[\frac{180(T - T_r)}{T_s - T_r}\right] + B_o$$

or

$$\text{Irradiance} = A_o T^2 + B_o T + C_o$$

where T is local time, T_r is local time of sunrise, and T_s is the local time for sunset.

Consider the observations plotted on Diagram 6.1, and fitted to both sine and second-order polynomial curves, with the observed irradiance, the estimated irradiance values from both regression curves and the residuals included in Table 6.1. The two regression curves derived using the data in Table 6.1 are

$$\text{Irradiance}_{\text{poly}} = -7.3929 \times (T - 12)^2 + 8.68 \times (T - 12) + 150.3$$

Table 6.1 Irradiance observations at Orange on 15 June 1986

Time	Observed Irrad. ($W m^{-2} m^{-1}$)	Polynomial Regression		Sine Regression	
		Irrad.	Residual	Irrad.	Residual
0900	59	57.7	+1.3	59.1	−0.1
1000	101	103.4	−2.4	102.1	−1.1
1100	134	134.2	−0.2	133.8	+0.2
1200	152	150.3	+1.7	151.2	+0.8
1300	152	151.6	+0.4	152.6	−0.6
1400	137	138.1	−1.1	137.8	−0.8
1500	110	109.8	+0.2	108.3	+1.7
r^2			11.79		5.79

$$\text{Irradiance}_{\text{sine}} = 150.3 \times \sin\left[\frac{T - T_r}{T_s - T_r}\right] + 3.02$$

where T_r is 0736 hours and T_s is 1731 hours.

If the residuals are sufficiently small then the derived regression curve is used to estimate the irradiance at the time of acquisition of radiance observations of the targets, so as to calculate target reflectance.

The main disadvantages of this method are:

- variations in irradiance between the regression estimate and the actual irradiance due primarily to cloud; and
- differences in calibration between the instruments.

The measurement of reflectance factors

In this method the observations of incident irradiance are replaced with observations of a control surface that has stable and known reflectance characteristics. Normally, both the target and control surfaces are measured using the same instrument and identical optical systems so that calibration errors have an equal effect on both observations. If the target and the control surfaces are measured with different instruments then different calibration factors will apply to the different observations. However, simultaneous observations of the control surface with both instruments provides the necessary relative calibration of the two instruments as is required by this method. The ratio of the target to control observations (for the one instrument) or calibrated observations (for different instruments) are the reflectance of the target relative to the control surface reflectance, or reflectance factors. As long as the reflectance of the control is stable then the method readily provides consistent measurements of reflectance factors for the specified control surface.

The reflectance of the targets can be calculated from:

$$r_{c,\lambda} = \frac{\pi L_{c,\lambda}}{E_{i,\lambda}}$$

$$f_{t,\lambda} = \frac{L_{t,\lambda}}{L_{c,\lambda}}$$

$$r_{t,\lambda} = \frac{\pi L_{t,\lambda}}{E_{i,\lambda}}$$

$$= \left[\frac{L_{t,\lambda}}{L_{c,\lambda}}\right] \times \left[\frac{\pi L_{c,\lambda}}{E_{i,\lambda}}\right]$$

where $r_{c,\lambda}$ is the reflectance of the control surface, $f_{t,\lambda}$ is the reflectance factor of the target, and $r_{t,\lambda}$ is the reflectance of the target, so that $r_{t,\lambda} = f_{t,\lambda}\, r_{c,\lambda}$.

With some instruments the radiance of the target and the control are measured simultaneously. This practice avoids errors due to variations in the incident radiation.

The main advantage of the method is that it can provide good values of target reflectance factors; the main disadvantage is its dependence on the control surface having consistent reflectance values for all of the observations being taken.

The control surface needs to:

- be lambertian or nearly lambertian;
- be not significantly degradable, in terms of reflectance value, with time or exposure to sunlight;
- be resistant to wear and tear in its use;
- have either known reflectance values, that are similar, or slightly greater to the reflectances expected, or reflectance values near to 100 per cent. The advantage of 'grey' control surfaces is that they create similar electronic responses as the target in the instrument, thereby minimizing one source of instrumental error. However the derived reflectance factors will be quite different to the actual reflectance of the targets, and so the reflectance factors will need to be converted into reflectance before they can be used. The advantage of 'white' control panels is that the reflectance factors are similar to actual reflectance and so they are easily checked for errors. A disadvantage is that users might be tempted to use these reflectance factors as actual reflectance values.

Various types of control surfaces are used, but the two most common are Halon powder and Halon paint. Halon powder is compressed under constant pressure of 14 pounds per square inch to form a uniform, reasonably lambertian reflectance surface with typical reflectance values as shown in Diagram 6.2. The main disadvantage of the powder is that it becomes dirty very easily, so that it has to be replaced frequently. To ensure that the surface formed is flat and of consistent compression, the halon has to be filled into a mould and compressed using a matt surface that will not significantly distort during application of the pressure. Thick matt glass is suitable for smaller control surfaces. **Halon paint** is used to cover a reasonably smooth surface to a depth of at least 0.1 mm. Suitable bases must be matt and have reasonably lambertian reflectance characteristics. One solution is to create a sandwich of heavy duty particle board between aluminium plates. The surface of the aluminium plates are etched with hydrochloric acid to provide the necessary matt surface.

6.2.3 Considerations in collecting field spectral data

Need for representative data

The reflectance of a vegetative canopy depends upon the reflective, transmissive and absorptive characteristics of all of the components in the canopy, and their area, location in the canopy and orientation as well as the reflective, transmissive and absorptive

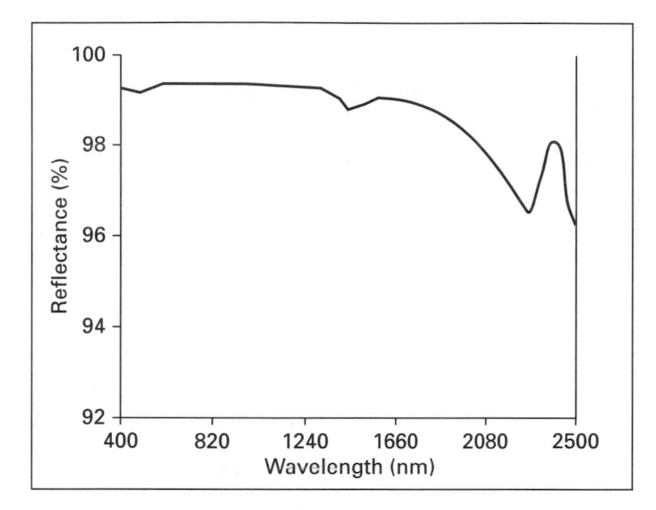

Diagram 6.2 Typical reflectance characteristics of Halon powder from Grove et al. *(1992).*

characteristics of the background soil or water layer. Because the distribution of vegetative components in a canopy will never be identical in different fields of view, and in fact may change over time due to such factors as wind, so the measured reflectance will never be identical at all points in the canopy. It is necessary to correct for this variability by using the mean of a sample of observations as being representative of the cover.

Need for lambertian control surfaces

Most of the reflectance data required in remote sensing is bi-directional in form, i.e. the dominant source of energy is from one direction, and the energy is recorded in another direction. Such reflectance is dependent upon the lambertian characteristics of the target and the control surfaces. The lambertian characteristics of the control must be measured so as to determine the range of solar zenith angles over which the control can be used. This is best done using an artificial light source to represent the source, and measuring the reflected radiation at set zenith angles, at a constant distance from the target. Variations in the measured radiance from the control is indicative of variations from the lambertian norm by the surface.

The effect of non-lambertian surfaces on reflectance

Target surfaces, particularly those that are a composite of different cover types, may not be lambertian. If this is the case then the radiance recorded by the sensor can change from image to image. It may be necessary to verify this hypothesis and determine the nature of the distribution. The best way to do this is to measure the radiance at various azimuthal and zenith angles.

Impact of clouds on measurement of reflectance

Clouds not only block solar radiation, but they can reflect radiation. When this occurs then the irradiance will decrease and increase over affected areas relative to that which would occur due solely to the direct solar beam. The changes in irradiance can create

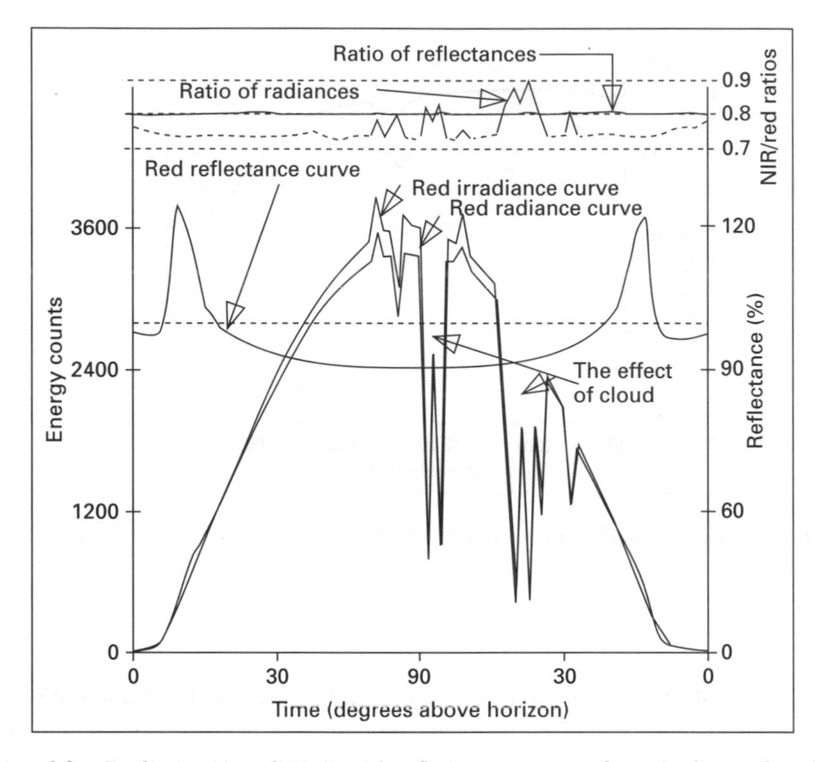

Diagram 6.3 Radiance, irradiance and reflectance curves for a red waveband recorded throughout the day, for a lambertian Halon target and curves of the ratios of (NIR/red) radiance and reflectance values using the same field data set for red and NIR wavebands.

significant spikes in the irradiance recorded at the surface (Diagram 6.3). It is therefore safest to take field spectral observations when there are no, or only a few clouds that are a long way from the Sun–surface–sensor geometric figure.

The data in Diagram 6.3 contains other interesting information. The irradiance and the radiance curves should have maximum values that follow a curve that peaks at solar midday, decreasing away from this point. The reflectance curve should be flat throughout the day, but in Diagram 6.3 it has peaks at low solar elevations indicating that the diffuse hemispherical objective acquiring the irradiance data is not collecting all of the incoming radiation at these low angles. Nor is the curve flat at higher angles, either because the surface is not truly lambertian, or because the objective is still affecting the incoming radiation. The NIR ratio curves show that the ratio of the reflectances is consistent throughout the day, as is expected from the definition of reflectance. The ratio of the (NIR/Red) radiances exhibits significant changes when there was cloud cover, indicating that the proportions of red and NIR energy changed under these diffuse light conditions.

6.2.4 Collection of other data with spectral data

If the spectral data are being collected to develop numerical models intended to estimate canopy physical parameter values then a number of matters need to be considered, including:

- the ability of the spectral data to estimate the desired parameters;
- the need for representative spectral and physical data;
- the impact of variations in canopy reflectance on derived models; and
- the conversion of field based models to airborne or space-borne applications.

Ability of spectral data to estimate selected physical parameters

The more closely linked the physical parameters are to canopy reflectance or radiance, the more likely that the reflectance or radiance data will provide good estimates of that parameter, as discussed in Chapter 5. Conversely, the more factors that affect the parameter of interest but do not significantly affect the reflectance or radiance from that surface, the less likely that the image data will provide good estimates of that parameter. Spectrometer data may be used to establish the level of correlation between selected spectral bands and the physical parameters of interest, and to develop a model to estimate the physical parameters from remotely sensed data in specific wavebands.

Similar considerations apply if the objective is to develop models estimating parameters that may be primarily influenced by thermal radiation, such as stress and moisture content, or radar imagery, such as biomass, moisture content and canopy structure.

Need for representative physical and spectral data

The physical and spectral data collected to investigate relationships between the two, must be compatible. First, it is essential to cover the full range of conditions that are likely to be met by the model. Consider the development of models designed to estimate pasture (grassland) canopy parameters. Canopies of green vegetation exhibit variations in reflectance in the NIR up to GLAI values of about 6 due to the transmissivity of green vegetation at these wavelengths (Diagram 6.7). Green vegetation shows little variation in red reflectance above GLAI values of about 3, and brown vegetation shows little variations in reflectance in both wavelengths above LAI values of about 3, because of the opacity of vegetation under these conditions. The spectral and physical data needs to be distributed across a domain that varies between 0 GLAI 6 and 0 BLAI 3, similar to that shown in Diagram 6.4. Similar considerations apply for estimating attribute values for other land-cover types.

In developing numerical models it is important that the samples are distributed across the domain, and sufficient samples are taken to ensure that the conclusions drawn are statistically valid. In general a minimum of 30 samples would be required for this purpose.

Second, the samples used must be of sufficient area to include representative proportions of vegetation, soil, shadow, and litter within the spectrometer field of view (FOV) so that the observations of reflectance will be representative of that cover type. If the herbage contains individual plant clumps then the sample area must be large enough to be representative of cover conditions across the surface. It is for this reason that samples of herbage will normally be larger than 1 m^2 in area, and samples of larger plants will need to be correspondingly larger again. The field of view of the instrument is specified as a cone angle, λ, so that the area covered can be calculated as a function of the distance from the sensor (Diagram 6.5).

If the distance from the target to the sensor is d then

$$\text{diameter of FOV} = 2d \tan\left(\frac{\eta}{2}\right)$$

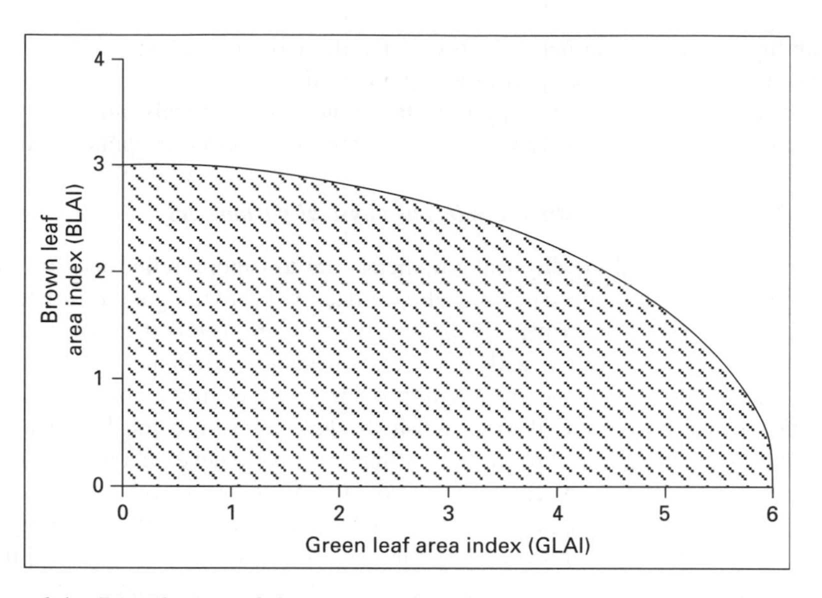

Diagram 6.4 Distribution of data required to develop a model to estimate GLAI and BLAI for pastures from spectral data.

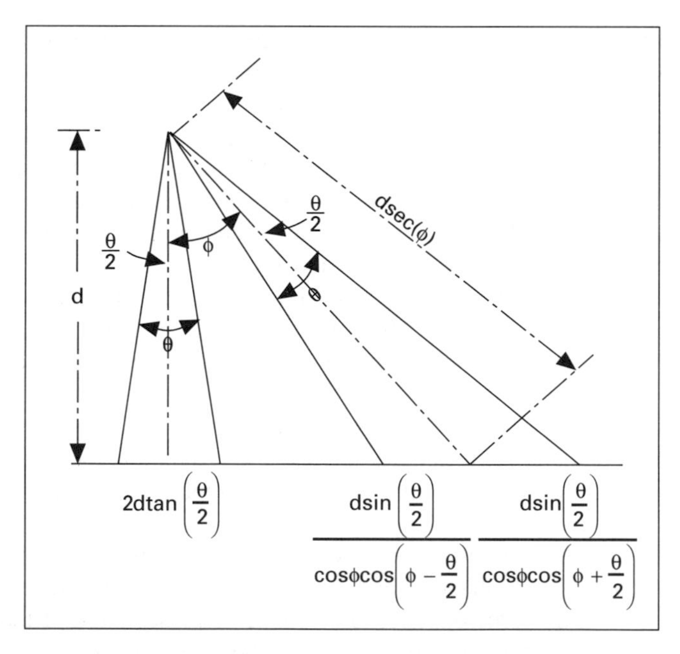

Diagram 6.5 Calculation of the area covered by an instrument of specified IFOV at distance d *from the target.*

where η is the solid angle defining the instrument's IFOV, and

$$\text{area of FOV} = \pi\left(d\tan\left(\frac{\eta}{2}\right)\right)^2$$

when the instrument is in the vertical position (Diagram 6.6). If the spectrometer is tilted by ϕ then

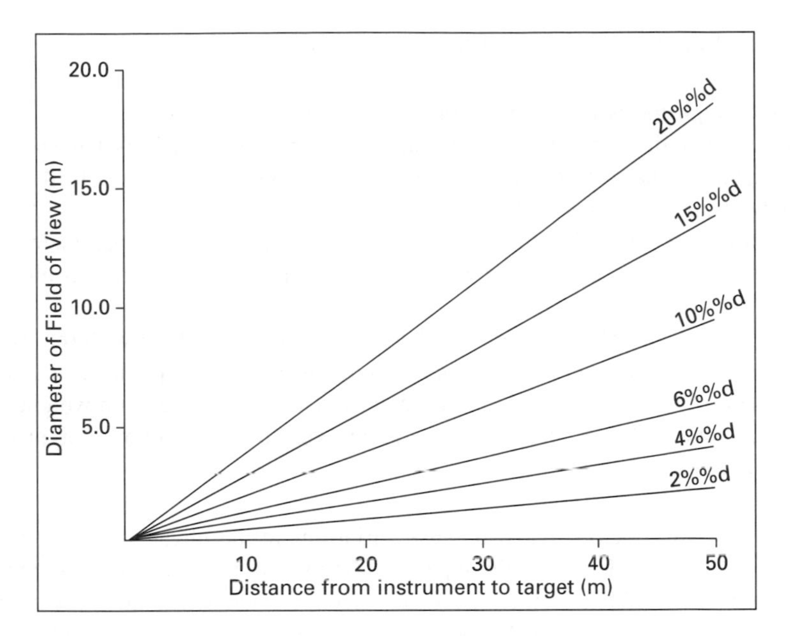

Diagram 6.6 Relationship between cone angle, distance from the sensor to the target and diameter of the FOV on the target for vertical observations of the target.

$$\text{diam}_{\text{FOV},\,\phi} = \text{d} \sin(\eta) \left\{ \sec\left(\phi - \frac{\eta}{2}\right) \sec\left(\frac{\phi + \eta}{2}\right) \right\}$$

in the direction of tilt. While

$$\text{diam}_{\text{FOV},\,90+\phi} = 2d \sec(\phi) \tan\left(\frac{\eta}{2}\right)$$

at right angles to the direction of tilt. Also,

$$\text{area of FOV} \approx \frac{\pi}{4} \{Diam_{\text{FOV},\,\phi} \times Diam_{\text{FOV},90+\phi}\}^2$$

Third the field parameters and spectral observations should be measured to similar levels of accuracy and resolution. The simplest field spectrometers can measure the radiance incident on the sensor to resolutions of about 1 per cent of full scale deflection, and standard errors of about 3 per cent of the full scale deflection. Many better quality field instruments can measure radiance to better than 1 per cent of full scale deflection with standard errors of less than 1 per cent. Before starting to collect field data, take a set of observations on a standard surface, to determine the resolution and standard error that can be achieved with the instrument. The field observations of canopy parameters should achieve similar resolutions and accuracies to that determined for the instrument. To achieve this may require measurement of a number of samples (or sub-samples) as discussed below, and relating the average values to each spectral observation since for n observations

$$T_{\text{Av}} = \frac{T_{\text{Ind}}}{n^{1/2}}$$

The usual method of collecting field data is as follows.

1. Clip and weigh the whole sample for fresh weight.
2. Shuffle or mix the sample and partition it into fractions, such as halves, and then halves again, or quarters. If this sub-sample is still too large then it is further shuffled and halved until the subsamples are suitable for measurement.
3. Sub-samples are selected to measure GLAI and the other relevant parameters. Different sub-samples may be used to measure the different parameters, to sort into species, etc.
4. Each of the sub-samples used are weighed, dried and reweighed, with the weights of the sub-samples and the whole samples being used to extrapolate to estimate the parameters for the whole sample. The standard measure of vegetation weight is of dry vegetation, that is the vegetation is dried in an oven for a minimum of eighteen hours at a temperature of 450°.

There are many factors that will cause variations in the physical parameters of the land-cover of interest. Measuring these parameters usually involves disturbing that land-cover in some way. If this is the case then the experimental design must incorporate sufficient training sites to cover this destructive collection of field data and ensure that an adequate temporal sequence of training sites is available.

In designing field data collection programmes, it is usual that some trade-off between resolution, accuracy and cost has to be made. The analyst should consider carefully the implications of this trade-off in preparing the experimental design. If the analyst is exploring new ideas that may, or may not be successful then a slight reduction in resolution and accuracy in the field data being collected may allow him to collect a mass of data that more exhaustively covers the full range of possible conditions more economically. Such data sets will allow the analyst to verify or to refute the hypotheses being tested quickly and economically, but will not necessarily allow the proposed model to be developed to the required level of resolution or accuracy. Having verified the concept, more expensive data sets can be used to fine tune the model. This approach has the advantage that development of the crude model will provide considerable insights into the processes being studied, and how the model will actually be used; these insights may well influence future experimental design. If the analyst is conducting the work to prepare for an operational programme then resolution and accuracy are of supreme importance. The analyst may choose to take very detailed observations over subsets of the conditions that will be met, and know that the analysis will be good within, but may be less accurate outside of, these subsets.

Impact of variations in canopy reflectance on derived models

The reflectance of a canopy will usually exhibit variations in response due to a variety of causes. Some of these, such as variations in canopy geometry and the impact of wind on the canopy, can be considered to be a random noise component that has little impact on the average taken for a set of observations. Other causes do not act in a random manner, but introduce a bias, either up or down, into the reflectance data. These causes include variations in atmospheric attenuation, and some surface conditions such as crop stress. Since these causes bias the reflectance values they will often reduce the accuracy of any derived models. It is important either to minimize or to correct for their effects as far as possible.

Systematic errors should also be corrected in the data. In practice, many systematic

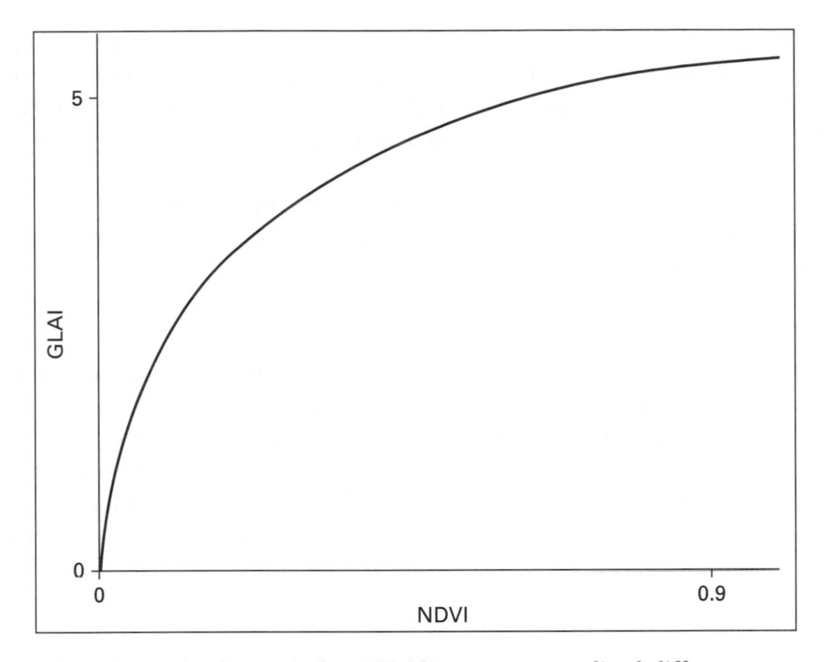

Diagram 6.7 Green leaf area index (GLAI) versus normalized difference vegetation index (NDVI). The diagram shows that NDVI changes with changes in GLAI until GLAI > 6.

errors are too small to justify the significant amount of work necessary to establish a credible correction mechanism. In consequence, they remain in the data and reduce the accuracy associated with estimates derived from the data. However, the analyst cannot make a final decision on this matter without collecting sufficient field data to:

- identify the significant sources of systematic error, so as to correct for them prior to using the data in the estimation model; and
- indicate the magnitude of the residual errors, as indicated by the variance of the data so corrected.

Implementation of models in airborne and space-borne systems

Models developed on the ground generally estimate physical parameters from measurements of reflectance data. Satellite and airborne data are derived from the radiances received at the sensor. To use models that have been developed on the ground requires that the airborne or space-borne sensor data be corrected from a measure of radiance to reflectance at the surface. This requires an atmospheric correction to the sensor data as discussed in section 5.3.

6.3 The use of field data in visual interpretation

Our general knowledge of the physical environment, its processes and how they are manifested in remotely sensed data, are based largely on conceptual models that are constructed from a mix of scientific and human observations. These models are made more relevant to specific conditions by the use of factual information on aspects of the environment. Not always is the process error-free; errors are made in the collection of data, its analysis and in making inferences to amend or to build on existing conceptual models. The process is therefore very much an empirical one of model building and evaluation, with modifications to the models only being accepted after exhaustive evaluation.

This process of conceptualization and quantification of models forms the basis of all decisions made in extracting information by visual interpretation. The process involves two tasks: identification and interpretation.

Identification

This is the process of identifying objects and features in the image or images. It is the simpler of the two tasks, only requiring sufficient information in the imagery to first discriminate the feature from other features and then identify the general geometric shape and characteristics of the feature. If the imagery and facilities used do not allow the routine identification of features then errors will occur in the interpretation. These errors **of non-identification** are called **Class A errors**. They require consideration of the:

- scale of the imagery;
- resolution characteristics of the imagery;
- contrast of the object relative to its surroundings;
- photographic processing;
- the interpretation facilities used and the physiological condition of the interpreter; and
- the knowledge levels of the interpreter and the psychological aptitude of the interpreter for the task.

The first five items are important components of any project design; what is always more difficult to assess is the physiological and psychological aptitude of the interpreter. However, the importance of these factors should not be underestimated, and procedures should be established, and interpersonal relationships maintained that are conducive to a good working environment.

The most reliable way to assess the suitability of an interpreter for a task is through the accuracy assessment. However, accuracy assessment is performed last after all of the work has been done. It is important, therefore, not to depend on this check but to build a positive interpretation environment for the interpreter.

Interpretation

The interpretation of imagery can be at different levels of complexity as discussed in Chapter 3. Whatever the level of complexity, all interpretation tasks may contain the following types of errors.

- A feature may be identified in the image, but not recognized as being important to the interpretation, or **errors of mis-identification** may occur. Errors of this type are called **Class B** errors.
- The feature may be identified and described to the satisfaction of the interpreter, using established conceptual models in the interpretation, but be erroneous. These **errors of mis-interpretation** are called **Class C** errors.
- The features may be identified, their importance recognized, but this importance not quantified to the satisfaction of the analyst. In this case, the analyst may either use field or auxiliary data to describe the feature satisfactorily, or he may isolate the feature as an unresolved object, creating **Class D errors**, or errors of **non-interpretation**.

A major reason why analysts should have considerable knowledge of the disciplines associated with the interpretation is to minimize Class B and Class C errors during the interpretation. Class B errors, arising when the importance of features are not recognized, are minimized through knowledge of the disciplines of interest and the consequent development of good conceptual models. Field data are of some use in reducing Class B during the interpretation process, but are primarily used to minimize Class C and Class D errors. The field data does this by facilitating the inductive and deductive processes that strengthen and improve the conceptual models employed by the analyst, thereby improving the accuracy and efficiency of the interpretation. The field data does this by identifying errors in the interpretation (Class C errors) that will lead to a re-evaluation of that interpretation, and similar interpretation elsewhere in the image. The field data will also explain features (Class D errors), leading to a re-evaluation and re-interpretation.

Conceptually Class C and Class D errors differ in that Class D errors have been recognized as a limitation of the data, facilities, knowledge base or interpretation skill of the analyst whereas Class C errors have not been recognized. Because the Class D errors have been recognized they will be visited and resolved in the field visit. On the other hand, Class C errors would normally be addressed only by driving through areas typical of the interpretation so as to assess the correctness of that interpretation and addressing errors, and their more global implications, when they are found.

To summarize then, Class A errors (or errors of non-identification) will be minimized by ensuring that the imagery used is of adequate resolution, the processing employed does not degrade the photography, the facilities used are of adequate optical and illumination quality and the interpreter is both interested and capable, physically and psychologically, of conducting the interpretation. Class B errors (or errors of mis-identification) are minimized primarily by ensuring that the analyst has adequate knowledge levels of the discipline and the characteristics of the area being interpreted. These errors can be further reduced by the proper use of field data. Class C errors (or errors of mis-interpretation) are minimized by ensuring that the interpreter has adequate knowledge levels of the discipline and the area being interpreted, has knowledge of and is skilled in the methods of extracting information from imagery, conducts adequate field work and is both physically and psychologically suited to the interpretation task. Psychological preparation for interpretation depends on many factors, but it can be affected by physical exhaustion, personal affairs and work conditions. Class D errors (or errors of non-interpretation) are minimized by conducting adequate field work during interpretation and by the knowledge levels of the interpreter.

The amount and type of field data required as part of visual interpretation is thus dependent upon the knowledge levels of the interpreter, the status of the conceptual models constructed by the interpreter and their relevance to the interpretation task at hand. It is therefore impossible to specify the amount of field data that would be required in any given situation. In general, the inexperienced analyst will require more field data than will the experienced analyst, and any analyst confronting new conditions will require more field data than when interpreting conditions similar to those met previously. The analyst has to judge for himself the amount of field data that is required, and the relative density of field data required in the different conditions being met in the interpretation. Some of this field data should be acquired before the conduct of the interpretation task, by moving through the area to gain an appreciation of the area and its conditions. This initial traverse through the area should not be restricted to too small a portion of the area, or too small a sample, as it is important for the analyst to sample the full range of conditions that might be met within the area. Sampling the full range of conditions allows the analyst to be confident that the conceptual models being constructed represent the range of conditions within the area. The general approach to the collection of field data is therefore to:

- drive or move through the area to assess the full range of conditions that will be met and to indicate the range of conceptual models that might need to be developed or used;
- conduct an interpretation of the imagery, which will readily identify many of the existing limitations of the conceptual models held by the analyst and identify confusions that will require either new models or refinements of existing models;
- conduct a field visit to the area, to establish other models as required, and to refine existing models;
- repeat this cycle until the analyst is satisfied with the interpretation results being achieved; and
- conduct a statistically based accuracy assessment of the interpretation results. If this accuracy assessment shows that the interpretation is not of an adequate standard then the analyst has to identify the dominating classes of errors so as to decide on an appropriate course of action.

6.4 The use of field data in the classification of digital image data

Classification was defined as partitioning the response domain into a number of discrete class sub-domains, where individual, or groups of these sub-domains represent different physical classes of land-cover or land use. Each spectral class domain is defined by means of an algorithm that depends upon a number of parameters for implementation. Field data are required to determine estimates of these parameters for each spectral class, or train the classifier. The training data must therefore be representative of all of the physical classes i.e. it is necessary to ensure that the field data covers the full range of conditions that are likely to be met in each class. The safest way to do this is to ensure that the field data adequately samples the whole area covered by each class.

Data must also be sufficient to derive the parameters for all spectral classes contained in the physical class. If the data for the physical class are a poor fit to the probability distribution that underpins the selected classifier algorithm then the data for the physical class may need to be partitioned into a number of spectral classes, as discussed in section 5.6. If this occurs then collection of field data will need to reflect the increased use of that data by the more classes.

6.4.1 Classification using the maximum likelihood criteria

The group of classifiers that use the maximum likelihood criteria are based on the assumption that the digital data for each spectral class are a close approximation to the normal distribution, with the mean and covariance arrays providing the class parameters in m-dimensional space. Classification can proceed in either of two ways: by classification then field verification or by collection of field data then classification.

Classification then verification

Training areas for the physical classes are defined by visual interpretation of the image data. If the physical class spectral data are not a good fit to the normal distribution then clustering is employed to create spectral classes within each physical class so as to classify the image into the spectral classes. Field data are used to determine exactly the physical representation of each spectral class by sampling across the area, and processing the data to create a confusion matrix as discussed under accuracy assessment in section 6.5, to identify the level of correspondence between the spectral and physical classes. If this process shows a high level of correspondence then the classification can be accepted as being ready for accuracy assessment. If significant proportions of individual spectral classes represent different physical classes then the classification work must be reviewed.

Field data collection then classification

In this approach, field data are collected for known samples of all physical classes so as to define physical class training data. The boundaries of these training areas are digitized so as to create training areas in the image data, to determine class statistics and analyse the goodness of fit of the data to the classifier algorithm. If analysis of the spectral distribution in the classes shows that the data are not a good fit to the normal distribution then the data are split into a number of spectral classes using clustering techniques. After the spectral classification the groups of spectral classes are assigned to their relevant physical classes in preparation for accuracy assessment.

6.4.2 Classification using the vector classifier

Unlike the maximum likelihood classifiers, the vector classifier assumes a linear domain for the class data, where changes in response within that domain reflect changes in physical conditions within the class. The class domain is separated from the domain of other classes by a threshold surface at some distance from the linear surface defined by the response values for the nodes used in the classification. The vector classifier is suitable for use where these criteria are met – criteria that frequently occur in the natural environment.

Field data are therefore required to:

- verify that the class data occupy a linear domain;
- assist in defining node response values; and
- develop models to estimate physical parameters from the proportion coordinates derived by the vector classifier.

It is therefore necessary to collect field data that is representative of the full range of class conditions that are likely to be met and assess the linearity of the reflectance data. If the data occupies a linear domain, use the observed data to assist in identifying node response values. Field spectral data should also be collected for conditions that represent node conditions such as full green or brown herbage canopy conditions. Finally, field data is used to develop models to estimate physical parameters from the proportion coordinates derived by the vector classifier. These models relate variations in physical conditions to the proportion coordinates. Examples of this are to convert vector classifier proportion coordinates into estimates of herbage LAI and biomass.

6.5 The stratified random sampling method

An important technique in assessing the accuracy of information derived from remotely sensed data is by the use of a stratified random sample survey. This method comprises the following components

- delineation of strata
- determination of the strata sampling fractions and the number of samples
- selection of samples
- measurement of samples
- comparison of the samples with derived information

each of which are now considered in turn.

6.5.1 Delineation of strata

The simplest way to conduct a sample survey for accuracy assessment is to select a set of unbiased samples across the area to be assessed. The unbiased set of samples is selected as a random sample from a population using a specific frame of reference which describes and contains the population in the random selection process. This frame of reference might be the range of map coordinates covering the area, a telephone listing of all property owners in the area, a graphical list (as a map) of all cadastral land parcels numbered sequentially from 1 to *n*, etc. In this process the whole frame of reference, whether it be the coordinates for the whole area, the complete telephone listing or the set of numbers for all cadastral land parcels, is considered as the one strata or domain which is treated as a unit. This is suitable if all classes to be assessed represent similar proportions of the total area, so that the random sampling will select samples from each class.

Generally, however, this condition is not met. Often some classes only represent a small fraction of the total area, the mix of classes varies significantly in different parts of the area classified, or different processing is employed in different parts of the area. When this occurs then the random sampling process will not select representative samples for all areas or classes. The solution is then to partition the total area into a number of strata, and conduct random sampling within each of these strata, to ensure that all classes are suitably represented.

Strata may be chosen on the basis of geomorphic, climatic, land-using and land parcel size conformity, with each strata representing areas which are relatively homogeneous in the chosen characteristic(s). In general, the strata will not be chosen on the basis of classes within digital classification since errors in the classification may introduce bias in the stratification. However, if different processing procedures are used in different parts of the area then these parts form strata boundaries because of differences in the classification accuracy in the different parts.

Once the strata have been selected they must be mapped over the frame of reference that will be used to select the samples. If coordinates are to be used as the basis for selecting the samples then the strata need to be mapped on an appropriate map. If the frame of reference is the listing of telephone numbers then the population of telephone numbers needs to be partitioned into the different strata.

Within the chosen strata the samples then still need to be selected in an unbiased way, and the simplest way is to use a random sampling process. The random samples can be chosen from various frames of reference within the strata, as long as the chosen frame of reference includes, but does not duplicate, all members of the strata population. Suitable frames of reference include those discussed above for random sampling.

6.5.2 Determination of the sampling fraction and the number of samples

The sampling fraction (SF) is the fraction of the whole area, or the frame of reference, that will be sampled. If the whole area, or the whole frame of reference, is sampled (i.e. SF = 1) then the samples will collect all of the information in the area and will represent

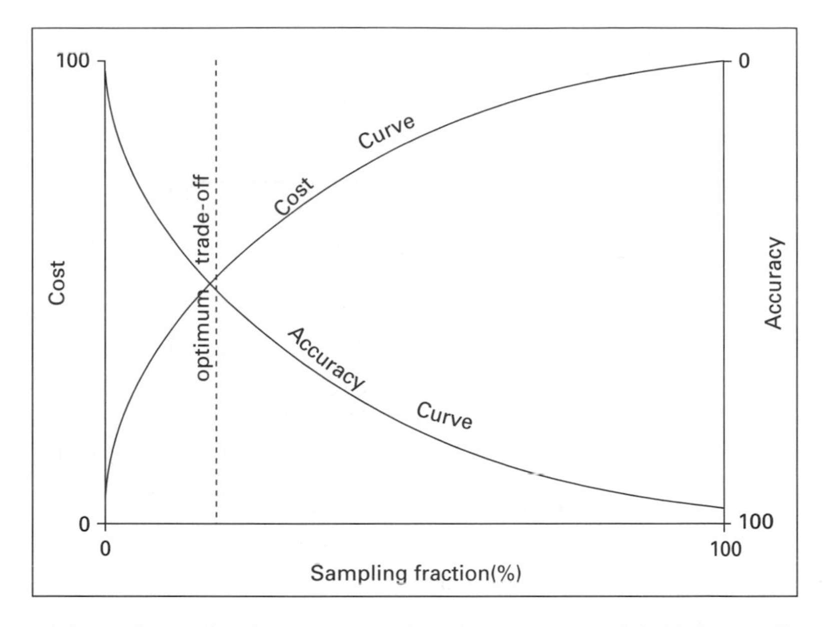

Diagram 6.8 Relationship between sampling fraction, cost of field data collection and accuracy.

the 'correct' situation. As SF is reduced from 1, so the proportion of the area being sampled is reduced; the probability that the sample will not be representative of the population increases but the cost of the sampling decreases (Diagram 6.8). The objective is to select a SF so that the samples are a reasonable representation of the population, whilst the cost of the sampling is minimized. The most suitable SF depends on the size and number of samples as well as their nature or shape. The SF could all be contained in the one, large sample, but this will lead to errors in the sampling. Generally the more, but smaller the individual samples, the better, as discussed below. For remote sensing a SF of 1 per cent is the largest that is usually required as this SF has been found to provide samples that are representative of the whole population at a realistic cost, when the individual samples are of 1 ha in size or smaller. The smallest SF to be used should be larger than 0.01 per cent as this SF has been found to provide a sample that is not representative of the population in some studies, at this sample size.

Once the SF has been selected this indicates the size of the area to be sampled. This area is then divided amongst the number of samples to be taken. In general the more, and the smaller, the samples, the better the results. In practice, the more samples, the higher the cost; also it is not realistic to have the samples smaller than the resolution of the mapping being assessed. With small samples it is often very difficult and expensive to locate the samples in the field and in the mapping, so the analyst needs to consider very carefully the question of sample size.

If point or area samples are chosen to be a specific size, then the minimum size is controlled by the spatial registration and resolution of the data. If the image data have been registered to a map with a standard error of $\pm s$, then 68 per cent of the time the actual pixel will be within $\pm s$ of the nominal or known position of the pixel, and 95 per cent of the time the actual pixel will be located within a distance of $\pm 2s$ of its nominal position. This means that the actual pixel position is within an area, at the 95 per cent confidence level, that is the pixel size $+2s$ on all four sides (Diagram 6.9), i.e. the pixel is located somewhere within a distance of (pixel size + $4s$) in each direction. Consider

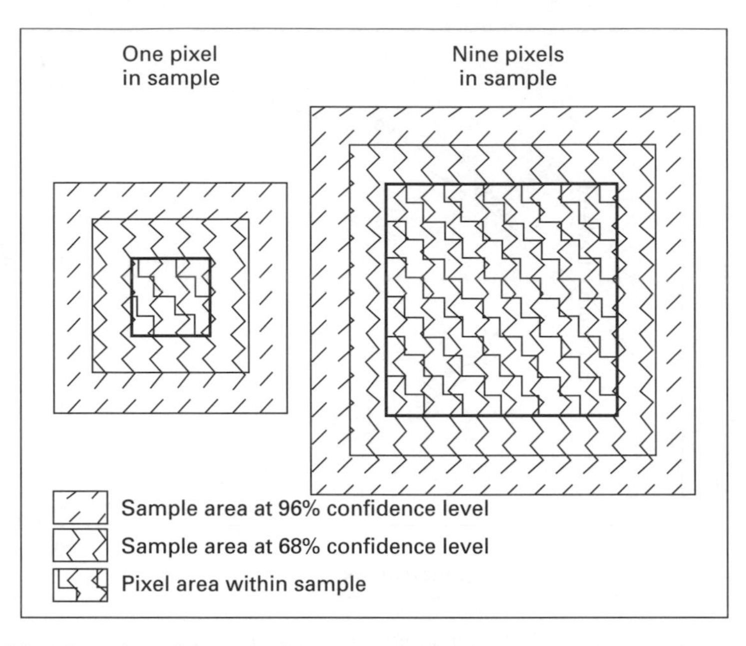

Diagram 6.9 Sampling for a pixel in scanner data.

a Landsat TM image that has been rectified with a standard error of ±20 m in both directions. A sample for a pixel will need to be of size 30 + 80 m at the 95 per cent confidence level, or an area of 110 m by 110 m. Clearly, whilst the actual pixel is likely to be within this area, so will other pixels; also the actual location of the pixel is unknown within this area. The samples of this size therefore must be chosen to be reasonably homogeneous in the physical parameters being measured.

The accuracy of the rectification has less impact on larger samples. Consider samples that are to be n by n pixels in size, then the sample will have sides of length ($n \times$ pixel size $+ 4s$) at the 95 per cent confidence level. Consider Landsat TM data where the pixel sizes are 30 m by 30 m, for a sample of 10 by 10 pixels, rectified to an accuracy of ±20 m s.e. The sample size will be a square where

$$\text{length of side} = [n \times \text{pixel size}] + 4s$$

where n is the number of pixels along, or across the side of a sample, and s is the standard error associated with rectification of the image. So

$$\begin{aligned}\text{length of side} &= 10 \times 30 + 4 \times 20\\ &= 380\text{ m}\end{aligned}$$

Landsat MSS is somewhat different since the separation of the pixels along a scanline is only 60 m, not the size of a pixel (80 m). If the samples are to be 10 by 10 pixels (of size 80 m by 80 m) and the rectification accuracy had a standard error, s, of 50 m, then the sample size (in m) along the scanlines will be given by

$$\begin{aligned}\text{size} &= [\text{pixel size} + (n - 1) \times \text{pixel interval}] + 4s\\ &= [80 + 9 \times 60] + 4 \times 50\\ &= 820\end{aligned}$$

Across the scanlines, we have

$$\begin{aligned} \text{size} &= [n \times \text{pixel size}] + 4s \\ &= [10 \times 80] + 4 \times 50 \\ &= 1000 \end{aligned}$$

For the 10 by 10 pixel sample with TM data only 80 m is due to the errors in position, representing 38 per cent of the sample area in comparison to the situation when sampling for one pixel where 93 per cent of the sample is due to registration errors, for the condition that the standard error is ±0.67 of the pixel size.

For TM and SPOT data the ratio of the field area sampled to the pixel based sample is given by

$$R = \frac{[nd + 2Cs]^2}{(nd)^2}$$
$$= \left[\frac{1 + 2Cs}{nd}\right]^2$$

where

s = standard error
C = multiple of p to set confidence range
d = pixel dimension
n = number of pixels in the sample

and the power of 2 is because areas are being used in R.

The relationship between this ratio and the proportion (s/d) is given in Diagram 6.10 when the envelope about the image sample is $\pm 2s$ (i.e. a 95 per cent confidence interval). This shows the inefficiency introduced into the collection of field data when (s/d) is large, due to low rectification accuracies, and when n is small. This is a particularly good example of the use of high resolution data that is registered to the coarser multispectral data being used in the analysis, to define field sample areas that are not much larger than the actual pixel size.

If only a small number of samples are chosen, say 10, then the number of errors discovered by those samples would also be small, say 0, 1 or 2. However the achievement of no errors does not imply that the total population is error free; this result may occur by chance since the number of samples is so small. Let the proportion of the classification that is in error be specified by p. The probability of discovering no errors when taking x samples from this population with p per cent actual errors is given by the binomial distribution as $(1 - p)^x$, and the probability of discovering y errors in x samples is ${}^xC_y \times p^y \times (1 - p)^{(x-y)}$.

Table 6.2 shows the probability of scoring no errors in samples of varying size taken from a population with a range of error proportions, p. This table indicates that no error sampling results can quite easily arise with small sample populations, even when the actual error rate is quite high. Taking the conventional probability level of 0.95/0.05, or 95 per cent/5 per cent, the table can be divided into two parts as has been done by the figures shown in bold. Above and to the left of these, the probabilities of obtaining error free sampling results are high, even when the actual errors are present in significant

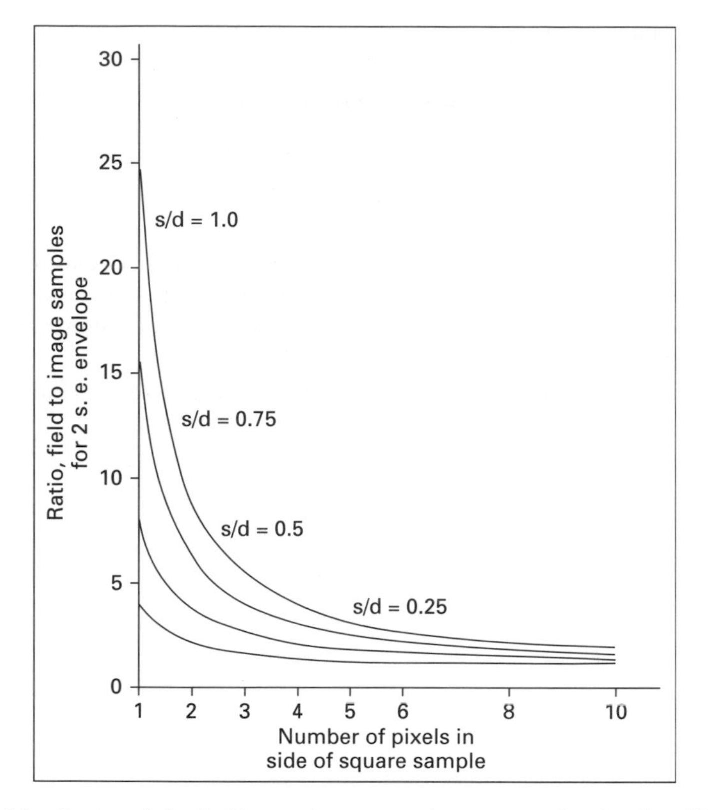

Diagram 6.10 Ratio of the field sample area to image sample size for different numbers of pixels in the image samples, for different rectification accuracies (different s/d *values).*

proportions. Below and to the right of them there is a small to negligible chance of getting an error free result from the sampling when there are actual errors in the population.

Thus if the permissible error rate in the image interpretation is set at say 85–90 per cent, as suggested in the US Geological Publication on land-use mapping (USGS Circular 671) or as required in an operational programme specification, then the sample size in each category or strata necessary for 85 per cent interpretation accuracy is at least 20 samples, and at least 30 samples for 90 per cent accuracy. The tables can thus be used to determine the minimum number of samples that are required for any given interpretation accuracy. Tables 6.3 to 6.5 inclusive provide more detailed calculations of the probabilities of scoring errors in samples of varying sizes with the specific interpretation accuracy levels of 85, 90 and 95 per cent respectively.

6.5.3 Selection of samples

It is necessary to consider both the type of samples to be used and how they will be selected. There are at least two types of samples that can be used: uniform sized or irregular ones.

Uniform sized samples are all of the one size and shape for a particular project or task, without consideration for existing physical or administrative boundaries. The minimum

Table 6.2 Probability of scoring no errors in samples of varying sizes from a population with a probability, p, *of real errors in each pixel*

p	$n = 5$	10	15	20	25	30	35	40	45	50	55	60
0.01	0.9510	0.9044	0.8601	0.8179	0.7778	0.7397	0.7034	0.6690	0.6362	0.6050	0.5754	**0.5472**
0.05	0.7738	0.5987	0.4633	0.3585	0.2274	**0.2146**	**0.1661**	**0.1285**	**0.0994**	**0.0769**	**0.0595**	0.0461
0.10	0.5905	0.3487	0.2059	**0.1216**	**0.0718**	0.0424	0.0250	0.0148	0.0087	0.0052	0.0030	0.0018
0.15	0.4437	0.1969	**0.0874**	0.0388	0.0172	0.0076	0.0034	0.0015	0.0007	0.0003	0.0001	0.0000
0.20	0.3277	0.1074	**0.0352**	0.0115	0.0038	0.0012	0.0004	0.0001	0.0000	0.0000	0.0000	0.0000
0.25	0.2372	**0.0563**	0.0134	0.0032	0.0008	0.0002	0.0000	0.0000	0.0000	0.0000	0.0000	0.0000
0.30	0.1681	**0.0282**	0.0047	0.0008	0.0001	0.0000	0.0000	0.0000	0.0000	0.0000	0.0000	0.0000
0.40	**0.0778**	0.0060	0.0005	0.0000	0.0000	0.0000	0.0000	0.0000	0.0000	0.0000	0.0000	0.0000
0.50	**0.0313**	0.0010	0.0000	0.0000	0.0000	0.0000	0.0000	0.0000	0.0000	0.0000	0.0000	0.0000

Figures in bold indicate approximate 0.05 probability level.

Table 6.3 Probability of scoring errors in samples of varying sizes from a population with real errors of 15 per cent, i.e. an interpretation accuracy of 85 per cent

Sample size	Number of errors					
	0	1	2	3	4	5
15	**0.0874**					
20	0.0388	0.1368				
25	0.0172	**0.0759**	0.1607			
30	0.0076	0.0404	0.1034			
35	0.0034	0.0209	**0.0627**	0.1218		
40			0.0365	0.0816		
45			0.0206	**0.0520**	0.0963	
50				0.0319	**0.0661**	0.1072
55				0.0189	0.0434	0.0781
60					0.0275	**0.0544**
65						0.0365

Figures in bold indicate approximate 0.05 probability level.

Table 6.4 Probability of scoring errors in samples of varying sizes from a population with real errors of 10 per cent, i.e. an interpretation accuracy of 90 per cent

Sample size	Number of errors			
	0	1	2	3
15	0.2059			
20	0.1216			
25	**0.0718**	0.1994		
30	0.0424	0.1413		
35	0.0250	0.0973		
40		**0.0657**		
45		0.0436	0.1067	
50		0.0286	0.0779	
55			**0.0558**	0.1095
60			0.0393	0.0844
65				**0.0636**
70				0.0470

Figures in bold indicate approximate 0.05 probability level.

size of the samples will need to consider the influence of data resolution as discussed earlier. They can be either circular, centred on a location, or rectangular as is suitable to the task. The samples are selected as a random sample within each strata.

A major difficulty with this type of sample is to identify the location and boundaries of the sample in the mapping being assessed, and on the ground. The samples may contain information on more than one class, or the condition of the land-cover within the sample may vary across the sample. If the samples contain more than one class then the

Table 6.5 Probability of scoring errors in samples of varying sizes from a population with real errors of 5 per cent, i.e. an interpretation accuracy of 95 per cent

Sample Size	Number of errors 0	1	2
45	0.0994		
50	0.0769		
55	**0.0595**		
60	0.0461	0.1455	
65	0.0356	0.1219	
70		0.1016	
75		0.0843	
80		0.0695	
85		**0.0572**	
90		0.0468	0.1097
100			0.0812

Figures in bold indicate approximate 0.05 probability level.

proportions of the classes need to be determined and carried through the accuracy assessment. If the assessment is on condition of the covers then variation in condition can create difficulties in the assessment. It may be necessary to select other samples that do not contain too much variation in condition within the individual samples.

This form of sample is very appropriate for assessing the accuracies of condition parameters at the resolution of the data, e.g. to assess mapping of vegetative biomass, cover, water quality, etc. For these assessments, the samples should lie within the one cover type for the assessment, and be reasonably uniform in cover condition to avoid, as much as possible, discrepancies between the actual and the measured cover due to discrepancies between the areas mapped and the area field sampled. Each sample would normally be designed to match one or more pixels in the mapping being assessed.

Uniform samples are also good for thematic classifications, such as soils, land systems and land-use mapping, particularly where the mapping does not follow (or may not be adequately represented by) samples that are constrained by administrative or physical boundaries. The samples would often be large relative to the resolution in the data, and may contain more than one class within each sample so that the samples need to be located and then mapped for comparison with the classification.

Samples of this type would normally be selected as a random sample from a regular grid across the area to be assessed. The random selection of sample sites can be done with the use of a table of random numbers (Table 6.6). The gridlines, both northings and eastings, are both numbered and a pair of random numbers specifies each coordinate point of a sample.

Irregular samples are those in which the sample boundaries are defined by physical or administrative boundaries, such as property or paddock areas. The main advantages of this type of sample are that the samples can be unambiguously located in the mapping and on the ground, and the sample can be readily revisited if required. Another advantage is that the individual fields will often contain the one land use or land-cover, but the boundary of the field will contain mixed pixels that need to be considered.

Irregular samples may have the disadvantage that they may be very large relative to

Table 6.6 Table of random numbers

1616	5704	8171	1746	5329	7346	4273	7763	6258	6059
9863	8952	7723	6108	6390	8038	4271	8570	0481	0550
0103	0935	0254	5196	9275	5829	2423	2519	8997	9129
2907	1634	4922	5296	8934	1711	0691	2438	5506	8359
7261	8054	7099	2464	1138	8365	2723	4037	8458	4853
7111	4182	7937	0045	9854	5289	2634	4013	6038	0886
6105	6618	7682	1118	6190	9063	7857	3206	3995	7594
8189	4234	0049	9753	3316	2691	5758	4248	5105	4827
1024	9084	2216	2696	5411	0196	5881	3797	8098	7281
1428	3343	0132	5839	1954	5657	2358	2487	7736	2097
3541	1789	8704	2832	1345	5903	9108	6924	8444	4283
0789	3687	9873	7764	7519	0561	1164	3175	4938	9660
2759	1558	1968	9547	2569	1190	2619	0740	8359	9095
9598	4552	2735	8681	1629	3760	3935	0524	4900	2907
1295	7272	8184	3658	0510	7050	3104	1267	7401	7290
3523	0668	5250	3955	9228	2889	6487	8000	8453	9797
8633	9573	8092	2649	5450	4121	0662	7391	3505	2137
0282	9623	1646	1551	6031	5527	8414	7158	9471	4835
4446	3496	3268	4822	4017	4325	3331	2626	5934	9900
0877	0719	9446	1751	0373	9989	2844	1687	5616	5609
6159	3708	0846	5676	2948	3387	7079	0380	9681	7968
6770	1801	6719	2949	5867	0856	2724	2070	4631	0432
2309	0879	1878	0032	8674	7855	5572	5854	7607	5373
8940	2639	7458	5955	8711	7406	4946	3194	8666	6697
8495	6642	9074	1371	0071	2441	6762	3892	3926	3029
5214	4902	1931	2815	5101	1909	9794	5243	2221	1766
8956	3141	3787	2816	6248	0184	4606	0439	9410	7621
6594	0593	0668	3472	7317	6534	0065	7578	2397	1304
1308	1575	0283	4826	5377	6296	5652	2826	1215	7553
0318	3357	1671	6027	1518	3932	3701	0586	2514	3541
1004	0095	8504	3280	1901	8503	2929	8004	2152	1476
2394	9728	6043	4225	2648	4813	3468	3922	7485	0325
3563	4290	9074	3317	5877	8336	7622	0089	6155	1317
4286	0336	4533	6077	7292	1076	2255	1100	3760	4773
6726	9287	0996	8537	8216	3901	7005	1266	1739	9934
9193	8856	3576	9735	1937	1466	0757	2441	0690	0772
3714	7335	3201	0794	7828	9033	7156	6377	8924	2428
0746	5058	0873	4297	2042	6468	4835	0438	2828	3694
9218	0946	9499	1741	2860	6794	2654	6370	8473	7661
0049	9843	3967	6840	4131	9228	4957	1555	1181	4189
0859	4141	3359	4328	1451	0271	2445	4157	2211	7979
6705	1954	3233	3468	2793	3935	6251	3555	4099	4619
2499	4806	9641	2125	2903	5771	9649	9474	9890	2152
6586	2746	7093	2739	6437	0163	2103	4378	1874	7707
5270	0320	8496	1437	5105	6399	8102	8456	1778	4845
3288	2993	5821	7105	6858	7908	8637	9876	7045	6623
5416	3940	9857	0205	6515	7323	5151	7506	3813	5168
9522	1859	5457	4422	7235	8124	1494	2404	4226	9214
9310	2794	9045	3933	5026	8846	9057	4047	7163	6259
1920	8520	1567	7803	3223	5059	2483	6499	1800	7850

Each digit is an independent sample from a population in which the digits 0 to 9 are equally likely, that is each has a probability of $\frac{1}{10}$.

the sampling fraction, and so the area sampled, to ensure that the number of samples is adequate, may be larger than is necessary.

The samples are normally chosen by first deciding on the sample units (e.g. paddocks or properties), identifying these units on a map and then numbering each unit on the map. These numbers are then used with a table of random numbers to select the samples. The paddocks or properties can be numbered in any convenient way provided it is, unbiased and the numbers on each sample unit are unique. It may be convenient to use telephone numbers if all properties have the telephone.

With samples of this type there may be two distinct forms of error to be considered. There are **errors in classification** itself, i.e. errors in assigning pixels that are of one class to that class. The frequency of this type of error will be determined by comparing the classifications within the sample. Accuracy assessment for this purpose can be done by defining the sample boundaries within the fields so as to avoid the mixed pixels around the border of the field. There are also errors due to **mixed pixels** along the boundary of areas of land-cover or land-use types. The errors associated with mixed pixels are likely to be quite different to the errors associated with the centres of fields, so that it may be necessary to consider these two sources of error quite separately. Digitization on the border of fields may not be suitable because slight errors in the location of this border in either mapping may introduce significant 'apparent' rather than 'real' errors into the accuracy assessment. A solution is to create samples that are larger than the fields and compare the areas of the field as stated in the field and the mapping.

6.5.4 Measurement of samples

The samples are located in the field and the parameters of interest are measured within the sample area. The method of measurement depends on the parameters to be measured: land use and land-cover, or condition parameters.

Land use and land-cover A typical mapping of land-cover or land use within a rectangular sample is shown in Diagram 6.11. For land-use or land-cover accuracy assessment the best method is to map the sample areas on the ground, aided by current, high-resolution aerial photographs. Visual interpretation of moderate scale aerial photographs, supported by extensive field visits would be the second-best choice. The mapped field information for a sample is then compared to the same area as mapped by the remotely sensed data, using either of two methods.

1. In the first method an overlay of one mapping onto the other is made as shown in Diagram 6.11, and a dot grid is laid on top. A count of the dots within each subsample area and for the total sample computes the proportion of the sample area covered by each subsample area. These proportions are then used in accuracy assessment (section 6.6).
2. Alternatively, we can digitize the field mapping to create a theme image file of the field data. This is compared, on a pixel by pixel basis, with the classification theme file derived from the remotely sensed data.

Condition parameters The sample area is defined and the physical attributes of interest are measured for a random distribution of points across the sample, so as to derive estimates of the parameters for the whole sample.

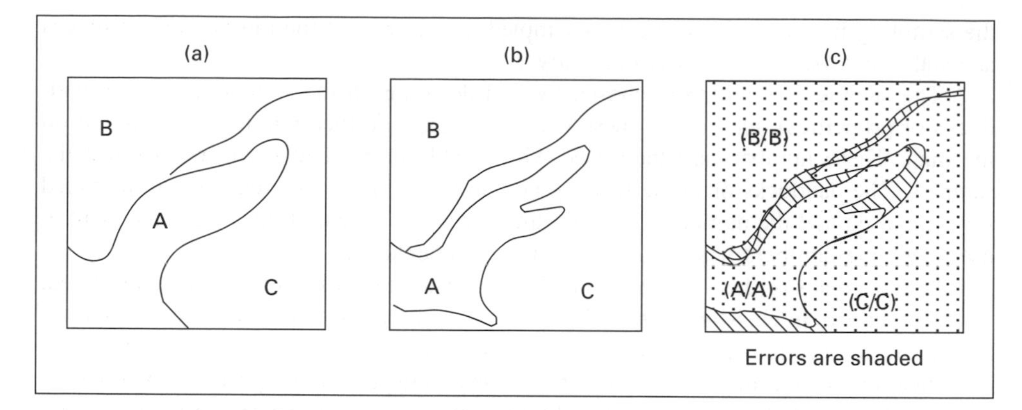

Diagram 6.11 Mapping of a sample area of a land-use map for accuracy assessment: (a) image classification; (b) field mapping; and (c) overlay of both.

6.6 Accuracy assessment

6.6.1 The role of accuracy assessment

Much has been made of the opportunities that exist for errors to arise in extracting information from remotely sensed data. Resource managers, being well aware of this characteristic, will want to weigh the value of this information relative to other information, in making decisions. They will want to decide when to use this information as the basis of final decisions, and when to use it as an indicator, to be supplemented by other information on which to base final decisions.

If resource managers weigh the value of this information too lowly then they will not be making the best use of the information available to them; it will represent a waste of expensive processing time and field data collection. If they weigh the information too highly then they may make erroneous decisions that might reflect on them, and on this source of information. It is therefore important to provide the users of this information with an objective means of properly weighing this information relative to other types of information that they might use.

A very good basis for weighting information is the accuracy assessment done on the information, as long as that accuracy assessment is relevant, representative of the population, easily understood and usable and it can be related to other information.

Accuracy assessment can assess many facets of the information derived from remotely sensed data. The resource manager, wishing to use the information for a specific purpose or purposes will find that some forms of accuracy assessment are more relevant than others. For example, in digital classification the manager may be interested in the accuracy of areas estimated from the classification. If so, the accuracy assessment should be of the accuracy of determining areas rather than the accuracy in identifying fields. Alternatively, the manager may be interested in the locational accuracy of the classification so as to relate this information to other types of information or data. In visual interpretation the manager may be interested in the positional accuracy of the interpreted class boundaries, or in the accuracy of classifying specific locations, and so on.

The easiest way to ensure that the accuracy assessment is representative of the population is to use a statistically valid and sufficient sample to conduct the accuracy assessment. The sample must therefore:

- cover the area of interest;
- represent all classes;
- be distributed in an unbiased way;
- be of sufficient size; and
- select appropriate sample types.

It is essential that accuracy assessment results be stated simply and in an unambiguous fashion. Simplification of accuracy assessment is often likely to be achieved by some compromise between theoretical precision and purity on the one hand and practical simplicity on the other. Consider the solution used in topographic mapping where the position of contour lines are stated as being '*x* per cent accurate *y* per cent of the time' where a contour is considered to be accurate if it is within ±(half a contour interval) from its 'true' position. Statistically, the contours are plotted with accuracies related to the ±(half contour interval) and the '*x* per cent accuracy *y* per cent of the time' rule gives a confidence interval.

The accuracy of information derived from remotely sensed data often needs to be related to the accuracy of other information. This is necessary to:

- allocate relative weights to the different sets of information in making decisions; or
- to derive the accuracy of information extracted from combinations of the parent information types.

The issues raised in relating one set of information to others has not been adequately addressed in the literature. The discussion in section 6.1 identified two complementary indicators of the quality of information derived from remotely sensed data: accuracy of information (6.1); and consistency of information (6.3).

6.6.2 Comparison of field data with classification results

Accuracy assessment of information extracted from remotely sensed imagery depends upon whether the information is concerned with classification or estimation.

Classification

There are three different aspects of classification that may require accuracy assessment.

1. Classification accuracy – the accuracy of classification of individual pixels or image elements.
2. Accuracy of identifying production units such as paddocks – a management unit can be delineated even though some of the pixels contained within the unit might be wrongly classified. The important criterion is that sufficient pixels are properly classified so as to assign the paddock to the proper class. It is thus a much less rigorous criterion for accuracy assessment than classification accuracy, yet it is adequate for many purposes.
3. Boundary definition – the accuracy of delineation of the boundaries of management units such as paddocks. The accuracy of delineation of the mixed boundary pixels will usually be different to the accuracy of classification of the internal pixels in an area of the one landclass. Yet this accuracy is important if the resource manager is concerned with areas, or with an accurate definition of the boundaries of the paddocks.

Table 6.7 Comparison of the pixel counts for the combinations of classes in two classification theme files in which one file is from image processing and one is derived from field data

Image Classification	Field Classes						
	Cropping	Pasture	Forest	Soil	Water	Urban	Total
(a) In one sample							
Cropping	82	–	–	–	–		82
Pasture	6	15	–	–	–		21
Forest	–	–	2	–	–		2
Soil	–	3	–	–	–		3
Total	88	18	2	–	–		108
(b) For all samples in one strata							
Cropping	1 421	43	–	5	–		1 469
Pasture	247	1 360	–	81	1		1 689
Forest	2	104	288	–	1		395
Soil	–	29	–	494	–		523
Water	–	1	3	–	104		108
Total	1 670	1 537	291	580	106		4 184
(c) For all samples for all strata							
Cropping	3 694	106	–	74	–	–	3 874
Pasture	432	11 872	106	234	6	1	12 651
Forest	36	327	1 096	–	1	–	1 460
Soil	59	221	–	736	–	–	1 016
Water	–	–	1	1	429	13	444
Urban	–	–	–	21	–	53	74
Total	4 221	12 526	1 203	1 066	436	67	19 519

Classification accuracy

Accuracy assessment of an image-based classification against field data will create a 'confusion' matrix (Table 6.7) which may contain either counts or proportions depending on how it is constructed. A confusion matrix can be constructed for each sample, similar to Table 6.7(a). The confusion matrix is a two-dimensional array, with one dimension being the image-based, and the other dimension being the field-based classes. Each of the classes that occur in both themes are represented by one location in that dimension of the confusion matrix. During comparison the class assignments in the two theme files for each pixel are extracted. The location of each of these classes on the axes of the confusion matrix is determined, and the point in the matrix at the intersection of these values in each dimension is incremented by one to indicate that another pixel has that particular combination of classes. The action is the same as the construction of a 2–D histogram, except that in this case the dimensions represent classes in theme files rather than values in wavebands. The results of this comparison is a confusion matrix of the form shown in Table 6.7.

The values in the matrix may:

1. indicate pixel counts, as is usually the case when the comparison is of digital theme files from either digital classification or digitization of visual interpretation as is shown in the example here; or
2. they may be the sum of the proportions of the sample areas as determined by the use of a dot grid when the information is derived by visual interpretation.

In Table 6.7(c) the total number of pixels of crop from the field data are 4221. If the values in the crop column are divided by this value then the resulting proportions are the accuracies of classifying crop. The accuracy of classifying crop, Pr{crop/crop} = 3694/4221 = 0.875. The error of omission, E_o, is the error of classifying cropping as another class, so that

$$E_o\{\text{pasture/crop}\} = 432/4221 = 0.102$$

This processing is done for each column in the matrix to give the error of omission for each class. The error of commission, E_c, or the error of classifying a pixel as crop when it is another class is the value in the row that contains the accuracy of classification (Diagram 6.12). The error of commission for crop is

$$E_c\{\text{crop/pature}\} = 106/12\,526 = 0.008$$

It can be seen that the errors of omission in relation to one class are the errors of commission in relation to a second class and vice versa. The probabilities have been computed from Table 6.7(c) and presented in Table 6.8. This table shows that the probabilities of correctly classifying a class are on the diagonal, the errors of omission in relation to that class are the other values in that column, and the errors of commission in relation to that class are the other values in the same row. Thus the accuracy of mapping cropping is 87.5 per cent and the main error of omission, mapping cropping as pasture, is 10.2 per cent whilst the largest error of commission is mapping soil areas as crop at 6.9 per cent.

The Central Limit Theorem states that the proportions in this table are average proportions and as such they are approximately normally distributed as long as more than 30 samples have been used to derive the proportions. The standard error associated with each proportion, p, can be calculated from the equation:

$$\text{s.e.}(p) = p(1 - p)$$

so that the variances associated with each proportion can be determined and used in the analysis of the propagation of error. The variances associated with the proportions in Table 6.8 are given in Table 6.9.

If probability confusion matrices similar to that shown in Table 6.8 are constructed for each sample, and for the average values for the strata then the consistency of these accuracies or probabilities can be calculated, as they can for the whole mapping.

The confusion matrix can also be used to determine the proportions of each class in the classification that belongs to each actual class, by dividing the values in each row by the sum in that row as shown for the data from Table 6.7(c) in Table 6.10, with variances in Table 6.11.

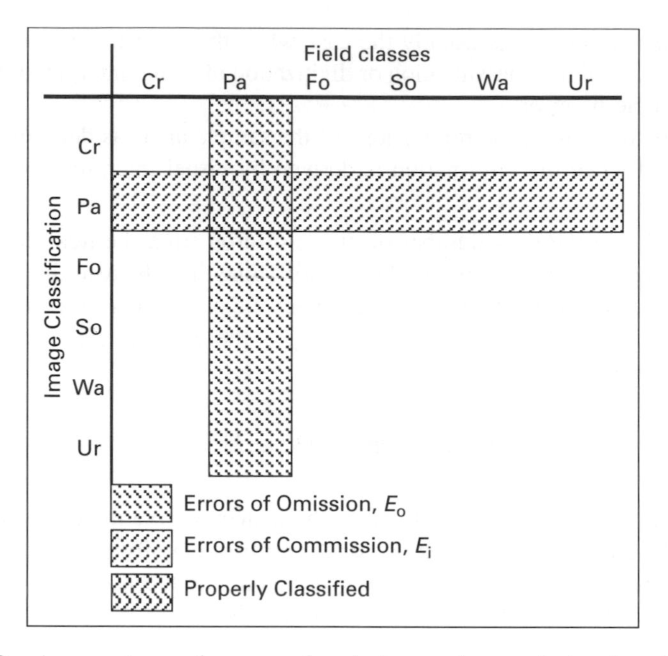

Diagram 6.12 Accuracies and errors of omission and commission in relation to the one class.

Table 6.8 Probabilities of occurrence in relation to total actual occurrence of a class

Image classes	Field classes					
	Cropping	Pasture	Forest	Soil	Water	Urban
Cropping	0.875	0.008	0.0	0.069	0.0	0.0
Pasture	0.102	0.948	0.088	0.220	0.014	0.015
Forest	0.009	0.026	0.911	0.0	0.002	0.0
Soil	0.014	0.018	0.0	0.690	0.0	0.0
Water	0.0	0.0	0.001	0.001	0.984	0.194
Urban	0.0	0.0	0.0	0.020	0.0	0.791
Total	1.0	1.0	1.0	1.0	1.0	1.0

Table 6.9 Variances associated with average proportions given in Table 6.8

Image classes	Field classes					
	Cropping	Pasture	Forest	Soil	Water	Urban
Cropping	0.1094	0.0079	0.0	0.0642	0.0	0.0
Pasture	0.0916	0.0493	0.0803	0.1716	0.0138	0.0148
Forest	0.0079	0.0253	0.0803	0.0	0.0020	0.0
Soil	0.0138	0.0177	0.0	0.2139	0.0	0.0
Water	0.0	0.0	0.0010	0.0010	0.0157	0.1564
Urban	0.0	0.0	0.0642	0.0	0.0	0.1653

Table 6.10 Probabilities of pixels classified as being in a class being in that and other classes

Image classes	Field classes						
	Cropping	Pasture	Forest	Soil	Water	Urban	Total
Cropping	0.95	0.03	0.0	0.02	0.0	0.0	1.0
Pasture	0.03	0.94	0.01	0.02	0.0	0.0	1.0
Forest	0.03	0.22	0.75	0.0	0.0	0.0	1.0
Soil	0.06	0.22	0.0	0.72	0.0	0.0	1.0
Water	0.0	0.0	0.0	0.0	0.97	0.03	1.0
Urban	0.0	0.0	0.0	0.28	0.0	0.72	1.0

Table 6.11 Variances of the proportions given in Table 6.10

Image classes	Field classes					
	Cropping	Pasture	Forest	Soil	Water	Urban
Cropping	0.0443	0.0266	0.0	0.0187	0.0	0.0
Pasture	0.0330	0.0378	0.0083	0.0182	0.0	0.0
Forest	0.0247	0.1738	0.1872	0.0	0.0	0.0
Soil	0.0847	0.1702	0.0	0.1996	0.0	0.0
Water	0.0	0.0	0.0022	0.0022	0.0326	0.0284
Urban	0.0	0.0	0.0	0.2033	0.0	0.2033

The data in Table 6.10 can then be used to estimate the area of each land use or land-cover within the total area mapped. If the classification results in estimated areas of crop, E_C, pasture, E_P, etc., then the actual area of crop, $\underline{\overline{A}}_C$, pasture, $\underline{\overline{A}}_p$, etc. are:

$$\underline{\overline{A}}_C = 0.95E_C + 0.03E_P + 0.00E_F + 0.02E_S$$

$$\underline{\overline{A}}_p = 0.03E_C + 0.94E_P + 0.01E_F + 0.02E_S$$

.

.

.

etc.

The confusion matrix, when converted into probabilities of the total number of pixels in that class indicate the average accuracy of classification and average error of omission and commission. It should also be seen that the sum of the accuracy of classifying a class, and the error of omission associated with that class should sum to 1.0, whereas the sum of the accuracy and the errors of commission will not necessarily sum to unity. Similar matrices can be constructed for each strata, to indicate the quality of the classification within the strata. If the overall accuracies are not satisfactory then analysis of the confusion matrices for the strata may indicate the major sources of error and hence where improvements need to be taken in the classification.

Table 6.12 Acccuracy assessment of 486 production units

Classification	Field results	
	Cropping	Not cropping
Cropping	172 (96.1)	11 (3.6)
Not cropping	7 (3.9)	296 (96.4)

* The percentages are given in brackets

Accuracy of identifying production units

For some purposes the resource manager is only interested in the accuracy of classifying production units, such as paddocks used for cropping. If this is the case then the occasional misclassification of pixels within the fields is of no account; as long as it does not affect the analyst's ability to identify the field properly. Typical examples could be where the manager requires information on field use, to assess trends in cropping and land use, or to identify particular cropping practices for some purpose. The extraction of this type of information requires reasonable resolution so as to define boundaries, but it does not require high resolution data within the fields; visual interpretation is very suitable for extracting this type of information under many conditions.

Accuracy assessment is very simple for this type of information as the accuracy assessment only has to compare actual with classified land use for an unbiased sample of production units. It is essential to define a sample of production units independently of the classification so that the accuracy assessment yields the accuracy as well as the errors of omission and commission. For example, a sample of 486 production units may yield the counts shown in Table 6.12.

Boundary definition

Mapping of boundaries on an image creates errors in the accuracy of location of that boundary due to a combination of interpretation, identification and cartographic errors. These errors are usually larger when attempting to delineate boundaries that are poorly defined on the image, for whatever reason. Thus the accuracy of defining the boundaries of soil groups on a soils image will usually be less accurate than will be the accuracy of defining the boundaries of cropping fields. The accuracy of defining boundaries can be assessed by the following methods.

1. Plot the locations of the boundaries in the field on one or more random traverses across the area and compare this with the intersections of the traverse with the mapped boundaries. The difference between the two, as a distance, can be computed as an average and a variance, taking differences in one direction as positive and as negative in the other direction. Traverses can require considerable field work and it can be difficult ensuring that they are representative.
2. Take a set of sample areas along the various boundaries, and measure the areas of each land-cover in each sample, both from the classification and from another source such as field mapping. The differences between the two, taken negative in one direction and positive in the reverse, are used in a similar manner to the distances to compute an average and a variance. The areas calculation will tend to average small idiosyncrasies that may occur in either of the mappings.

3. Overlay the mappings in a geographic information system and extract the areas that are within the overlap as errors of omission and commission and the areas that are classified the same in both mappings as being correctly classified.

6.6.3 The use of field data in estimation

Estimation of the values of physical environmental parameters from remotely sensed data will usually be done for each pixel in turn. The samples used to assess the accuracy and consistency of the estimation will therefore be related to single, or groups of, pixels. They must be located as accurately as possible relative to the individual or groups of pixels. Field data will collect information on the parameters within these samples for comparison with the estimates for the same parameters from the satellite data. Sets of pairs of values, from the field data, and derived from the satellite data, for sets of samples, can then be used either in regression analysis to analyse the accuracy of the satellite data in estimating the selected physical parameters or to derive estimates of accuracy and consistency.

Consider mapping the density of woody cover as discussed in Chapter 9. The vector classifier was used to process satellite data (Photo 9.6) to derive estimates of the proportion contribution to response of woody canopy cover in both images. Aerial photographs (Photo 9.8), were acquired for a transect across the area mapped, located so that 1 ha samples could be readily identified on the aerial photography and in the mapping from the satellite data. The 1 ha samples were also chosen to ensure that the full range of cover conditions was represented in the samples. A set of 65 samples was used in the analysis. The percentage woody cover was measured in each 1 ha sample on the aerial photographs using a dot grid and compared with the estimate of contribution to response due to woody cover from the satellite data. The two sets of data were then compared in a linear regression analysis so as to derive a model to estimate percentage woody cover from the proportion (woody cover) data derived by the vector classifier.

7
Geographic information systems

7.1 Introduction

7.1.1 The role of geographic information in resource management

Consider the situation where a resource manager has a range of spatial information, including elevation data, slope and aspect, soil classes, individual site soil tests, current land-ownership, drainage and water features, current and historical land use and recent land-cover status (Photo 7.1). Different subsets of this information are used to address different issues. To promote the use of a particular fertilizer requires information on land-use history, soils and ownership, to locate those farmers that are the target client group. Alternatively, the manager may wish to select a route for a road, power or telecommunications line through an area, where the cost of construction is a function of soil type, topography, water obstacles and land values, as could be assessed from land use. Integration of these factors will indicate potential corridors that can then be assessed in more detailed analysis. Clearly each management issue will use a different subset of the available information and there will be overlaps in the subsets required to address the different questions.

Each type of spatial information could be conceptually viewed as a layer of information on that aspect of the environment. The set of information layers create an overlay for the area, much like a 'dagwood sandwich'. In some of these layers, each unique sub-area may have only one attribute value, e.g. an elevation layer may have only the one attribute value of elevation for each location in the overlay. In other layers there may be multiple attribute values. The layer of land ownership may have (a) the name of the owner, (b) the address of the block, (c) the name of the lessee, and (d) the tax or valuation number of the block. The layer of soil classes may have attribute values within each sub-area of soil class name, soil association, typical soil colour and profile or soil chemistry information.

The whole set of information is called a database, where the database consists of spatial and attribute data. The mix of spatial and attribute data in the database will vary from situation to situation. Subsets of the information in the database are integrated together in some way to create management information or information designed to address a specific management issue.

These spatial and attribute databases are the core of a geographic information system (GIS) which could be conceived as consisting of those components shown in Diagram 7.1. A GIS is defined as a means of storing, retrieving, analysing, and displaying spatially related sets of resource data so as to provide management information or to develop a better understanding of environmental relationships.

A manual GIS involves creating a base map and one or more layers on clear film to overlay over the base map. The combination of layers can thus indicate areas affected by different combinations of values within the GIS. Thus a land-use map, overlain with a layer depicting areas subject to flooding, allows a planner to assess the potential costs of flooding within the study area. Whilst a manual GIS can include many layers, the more layers that are used, the more difficult becomes the interpretation. In addition, manual GIS techniques are:

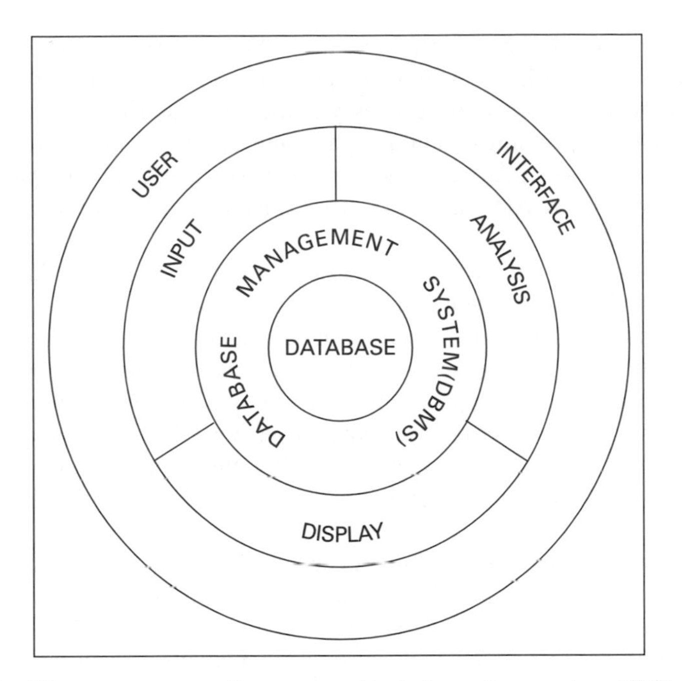

Diagram 7.1 The components of a geographic information system (GIS).

- labour intensive – requiring that the layers all be at the same scale on the same projection;
- difficult to use for the computation of statistics since measurement of all areas has to be done manually; and
- not suitable for creating new complex combinations of old layers since manual GIS techniques do not include numerical and some logical processes to combine layers.

In consequence, manual GIS techniques have never been used extensively as an operational tool by resource managers.

Computer-based GIS have their origins in the early days of computers. Many of the systems developed in the 1960s were designed to address a specific task. For example, the Universities of Minnesota and Michigan both developed GIS to analyse environmental data, and to assist in the management of their states' natural resources. Both of these systems started as punched-card batch-oriented systems that depended on gridcell or raster data as the form of the spatial data that they processed. These systems suffered from the difficulty they had of creating large files since each layer had to be stored as a grid consisting of n lines, each of m cells. Very fine resolution raster files (that is small cell size), meant a large number of cells, incurring high storage and processing costs and creating difficulties in printing hard-copy products because of the finite number of lines, and cells per line, that could be handled by printing and display devices. Reducing the storage and processing costs to acceptable levels meant sacrificing resolution in raster files by adopting a larger cell size.

Other early developments were more concerned with drawing architectural or engineering drawings, or in the production of line maps, such as topographic maps. Systems with this type of focus started with linear features that were converted into line or vector data, where lines are retained in the computer as strings of (x, y) coordinates. Storage of vector data is more accurate, and at a higher resolution than that of raster data because there is little limit on the recorded (x, y) values. Vector-oriented systems evolved into either computer-aided design (CAD) or computer-aided mapping (CAM) software. Generally these systems developed most quickly in the 1960s and 1970s because the vector data was much less demanding of storage space, and easier to display. From these origins came the CAD software of Autocad and the CAM software of Intergraph and Arc/Info, although both systems (CAD and CAM) have since evolved so as to attract a broader range of clients.

CAD software is based on three-dimensional plane cartesian coordinates. It is usually very good at the development and presentation of objects such as engineering components and architectural buildings. However, it does not usually provide for transformation between projections, or for the analysis of spatial relationships between objects. Early CAM software focused more on mapping from photogrammetric machines, map editing and the transformation of data between map projections. The high positional accuracy of vector data was seen as a major advantage for this type of work. CAM software was generally not as good at depicting engineering objects or the analysis of spatial relationships.

The analysis of spatial relationships is concerned with describing and understanding the relationship between, and interdependence of objects in the GIS, e.g. the relationship between an area and the arcs that surround it so as to be able to study circumference–area relationships. Another application concerns the relationship between a linear feature, such as a road, and area features adjacent to it. This type of relationship allows the analyst to assess things like land uses along a road or powerline, and hence estimate demand on that linear feature, either as total demand or as demand per unit distance. There are many other spatial relationships that need to be amenable to analysis so as to provide important management information.

In the 1980s the cost of display and storage devices dropped significantly. This has allowed the raster based systems to compete with the vector-based systems on more equal terms. In consequence, the 1980s saw the emergence of a number of raster-based systems, including SPANS, GRASS and EPPL7, designed primarily for the analysis of spatial relationships.

GIS can be used for a range of purposes that can be grouped into two broad categories: linear and area feature analysis.

- If the resource manager is concerned with linear features (such as transport and service corridors, land administration and ownership) then a GIS that is strong at vector-based analysis will often be most suitable. Such analysis includes estimation of demand on linear supply features, management of length, analysis of network efficiency and optimization of a network, and maintenance of records of landownership.
- If the resource manager is concerned with area features (such as land use, geology, soils, agriculture, forestry, etc.) then a GIS that is strong at raster-based analysis can often be most suitable. Such analysis includes overlaying, comparing, modelling, and estimation.

There are various terms that have been developed to describe systems designed to address specific GIS applications, such as:

LIS	land information system – concerned primarily with the management and administration of land parcels;
FIS	facilities information system – concerned primarily with the management of transport, communication and service facilities, such as roads, railways,

Diagram 7.2 Examples of vector data – the topographic and drainage data as taken from a topographic map for inclusion in the GIS: (a) contours; (b) soil type; (c) land-ownership; and (d) drainage.

sewage, water, power and telephone lines, although they may also be concerned with area features such as buildings, and their characteristics, and their distribution; and

NRMIS, RIS Natural Resource Management Information Systems – concerned primarily with the management of natural and other area resources such as water, soil, and vegetation resources.

Both LIS and FIS usually depend on vector data because of its greater positional accuracy. This book is primarily concerned with area resources, and hence with NRMIS. However, proliferation of terms is not helpful unless the new term describes an idea, activity or object in ways that cannot be achieved using the existing language. In this case most of these terms are probably unnecessary, so the term GIS is used here to describe the software/hardware systems shown in Diagram 7.1.

There are four input forms by which data can be supplied to a GIS:

1. data provided as hard-copy or analogue maps which starts as linear or vector data as shown for the topographic data in Diagram 7.2;

2. raster data, such as scanned maps or photographs and remotely sensed data, in the form of raster or grid cell data;
3. point data, generally collected in the field as a set of specific attributes, at a location at a specific time; and
4. record data, usually as vector, point or attribute data in a relational database file of attributes of specific resources in the area covered by the GIS.

Generally all of the data in a GIS need to be converted into the same form before analysis, although in some software, and for some tasks, this is not necessary. The data in Photo 7.1 were all converted into raster form for the analysis shown in Photo 7.2 and discussed in section 7.5. Data does not have to be converted into raster form for use in a GIS; some forms of analysis are best done with vector data, some with raster data and some analysis can be done with either type of data. The form chosen will depend on the task at hand, the form the data is in to start with and the GIS system being used.

The resource manager thus has at his disposal a library of information, subsets of which are used to derive different management information. But how does the resource manager bring these sets of information together so as to provide the required information in an easily usable form?. The manager may wish to ask questions of the form: 'Where are the areas of northerly aspect with annual average rainfall of a–b per annum, of soil types X (with slopes in the range $1{:}x(1)$ to $1{:}x(2)$) or Y (with slopes in the range $1{:}y(1)$ to $1{:}y(2)$), that are within distance d of roads within a designated area? Which properties have more than C ha in area that fit these conditions (since these properties have conditions suitable for growing a particular crop, potentially in a commercially viable quantity)?' Extracting this information from analogue maps is extremely time-consuming. More complex questions rapidly become too costly even to contemplate using paper maps. Yet these types of questions are valid if we wish to ensure that our resources are managed most effectively.

7.1.2 The role of geographic information in resource maintenance

Let us consider resource management from the perspective of resource maintenance. Depletion of physical resources rarely occurs as large losses over very short time periods, but more usually as small but frequent incremental losses. In the area covered by the data in Photo 7.1, there may be anecdotal information of significant soil erosion. Is the soil erosion widespread? Is the erosion loss increasing or decreasing, and at what rate? What is the likely cause of this erosion, and what is the best way of rectifying the problem?

In general, the factors that cause environmental degradation, such as soil erosion, are well known. However, the magnitude of the contribution of each of the factors, in each specific situation, is rarely known with any accuracy. The detailed information necessary to conduct management and operational control tasks is thus rarely available to resource managers and hence they cannot determine the best way to address specific resource issues.

Management of degradation is integrally involved with community perceptions on the extent and the severity of the problem. Degradation usually occurs in small increments that are unevenly distributed over time and space; their impact are neither uniform nor readily recognized, delaying community awareness of their severity and extent. Even at early stages of resource degradation some land-users will recognize the onset of significant environmental problems, but many will not; the degradation will need to be well advanced, with evidence of its extent and severity obvious to most people before general community acceptance of the need for action will be gained.

The causes of resource degradation are often removed, either spatially or temporally, from the effects of the degradation. This separation of cause from effect makes it more difficult to gain community acceptance of the causes and the proposed remedies. Resource managers who are suffering resource degradation will thus often have some difficulty convincing other resource managers, particularly those who are causing the resource degradation, of the seriousness of the problem and the need for implementation of suitable remedies. A GIS is a very good tool for changing community perceptions on the severity of the resource degradation, and gaining community acceptance for proposed remedial actions by providing relatively objective and quantifiable spatial and temporal information. The GIS can also be used in a non-confrontationist way to assist in resolving degradation issues.

It is in the best interests of both the individual land-user and the community to rectify degradation issues as early as possible. To achieve this it is necessary to convince the majority of land-users that action needs to be taken well before the full implications of the resource degradation are obvious. Prediction of future resource conditions is not an activity that the resource manager can estimate with any degree of confidence or reliability without predictive models and the quantitative data necessary to drive these models. Predictive models provide an indication of future conditions, within the constraints of the data used to create, and drive, these models, that would not otherwise be available to the resource manager. Initially our understanding of environmental processes are likely to be simplistic, leading to rather coarse or crude models that may take into account only the first-order influences on resources. Despite this limitation, the models can still indicate worst- and best-case scenarios if current management practices persist. The models should thus be used as another facet of the information available to the resource manager.

Even if models become significantly more accurate, there are many random or apparently random activities in the environment that will affect the accuracy of any prediction. Models are always going to be subject to these influences and so their role will be as an information support for, and not a replacement of, the resource manager. The impact of random processes on models, and the role of these models in the decision-making process may influence the level of accuracy that it is reasonable to strive for in the development of the models. The characteristics of models, their assumptions, processes and the impacts they predict on resources will need to be integrated into the manager's decision-making process. Models will be more readily accepted if their assumptions and predictions are displayed in an easily understood and usable form so that the resource manager can easily integrate this type of information into the decision-making process.

7.1.3 The conversion of data to management information

The conversion of data into management information follows the path (Diagram 7.3):

1. preprocessing, including acquisition and calibration, of all source data;
2. extraction of raw information from the data;
3. creation of derived information from the raw information, by use of models and analysis in the GIS;
4. presentation of the derived management information; and
5. development of decisions using this derived information.

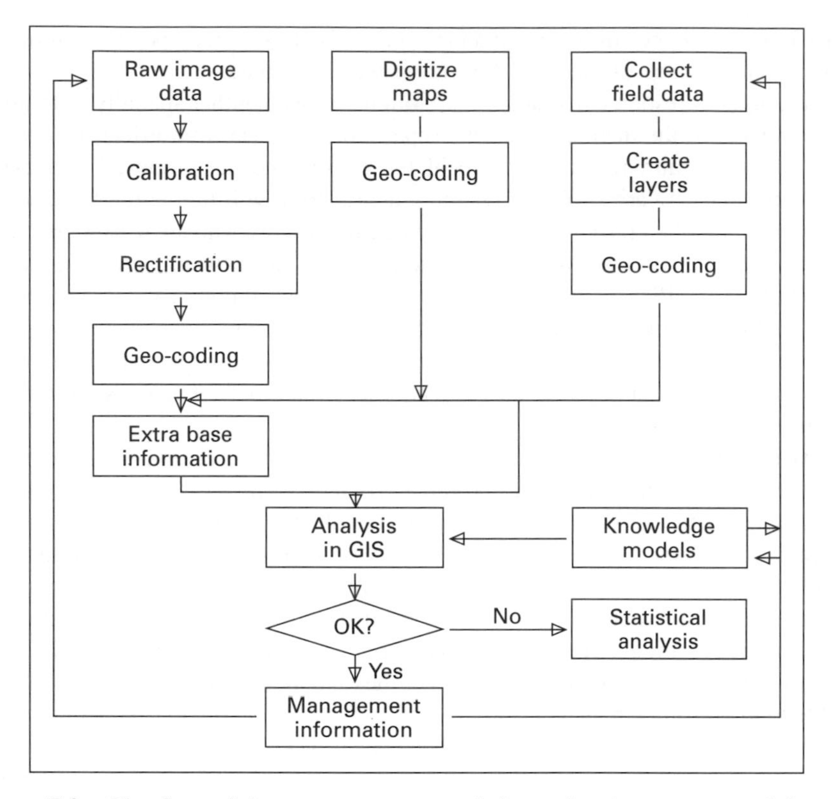

Diagram 7.3 The flow of data to management information in a resource information system.

Preprocessing is defined as the preparation and entry of data into the GIS. The complementary methods of preprocessing data are:

- calibration and rectification of remotely sensed imagery;
- digitization of analogue maps to create vector file data;
- acquiring data from existing files, such as survey files and marketing information files, that will create either spatial vector data or relational database files; and
- acquiring data from field data observations or point observations such as observations of soil chemical composition or fertility, the depth to the water table, etc.

The resulting raw data can be in the form of point, vector or raster files.

Point files often constructed as a relational database file, contain point coordinates (position and time) and values in a number of attributes at that location. Thus soil tests, of pH, exchangeable cations, nitrogen content, etc. for which the coordinates are known can be constructed as a database file, for use as input to a GIS. If an image is required of one of the parameters stored in the database, then the database software is used to create a file containing the point or observation coordinates and that parameter value for each point or observation in turn. This file is then fed into the GIS where the point observations are used in interpolation to estimate values either along contour lines, thus creating vector files, or for all cells in a grid, thus creating raster files.

Vector files are files that contain the coordinates (x, y) of points along lines, thereby defining the shape of the lines, and attribute data about the lines, such as connecting lines and associated polygons. The lines may be stored as arcs, or line segments, that join nodes, or points of junction of two or more lines. The relation of these arcs to nodes and to the polygons contained by one or more arcs is retained in the file as spatial relationship data. Vector data often come from the visual interpretation of remotely sensed images or where the information is extracted by digitization of maps. They can be converted into raster files using painting algorithms.

Raster files contain data that are in the form of a large grid array, consisting of a number of lines, with each line consisting of a number of cells. Generally raster data are defined in terms of rows and columns, or lines and pixels, with each data record consisting of the data for the cells in the row or line. The data may consist of a number of layers, with each layer consisting of rows and columns, with one value per cell. This value can be linked to values in an attribute file. Raster data usually come from the digital processing of remotely sensed images, scanning of hard-copy maps, conversion of vector files and the interpolation of an irregular net of point observations or field data contained in a point file. Analysis of spatial data can be done using either vector or raster data, with both types of data having advantages and disadvantages for different types of processing.

After preprocessing the data are analysed or processed to **create raw information**; the interpretation of remotely sensed images using visual interpretation or digital image processing creates raw information, as does the conversion of digitized map data into raster data. Raw information is defined as that information that is derived directly from data. Thus raw information forms the individual layers within a GIS. Typical examples of raw information are land-cover and use.

Sometimes this raw information is required directly by resource managers, e.g. they may use information on the area under crop, or land-use maps on their own for a specific purpose. However resource managers will normally integrate this information with other information to provide the information that they require to make management decisions. In simple cases the raw information sets can be overlaid to derive the required information. More usually however, more complex combinations of raw information need to be numerically and/or logically integrated to derive the required management information. This **derived management information** will usually be in the form of tabular, or statistical, information, and maps. Facilities used in the **presentation of information**, and methods of presentation, are discussed in section 7.6.

7.1.4 Functions of a GIS within an organization

A GIS can be used for a variety of purposes, including:

- a decision support system within the organization, by providing the information necessary to the managers of the organization in the making of resource management decisions;
- management of the spatial data belonging to the organization, typically the maps of mapping organizations, including the revision and updating of those maps and data;
- better management of corporate resources where that management requires spatial information; and
- investigation of the statistical correlation between resource parameters for the purpose of better understanding environmental interrelationships.

The nature of the GIS required by an organization depends on the relative importance of these applications, as well as the characteristics of the applications within the organization.

Decision support system

A decision support system is a computer-based system that transforms and presents information in a form and manner designed to assist a manager make specific resource management decisions. The system usually employs models of environmental and decision-making processes to transform the information and then present it in a manner and form best suited to the particular decision-making process. Decision support systems can take many forms of which a GIS is only one type.

The most important issue to be addressed in this role of a GIS is whether the GIS is to be used mainly for the management of features that are primarily linear or area in character. Linear character features include many cultural and administrative features such as:

- transport, energy and service features, such as powerlines, telecommunications, water and sewage pipes, etc.; and
- land-ownership, or cadastral and land administration data which can be used as either a linear feature, representing the boundary or perimeter, as is usually the case, or as an area feature. It is usually treated as a linear feature because of the much greater positional accuracy of vector data, and smaller data storage requirements.

Area character features include many physical environmental and land-using features such as:

- geologic associations, soil classes and soil conditions;
- land-cover, land use and land-cover condition;
- hydrologic and meteorologic information; and
- landform and topography.

The management of linear features is usually concerned with questions of: identity; context, or what is in the vicinity, and proximity or how close; length; intersections; and capacity. Therefore, the management of linear features is often best done with GIS that are strongest in the use of vector spatial data.

The management of area features are usually concerned with questions of: identity; area; distribution and statistics of feature type, quality, quantity or volume and amount or production; correlations and predictions. So, the management of area features is often best done with GIS that are strongest in the use of raster spatial data.

The nature of resource management within an organization will also affect the selection of a GIS. If the management style is centralized then the GIS is likely to be centralized, often with technical staff running the system; a distributed style of management is likely to implement a distributed GIS, in which case the GIS will be used by individual, often junior, resource managers, many of whom will have limited experience at using computer systems and none will have readily available support services. There are thus alternative arrangements or structures for the establishment of a GIS in an organisation.

1. In a **centralized GIS** the preponderance of the GIS hardware, software, support and users are at the centralized location. A centralized GIS is appropriate where the main

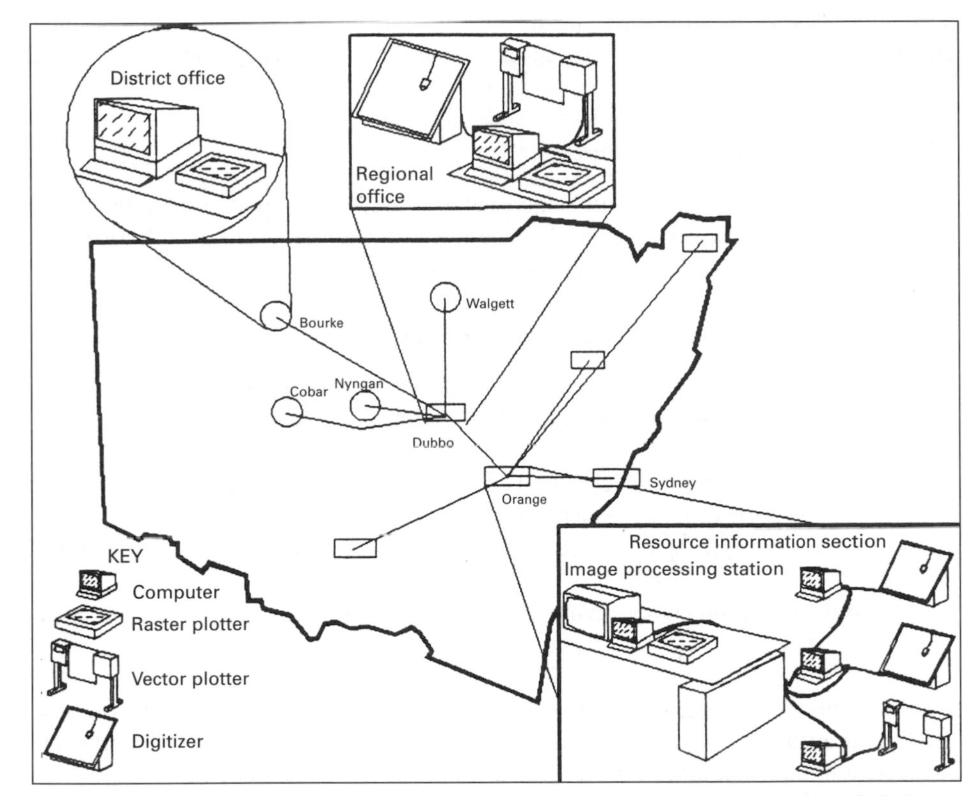

Diagram 7.4 Typical structural arrangements for an operational, distributed GIS, based on the needs of the Department of Agriculture in New South Wales, Australia.

functions of the organization are either the management of information or where it is critical that the organization be able, quickly and efficiently, to analyse the GIS and other data and to respond to issues.

2. In a **distributed GIS** (Diagram 7.4), there will usually be a centralized core and a network of distributed users. The centralized core provides:

   (a) the main databases required by the users, either purchased or created by the core group;
   (b) a skills base to support the distributed users with advice, training and system developments;
   (c) a point of contact with vendors of GIS hardware and software, and the suppliers of databases; and
   (d) a focus for the needs of the end users for databases, facilities and system capabilities.

A distributed GIS is most appropriate as a decision support tool for managers of physical resources. In this case, where the decisions are being made by resource managers 'in the field', the GIS needs to be readily accessible to them if it is to be used effectively.

Management of organizational data

This is an administrative function required of all GIS but rarely the primary use of the GIS. This function will involve both vector and raster data; they need to be capable of accepting, storing, reformatting, listing, windowing, and mosaicing together files of adjacent areas, as well as registering layers of the same area. They may not be strong at the analysis of both data types.

Management of corporate resources

This is a similar function to a GIS being used as a decision support system, but supports managers higher up the organizational structure and less likely to use the GIS themselves. The system is more likely to be centralized and implemented by technical staff.

Investigation of the statistical correlation between resource parameters

This role of a GIS is likely to be primarily utilized to understand or investigate macro level relationships in the environment. Investigation of these correlations can:

1. indicate causal relations which will usually result in correlation, whereas correlation does not necessarily indicate causal relations;
2. indicate the strength of these causal relations – as a function of the correlation between the parameters, and
3. lead to the development of environmental models, from the work done as a result of 1 and 2 above.

In summary, the role or purpose of a GIS can be defined as follows: the purpose of a GIS is to manage and process raw information so as to present the derived products as management information to the resource manager so as to facilitate his management of natural or physical resources.

7.2 Components of a geographic information system

7.2.1 Functions of a GIS

A GIS should be able to effect several functions (Diagram 7.5).

Management of database files on resource attributes File management includes storage, retrieval, editing or updating, adding to and display of both the files and information on file characteristics. The files may be both spatial files or files that contain information on the location (in position and time) of each attribute value, and attribute files or files that contain values on multiple sets of attributes.

Editing of contents of both spatial and attribute files For spatial files this includes the following.

- **Mosaicing** is the joining together of two or more spatial files covering different areas in the same attribute or parameter so as to create a file that covers all of the areas covered in the original files (Diagram 7.5(b)).
- **Windowing** is the process of extracting portion of a spatial file so as to create a new file of a subset of the original file. Windowing is usually of rectangular subsets of the original file (Diagram 7.5(c)).
- **Masking** is the process of extracting portion of a spatial file within an irregular boundary that is defined on a mask file where different values in the mask file represent areas

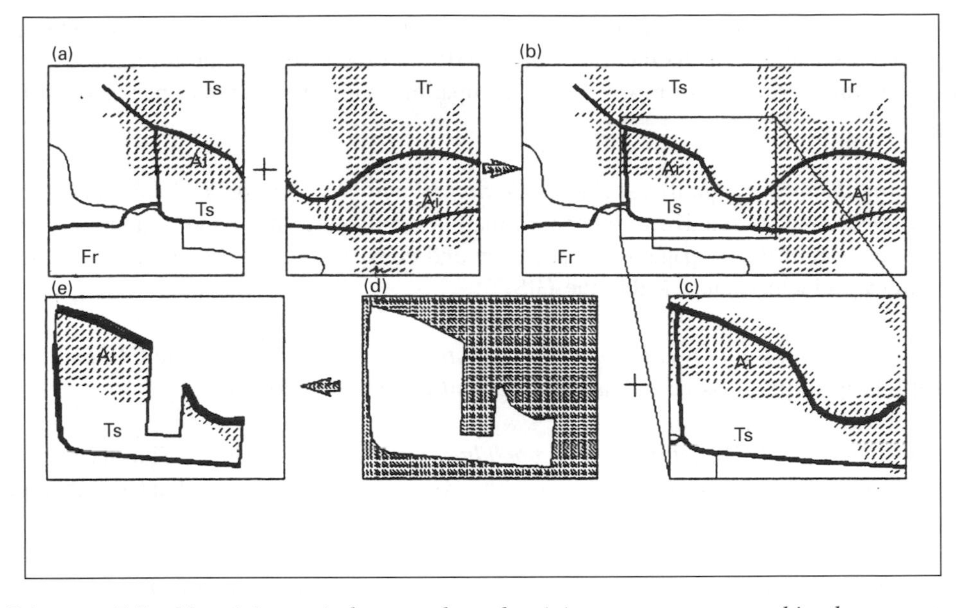

Diagram 7.5 Mosaicing, windows and masks: (a) two maps are combined to create (b) a mosaic, from which (c) are subsetted a window that can be (d) masked to create (e) a map of a specific area.

that are either outside or inside the area of interest (Diagram 7.5(d)). A mask can be of any type, including:

- administrative areas such as a shire or property;
- stable physical environmental regions such as soil types or geologic associations; and
- variable physical factors, such as water surfaces, specific land uses, and cloud areas.

• **Registering** of spatial files is combining files from different databases into the one database. It will often require conversion of spatial files to different projections or pixel sizes (for raster files), conversion of codes or symbols to a common standard and windowing and mosaicing when the files do not cover exactly the same area.

Editing of attribute files involves standard database processes of merging, editing and transforming files.

Conducting numerical and logical processing of the database files This is necessary to derive resource management information, or information of direct interest to the resource manager. This analysis may involve the logical or numerical integration of data, e.g. 'cropping in the last three years', 'cropping on soil type *X*' or estimation of soil loss. It may also involve analysis of spatial relationships including proximity of features, areas, lengths, route distances and trajectories or profiles. The GIS should be capable of accessing, using and then exiting from software modules designed to efficiently conduct specific modelling or integration tasks.

Displaying the derived information in a manner that is compatible with the requirements of the resource manager This includes the display and preparation of spatial information as maps or images, and statistical information as tables, graphs or statistics.

Facilitating the statistical analysis of the data in the GIS This generally involves conducting analysis on the databases in the GIS and then sending the resulting data to files that are formatted in a way that is compatible with the statistical packages to be used.

Allowing the manager to assess the weight that he will place on the derived information This 'weighting' is relative to other information available to him, and is done through consideration of the propagation of accuracy and consistency, preferably displaying the accuracy and consistency within the GIS.

Allowing integration of predictive models into the analysis This refers to GIS data and the plotting of progressive estimates as derived from those models.

Allowing the use of GIS databases and capabilities within expert systems Thus the system can be partially automated and various functions and processes involved in resource management optimized.

Providing linkage between databases, particularly physical resource and economic databases This allows the manager to address 'what if' type questions and to convey decisions and actions to other databases for evaluation of their implications to other managers.

Ensuring user friendly interfaces This is necessary because the resource manager will often need to interface closely with the GIS and the derived information will then be used by various types of resource managers rather than by technically skilled people.

At present most existing GIS systems can adequately meet only some of the requirements discussed above.

7.2.2 Structure of a GIS

In the generalized structure of a GIS as shown in Diagram 7.1, the core of the system is the database, which contains the 'raw' information as spatial and attribute data files, defined as a shared collection of interrelated, spatially referenced, data designed to meet the needs of the organization or the user. The database is created, maintained and accessed by a software system called the Database Management System (DBMS), which provides facilities to store, retrieve, modify and delete data and files from the system. In some GIS systems the DBMS is transparent to the user, i.e. the tasks are performed by the DBMS without the user being aware of the activity taking place, or of the processes involved. In other GIS systems the operator is aware of, and can gain access to the DBMS so as to modify the operation of the DBMS or to direct the system to conduct specific tasks.

A GIS usually contains four components: input, storage, analysis and output, each of which are now considered in turn.

7.3 Input to a GIS

Spatial data can be input to a GIS in four ways: conversion of analogue maps (section 7.3.1); interpolation from a network of point observations (section 7.3.2); processed results of image processing (section 7.3.3); and/or drawn from existing databases (section 7.3.4).

7.3.1 Conversion of analogue maps

Conversion of analogue maps is done either by manual digitizing directly from the map, or by scanning the map with a scanning device, and then identifying the linear features to be edited and retained by use of semi-automated digitizing and line following techniques. Both processes create vector data as the final output, although scanning creates raster or gridcell data as an intermediate step, with the gridcell data having a local origin and orientation.

Manual digitizing

A digitizing table consists of a table and controller attached to a computer. The table contains a fine electrical graticule of parallel horizontal and vertical grid of wires embedded within the table, usually at a resolution of about 0.05 mm to 0.2 mm, representing the finest resolution data that can be recorded from the table. The controller consists of a cursor and controller keys. The cursor is a glass or perspex window containing an etched cross-hair and electrical loop. The controller handle contains between four and sixteen cursor keys that, by sending different signals to the computer, can control the operation of the computer. As the cursor is made to follow a line by the operator, the circuit containing the cursor and table is completed by pressing the appropriate cursor button. The intersection within the graticule of wires in the table closest to the cursor is identified. These intersections are recorded in the computer as a string of digitizer (x, y) coordinates along the line being followed by the cursor. As the data are digitized they are displayed on a graphics screen so that the operator can see how the work is progressing.

The digitizer coordinates recorded in this way need to be converted into map coordinates before use, since the digitizer coordinates will change if the map is positioned differently on the table. The digitizing is started by recording the digitizer coordinates of at least two, and preferably four control points. The control point coordinates and the digitizer coordinates are related through a rotation, translation, and change in scale if the control point coordinates are in a cartesian coordinate system. In this case the transformation is of the form

$$\text{CONTROL X} = \text{A} + \text{B} \times (\text{digitizer } x) - \text{C} \times (\text{digitizer } y)$$

$$\text{CONTROL Y} = \text{D} + \text{C} \times (\text{digitizer } x) + \text{B} \times (\text{digitizer } y)$$

Whilst this transformation requires a minimum of two control points because of the four unknowns (A, B, C and D), it is good practice to use four points located near the corners of the area being digitized so as to have redundancies in the data for error checking. The best control points are grid intersections as these are easily seen on the map, and their coordinates are precisely known. With most topographical map series the grid coordinates are cartesian coordinates and so can be used in this way quite satisfactorily. The digitizer coordinates are transformed into the same projection as the map coordinates (Chapter 3).

The strings of map coordinates (x, y), are then stored as vector data in which the vectors represent either points, linear features or the boundaries of area features.

Digitizing can normally be done in three modes: point mode, stream mode, or continuous mode. In **point mode**, once the cursor is positioned over a point, the relevant cursor button is pressed to record the coordinates for that point. The method is slow but accurate. It is suitable for digitizing control points, individual point features such as houses, and linear features that consist of straight stretches, such as land-ownership

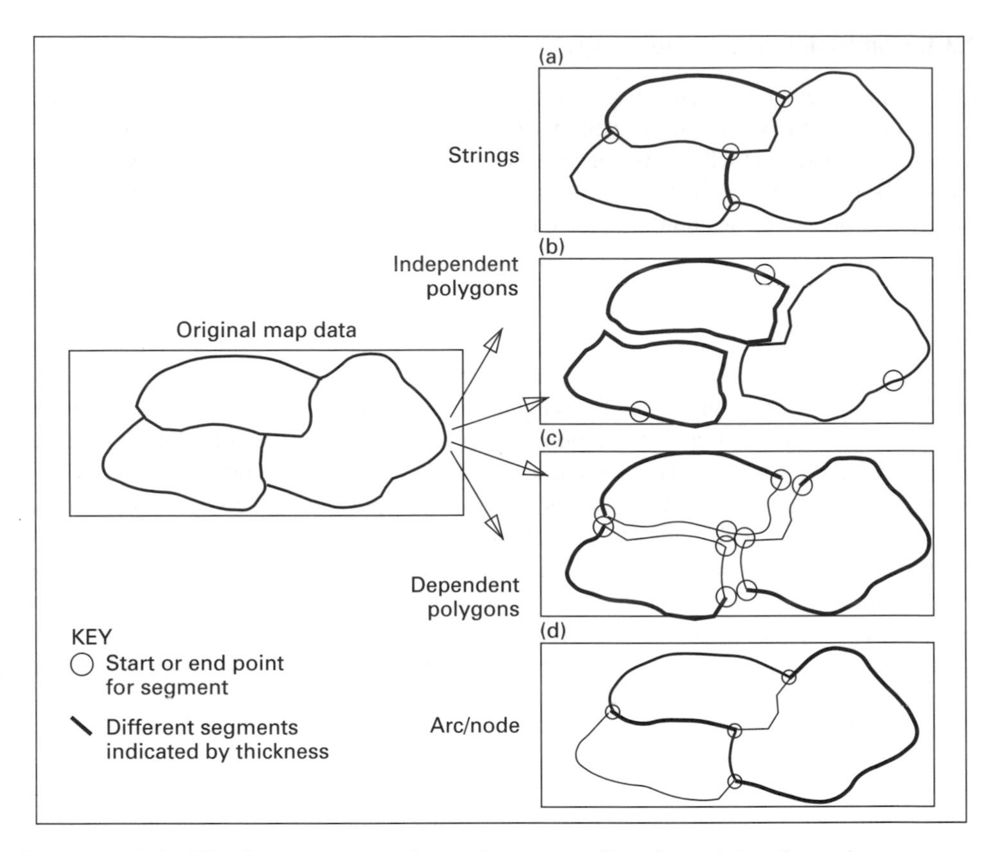

Diagram 7.6 The four methods of creating vector line data: (a) independent polygons; (b) dependent polygons; (c) arc nodes; and (d) strings.

boundaries, by digitizing just the changes in direction along the boundary. Digitizing by point mode leads to the most efficient storage of data. In **stream mode**, once the cursor is correctly positioned the appropriate cursor button is pressed and points are digitized at a constant time rate or frequency. With some systems the time rate can be adjusted, to allow for differences in operator experience and variations in the complexity of the information being digitized. This approach often leads to excessive storage of data, but it allows the experienced operator to vary his digitizing pace with variations in the complexity of the information being digitized. It is a good way for experienced operators to digitize irregular features such as creeks and contours. **Continuous mode** is similar to stream mode except that points are digitized at a set spatial interval rather than a set time interval. This mode does not allow for variations in the complexity of the information, but can reduce the amount of data stored by inexperienced staff.

Vector data can be stored in five ways (Diagram 7.6): Points, Strings, Independent Polygons, Dependent Polygons, and Arc-Nodes. For **points**, each is stored independently in the data file. Attribute value(s) may be either recorded with the coordinates in the file, or they may be recorded once for a set of site coordinates, for example the sites of all houses. With **strings**, the data for each line or vector are created once, often in a way most convenient to the person creating the data. It ignores the relationship between the boundaries being created and the entities that the boundaries define as displayed in Diagram 7.6(a). This type of vector data is suitable for line features or for mapping where the boundary information is depicted on a map and the observer is required to draw inferences on any relationship that exists between the line boundaries and the enclosed areas. It is unsuitable for digital analysis of the data where the computer needs to enquire of the attributes for the areas defined by each boundary.

With **independent polygons**, the full boundary of each area or each line is digitized independently of each other line or boundary (Diagram 7.6(b)). Most lines or boundaries are digitized twice as they are used for two areas, increasing storage, incurring costs due to the double digitizing and creating errors and confusion because the boundaries will not be digitized in exactly the same place both times. It is *not* a recommended way of creating vector data. With **dependent polygons**, however, the line segments are created once, and used for all abutting polygons, but the data are stored as complete polygons (Diagram 7.6(c)). The approach eliminates the duplication and errors that will occur with independent polygons, but it incurs storage costs because of duplication in the storage of data.

With **arc–node storage**, the arcs between segment joins or nodes are digitized once and stored as separate entities (Diagram 7.6(d)). Polygon information is stored either as a set of nodes that specify the arcs, or as lists of arcs required to construct the polygon. This approach does not incur the errors or costs of double digitizing; it minimizes storage costs, but it does incur some additional processing costs in reconstructing the polygons. Arc-node storage is suitable for storing all three types of data: point, line and boundary. It is thus the best of the currently used methods of storing vector data.

Scan digitizing

A scanning device senses the density in a strip or scanline across the photograph or map, and converts that density into a digital value for each of a finite number of elements, or pixels, across the line. A series of lines are recorded to scan digitize the whole of the image. Conceptually the process is the same as scanning the environment with an imaging scanner. In scanning a photograph, the density within the field of view controls the amount of energy that enters the scanning system. This density depends on the range of densities in the photograph and the illumination of the photograph.

Scanning of maps or photographic images can be done using systems based either on a television camera or on a scanning densitometer. The first are quick and cheap, but provide relatively coarse data. Small format scanning densitometers are also inexpensive, but large format high resolution scanners are expensive. High resolution scanning densitometers generally use apertures of 25 μm or 50 μm, indicating the resolution of the data at image or map scale.

The scan digitized data is in raster form (Diagram 7.7) with the size of the grid cells being set by the scanning device and, in the case of video camera systems, by the distance between the scan digitizer and the image. The best resolution that can be achieved with video camera systems is about 1000 values per line, for about 1000 lines. The resolution of the digitized data is the area covered divided by this resolution.

The raster data produced by a scan digitizer has pixels of a specific size that may not be at the desired location or orientation. The scanned data can then be dealt with in either of two ways depending on whether raster or vector data is required. If raster data are required then the data needs to be resampled as discussed in Chapter 5. If vector data are required then the scanned image can be interrogated using line following software to identify the linear and point features in the scanned data. Line following routines can be confused, particularly if the scanned data contains a lot of noise. In consequence the line following is usually followed with editing to clean up the final record, and to add text and other details not included in the line following phase of the work.

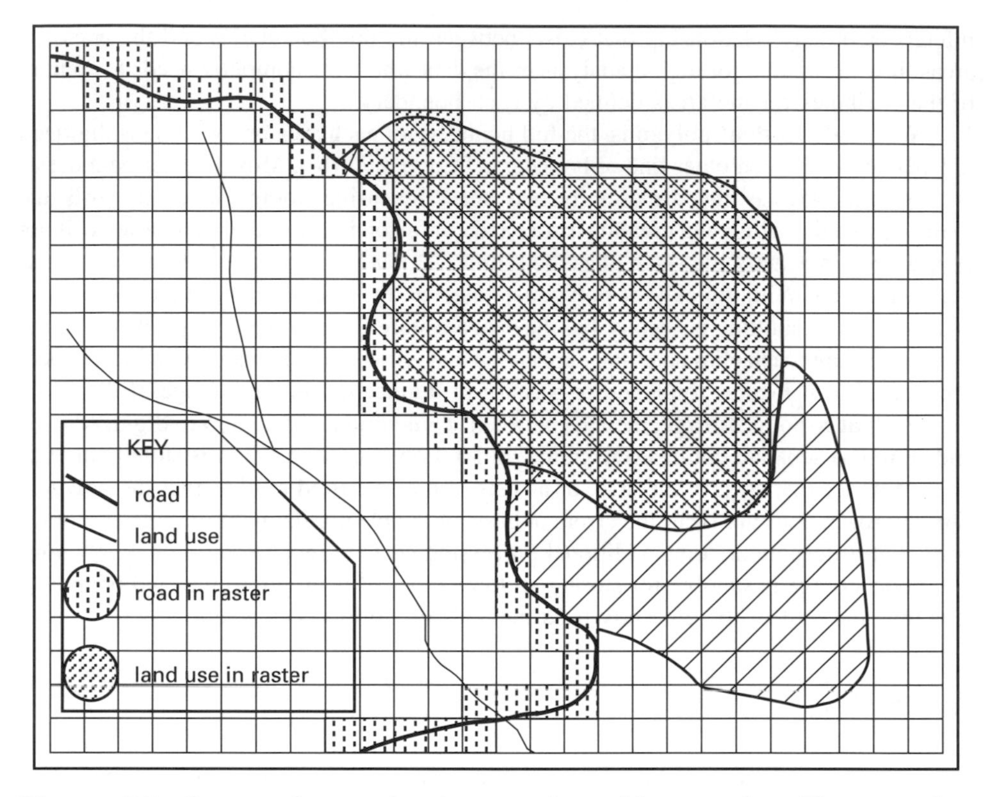

Diagram 7.7 Storage of raster data in comparison with vector data. The vector data would be stored as [attribute, (x_1, y_1), (x_2, y_2), . . .] for each arc or polygon. The raster data are stored as strings of attribute values per data record, with one value for each cell in the raster.

7.3.2 Point observations

Point observations, at a time, must include coordinate and time data as well as other types of data, or attribute data, such as temperature, humidity, wind direction and velocity for climatic data, or soil pH, humus content, moisture content and friability for soils. These data are usually stored in database files that can be interrogated to extract the information to be included in the GIS by writing these data to a separate file. This extracted data file can then be input to the GIS. The GIS may use the data as a set of triangulated independent networks (TIN) of data points to interpolate between the point observations to either create contours as vector data files, or to calculate gridcell values for raster files.

Interpolation between point observations is done by creating a mathematical surface through the known points and interpolating to estimate values at the pixel centres within the area covered by the surface. The simplest surface is linear interpolation as a plane of the form:

$$\text{estimate} = A_0 + B_0X + C_0Y$$

which could be constructed within triangles of points across the surface (Diagram 7.8). Planar models of this type can be suitable if:

1. the points are observed on changes in gradient, as can be done for parameters that can be observed, such as ground elevations;

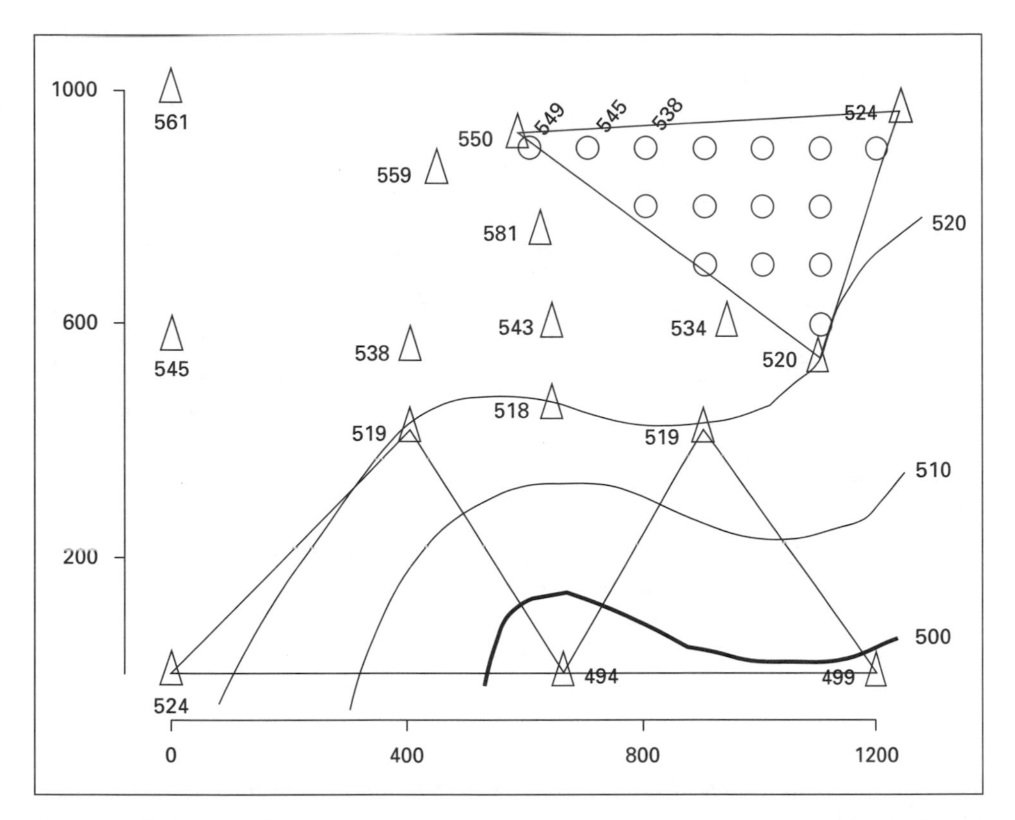

Diagram 7.8 Distribution of spot heights used to create a digital terrain model, and estimated gridcell values using linear interpolation within triangles of points.

2. the distribution of points forms well conditioned triangles, i.e. none of the internal angles should be less than 30°; and
3. the unknowns (A, B, C) are solved to the number of decimal places necessary to ensure that rounding-off errors do not significantly affect the final values.

More complex surface shapes can be used including: second- or third-order polynomial curves; chebyshev polynomial curves; and/or spline curves. Whilst these more complex shapes will require more points for their solution, they are still created within sub-image areas and used to estimate cell values within the area used to derive the parameters of the surface. All of these surfaces are continuous in that they define a continuous figure within the domain within which they are derived.

If there are discontinuities in the real world then the numerical surface may be a poor fit to the actual surface at that point. There are many causes of discontinuities in nature; impervious rock layers can form barriers to the flow of ground water, and hence cause discontinuities in the elevation of the ground water surface. Changes in soil types or associations can cause discontinuities in the chemical composition of the soil. When discontinuities are known or suspected then the discontinuity should be used as the edge of surface-fitting models by observing the parameters of interest along the discontinuity. Sometimes discontinuities can be identified if surfaces are fitted to points by use of least squares such that the residuals from the adjustment will be large, and of opposite sign, in the vicinity of discontinuities.

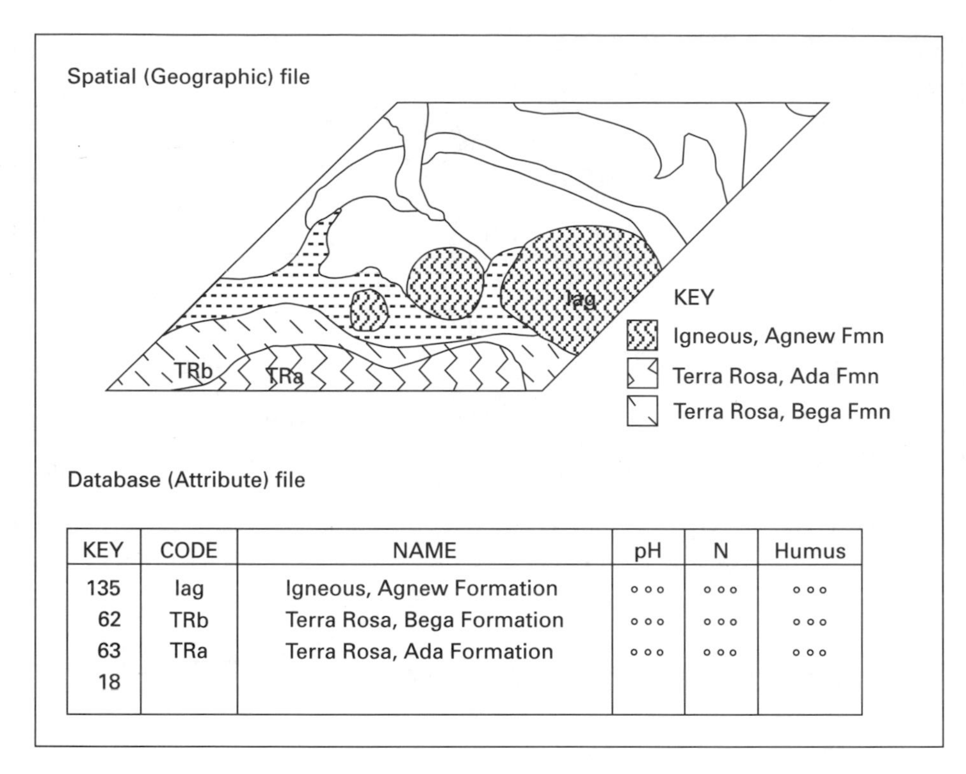

KEY	CODE	NAME	pH	N	Humus
135	Iag	Igneous, Agnew Formation	∘∘∘	∘∘∘	∘∘∘
62	TRb	Terra Rosa, Bega Formation	∘∘∘	∘∘∘	∘∘∘
63	TRa	Terra Rosa, Ada Formation	∘∘∘	∘∘∘	∘∘∘
18					

Diagram 7.9 Illustration of the two types of files in a GIS used for the analysis of soil databases.

The information contained in both vector and raster spatial files usually refers either to one attribute of the surface (such as soil pH, elevation or biomass), or to a key parameter that represents specific values in a family of attributes. For example, the key parameter may be coded to represent one soil class that has specific values in soil colour, Ph, silica and clay content. If the information only refers to one attribute then the values for that attribute will normally be held in the spatial files themselves. If the information is of a key parameter then storing each attribute as values in a GIS layer will incur significant storage costs. Most GIS systems store the key parameter in the spatial database file and link this key to the other attributes stored either in simple tables or in relational database files. The GIS uses the key to derive spatial files for values in the desired attributes in preparation for subsequent processings.

Consider the case of soils, where those soils may be categorized by the assigned soil names (Diagram 7.9). Each name would refer to soils that have many common characteristics, often in soil colour, Ph, chemical constituents, friability and structure. Storing the soils data in this way significantly reduces storage costs. An image of soil Ph can be extracted by running the GIS file containing the key attribute through a table so as to convert the soil name values to Ph values.

The key attribute must be unique for each set of parameter values, for all parameters that are represented by that key attribute. In the example shown in Table 7.1, the name 'residential' represents more than one land-use class; it is not unique and hence it cannot be used as the key attribute. If none of the stored parameters are sufficiently unique to use as the key parameter then it will be necessary to create one.

Table 7.1 Illustration of a GIS database for land use in which a unique key attribute had to be constructed for the database. (The Land-use categories used in the table are from the USGS Classification given in Chapter 4)

Keycode	Land use (Level two)	Land use (Level three)	% Built up	Population density
103	residential	111	30	15
121	residential	111	20	10
125	residential	113	70	40
102	residential	112	50	30
129	residential	111	30	20
154	commercial	121	80	20

Table 7.2 Relationship between scale, resolution (pixel size), size of coverage and number of cells in each file area.

Typical scale of use	Resolution (cell size)	Format (map size)	Approximate number gridcells in file
1:1000 000	1 km × 1 km	6°Long × 4°Lat	260 000
1:250 000	100 m × 100 m	1.5°Long × 1°Lat	1.6×10^6
1:100 000	50 m × 50 m	30′Long × 30′Lat	1.1×10^6
1:100 000	25 m × 25 m	30′Long × 30′Lat	4.3×10^6
1:50 000	25 m × 25 m	15′Long × 15′Lat	1.1×10^6
1:25 000	10 m × 10 m	7.5′Long × 7.5′Lat	1.7×10^6

7.3.3 Image processing

Raster files derived from image processing may require reformatting before they can be used or analysed in the GIS. It may be necessary to change the file formats to those that can be used in the GIS. Most GIS systems use a unique internal data format, designed to optimize processing in the GIS. However all GIS systems can communicate (import/export datafiles) with other GIS through a number of accepted formats, and can convert data in these formats into the internal data format preferred by the GIS. It may also be necessary to change the file boundaries to fit those that are conventionally accepted in the GIS. Whilst most GIS will accept a range of file sizes and locations, the organization may have adopted, as a convention, specific boundaries to the GIS files, e.g. map sheet formats may have been adopted. If this is the case then the data will need to be reformatted into these files sizes for inclusion in the GIS.

Raster files derived from remote sensing have to be rectified (section 5.4) and geocoded before they can be used in a GIS as this allows the different data sets to be registered one with the other. The standard resolution and coverage of GIS files is given in Table 7.2, although other resolutions may be adopted for specific purposes.

Information derived from remotely sensed data by visual interpretation is defined by linear boundaries that require digitization to create vector files. Digital image processing results in raster files being created.

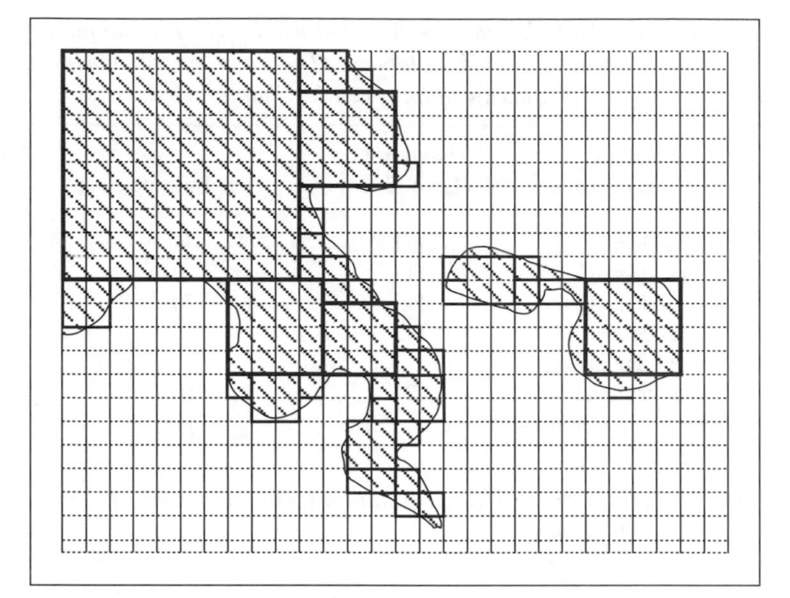

Diagram 7.10 Quadtree data structures.

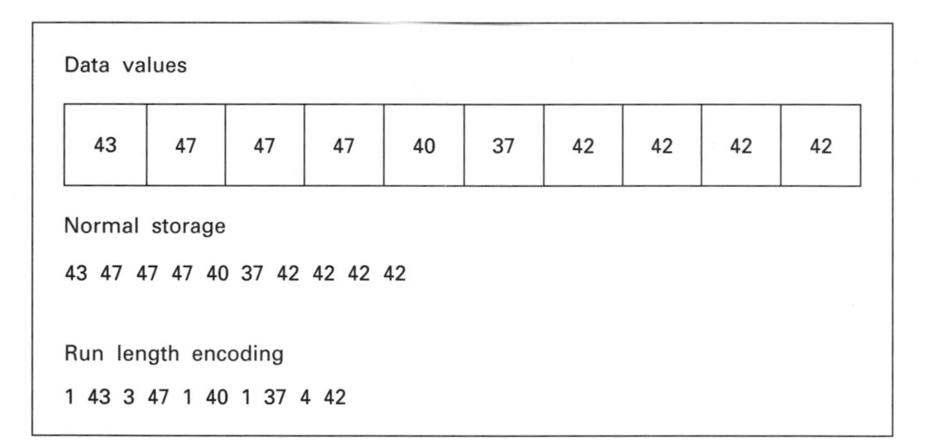

Diagram 7.11 Run length encoding.

7.3.4 Data drawn from existing files

Most forms of existing files, such as survey and seismic data, create either point or vector data that can be dealt with in the manner discussed earlier.

7.4 Data storage issues

The data storage devices, and the file types, for GIS files, are the same as for other file types. However GIS files, particularly raster files, can be very large. Since there is often redundancy in the information contained within the GIS files, there is considerable scope to reduce data storage needs by the use of data compression techniques.

Inspection of the GIS data files in Diagrams 7.10 and 7.11 will show that many of the cells in the GIS layers have the same values, creating data redundancies since the data can be reconstituted from a much smaller data set. Compression of the data to eliminate data redundancies will reduce the storage requirements for the data. Whilst there are many techniques for data compression, two are commonly used within GIS systems. These are quadtree data structures and run length encoding.

Quadtree data structures

Here, the raster data are analysed to find the largest square that is a multiple of 1, 2, 4, 8, . . ., 2^n cells within each area of the same values in the layer. Once the largest square has been found it replaces all of the equivalent individual cells. The area that remains around this first (and largest) cell is then analysed again on the same basis. In this way an area of the same values are stored as a series of cells of varying size, but with sizes that are related to each other. This approach significantly reduces data storage without affecting the quality of the information.

Run length encoding

If a line of data contains a run of contiguous cells with the same values then storage savings can be made if the run of values is replaced with the value, and the number of occurrences. Storage of the data includes the number of occurrences of a value, and the value, at each change in value along the line of data (Diagram 7.11). It can save a lot of space when there is high redundancy, but will cause higher storage costs if there is little data redundancy.

7.5 Analysis of data in a GIS

Data may be analysed using a GIS for three functions: to manage the databases; to create management information; and to develop better models and methods. This can be done in four ways: administrative processing of the data; combining sets of information; statistical analysis; and modelling.

The command names, functions and the method by which they are used to process data within a GIS will vary from system to system. GIS software needs to deal with spatial and attribute data in ways that are technically complex, but which, to the user, must be easy to learn. The software must minimize the chances of incurring errors either in the processing or in the input of data or commands by the user, and yet be accurate, effective and efficient in the conduct of the task and in the use of resources.

Achieving these requirements with the complex processes required to handle spatial and attribute data in a GIS is not easy. In consequence, the different GIS systems have resolved these complexities in quite different ways. Some software designers attempt to make the learning and use of the GIS easier with the use of windows, while others make extensive use of menus. Similarly the command names used in the different GIS systems can be quite different for the conduct of similar tasks, just as the method of executing a task can be quite different between software systems. All GIS systems also have some functionality unique to that particular package.

Because of these differences it is impossible here to cover the commands and their methods of implementation as are available in all the different GIS systems. Instead this text has chosen to use a low cost, raster-based GIS as the basis of the examples given. This GIS package is easy to use, and would be readily accessible to most students and institutions since it runs on standard 80286 and higher, PC compatible, computers.

The software selected uses menus and help windows as well as commands as the means of directing processing. The commands given in Table 7.3 are many of the

Table 7.3 Typical command capabilities of GIS in the processing and analysis of spatial files

Command	Meaning (subject to parameters to be specified)
ALIGN	Match two spatial files by changing the coordinates.
ASPECT	Derive aspect data from elevation data.
BORDER	Draw borders around homogeneous areas in raster files.
BUFFER	Map areas within a given distance of a feature.
CLASSIFY	Classify spatial data file into new classes.
CLUSTER	Identify groups of contiguous data cells.
COLOUR	Edit the colour palette used to display a file.
COPY	Make a copy of a spatial file.
COUNT	Count the number of cells, and percentages, for each combination of values in one or more sets of spatial files.
DELETE	Delete files from the disk.
DIRECTORY	List the available spatial files.
EDGE	Detect edges in raster files.
EDIT	Interactively edit the data values in a spatial file.
ENDLOG	End a log of actions as a file that can be subsequently reused.
EVALUATE	Numerically and logically evaluate combinations of spatial files.
EXPORT	Take vector, raster or attribute data out of the GIS in various formats.
FILTER	Filter spatial files within a window.
HEADER	Edit the header record of spatial files.
HELP	Ask for help with the available commands.
IMPORT	Bring in raster, vector or attribute data into the GIS in different formats.
MAPLIST	List the header record of spatial files.
MOSAIC	Join two spatial files.
NEIGHBOUR	Test neighbour cells for a specified value.
PLOT	Plot specified spatial file.
PRINT	Print specified spatial file.
QUIT	Exit from the GIS.
RADIUS	Calculate the distance from a cell to the nearest different cell of specified type.
REFLECT	Calculate reflected energy given slope and aspect.
RENAME	Change file name.
REPLAY	Replay a log file.
RESAMPLE	Resample spatial files to overlay.
SLOPE	Calculate slope from height data.
SMOOTH	Curve fit discontinuous data.
STARTLOG	Start a log of actions in the GIS, to create a record.
TREND	Create raster trend surface using (x, y, z) data as input.
WINDOW	Create new files of subareas from original spatial files.

commands used in this software, and illustrate many typical functions within a GIS, as well as some atypical functions. Where commands are mentioned they appear in small capitals e.g. DELETE.

7.5.1 Administration of the databases

The user will want to be able to manage his files by being able to:

- list files, or subsets of files using user specified constraints or parameters;
- view and edit the contents of the file – files generally consist of file administrative or

descriptive information in header and/or trailer records with the actual data in data records and so different types of commands may be required when looking at each type of record; and
- DELETE, RENAME and COPY files.

The user will want to be able to resize files by being able to:

- create subsets of some files by WINDOWing portion of the file;
- join files together or MOSAIC files;
- focus on irregular areas for subsequent analysis by masking out irrelevant parts of the files – e.g. the user may be interested in deriving information for irregular areas like a National Park, areas of water surface as defined on a remotely sensed image, a property, Shire or the catchment area for a river or a grain silo; and
- incorporate map coordinates into the files so as to analyse the spatial relationship between files, and properly overlay files.

7.5.2 Combining sets of information

A GIS contains both spatial and attribute files. A contiguous area is a group of connected cells of the same class in raster files and the area bounded by a vector in vector files. The spatial files for an area need to be registered so that they can be analysed or combined using numerical and/or logical processes. If the spatial files contain a key parameter then this can be replaced in the analysis with the values for a selected attribute as taken from attribute files.

The logical commands that can be used normally include those listed below. In these logical commands, both *A* and *B* can be numerical, character or logical variables or expressions, but usually both have to be of the same type, or be converted into a type or types that are comparable.

[*A* AND *B*] giving TRUE as the result if both occur and FALSE if either one, or neither, occur.
[*A* OR *B*] giving TRUE if either occur and FALSE if both, or neither, occur.
[*A* XOR *B*] '*A* and/or *B*', giving TRUE if both, or either, occur and FALSE if neither occur.
[*A* EQ *B*] 'A equal to B', giving TRUE if $A = B$, otherwise giving FALSE. Can also be used in thc form, $[A - D]$.
[*A* LT *B*] '*A* less than *B*', giving TRUE if *A* is less than *B*, otherwise giving FALSE. The alternative form is $[A < B]$.
[*A* LE *B*] '*A* less than or equal to *B*', giving TRUE if *A* is less than or equal to *B*, otherwise giving FALSE. The alternative form is $[A \quad B]$.
[*A* GT *B*] giving TRUE if *A* is greater than *B*, otherwise giving FALSE. The alternative form is $[A > B]$.
[*A* GE *B*] giving TRUE if *A* is greater than or equal to *B*, otherwise giving FALSE. The alternative form is $[A \geq B]$.
[*A* NE *B*] giving TRUE if *A* is not equal to *B*, otherwise giving FALSE. The alternative form is $[A <> B]$.

These logical processes may be used to test for a condition, and direct processing into different activities depending on the results of the test. One way to do this is to use a set of commands within the GIS to conduct this task, of the form

IF (test) THEN (do one thing)
ELSE (do something else) END

Consider the example

IF $(2 * A - 5)$ GT $(3 * B - A)$ THEN C: $= 2 * A - 5$
ELSE $C: = 3 * B - A$
END

which assigns (: =) the value $(2 \times A - 5)$ to new parameter C when $(2 \times A - 5)$ is greater than $(3 \times B - A)$ and assigns the value $(3 \times B - A)$ to C under all other conditions. The END statement indicates the end of that IF test.

IF $(A$ GT $B)$ XOR $(A$ GT $100)$ THEN $C := 0$
ELSE $C := (B - A)/(B + A)$
END

assigns the value 0 to C when A is greater than B and/or A is greater than 100, and assigns it a computed value in the remainder of the domain.

With some GIS this sort of test and the processing directions are set up in a different way to the examples given here.

The numerical commands, that can be applied to numerical variables A and B can include those listed below. With some systems more sophisticated commands may also be possible including the trigonometric and logarithmic functions.

[A = constant]	Assigns constant value to A
[$A + B$]	Addition
[$A - B$]	Subtraction
[$A * B$]	Multiplication
[A / B]	Division
[$A ** B$]	A to the power B
MIN (A, B)	Smaller of values A and B
MAX (A, B)	Larger of values A and B

Numerical combinations give a numerical result whereas logical combinations give either a 1 or 0 result. However logical combinations are often incorporated within a test and assignment command as shown above, so that numerical values can be assigned, depending on the result of the logical combination.

A complication that arises is that the processing to be done can depend on the values of some intermediate variables. This is solved by the use of intermediate variables in the processing, e.g.

INTERMEDIATE := (calculation using values in the source data)
IF *INTERMEDIATE* < 0 THEN
do one branch of processing of the data
ELSE
do another branch of processing of the data
END

This is particularly important when the IF tests and the numerical processing are interdependent and depend on multiple files since under these conditions it can be very difficult to conduct the analysis in stages. However, conducting the analysis in stages is often a good idea as it simplifies the analysis task, reducing the risk of errors.

1. **Define files**. Specify the number and then the file names of both the input and output files. Each file will be given a local name for processing, usually by the software.
   With some GIS, when the user is asked which files that he wishes to use, the GIS automatically assigns local names to the files. If this occurs the user must be familiar with the naming convention adopted if he needs to use these later to specify parameters in the processing.
2. **Set up the analysis commands**. Sometimes the analysis commands operate in an identical way on all pixels or regions. Such a command is called a global command. When this is the case the global command need only be set up in a simple way for the GIS to analyse all pixels or regions automatically. In other cases the assignment or task to be done can depend on the values in the files, on the processing results or the interrelationship between the values in some of the files. In this case the processing route can vary, and so the alternatives need to be set up in a suitable way. This is achieved by creating a loop of instructions where the loop would normally be completed for each pixel or region, although the actual path through the loop may vary from pixel to pixel or region to region. (There are some commands that can operate as either global or loop commands, whereas other commands can operate in only one of the two modes.)
3. **Execute the commands** whilst maintaining a detailed record of the actions that have been taken in an operational diary. GIS operations readily create data files that can be very large. It is essential that the analyst maintain a strict record of the work done, the files created and their purpose. It is also essential to establish a regular procedure of 'housekeeping' to delete older and unneeded files or copy the hard disc files to tape or floppy disc if they are not needed in the near future. When the analyst has completed a GIS task it is also a good rule to create, and maintain, a text file that records and describes the function of each layer in the GIS.

Consider the GIS (Diagram 7.12) established by the New South Wales Department of Agriculture. The GIS consists of the following layers:

1. **dynamic information layers** or layers of information that change at a relatively fast rate so that more than one layer of this information may be contained in the GIS, each layer recording the information at a date, or for a defined period – in this example the GIS contains agricultural parcels that have been cropped in a sequence of seasons;
2. **stable information layers** or information that does not change significantly in the expected life of the GIS – geographic parcels of attributes of the area that affect crop production, including soils, elevations, slopes and aspect;
3. **administrative features** relevant to the use of the GIS such as property boundaries, so as to identify individual land managers, Shire boundaries since most agricultural statistics relate to these administrative areas, and the area of responsibility of the agronomist; and
4. **orientation features** to assist in navigation and orientation by the user.

This GIS was established primarily to support the activities of agronomy advisers, where they are concerned with promoting practices that improve agricultural productivity and maintain agricultural resources. The data in the GIS allows the agronomist to identify farmers who are following current recommendations, and those who are not. One of the first things that the agronomist might do is determine why particular farmers are not following the recommended rotations since the reasons given could lead to identification of better practices than those currently being promoted. The GIS could also be used by

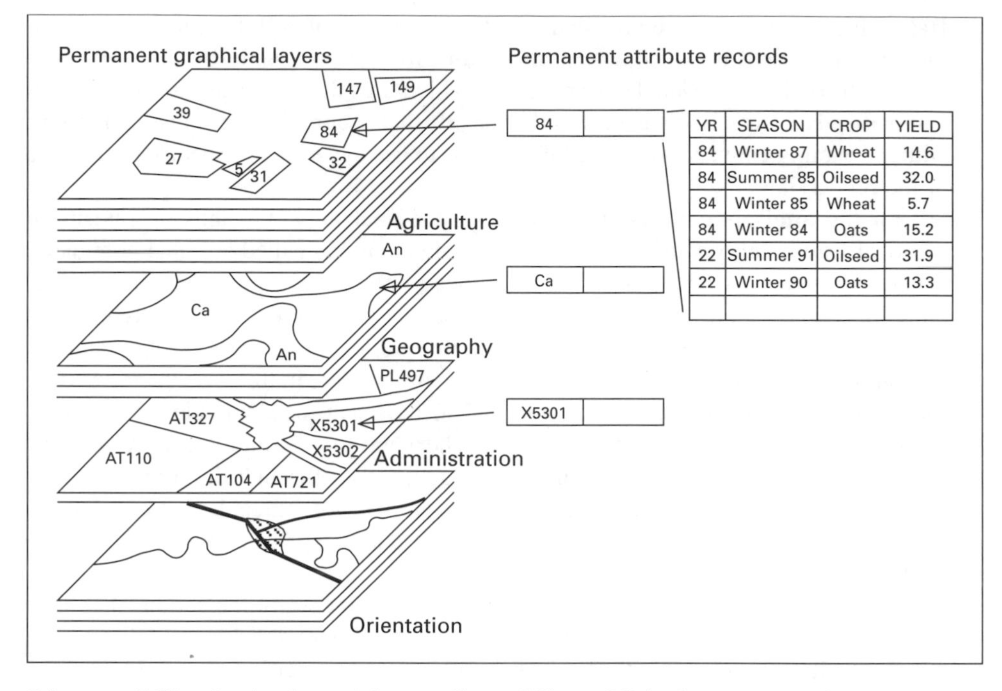

YR	SEASON	CROP	YIELD
84	Winter 87	Wheat	14.6
84	Summer 85	Oilseed	32.0
84	Winter 85	Wheat	5.7
84	Winter 84	Oats	15.2
22	Summer 91	Oilseed	31.9
22	Winter 90	Oats	13.3

Diagram 7.12 A mixed spatial – attribute GIS established to support the management of agricultural resources.

the agronomist to analyse the relationship between yield and the various causal factors. If the attribute database related to the cropping parcels contains yield or production information then the agronomist can relate this to soil, geomorphological and climatic data in the GIS to see if a relationship can be identified.

7.5.3 Statistical analysis

Statistical analysis of GIS data will most usually involve estimation of statistical parameters, testing of goodness of fit of data to a specific probability distribution, correlation analysis between various parameters, and regression analysis. Most GIS do not contain statistical tools, but can create files that can be used in standard statistical packages. This is usually done in either of two ways; by creating count or n-dimensional histogram files or by analysis of the spatial data files themselves.

In the first case the m spatial files, representing the m parameters that are to be analysed, are scanned to create an m-dimensional histogram file. In the second case, the spatial files may need to be converted into a suitable file structure for use in the statistical program before analysis. The first case loses the spatial locational information in the data before statistical analysis, but has the advantage that it significantly condenses the data and hence will significantly speed up the statistical analysis. Generally spatial location data is not required in statistical analysis. The second case retains the spatial locational information, and as such this case must be used when this information is critical to the analysis.

Correlation analysis can assist resource managers to:

- identify or verify the primary factors that influence the state of specific resources or aspects of the environment;

- assess the relative importance of the causal parameters; and
- develop a relationship between these parameters and the resources or aspects of the environment that are of interest.

If the correlation analysis yields a significant level of correlation then the analyst might use this data to create a regression model to estimate values of the dependent parameters from values of the independent parameters. Such models can be used either to estimate values in the dependent parameter, or to predict future values for the dependent parameter under specific conditions of the independent parameters.

7.5.4 Modelling

Modelling activities can be of two forms: estimation and prediction models. Estimation models, discussed in Chapter 5, designed to convert remotely sensed data into estimates of one or more physical parameters, may need to be implemented within a GIS environment because the models may utilize a range of data types. Predictive models will normally be implemented within a GIS for the same reasons.

GIS databases are important for both the creation and implementation of many models. Models may be created as a result of statistical analysis. Once the models have been developed then the GIS provides the numerical data necessary for implementation.

A model of erosion risk could be derived using the universal soil loss equation (USLE) to estimate soil loss due to water erosion, where

$$A = 0.224RKLSCP$$

for which

A = soil loss per unit area
R = rainfall and run-off factor
K = soil erodibility factor
L = slope length factor
S = slope steepness factor
C = cover and management factor
P = conservation practices used factor

This model is discussed in detail in Chapter 9 as an application of modelling in a GIS. Here it will be treated in a simplified way to illustrate the concept of implementing a model within a GIS. The USLE can be applied in the following steps.

1. **Preparation of Erosion Potential Maps** Soil mapping units were categorized in terms of soil erodibility (element K in the USLE). Because this factor is essentially unchanging it can be retained as a permanent entry in the GIS. The digitized soil maps are retained as the generalized soils information in the GIS (Photo 9.9). For this application, the classes are converted into a soil erodibility factor, K, using a combination of field visits and experience with the soils data, where the conversion is discussed in detail in section 9.6.
2. **Slope Factor** Slope maps are usually derived either by analysis of the elevation data in the GIS, or by manually mapping the slope into classes using the analogue map, and digitizing these classes into the GIS. The slope values are used to create a slope factor layer in the GIS using the conversion discussed in section 9.6. The slope data

is also used to estimate the slope length, i.e. the length of the slope from the start of the slope to the point or cell being analysed in the GIS.

3. **Cover and Management Factor** Land-cover changes are affected by both management and climatic factors. It is important to assess the amount of cover that can be expected at the time when the area is most at risk from erosion. Satellite imagery was used to assess cover at the time of acquisition, and land-use in that season. The cover was estimated by calculation of the NDVI, and converted into LAI by the use of field data collected for the purpose. The cover factor at the time of maximum risk was then estimated by extrapolation for each land-use class, using field data to relate land-use and cover to the cover factor, C.
4. **Rainfall and Run-off Factor** The rainfall erosivity can be estimated using various equations, such as $R = 6.28p^{2.17}$ where p is the total rainfall (in cm) for a 1 in 2 year, 6-hour event. For the small area of this watershed the factor will be constant and can be ignored to give relative erosion risk maps. For larger areas where the variability in rainfall cannot be ignored, site records of rainfall can be used to construct a map of rainfall (in cm) for a 1 in 2 year, 6-hour event.

The USLE can now be applied by the product of the three parameters that are stored in the GIS, calculating the erosion risk for each cell in turn.

$$\text{Annual soil loss} = 0.224R(KL/S)(CP)$$

where the annual loss will be in tonnes $ha^{-1}\ yr^{-1}$.

This mapping can be compared either with field estimates of erosion risk, or with mappings of actual erosion. The field data can be entered into the GIS to conduct the statistical analysis of the correlation between actual and estimated erosion. If this comparison does not produce an acceptable level of correlation then either more detailed models incorporating other parameters may need to be considered and/or more accurate data measurement techniques may need to be used. It may be necessary to develop better models, with the development influenced by detailed correlation analysis to identify the parameters of interest.

Sometimes the model may need to be built into an expert system, to simplify the process of estimating model parameter values, or the process of utilizing the model in the management of resources. An important component of such models built into expert systems will be the facility to have modelling influence decisions by ensuring that the resource manager understands the results that are derived from the model, and that the model outputs are in a form suitable for the decision-making processes involved. This will often mean simplifying the output from the model into classes or levels that are compatible with the needs of the resource manager. For example, an expert system looking at the economic return of a crop may map the area into only three classes: poor investment; marginal to fair investment with other crops; and good to excellent investment – in which case most of the activity is likely to focus on the third class in the short-term with one strategy, expand to the second class areas with a second strategy, and ignore the areas in the first class in relation to that crop.

Use of the decisions to influence feedback mechanisms is also important. Where possible, include the proposed decisions in the expert system so as to provide the manager with information on the likely impact of those decisions on the resources of interest, even if these predictive capabilities are rather crude. If the expert system prediction indicates a potential problem then the manager can decide to expend some resources to assess the situation in some detail, whereas he is unlikely to do this without that predictive capability. Clearly such predictive capability is of no use once it becomes discredited; the predictions need to be satisfactory in providing a reasonable indication of potential impacts.

7.5.5 Example application of a GIS

Consider the case of the resource managers of an area who have access to a GIS with the data sets as shown in Diagram 7.13 and depicted in Photo 7.1. There are a total of eighteen overlays in the original data; Elevation, Soils, Ownership, Woody Cover, Drainage, Soil Erosion and Cropping in each year, 1980–1991 inclusive.

The data sets are oriented about, but are not exclusive to agricultural management issues. Many of them are useful for other purposes; even the cropping information may be important in resolving various environmental issues. These overlays will now be used to address a range of tasks, to illustrate, in a simple way, the role of GIS in resource management.

The first question that a resource manager needs to address is where can he gain access to the information he requires in a cost effective way. Generally topographic, drainage and catchment area information will be digitized from existing topographic maps, and converted from vector files into the raster files as depicted. The soil map data will be acquired in the same way from soils maps. Ownership information can also be acquired by the digitization of maps; up-dated with sales since publication of the map if necessary.

Information on soil erosion will come either from field mapping, or from mapping on up-to-date large scale aerial photography. The density of woody cover can be measured by digital processing of remotely sensed data, and the cropping will come from either visual interpretation or digital classification of remotely sensed data.

Consider an agricultural adviser moving into the area who is told of the crop rotations that are promoted and those that are practised. Crop rotations have a significant effect on soil condition and hence on productivity and degradation. They are an important technique in maintaining agricultural resources whilst gaining high yields. Promotion of better practices is an important issue in agricultural production. How can the adviser objectively assess what rotations are being used in this area?

The cropping layers contain the information that the adviser needs on the crop rotations that are practiced in the area. A GIS can analyse these layers to identify all of the unique combinations of values that occur in the cropping layers and place these combinations in a table file. The first step in this analysis might be to determine the cropping frequency in each cell, by adding together the cropping layers, to give the results shown in Table 7.4, shown in Diagram 7.14, Layer 5 and Photo 7.2. This cropsum overlay indicates that no cells that have been cropped more than 10 times in the period 1980–1991 inclusive. Diagram 7.14, layer 5 shows that there are a number of distinct blocks that have been cropped for each frequency, except for the 3 crops in 12 years (that is for 0, 5, 6, 8 and 10 crops in the 12 years). Now a cropsum of 6 crops in 12 years can be achieved by rotations of 1 in 2 years, 2 in 4 years, and 3 crops in 6 years, so that the summary does not provide definitive information on all crop rotations.

With any rotation, say 2 crops in 4 seasons, with the land resting for the other 2 seasons, the actual number of crops recorded in a 10-year period will depend on what part of the cropping cycle is occurring when the recording period starts. For 12 years, cropping 2 out of 4 will always yield 6 years of cropping as occurs for 15.11 per cent of the area (Table 7.4). If the cropping is 2 years out of 3 then these areas would record about 8 years of cropping, as occurs for 32.44 per cent of the area.

(1) Layers 1–3: Elevation (metres), soil types and land ownership or properties. The elevations are in the range 11–95 m. There are six soil types; 1 = elevated sandy, shallow soil; 2 = thin sandy soil on the rocky escarpment; 3 = alluvial flats; 4 = sub escarpment eroded soils; 5 = alluvial slopes; and 6 = sub escarpment soils, deep profile soil. There are seven properties in the sample area.

Topography														
92	91	88	87	75	60	39	28	26	28	27	25	23	22	21
93	90	85	79	68	50	31	24	23	26	27	27	25	23	22
95	89	79	75	58	40	26	22	23	27	31	29	27	25	24
93	82	68	57	42	31	24	20	23	27	30	30	28	27	26
86	67	55	48	39	30	23	18	22	26	29	30	27	25	24
69	54	47	41	32	28	23	17	22	26	28	30	28	29	27
52	46	44	38	31	28	23	17	22	25	28	30	31	29	28
45	40	37	34	29	26	22	16	21	24	26	28	29	28	27
39	35	36	30	29	26	22	16	21	23	25	27	28	27	25
30	29	27	25	25	22	18	15	18	19	20	21	23	25	23
26	25	22	18	19	17	15	14	15	15	14	15	17	18	18
25	19	17	16	15	15	14	13	13	13	12	12	13	14	15
22	22	21	19	18	18	17	16	15	14	14	12	12	11	12
25	26	29	27	25	26	25	26	25	22	18	14	13	12	11
28	30	34	31	33	35	37	34	30	27	24	20	16	14	12

Soil type														
1	1	1	1	2	2	4	4	5	5	6	6	6	5	5
1	1	1	2	2	2	4	5	5	6	6	6	6	6	5
1	1	2	2	2	2	4	5	5	6	6	6	6	6	6
1	2	2	2	2	4	5	5	5	6	6	6	6	6	6
2	2	2	2	2	4	5	5	5	6	6	6	6	6	6
2	2	2	2	4	4	5	3	5	6	6	6	6	6	6
2	2	2	4	4	4	5	3	5	6	6	6	6	6	6
2	2	4	4	4	5	5	3	5	6	6	6	6	6	6
2	4	4	4	4	5	5	3	5	4	6	6	6	6	6
4	4	4	5	5	5	5	3	5	4	4	4	4	4	4
5	5	5	5	5	5	5	3	3	5	4	4	4	4	4
5	5	5	5	5	5	3	3	3	3	3	3	3	5	5
5	5	5	5	5	5	5	5	5	5	5	3	3	3	3
5	4	4	4	4	4	4	4	4	4	5	5	3	3	3
4	4	4	4	4	4	4	4	4	4	4	4	5	5	5

Ownership														
1	1	1	1	1	1	1	1	1	1	6	6	6	6	6
1	1	1	1	1	1	1	1	1	6	6	6	6	6	6
2	1	1	1	1	1	1	1	1	6	6	6	6	6	6
2	2	2	2	2	1	1	1	1	6	6	6	6	6	6
2	2	2	2	2	2	2	2	6	6	6	6	6	6	6
2	2	2	2	2	2	2	6	6	6	6	6	6	6	6
2	2	2	2	2	2	2	6	6	6	6	6	6	6	6
3	3	3	2	2	2	2	6	6	6	6	6	6	6	6
3	3	3	3	3	3	5	5	5	5	5	5	5	5	5
3	3	3	3	3	5	5	5	5	5	5	5	5	5	5
3	3	3	3	3	5	5	5	5	5	5	5	5	5	5
3	3	3	3	3	5	5	5	5	5	5	5	5	5	5
3	4	4	4	4	4	4	4	4	4	4	4	7	7	7
4	4	4	4	4	4	4	4	4	4	4	4	7	7	7
4	4	4	4	4	4	4	4	4	4	4	4	7	7	7

(2) Overlays 4–6: Slope, aspect and percentage woody cover. Slope – each digit represents percent slope. The aspect is 0 (north) and increases by units of one every 45° in the clockwise direction, so that east is 2. The woody cover is in percentages cover, so that 15 represents 15% woody cover on that pixel.

Slope														
1	4	6	13	17	21	11	4	3	3	2	2	1	1	0
3	7	9	14	19	19	7	2	0	3	1	2	2	1	1
9	14	15	20	19	14	4	2	2	4	4	2	2	2	2
18	19	14	15	11	7	4	2	3	4	3	2	2	2	2
22	14	9	11	9	7	5	1	4	4	3	3	2	1	0
17	8	6	9	4	5	6	0	5	4	2	2	2	4	3
8	6	7	7	3	5	6	1	5	3	3	2	3	1	1
7	5	4	5	3	4	6	0	5	3	2	2	1	2	2
9	6	9	5	4	5	6	1	5	4	5	6	5	2	2
4	4	6	7	6	5	3	1	3	4	6	6	6	7	5
4	6	5	2	4	2	1	1	2	2	2	3	4	4	3
6	2	1	1	0	1	1	0	0	1	0	0	1	3	3
2	3	4	3	3	3	3	3	2	1	2	0	1	0	1
3	4	8	8	7	8	8	10	10	8	4	2	1	1	0
3	4	5	4	8	9	12	8	5	6	7	6	3	2	1

Aspect														
3	3	3	3	3	2	2	4	4	5	2	2	2	2	0
2	3	3	3	3	2	2	4	5	6	1	1	1	1	0
3	3	3	3	3	2	3	4	5	6	0	1	1	1	0
3	3	3	2	2	2	3	4	5	5	6	1	1	1	0
3	3	3	3	2	2	2	4	6	6	6	2	2	2	2
4	4	3	2	2	2	2	4	6	6	5	1	1	0	0
3	4	3	2	3	2	2	4	6	6	6	5	0	1	0
3	4	3	2	2	2	2	4	6	6	5	5	3	3	4
4	4	4	4	3	3	2	4	6	4	4	4	4	3	4
4	3	3	4	4	4	4	4	4	4	4	4	4	4	4
3	4	4	3	4	3	3	4	4	3	4	4	4	4	4
2	3	3	3	3	3	3	4	4	3	3	4	4	4	3
1	1	0	0	0	0	0	0	0	1	0	0	2	3	4
0	0	0	0	0	0	0	0	0	0	1	0	1	0	2
0	0	7	1	0	0	0	0	1	1	1	0	0	1	0

Woody Cover														
10	5	15	20	90	90	65	0	0	0	0	0	5	0	0
15	10	25	85	90	75	85	0	5	0	5	0	0	0	5
5	10	70	90	55	50	50	0	5	0	5	0	0	5	10
15	80	85	90	60	35	40	5	0	0	0	5	0	0	0
75	75	90	85	70	35	5	0	5	0	10	0	0	0	5
65	65	90	85	75	40	10	0	0	0	5	0	5	0	0
50	55	85	90	95	45	0	0	5	10	0	0	0	0	0
25	20	35	75	60	15	5	0	5	0	5	0	5	5	0
10	5	15	25	15	5	5	5	0	0	5	0	5	5	0
10	5	5	0	5	5	0	5	15	0	0	0	5	5	5
15	5	0	0	5	0	0	5	5	10	10	15	10	0	5
0	5	5	5	0	0	0	0	5	5	5	0	5	10	5
5	50	65	65	75	55	65	55	60	45	50	35	0	0	0
25	65	95	90	65	75	75	70	75	70	65	55	0	0	5
30	45	80	75	85	25	25	45	90	85	80	75	0	0	0

(3) Overlays 7–9: Cropping in 1980, 1981 and 1982. In each cropping overlay 1 represents cropping in that year, and 0 represents no cropping.

1980														
0	0	0	0	0	0	0	1	1	1	0	0	0	1	1
0	0	0	0	0	0	0	1	1	0	0	0	0	1	1
0	0	0	0	0	0	0	1	1	0	0	0	0	1	1
0	0	0	0	0	0	0	1	1	0	0	0	0	1	1
0	0	0	0	0	0	1	1	1	1	1	1	1	1	1
0	0	0	0	0	0	1	1	1	1	1	1	1	1	1
0	0	0	0	0	0	1	1	1	1	1	1	1	1	1
0	0	0	0	0	1	1	1	1	1	1	1	1	1	1
0	0	0	1	1	1	1	1	0	0	0	0	0	1	1
0	0	0	1	1	1	1	1	0	0	0	0	0	1	1
0	0	0	1	1	1	1	1	0	0	0	0	0	1	1
0	0	0	1	1	1	1	1	1	1	1	1	1	1	1
0	0	0	0	0	0	0	0	0	0	0	0	0	0	0
0	0	0	0	0	0	0	0	0	0	0	0	0	0	0
0	0	0	0	0	0	0	0	0	0	0	0	0	0	0

1981														
1	1	1	0	0	0	0	0	0	0	1	1	1	1	1
1	1	1	0	0	0	0	0	0	1	1	1	1	1	1
0	1	0	0	0	0	0	0	0	1	1	1	1	1	1
0	0	0	0	0	0	0	0	0	1	1	1	1	1	1
0	0	0	0	0	0	1	1	1	1	1	1	1	1	1
0	0	0	0	0	0	1	1	1	1	1	1	1	1	1
0	0	0	0	0	0	1	1	1	1	1	1	1	1	1
1	1	1	0	0	1	1	1	1	1	1	1	1	1	1
1	1	1	0	0	0	1	1	0	0	0	0	0	0	0
1	1	1	0	0	1	1	1	0	0	0	0	0	0	0
1	1	1	0	0	1	1	1	0	0	0	0	0	0	0
1	1	1	0	0	1	1	1	1	1	1	1	1	0	0
1	0	0	0	0	0	0	0	0	0	0	0	1	1	1
0	0	0	0	0	0	0	0	0	0	0	0	1	1	1
0	0	0	0	0	0	0	0	0	0	0	0	0	0	0

1982														
0	0	0	0	0	0	0	0	0	0	1	1	1	1	1
0	0	0	0	0	0	0	0	0	1	1	1	1	1	1
0	0	0	0	0	0	0	0	0	1	1	1	1	1	1
0	0	0	0	0	0	0	0	0	1	1	1	1	1	1
0	0	0	0	0	0	0	0	0	0	0	0	0	0	0
0	0	0	0	0	0	0	0	0	0	0	0	0	0	0
0	0	0	0	0	0	0	0	0	0	0	0	0	0	0
1	1	1	0	0	0	0	0	0	0	0	0	0	0	0
1	1	1	1	1	1	0	0	1	1	1	1	1	0	0
1	1	1	1	1	0	0	0	1	1	1	1	1	0	0
1	1	1	1	1	0	0	0	1	1	1	1	1	0	0
1	1	1	1	1	0	0	0	0	1	1	1	1	0	0
1	0	0	0	0	0	0	0	0	0	0	0	1	1	1
0	0	0	0	0	0	0	0	0	0	0	0	1	1	1
0	0	0	0	0	0	0	0	0	0	0	0	1	1	1

(4) Overlays 10–12: Cropping in 1983, 1984 and 1985. The overlays are depicted in the same manner described in (c) above.

```
1983                            1984                            1985
0 0 0 0 0 0 0 1 1 1 0 0 0 0 0 + 0 0 0 0 0 0 0 1 1 1 1 1 1 1 1 + 1 1 1 0 0 0 0 0 0 0 1 1 1 1 1
0 0 0 0 0 0 0 1 1 0 0 0 0 0 0 + 0 0 0 0 0 0 0 1 1 1 1 1 1 1 1 + 1 1 1 0 0 0 0 0 0 1 1 1 1 1 1
0 0 0 0 0 0 0 1 1 0 0 0 0 0 0 + 0 0 0 0 0 0 0 1 1 1 1 1 1 1 1 + 0 1 0 0 0 0 0 0 0 1 1 1 1 1 1
0 0 0 0 0 0 0 1 1 0 0 0 0 0 0 + 0 0 0 0 0 0 0 1 1 1 1 1 1 1 1 + 0 0 0 0 0 0 0 0 0 1 1 1 1 1 1
0 0 0 0 0 0 0 0 1 1 1 1 1 1 1 + 0 0 0 0 0 0 1 1 1 1 1 1 1 1 1 + 0 0 0 0 0 0 1 1 0 0 0 0 0 0 0
0 0 0 0 0 0 0 1 1 1 1 1 1 1 1 + 0 0 0 0 0 0 1 1 1 1 1 1 1 1 1 + 0 0 0 0 0 0 1 0 0 0 0 0 0 0 0
0 0 0 0 0 0 0 1 1 1 1 1 1 1 1 + 0 0 0 0 0 0 1 1 1 1 1 1 1 1 1 + 0 0 0 0 0 0 1 0 0 0 0 0 0 0 0
0 0 0 0 0 0 0 1 1 1 1 1 1 1 1 + 1 1 1 0 0 1 1 1 1 1 1 1 1 1 1 + 1 1 1 0 0 1 1 0 0 0 0 0 0 0 0
0 0 0 1 1 1 1 1 1 1 1 1 1 0 0 + 1 1 1 0 0 0 1 1 0 0 0 0 0 1 1 + 1 1 1 1 1 1 1 1 0 0 0 0 0 1 1
0 0 0 1 1 1 1 1 1 1 1 1 1 0 0 + 1 1 1 0 0 1 1 1 0 0 0 0 0 1 1 + 1 1 1 1 1 1 1 1 0 0 0 0 0 1 1
0 0 0 1 1 1 1 1 1 1 1 1 1 0 0 + 1 1 1 0 0 1 1 1 0 0 0 0 0 1 1 + 1 1 1 1 1 1 1 1 0 0 0 0 0 1 1
0 0 0 1 1 1 1 1 1 0 0 0 0 0 0 + 1 1 1 0 0 1 1 1 1 1 1 1 1 1 1 + 1 1 1 1 1 1 1 1 1 1 1 1 1 1 1
0 0 0 0 0 0 0 0 0 0 0 0 0 0 0 + 1 0 0 0 0 0 0 0 0 0 0 1 1 1 1 + 1 0 0 0 0 0 0 0 0 0 0 0 1 1 1
0 0 0 0 0 0 0 0 0 0 0 0 0 0 0 + 0 0 0 0 0 0 0 0 0 0 0 1 1 1 1 + 0 0 0 0 0 0 0 0 0 0 0 0 1 1 1
0 0 0 0 0 0 0 0 0 0 0 0 0 0 0 + 0 0 0 0 0 0 0 0 0 0 0 0 0 0 0 + 0 0 0 0 0 0 0 0 0 0 0 0 0 0 0
```

(5) Overlays 13–15: Cropping in 1986, 1987 and 1988. The overlays are depicted in the same manner described in (c) above.

```
1986                            1987                            1989
1 1 1 0 0 0 0 0 0 0 0 0 0 1 1 + 0 0 0 0 0 0 1 1 1 1 1 1 1 1 1 + 0 0 0 0 0 0 0 1 1 1 1 1 1 1 1
1 1 1 0 0 0 0 0 0 0 0 0 0 1 1 + 0 0 0 0 0 0 1 1 1 1 1 1 1 1 1 + 0 0 0 0 0 0 0 1 1 1 1 1 1 1 1
0 1 0 0 0 0 0 0 0 0 0 0 0 1 1 + 0 0 0 0 0 0 1 1 1 1 1 1 1 1 1 + 0 0 0 0 0 0 0 1 1 1 1 1 1 1 1
0 0 0 0 0 0 0 0 0 0 0 0 0 1 1 + 0 0 0 0 0 0 1 1 1 1 1 1 1 1 1 + 0 0 0 0 0 0 0 1 1 1 1 1 1 1 1
0 0 0 0 0 0 0 0 1 1 1 1 1 1 1 + 0 0 0 0 0 0 0 1 1 1 1 1 1 1 1 + 0 0 0 0 0 0 1 1 0 0 0 0 0 0 0
0 0 0 0 0 0 0 1 1 1 1 1 1 1 1 + 0 0 0 0 0 0 1 1 1 1 1 1 1 1 1 + 0 0 0 0 0 0 1 0 0 0 0 0 0 0 0
0 0 0 0 0 0 0 1 1 1 1 1 1 1 1 + 0 0 0 0 0 0 1 1 1 1 1 1 1 1 1 + 0 0 0 0 0 0 1 0 0 0 0 0 0 0 0
0 0 0 0 0 0 0 1 1 1 1 1 1 1 1 + 0 0 0 0 0 0 1 1 1 1 1 1 1 1 1 + 0 0 0 0 0 1 1 0 0 0 0 0 0 0 0
0 0 0 1 1 1 1 1 1 1 1 1 1 0 0 + 1 1 1 0 0 0 1 1 1 1 1 1 1 0 0 + 1 1 1 1 1 1 1 1 0 0 0 0 0 1 1
0 0 0 1 1 1 1 1 1 1 1 1 1 0 0 + 1 1 1 0 0 1 1 1 1 1 1 1 1 0 0 + 1 1 1 1 1 1 1 1 0 0 0 0 0 1 1
0 0 0 1 1 1 1 1 1 1 1 1 1 0 0 + 1 1 1 0 0 1 1 1 1 1 1 1 1 0 0 + 1 1 1 1 1 1 1 1 0 0 0 0 0 1 1
0 0 0 1 1 1 1 1 1 1 1 1 1 0 0 + 1 1 1 0 0 1 1 1 1 1 1 1 1 0 0 + 1 1 1 1 1 1 1 1 1 1 1 1 1 1 1
0 0 0 0 0 0 0 0 0 0 0 0 0 0 0 + 0 0 0 0 0 0 0 0 0 0 0 0 1 1 1 + 0 0 0 0 0 0 0 0 0 0 0 0 1 1 1
0 0 0 0 0 0 0 0 0 0 0 0 0 0 0 + 0 0 0 0 0 0 0 0 0 0 0 0 1 1 1 + 0 0 0 0 0 0 0 0 0 0 0 0 1 1 1
0 0 0 0 0 0 0 0 0 0 0 0 1 1 1 + 0 0 0 0 0 0 0 0 0 0 0 0 0 0 0 + 0 0 0 0 0 0 0 0 0 0 0 0 0 0 0
```

(6) Overlays 16–18: Cropping in 1989, 1990 and 1991. The cropping overlay is as described in (c) above.

```
1989                            1990                            1991
0 0 0 0 0 0 0 0 0 0 0 0 0 0 0 + 1 1 1 0 0 0 0 0 0 0 1 1 1 1 1 + 1 1 1 0 0 0 0 1 1 1 1 1 1 1 1
0 0 0 0 0 0 0 0 0 0 0 0 0 0 0 + 1 1 1 0 0 0 0 0 0 1 1 1 1 1 1 + 1 1 1 0 0 0 0 1 1 1 1 1 1 1 1
0 0 0 0 0 0 0 0 0 0 0 0 0 0 0 + 0 1 0 0 0 0 0 0 0 1 1 1 1 1 1 + 0 1 0 0 0 0 0 1 1 1 1 1 1 1 1
0 0 0 0 0 0 0 0 0 0 0 0 0 0 0 + 0 0 0 0 0 0 0 0 0 1 1 1 1 1 1 + 0 0 0 0 0 0 0 1 1 1 1 1 1 1 1
0 0 0 0 0 0 1 1 0 0 0 0 1 1 1 + 0 0 0 0 0 0 0 0 1 1 1 1 1 1 1 + 0 0 0 0 0 0 0 0 1 1 1 1 0 0 0
0 0 0 0 0 0 1 0 0 0 0 0 1 1 1 + 0 0 0 0 0 0 0 1 1 1 1 1 1 1 1 + 0 0 0 0 0 0 0 1 1 1 1 1 0 0 0
0 0 0 0 0 0 1 0 0 0 0 0 1 1 1 + 0 0 0 0 0 0 0 1 1 1 1 1 1 1 1 + 0 0 0 0 0 0 0 1 1 1 1 1 0 0 0
0 0 0 0 0 1 1 0 0 0 0 0 1 1 1 + 1 1 1 0 0 0 0 1 1 1 1 1 1 1 1 + 1 1 1 0 0 0 0 1 1 1 1 1 0 0 0
0 0 0 0 0 0 0 0 0 0 0 0 0 1 1 + 1 1 1 1 1 1 1 1 1 1 1 1 1 0 0 + 1 1 1 1 1 1 1 1 1 1 1 1 1 0 0
0 0 0 0 0 0 0 0 0 0 0 0 0 1 1 + 1 1 1 1 1 1 1 1 1 1 1 1 1 0 0 + 1 1 1 1 1 1 1 1 1 1 1 1 1 0 0
0 0 0 0 0 0 0 0 0 0 0 0 0 1 1 + 1 1 1 1 1 1 1 1 1 1 1 1 1 0 0 + 1 1 1 1 1 1 1 1 1 1 1 1 1 0 0
0 0 0 0 0 0 0 0 0 0 0 0 0 1 1 + 1 1 1 1 1 1 1 1 1 1 1 1 1 0 0 + 1 1 1 1 1 1 1 1 1 1 1 1 1 0 0
0 0 0 0 0 0 0 0 0 0 0 0 0 0 0 + 1 0 0 0 0 0 0 0 0 0 0 0 1 1 1 + 1 0 0 0 0 0 0 0 0 0 0 0 1 1 1
0 0 0 0 0 0 0 0 0 0 0 0 0 0 0 + 0 0 0 0 0 0 0 0 0 0 0 0 1 1 1 + 0 0 0 0 0 0 0 0 0 0 0 0 1 1 1
0 0 0 0 0 0 0 0 0 0 0 0 0 0 0 + 0 0 0 0 0 0 0 0 0 0 0 0 1 1 1 + 0 0 0 0 0 0 0 0 0 0 0 0 0 0 0
```

Diagram 7.13 GIS overlays, three overlays across, each overlay being 15 × 15 cells in size, with each cell being 50 m × 50 m in size.

Table 7.4 Summary of cropping, 1980–91; count of cells and percentage of area cropped

Number of years cropped	Cell count in class	Percentage of area
0	76	33.78
3	3	1.33
5	15	6.67
6	34	15.11
8	73	32.44
10	24	10.67

A total of 10.67 per cent of the area has been cropped for 10 years; these areas may be continuous cropping that have had the occasional break due to bad weather or some other reason. There are also some areas that have been cropped a few times; this may be opportunity cropping, again largely depending on the weather.

With any rotation there will be discontinuities for a whole variety of social, economic and physical reasons, so there will be patterns in the cropping of some areas that do not fit a regular pattern all of the time. Also, if the number of years being analysed are not a multiple of the number of years in the rotations being used then the summation will yield different numbers of years cropped depending on when the monitoring started in the cropping cycles of the different fields.

In this case the dominant rotations are 2 years out of 3 and 2 years out of 4. Twelve is a multiple of both cycles (3 and 4 years respectively), so, that it is an appropriate number of years to use. Monitoring of crop rotations should use a monitoring period at least twice as long as the crop cycles of the rotation practices adopted in the area. Thus 6 years would be suitable for the 2 crops in 3 years, but not for the 2 crops in 4 years rotation. However, in the study area there is a significant amount of cropping that occurs both on a more and less continuous basis.

There are a number of ways of mapping the crop rotations for the area.

1. Run a temporal template for each rotation through the data. Create a cropping pattern template for each crop rotation, for example the template for crop rotation 2 years in 4 would be of the form $\{C, C, F, F\}$. Compare the template with the cropping data set to determine whether the template matches the actual cropping, in which case the cell is supporting that crop rotation.
2. Classify the cropping data. Consider each season's image as a dimension of the data, and use these 12 dimensions to classify the data, using the techniques discussed in Chapter 5.
3. Transform by an arithmetic progression and sum the data. Consider those areas with the same cropsum value. Multiply the values in each season in turn by the sequential values in an arithmetic progression so that the sum of the previous values in the progression sum to less than the next value in the progression. A suitable progression when the data in the GIS is either 0 or 1 is 1, 2, 4, 8, 16, 32, 64, 128 for eight seasons. Each rotation practice will provide a unique sum value.
4. Implement a set of logic tests to create the equivalent of templates to run through the data.

The results of this analysis would be identification of the fields with each particular rotation practice, as shown in Diagram 7.14, layer 4 and summarized in Table 7.5.

Information on actual practices leads into a number of management decisions. Do the better practices need to be promoted or are they being widely adopted? If the level of adoption is low, is the level rising? Analysis of subsequent data sets can be used to address this question.

Another issue may be soil degradation. Is there cropping on sloping soils and is it causing erosion? Is erosion related to particular soil, slope or cropping classes? To answer these questions it is necessary to study the correlation between the parameters that may be involved. Tables 7.6 to 7.8 show count tables for combinations of erosion and soil, slope and cropping classes respectively.

Analysis of the tables suggests that soil classes 2 and 4 are most at risk from erosion, that all slope classes are at risk with most of the erosion occurring on slope class 1 and that the rotation practice of 2 crops every 3 years appears to be causing the most erosion, but that some erosion is occurring on all of the rotation practices not on the alluvial soils in the valleys. From this analysis and other considerations the agronomist may decide to promote crop rotations of 2 crops every 4 years instead of every 3 years. He may wish to address those farmers that are using the 2 crops in 3 years rotation as part of his campaign to change the practices in the district. The crop rotation information can be integrated with the ownership information to give him a listing of the farmers that are to be approached to attend that seminar. The farms that meet this condition are mapped in Photo 7.2 and shown in Diagram 7.14, layer 6 from analysis in the GIS.

Tables 7.6 to 7.8 indicate that erosion is starting to become a problem, and that it will probably get worse. There are some areas that are at risk from erosion, but further work is required to obtain a much better estimate. One way to do this is to estimate the erosion risk by the use of the USLE. The GIS, with suitable field data, contains all of the information necessary to estimate relative erosion risk using the USLE.

As an exercise, one could attempt to answer all of the questions above by the manual processing of the overlays in the GIS. Some of the tasks can be done manually without too much trouble, but others are quite intimidating, and this is for a sample area of only 225 cells. Complex calculations, involved in modelling and estimation must be done by the use of a computer-based GIS.

These values can be used to create new layers in the GIS as are shown in Photo 7.2 and Diagram 7.14.

7.6 The display of GIS data and information

GIS data and information need to be displayed in both statistical and map forms. Typical methods of display are: hard-copy maps, video maps and images, and statistical tables and graphs.

A resource manager that uses a GIS will often need to create a hard-copy map to use as the basis of directing work, or discussing programmes. Such a user needs to be able, quickly and cheaply, to produce such hard-copy maps. Inkjet plotters are a usual means of their production. Generally all GIS workstations need a facility to produce raster and/or line based hard-copy maps easily, quickly and cheaply. The facilities required to display GIS maps are similar to those used to display remotely sensed data and products.

7.7 Implementation of a GIS

The process of designing a GIS for a particular application consists of five main steps each of which are considered in the next five sections.

(1) Layers 1–3: Slope (3 degree ranges), woody cover (10% range classes) and property boundaries. The slope classes are 0 = 0–3°, 1 = 3–6°, 2 = 6–9°, 3 = 9–12° and 4 = 12–15°. The woody cover classes represent a range of 10% change in cover condition, so that 5 = 50–90% woody cover. The property boundaries are marked as value 1.

Slope																**Woody cover**																**Property boundaries**														
0	0	1	2	3	3	2	0	0	0	0	0	0	0	0	+	1	0	1	2	9	9	6	0	0	0	0	0	0	0	0	+	0	0	0	0	0	0	0	0	0	0	1	0	0	0	0
0	1	1	2	3	3	1	0	0	0	0	0	0	0	0	+	1	1	2	8	9	7	8	0	0	0	0	0	0	0	0	+	1	0	0	0	0	0	0	0	0	1	0	0	0	0	0
1	2	2	3	3	2	0	0	0	0	0	0	0	0	0	+	0	1	7	9	5	5	5	0	0	0	0	0	0	0	1	+	0	1	1	1	1	0	0	0	0	1	0	0	0	0	0
3	3	2	2	2	1	0	0	0	0	0	0	0	0	0	+	1	8	8	9	6	3	4	0	0	0	0	0	0	0	0	+	0	0	0	0	0	1	1	1	1	1	0	0	0	0	0
4	2	1	2	1	1	0	0	0	0	0	0	0	0	0	+	7	7	9	8	7	3	0	0	0	0	1	0	0	0	0	+	0	0	0	0	0	0	0	0	1	0	0	0	0	0	0
3	1	1	1	0	0	1	0	0	0	0	0	0	0	0	+	6	6	9	8	7	4	1	0	0	0	0	0	0	0	0	+	0	0	0	0	0	0	0	1	0	0	0	0	0	0	0
1	1	1	1	0	0	1	0	0	0	0	0	0	0	0	+	5	5	8	9	9	4	0	0	0	1	0	0	0	0	0	+	1	1	1	0	0	0	0	1	0	0	0	0	0	0	0
1	0	0	0	0	0	1	0	0	0	0	0	0	0	0	+	2	2	3	7	6	1	0	0	0	0	0	0	0	0	0	+	0	0	0	1	1	0	1	1	1	1	1	1	1	1	1
1	1	1	0	0	0	1	0	0	0	0	1	0	0	0	+	1	0	1	2	1	0	0	0	0	0	0	0	0	0	0	+	0	0	0	0	0	1	1	0	0	0	0	0	0	0	0
0	0	1	1	1	0	0	0	0	0	1	1	1	1	0	+	1	0	0	0	0	0	0	1	0	0	0	0	0	0	0	+	0	0	0	0	0	1	0	0	0	0	0	0	0	0	0
0	1	0	0	0	0	0	0	0	0	0	0	0	0	0	+	1	0	0	0	0	0	0	0	0	1	1	1	1	0	0	+	0	0	0	0	0	1	0	0	0	0	0	0	0	0	0
1	0	0	0	0	0	0	0	0	0	0	0	0	0	0	+	0	0	0	0	0	0	0	0	0	0	0	0	0	1	0	+	0	1	1	1	1	1	1	1	1	1	1	1	1	1	1
0	0	0	0	0	0	0	0	0	0	0	0	0	0	0	+	0	5	6	6	7	5	6	5	6	4	5	3	0	0	0	+	1	0	0	0	0	0	0	0	0	0	0	0	1	0	0
0	0	1	1	1	1	1	1	1	1	0	0	0	0	0	+	2	6	9	9	6	7	7	7	7	7	6	5	0	0	0	+	0	0	0	0	0	0	0	0	0	0	0	0	1	0	0
0	0	0	0	1	1	2	1	0	1	1	1	0	0	0	+	3	4	8	7	8	2	2	4	9	8	8	7	0	0	0	+	0	0	0	0	0	0	0	0	0	0	0	0	1	0	0

(2) Layers 4–6: Crop rotations, summation of cropping and property boundaries with 2 in 3 year crop rotation cycle. The crop rotations are 0 = none, 14 = 1/4, 23 = 2/3, 24 = 2/4, 25 = 2/5, 99 = 2/4 or 5, 98 = 2/3 or none, 56 = 5/6 and 67 = 6/7 where *x*/*y* means *x* yrs in *y* yrs. The values are the number of crops grown in the twelve years that have been monitored. The property boundaries are marked as value 1 with the 23 crop rotation class.

Rotations																**Crop summation**																**2in3yr rotation**														
25	25	25	0	0	0	0	24	24	24	23	23	23	56	56	+	5	5	5	0	0	0	0	6	6	6	8	8	8	10	10	+	0	0	0	0	0	0	0	0	0	0	1	4	4	0	0
25	25	25	0	0	0	0	24	24	23	23	23	23	56	56	+	5	5	5	0	0	0	0	6	6	6	8	8	8	10	10	+	1	0	0	0	0	0	0	0	0	1	4	4	4	0	0
0	25	0	0	0	0	0	24	24	23	23	23	23	56	56	+	0	5	0	0	0	0	0	6	6	6	8	8	8	10	10	+	0	1	1	1	1	0	0	0	0	1	4	4	4	0	0
0	0	0	0	0	0	0	24	24	23	23	23	23	56	56	+	0	0	0	0	0	0	0	0	6	6	8	8	8	10	10	+	0	0	0	0	0	1	1	1	1	1	4	4	4	0	0
0	0	0	0	0	0	24	24	23	23	23	23	23	23	23	+	0	0	0	0	0	0	6	6	8	8	8	8	8	8	8	+	0	0	0	0	0	0	0	0	1	4	4	4	4	4	4
0	0	0	0	0	0	24	23	23	23	23	23	23	23	23	+	0	0	0	0	0	0	6	8	8	8	8	8	8	8	8	+	0	0	0	0	0	0	0	1	4	4	4	4	4	4	4
0	0	0	0	0	0	24	23	23	23	23	23	23	23	23	+	0	0	0	0	0	0	6	8	8	8	8	8	8	8	8	+	1	1	1	0	0	0	0	1	4	4	4	4	4	4	4
98	98	0	0	0	24	24	23	23	23	23	23	23	23	23	+	6	6	6	0	0	6	6	8	8	8	8	8	8	8	8	+	0	0	0	1	1	0	1	1	1	1	1	1	1	1	1
23	23	23	23	23	23	67	67	24	24	24	24	24	99	99	+	8	8	8	8	8	8	10	10	6	6	6	6	6	5	5	+	4	4	4	4	4	1	1	0	0	0	0	0	0	0	0
23	23	23	23	23	67	67	67	24	24	24	24	24	99	99	+	8	8	8	8	8	10	10	10	6	6	6	6	6	5	5	+	4	4	4	4	4	1	0	0	0	0	0	0	0	0	0
23	23	23	23	23	67	67	67	24	24	24	24	24	99	99	+	8	8	8	8	8	10	10	10	6	6	6	6	6	5	5	+	4	4	4	4	4	1	0	0	0	0	0	0	0	0	0
23	23	23	23	23	67	67	67	67	56	56	56	56	99	99	+	8	8	8	8	8	10	10	10	10	10	10	10	10	5	5	+	4	1	1	1	1	1	1	1	1	1	1	1	1	1	1
0	0	0	0	0	0	0	0	0	0	0	0	23	23	23	+	6	0	0	0	0	0	0	0	0	0	0	0	8	8	8	+	1	0	0	0	0	0	0	0	0	0	0	0	1	4	4
0	0	0	0	0	0	0	0	0	0	0	0	23	23	23	+	0	0	0	0	0	0	0	0	0	0	0	0	8	8	8	+	0	0	0	0	0	0	0	0	0	0	0	0	1	4	4
0	0	0	0	0	0	0	0	0	0	0	0	14	14	14	+	0	0	0	0	0	0	0	0	0	0	0	0	3	3	3	+	0	0	0	0	0	0	0	0	0	0	0	0	1	0	0

(3) Layers 7–9: Factors used in calculating erosion risk using the USLE, multiplied by 100 for storage in the GIS.

Soil factor (*100)																**Slope factor (*100)**																**Cover factor (*100)**														
30	30	30	30	75	75	55	55	25	25	20	20	20	25	25	+	5	5	10	20	40	40	20	5	5	5	5	5	5	5	5	+	30	30	19	10	0	0	0	40	40	40	51	51	51	52	52
30	30	30	75	75	75	55	25	25	20	20	20	20	20	25	+	5	10	10	20	40	40	10	5	5	5	5	5	5	5	5	+	19	30	11	0	0	0	0	40	40	51	51	51	51	52	52
30	30	75	75	75	75	55	25	25	20	20	20	20	20	20	+	10	20	20	40	40	20	5	5	5	5	5	5	5	5	5	+	16	30	0	0	1	2	2	40	40	51	51	51	51	52	52
30	75	75	75	75	55	25	25	25	20	20	20	20	20	20	+	40	40	20	20	20	10	5	5	5	5	5	5	5	5	5	+	10	0	0	0	1	4	4	40	40	51	51	51	51	52	52
75	75	75	75	75	55	25	25	25	20	20	20	20	20	20	+	80	20	10	20	10	10	5	5	5	5	5	5	5	5	5	+	0	0	0	0	0	44	0	40	51	51	51	51	51	51	51
75	75	75	75	55	55	25	5	25	20	20	20	20	20	20	+	40	10	10	10	5	5	10	5	5	5	5	5	5	5	5	+	0	0	0	0	0	44	0	51	51	51	51	51	51	51	51
75	75	75	55	55	55	25	5	25	20	20	20	20	20	20	+	10	10	10	10	5	5	10	5	5	5	5	5	5	5	5	+	2	1	0	0	0	24	0	51	51	51	51	51	51	51	51
75	75	55	55	55	25	25	5	25	20	20	20	20	20	20	+	10	5	5	5	5	5	10	5	5	5	5	5	5	5	5	+	19	32	12	0	12	54	0	51	51	51	51	51	51	51	51
75	55	55	55	55	25	25	5	25	55	20	20	20	20	20	+	10	10	10	5	5	5	10	5	5	5	5	10	5	5	5	+	51	51	32	19	32	51	52	52	40	40	40	40	40	34	34
55	55	55	25	25	25	25	5	25	55	55	55	55	55	55	+	5	5	10	10	10	5	5	5	5	5	10	10	10	10	5	+	51	51	51	51	51	52	52	52	25	40	40	40	40	34	34
25	25	25	25	25	25	25	5	5	25	55	55	55	55	55	+	5	10	5	5	5	5	5	5	5	5	5	5	5	5	5	+	32	51	51	51	51	52	52	52	40	40	40	25	40	34	34
25	25	25	25	25	25	5	5	5	5	5	5	5	25	25	+	10	5	5	5	5	5	5	5	5	5	5	5	5	5	5	+	51	51	51	51	51	52	52	52	52	52	52	52	52	34	34
25	25	25	25	25	25	25	25	25	25	25	5	5	5	5	+	5	5	5	5	5	5	5	5	5	5	5	5	5	5	5	+	51	2	0	0	0	1	0	1	1	2	2	4	51	51	51
25	55	55	55	55	55	55	55	55	55	25	25	5	5	5	+	5	5	10	10	10	10	10	10	10	10	5	5	5	5	5	+	6	0	0	0	0	0	0	0	0	0	0	1	51	51	51
55	55	55	55	55	55	55	55	55	55	55	55	25	25	25	+	5	5	5	5	10	10	20	10	5	10	10	10	5	5	5	+	6	2	0	0	0	6	6	2	0	0	0	0	28	28	28

(4) Layers 10–12: Estimated soil erosion by the USLE, comparison of predicted erosion against current erosion and overlays of property boundaries on the comparison. Erosion risk: 0 = small risk, 1 = some risk, 2 = moderate risk and 3 = significant risk. Comparison shows: 0 = compatibility between prediction and actual erosion; 1 = prediction significantly lower then actual erosion; and 2 = actual erosion significantly lower than prediction.

```
USLE prediction                  Prediction vs. actual            Properties + difference
0 0 0 0 0 0 0 0 0 0 0 0 0 0 0 +  0 0 0 0 0 0 0 0 0 0 0 0 0 0 0 +  0 0 0 0 0 0 0 0 0 0 1 0 0 0 0
0 0 0 0 0 0 0 0 0 0 0 0 0 0 0 +  0 0 0 0 0 0 0 0 0 0 0 0 0 0 0 +  1 0 0 0 0 0 0 0 0 1 0 0 0 0 0
0 1 0 0 0 0 0 0 0 0 0 0 0 0 0 +  0 0 0 0 1 1 0 0 0 0 0 0 0 0 0 +  0 1 1 1 1 1 0 0 0 1 0 0 0 0 0
1 0 0 0 0 0 0 0 0 0 0 0 0 0 0 +  0 0 0 0 0 0 0 0 0 0 0 0 0 0 0 +  0 0 0 0 0 1 1 1 1 1 0 0 0 0 0
0 0 0 0 0 0 0 0 0 0 0 0 0 0 0 +  1 0 0 0 0 0 0 0 0 0 0 0 0 0 0 +  1 0 0 0 0 0 0 0 1 0 0 0 0 0 0
0 0 0 0 0 0 0 0 0 0 0 0 0 0 0 +  1 0 0 0 0 0 0 0 0 0 0 0 0 0 0 +  1 0 0 0 0 0 0 1 0 0 0 0 0 0 0
0 0 0 0 0 0 0 0 0 0 0 0 0 0 0 +  0 0 0 0 0 0 0 0 0 0 0 0 0 0 0 +  1 1 1 0 0 0 0 1 0 0 0 0 0 0 0
1 0 0 0 0 0 0 0 0 0 0 0 0 0 0 +  1 1 0 0 0 0 0 0 0 0 0 0 0 0 0 +  1 1 0 1 1 0 1 1 1 1 1 1 1 1 1
3 2 1 0 0 0 1 0 0 0 0 0 0 0 0 +  0 2 1 0 0 0 2 0 0 0 0 0 0 0 0 +  0 2 1 0 0 1 1 0 0 0 0 0 0 0 0
1 1 2 1 1 0 0 0 0 0 2 2 2 1 0 +  2 2 2 2 2 0 0 0 0 0 2 2 2 0 0 +  2 2 2 2 2 1 0 0 0 0 2 2 2 0 0
0 1 0 0 0 0 0 0 0 0 0 0 0 0 0 +  0 2 0 0 0 0 0 0 0 0 0 0 0 0 0 +  0 2 0 0 0 1 0 0 0 0 0 0 0 0 0
1 0 0 0 0 0 0 0 0 0 0 0 0 0 0 +  2 0 0 0 0 0 0 0 0 0 0 0 0 0 0 +  2 1 1 1 1 1 1 1 1 1 1 1 1 1 1
0 0 0 0 0 0 0 0 0 0 0 0 0 0 0 +  0 0 0 0 0 0 0 0 0 0 0 0 0 0 0 +  1 0 0 0 0 0 0 0 0 0 0 0 1 0 0
0 0 0 0 0 0 0 0 0 0 0 0 0 0 0 +  0 0 0 0 0 0 0 0 0 0 0 0 0 0 0 +  0 0 0 0 0 0 0 0 0 0 0 0 1 0 0
0 0 0 0 0 0 0 0 0 0 0 0 0 0 0 +  0 0 0 0 0 0 1 0 0 0 0 0 0 0 0 +  0 0 0 0 0 0 1 0 0 0 0 0 1 0 0
```

Diagram 7.14 GIS overlays, three layers across, each overlay being 15 × 15 cells in size, with each cell being 50 m × 50 m in size.

Table 7.5 Table of all unique combinations of cropping that occurred in the study area during the years 1980–89, the count of the number of cells that contained this particular combination and the crop rotation that the combination represented

Cropping layers												
80	81	82	83	84	85	86	87	88	89	Count	Class assignment	Comment
0	0	0	0	0	0	0	0	0	0	76	0	Never cropped
0	0	1	0	0	0	1	0	0	0	3	14	Cropped 1yr in 4
0	0	1	1	0	0	1	1	0	0	15	24	Cropped 2yr in 4
0	1	0	0	0	1	1	0	0	0	7	25	Cropped 2yr in 5
0	1	1	0	1	1	0	0	0	0	4	230	Crop 2/3 then not
0	1	1	0	1	1	0	1	1	0	33	23	Cropped 2yr in 3
1	0	0	0	1	1	0	0	1	1	8	245	Cropped 2/(4 or 5)
1	0	0	1	1	0	0	1	1	0	9	24	Cropped 2yr in 4
1	0	1	1	0	1	1	0	1	0	9	23	Cropped 2yr in 3
1	1	0	0	1	1	0	0	1	1	6	24	Cropped 2yr in 4
1	1	0	1	1	0	1	1	0	0	19	23	Cropped 2yr in 3
1	1	0	1	1	0	1	1	0	1	12	23	Cropped 2yr in 3
1	1	0	1	1	1	1	1	1	0	12	67	Cropped 6yr in 7
1	1	1	0	1	1	1	1	1	0	12	56	Cropped 5yr in 6

Table 7.6 Two-way count of erosion versus soil classes in the study area

Erosion classes	Soil classes 1	2	3	4	5	6
None 0	8	21	21	47	63	49
Minor 1	2	5	0	6	0	0
Moderate 2	0	1	0	1	0	0
Severe 3	0	1	0	0	0	0

Table 7.7 Two-way count of erosion versus slope classes in the study area

Erosion classes	Slope classes 0	1	2	3	4
None 0	154	40	9	6	0
Minor 1	1	5	3	3	1
Moderate 2	0	2	0	0	0
Severe 3	0	1	0	0	0

Table 7.8 Two-way count of erosion versus crop rotations practised in the study area

Erosion classes	Crop rotations None	1in4	2in3	2in4	2in5	5in6	6in7	2in3 or 0	2in4 or 5
None 0	70	3	69	28	6	12	12	2	7
Minor 1	6	0	2	2	1	0	0	1	1
Moderate 2	0	0	1	0	0	0	0	1	0
Severe 3	0	0	1	0	0	0	0	0	0

Table 7.9 Conversion of GIS layer values to cover factors suitable for use in the USLE

Crop rotation contribution	Contribution of woody cover to cover factor 0–10% (0.8)	10–20% (0.5)	20–30% (0.3)	30–40% (0.2)	40–50% (0.13)	50–60% (0.07)	60–70% (0.02)
None (0.2)	0.16	0.10	0.06	0.04	0.03	0.01	0.00
1in4 (0.35)	0.28	0.18	0.11	0.07	0.04	0.02	0.01
2in3 (0.64)	0.51	0.32	0.19	0.13	0.08	0.05	0.01
2in4 (0.5)	0.40	0.25	0.15	0.10	0.07	0.04	0.01
2in5 (0.38)	0.30	0.19	0.11	0.08	0.04	0.03	0.01
5in6 (0.65)	0.52	0.33	0.20	0.13	0.08	0.05	0.01
6in7 (0.66)	0.53	0.33	0.20	0.13	0.08	0.05	0.01
2in3 or 0 (0.64)	0.51	0.32	0.19	0.13	0.08	0.05	0.01
2in4 or 5 (0.43)	0.34	0.22	0.13	0.09	0.05	0.03	0.01

7.7.1 Analyse information needed to make decisions

The objectives of this phase are set out below:

- Identify the types of information required by the different users. Users can usually identify the information currently being used, but they can rarely identify the optimum information set for managing specific resources and hence they can not identify how spatial information can improve on current management practices. The GIS specialist needs to be able to assist managers identify their information needs.

- Identify the characteristics of the information required, its resolution, accuracy, timeliness and frequency.
- Assess how the information will be used as this will affect the types of facilities, staffing resources that will be required and staff training requirements. If the GIS is to be used primarily by field staff then the GIS software will usually be located on machines that will also be used for other purposes, so the software must be used on hardware that is compatible with other institutional functions. The field staff are likely to use GIS for only part of the time so that the software must be user friendly and require minimum training. On the other hand if the GIS is to be used primarily in a central location then it may be used by dedicated staff, significantly reducing staff training and familiarization requirements.
- Identify responsibilities in the collection of information. To address these issues it is usual to:

1. conduct interviews seeking information and to influence attitudes on the nature and relevance of GIS to the management of resources within the institution;
2. review available documentation on institutional functions, responsibilities and capabilities so as to ensure that all relevant groups in the institution are identified and their needs considered – this analysis will also assist in anticipating needs for GIS that might not be identified by the management in the institution;
3. break this information down into component parts, regroup into blocks of interdependent needs and functions so as to create a comprehensive list of information needs without duplication and then analyse the characteristics of the information required to identify conflicts in resolution, accuracy, timeliness or frequency, resolving these differences through either discussion or acceptance of differing needs that will have some impact on the information input into the GIS, or processing in the GIS; and
4. develop an application/task matrix that shows all of the data requirements for the GIS by responsibility in its collection against the use of that data or application.

7.7.2 Categorize and evaluate existing databases

The interviews and available information will also:

- identify all existing databases including the information contained in them, its formats (analogue, digital, scale or resolution, coverage, data format, informational resolution), institutional responsibilities in the updating, maintenance, access and use of the GIS;
- develop a flowchart of current data collection, decision-making and decision implementation processes so as to rationalize data collection, particularly where there are overlaps, identify staff requirements for access to GIS facilities and/or information derived from the GIS;
- construct a data dictionary to define all data types;
- construct a data model describing the way data flows from the collector of the information, through the collecting organizations to the GIS for integration to create management information – the data model must include the flow of information to resource managers; and
- catalogue and evaluate current data and its maintenance.

7.7.3 Specify new database

Use the data model to define the new database by:

1. specifying the information to be contained in the database and the characteristics of this information;
2. identifying those responsible for the collection of each data item in the database and specifying how they are to acquire and inserting this data into the database;
3. identifying all recipients of each data item in the database and specifying how they are to receive and analyse this data;
4. specifying the processes that are to be put in place for the supervision and maintenance of the database, including charges for accessing the database;
5. identifying all data formats required, including those to communicate with other institutions and to use within individual organizations; and
6. specifying all output types (analogue photographic and paper image, digital) by user and define capabilities required by each user from user requirements.

Once this process is complete then the elements required for the GIS can be specified.

7.7.4 Specify system elements

The next step is to develop a full specification for the elements in the proposed GIS.

1. Construct a flowchart of data flows, processing steps, storage stages and output stations for the GIS. This will differ from the data model in that the data model is a conceptual model whereas this model or flowchart will need to consider all of the activities and actions that will occur in an operational system. The data model may show land-use information going from institution A to organizations A, B and C. This flowchart may, for example, show that this information has to be reformatted to send to C, or may be packaged with other types of information to go to B.
2. Identify software requirements.
3. Specify hardware requirements.
4. Define inter and intra agency arrangements including responsibilities, agreements, facilities and financial arrangements; and
5. Identify staff requirements, including training of both operators and users and staff to support the GIS.

7.7.5 Develop plan of implementation

1. Preparing a schedule for implementation that identifies all tasks, establishes their sequence, defines prerequisites in these activities and specifies the time, funds and resources required for each task.
2. Partition the schedule into phases and prepare a budget for each phase.
3. Identify responsibilities in the conduct and management of each phase of implementation.

8

Resource management information systems

8.1 Introduction

So far, this book has dealt with methods of acquiring and analysing information. The methods of acquisition include the digitization of maps, collection of field data and the analysis of remotely sensed images. Other sources of information that have been mentioned include statistics, records and other types of observations. All of these sources of information have a role to play in information systems for resource management.

The acquired information needs to be properly managed, analysed, displayed and maintained to provide management information, with many of these tasks being conducted within a GIS which should also be capable of supporting statistical analyses, facilitating the use of predictive models and conveying to the potential user the quality of the information being displayed.

The human element is the third component in RMIS. Who makes decisions and how are those decisions made? What type of information is required by managers and how must it be presented? How are the decisions best conveyed to those who must implement them, and what role does the information system play in conveying the decisions, and relevant supporting information, to the working staff? In short, how should an institution's information base be structured and operated so as to assist that institution achieve its objectives?

A RMIS is defined as a functioning information system, consisting of facilities, staff and data, designed to support the proper management of physical resources by the provision of appropriate information to decision makers, to facilitate their understanding of the conditions relevant to making decisions.

This definition does not restrict a RMIS to spatially extensive data such as images and maps, but it does envisage that these data types will be an important component of such systems because of the importance of spatial information in the management of physical, environmental resources.

The 'information system' consists of the technical facilities, data and staff necessary to implement and operate it. As a system it will consist of an integrated set of components that must include:

- the acquisition of data, its conversion into information and the acquisition of information;
- the storage, maintenance and ongoing management of the information base, including its auditing, editing and updating, verification and assessment, provision of backup and documentation, (both technical and non-technical) on the system and its functions;
- the analysis and presentation of derived information, as statistics and maps, to resource managers;
- the facilities necessary to conduct these tasks;
- training and user support as is necessary to ensure that the information is properly and efficiently used;
- integration of the information system into the resource management structure by adaption of the information systems to that structure where the implementation and use of the information system will also influence the way management decisions are made, i.e. the management style will adapt to this new information source; and

- system management to ensure the effective and efficient management of the whole information system – staff, facilities, information flow and utilization.

The information system is an integral component of the management structure of the organization. It should be tailored to the needs of the organization. To execute this design requires a good appreciation of the role of information systems in organizations, how to design and implement information systems, and in the role of information in the management of resources.

Information may be used to make decisions and then used to convey background information on those decisions to staff who must implement the decisions. It is essential that the facilities necessary to convey the information to where it is required, when it is required, be included in the system. Similarly the level of training and user support required will clearly depend on the role of the recipients in either the conduct of the RMIS or the use of the information in the decision-making process, and their skills levels prior to the training.

Fundamental to the design of a RMIS is an understanding of how managers make decisions. It is a crucial component in ensuring that RMIS provide the optimum information set, in the right form, at the right time, to the appropriate resource manager concerned with the particular resource management decision.

The objective of this chapter is thus to impart an appreciation of the characteristics of information itself, its role in resource management and the impact of these information systems on the structure of resource management, so that the student can design and implement better operational resource information systems. To do this the chapter is structured so as to:

1. review the characteristics of current management practices;
2. analyse the nature of information;
3. analyse the nature of resource management decision-making;
4. consider the characteristics of resource management; and
5. develop principles governing the design of resource management information systems.

8.2 Current management methods

8.2.1 The goals of resource managers

The goals of most renewable resource management decisions are concerned with resolving the two issues of the short-term management of resources so as to maintain or improve productivity and the long-term maintenance of resources. The importance of both of these issues will often be stated quite explicitly in the goals or objectives of most government agencies concerned with the management of renewable resources. Many private corporations, including farmers and other land-users, also accept the importance of both issues, and would strive to achieve both objectives in the use of rural resources, even though the relative importance of each issue will vary from person to person and from situation to situation. There is considerable support in the community for setting both goals as objectives in the management of resources.

There is also considerable knowledge and understanding of the processes that cause resource degradation in our society. Most people can describe, in general terms, the processes that cause soil erosion and most other expressions of land degradation. Most land managers have an understanding of the processes of land degradation, including those forms of degradation that they are instrumental in causing in their day-to-day management.

Despite the recognized importance of maintaining resources and general knowledge of the processes that lead to resource degradation, there is clear evidence of extensive, serious and ongoing resource degradation in most countries. The Murray-Darling Basin Ministerial Council in Australia estimates that environmental damage due mainly to salinization, soil acidity and erosion, weeds and loss of soil fertility incur costs of A$200 m, annually in the Murray–Darling Basin (1989). Woody shrub invasion of the rangelands of the basin have been estimated to incur costs of A$50 m annually in lost sheep and cattle production, and higher costs of production (1988). Other examples of resource degradation can be found in Australia and in most other countries.

Resource degradation costs can be very high; in the end the degradation results in the collapse of the rural industries using that land, causing a collapse of the local economies that depend on the industry as a major source of income. Salinization of the Euphrates Valley was a major cause of the destruction of the civilizations that were flourishing in that valley at about 1000BC. Recent archaeological work suggests that resource depletion terminated early indian civilizations in the south west of the USA in about AD1300.

The effects of resource degradation on the economy of an area can occur again. In fact serious, and possibly permanent, land degradation is more likely now than at previous times, for two reasons. First, there is much greater utilization of land now than in the past because of both population, and economic pressures. Both pressures cause land to be used up to and beyond its long-term maximum potential – marginal lands are used – and the land is not given the chance to recuperate as it did under earlier practices. Second, the research effort that has been conducted over the last 50 or so years has changed the way that lands are managed and used, placing quite unknown and unpredictable pressures on the soil and the land.

Problems arise as the land is used beyond its long-term potential – losses in soil fertility and increasing weed infestations being notable examples. A common response in our research effort is to find ways of managing the degradation after it has occurred, incurring costs that raise the overall costs of production. Raising the costs of production can make the area a marginal producer compared to other producers, reducing the profit margins of the land-users. Ultimately the profit margins become too small for land-users to try other practices, they become locked into an industry in decline. The result is increased degradation as they put even more pressure on the land to produce. Ultimately either the industry becomes uneconomic or the resource collapses, with the attendant impact on income. Clearly resource maintenance is an important objective for all communities.

The nature of resource degradation and depletion, despite our knowledge and community support for resource maintenance, suggests that a major cause of degradation may be due to current management practices rather than the intentions of specific sectional groups in the community.

8.2.2 The nature of current management practices

Current management practices have the characteristics of being dependent on human observation, which are qualitative and subjective. They cannot be readily cross-checked as the situation often changes between those at the time of observation and those at the current time. They are often subject to bias by the observer, and to human error. One response of management to these characteristics is to incorporate checking or validating mechanisms that slow the whole decision-making process. Another difficulty is the inability of management to make decisions relating to the degradation of resources from the unverified observations of one, or a few persons. These field observations require general community acceptance before decisions based on them will be politically acceptable in the community.

Current practices do not contain permanent quantitative records of many aspects of resource status. In consequence, it is usually difficult or impossible to analyse changes over time in a statistically valid way. It is often difficult to identify trends, establish their magnitude and have these observations accepted by the community before a serious loss of resources has occurred. There is thus no scope to anticipate problems as those problems are starting to arise. Nor is there scope to use predictive models to assess the potential impacts of proposed actions, and hence assess options prior to the implementation of decisions.

An impact of both of these characteristics is that current management practices will tend to underestimate the level of degradation and will have neither the facilities to remedy this deficiency nor the resources to locate and quantify the magnitude of major problem areas. Thus there are no facilities to prioritize problem areas and weight them against other issues that are competing for scarce resources.

Current practices are also discipline-based so that the potential impact of decisions on other aspects of the environment are largely ignored.

Current management practices mean that serious resource degradation will always occur before adequate remedies will be politically acceptable and then instituted. However, they do appear to be suitable for making short-term decisions, primarily focusing on issues of increasing productivity, because they:

- are flexible;
- are relatively quick and easy to implement; and
- have a narrow (discipline-based) focus.

They are not suited to resolving long-term issues because they do not:

- have predictive models to assist managers to assess potential impact or the quantitative databases that are required to build these models;
- have the databases necessary to identify trends as they occur nor the facilities necessary to identify those trends and then galvanize community concern and action;
- address cross-discipline implications of proposed actions; nor
- have the facilities or infrastructure to resolve the conflicts that will arise as competing groups wish to utilize resources in different ways.

8.2.3 The characteristics of management systems suitable for addressing both goals of resource managers

These limitations mean that current management practices will need to adapt if they are to be able to address the two issues of increasing productivity whilst ensuring the maintenance of resources.

Adequate quantitative databases are needed to identify trends and assess relative impacts so as to set priorities. To do this requires databases of the relatively stable resources such as geology, topography and soils as well as information on key dynamic parameters of the vegetation, water, moisture and climate from monitoring of those resources.

Development and use of predictive models are needed because resource management systems that include resource inventory and monitoring information are not adequate in themselves as the data used in those systems is after the event in contrast to future-oriented decisions. Resource managers require information systems that assist them assess the potential impact of proposed decisions; they require predictive models as part of the information system.

Predictive models require good resource information to derive and subsequently drive the models; resource inventory and monitoring information is a prerequisite to their use. The better the information base, the better the models are likely to be in estimating potential future conditions. However, predictive models cannot predict future conditions *per se*, but can only predict conditions that might exist if the assumptions on which the model are based apply in the future.

It is necessary to also incorporate consideration of cross-discipline impacts of decisions on resources. Government agencies in most countries are designed to manage one aspect of the environment; they are discipline- rather than regionally oriented. The same applies to most commercial institutions. Yet the living flora and fauna respond or adapt to many aspects of the environment directly, and to all aspects indirectly, and it is in these interactions that many forms of degradation arise or are seriously exacerbated. These interactions may affect other aspects of the environment directly (primary reaction) or indirectly (secondary or tertiary reactions). The interactions that result from an action will have a ripple effect through the environment so that they will ultimately have an effect on the resources of interest to the manager. This effect necessitates a need for communication between all resource managers so as to assess the potential effects of proposed actions, and to arbitrate between conflicting demands. Consequently mono-disciplinary oriented management will continue to cause degradation.

Mechanisms to resolve conflicts that arise due to competing demands on resources are required. This will become very important as the actions at one place can have effects at other places and times; the beneficiary of the initial action will often be quite different to those who suffer degradation at other places or other times.

8.2.4 Considerations in the design of resource information systems suitable for meeting both goals of resource managers

The long-term management of resources must take into account, as far as is possible, the many and varied environmental interactions that may occur and affect the resource. There are three main consequences of this process on the design and use of predictive models.

1. Predictive models are indicators of probable future conditions if specific conditions persist, rather than explicit statements of conditions that will occur. Models will strive initially to produce short- and medium-term predictions of first- and second-order effects. They will subsequently undergo extensive refinement designed primarily to improve the accuracy and reliability of these predictions rather than providing longer-term predictions, because of the implications of the butterfly effect on longer-term predictions.
2. It will often be necessary to monitor both the interactive processes and their impacts if the environment is properly to be understood and managed, even at the primary and secondary levels.
3. The trade-off between increasing marginal costs of prediction and decreasing marginal benefits is likely to see the development of different types of models for different managers. Strategic predictions will be relatively coarse, have a long-term perspective, but be relatively low cost. Resource management models are likely to have a short- to medium-term perspective, but be highly accurate, reliable and high cost models.

The normal range of variability in resource conditions can mask trends in resource status. It is therefore necessary to monitor adequately these interactions and stresses and their effects over sufficient time, and at sufficient resolution, to identify trends. RMISs are therefore likely to include:

- a method of monitoring key environmental parameters to provide a permanent, quantifiable, objective and geo-referenced temporal record of those key parameters;
- a method of analysing this data to identify trends and to analyse correlations;
- modelling of processes to assist in identifying causal relations, predicting the effects of proposed actions and assessing the relative costs and effects of various actions; and
- integration of the RMIS into the management structure of the organization.

These specifications do not indicate how the information is to be used, by whom and hence the resolution, accuracy, reliability and timeliness requirements of the information. To set these specifications, it is necessary to understand information theory and the role of information in decision making.

8.3 The nature of information

We use information continuously in our daily lives, in assessing conditions and making decisions, yet most people would have difficulty in defining precisely what is meant by the term information, or describing how they use it. There are many reasons for this situation; information is not a tangible, or 'hard' asset; many components of information are subjective, involving heuristics, trial and error learning, intuition and common sense in addition to logic in an amalgam that is impossible to define; and the way the information is used depends on the user, the decision environment and the decision itself, the characteristics of other available information sets and other factors. In consequence, most people, when asked what information was necessary in making a specific decision, could not specify the full set of information that was used, nor the importance of, or weight given to, the information used in making the decision.

Here, information is defined to be part of a progression that starts with data, which is transformed into information, leading to better knowledge or understanding of the conditions surrounding the issue that requires a decision.

8.3.1 Data

Data are defined as the raw observations made about an object or surface, and can be of four types:

1. **nominal** data are classes of data which do not abide by the normal numerical rules of mathematics, being usually descriptive titles for objects, such as soil names;
2. **ordinal** data, where the classes are ranked into an order, but the relativities in that order are not known – soils that are given codes for their characteristics can fall into this category, e.g. the soil colour may be categorized using the normal Munsell soil colour charts, so that the soils can then be ranked by their Munsell classification;
3. **interval** data which ranks objects and assigns relative ranking values, so the rules of numerical mathematics can be applied, but there is no datum or reference start point for the data, i.e. the data are relative to the other ranked values–if the Munsell soil colour chart values are quantified as to the degree of yellowness of each Munsell value

then the resulting data will be interval data, as will the observations taken by a field spectrometer before they have been calibrated, and for satellite data before calibration; and
4. **ratio** data which ranks, scales and calibrates the data to a base value, e.g. reflectance data.

Data can sometimes be used directly as information, but it is more usual for the data to be processed to create information. Even in the case of a resource manager using an image, he usually will interpret that image to extract the information required. Data are usually processed to create information for several reasons.

- The data are of physical attributes of the environment that may be related to, but are not usually the actual parameters required by the resource manager. Thus the data may be the radiance from the surface, or the price of individual lots of animals sold at market, whereas the manager may want information on land-cover condition or trends in market price for particular grades of stock.
- The data may be far too detailed for human consumption and use. Thus image data are far too detailed for many purposes and are simplified by the process of classification. Various statistical processes can also be used to simplify the data, e.g. the prices of all lots of the different grades of animals sold may be brought together and averaged as a first step in providing price trends for stock grades at market.
- Data are past-oriented since they are actual observations or measurements conducted at a previous time, whereas decisions are future-oriented. The past-oriented data needs to be transformed into present- or future-oriented information or knowledge.
- Data can be context free in contrast to information. Quantitative data as observed by a machine such as a satellite sensor, are relatively context free. The data values are influenced to some degree by the observing conditions and by the sensor specifications, but they are independent of context in terms of resource decisions and are thus free of bias in relation to resource management.

  Other types of quantitative data, in which the individual measurements are guided by human decisions, may not be context free in relation to resources since the data collected can be influenced by the purposes for which the data are to be used. Independence, or lack of bias, is best assured by collecting the data in accordance with a proper sampling process.

The objective of converting data into information is to create the knowledge base necessary to make resource management decisions. Data analysis is thus concerned with the following.

- It generalizes from the data, replacing the very detailed and large volume of data with a much smaller number of classes of information that can then be understood and used by the human manager. An example of this activity is the conversion of digital remotely sensed data into land-cover classes.
- It estimates physical parameters by transformation of the data. This process does not necessarily reduce the volume of information, but converts the data into a parameter that is readily understood, and usable to the manager. An example of this process is to derive estimates of herbage biomass from remotely sensed data.
- It filters the data to minimize the impact of local events and anomalies, so as to identify trends over time. Filtering can be used for two purposes; identification of longer-term trends and local anomalies.

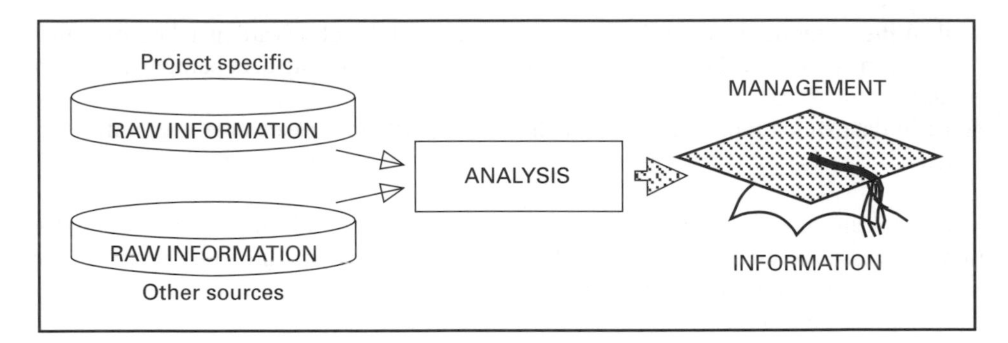

Diagram 8.1 Raw information needs to be combined to create management information.

- It transforms the data from the past-oriented data into present- or future-oriented information. This task is implicit in most of the processing done on data; the processing strives to eliminate the dependency of the derived information on the conditions at the time the data was acquired. Thus trends over time cover a longer period than the original data, and can be extrapolated beyond that period as long as specific boundary conditions are maintained. There are many forms of this type of transformation; converting image data into land-cover or land-use classes conducts a transformation from data at the date of acquisition, to information that covers a season or more, and may be indicative of conditions in other seasons. Models might be used to estimate current or potential future pasture conditions, using estimates of pasture biomass at the image date and extrapolating using actual or estimated meteorological conditions.

A manager generally requires combinations of sets of information to give the information required to make management decisions. A two-stage process of data analysis is involved. The potential, or 'raw', information is extracted from data in the first stage of the information analysis process. This activity includes image processing, the collection of field data and the digitization of map data. If this raw information is generally usable by a number of resource managers, then its derivation in a collaborative way amongst those potential users has two advantages.

1. It shares the costs in information extraction, particularly in project design, use of skilled staff, data analysis and processing to reduce the costs to the individual user. One of the most expensive components of a RMIS is deriving the raw information. This cost can be considerably reduced, per user, if the extraction process is shared.
2. It provides consistent base information to all users. The proper management of the environment is going to require integrated management where the impacts of proposed decisions on all resources are taken into account. As soon as the different managers start to address this need, they will require consistency in their information base. Collaboration in the collection of the information will ensure that all managers have a consistent database and of a known accuracy as long as accuracy assessment has been conducted.

To achieve these objectives requires that the derived raw information be in suitable building blocks for the information required by many users. The more users that can be satisfied with the information collected on a collaborative basis, the cheaper that will be per user, and in consequence the cheaper the whole information system to users.

8.3.2 Information

There is still much to be learnt about both the information requirements of resource managers and how those managers actually use information in the making of decisions. It is an area that requires considerably more investigation and research than has been conducted so far. Galliers (1987) states, 'I almost called this introductory section of the book, "on the elusive and illusive nature of information", given its inherent qualities of often being difficult to define on the one hand and deceptively easy to underestimate on the other. Major messages that will often be gleaned are that the difficulty should not be underestimated and that information has real meaning only in a given situation or context.'

Galliers quite explicitly describes the problem confronting those attempting to understand information and its roles, for the purposes of developing better RMISs. It is not so much one of defining the nature of information, but rather coming to grips with the role of information in the actual process of decision-making in contrast to the espoused process. What this distinction between actual and espoused decision-making processes means is that there is a recognized, admitted or espoused process that is often quite different to the actual process. Different components of the information base are given different weights in making decisions. Information is sometimes used to support the implementation of a predetermined decision, rather than the making of the decision itself. Sometimes information is manipulated to suggest a course of action that is predetermined by the manager, or some other person.

The issue then revolves around the need to understand how often, and under what conditions, people make decisions based on rational processes using independent information, in contrast to the making of decisions based on irrational or subjective processes without significant support from independent information. It is important to emphasize the constraint in the last phrase, 'without significant support from independent information'. Society accepts that many decisions must be based on subjective processes, but the decision maker is required to accumulate all the available information that it is realistic, or cost effective, to collect, and have considered this information before making the decision.

Making decisions without the benefit of the available information may be an irrational process, but it is often also based on the resource manager having a hidden agenda and desiring to make a decision that supports the objectives of this agenda. Hidden agendas need not be inimical to the interests of the resources being managed; but they are generally seen as being a negative factor since they cannot be adequately taken into account by others who are either involved in the decision-making process or affected by that process.

This issue means that the perceptions, desires and opinions of resource managers will influence discussions that are held with them when attempting to identify their information needs as part of the process of designing a RMIS.

Information can be defined as that collection of facts and opinions that improves the knowledge levels of the recipient in such a way that he or she is better able to undertake a particular activity or make a particular resource management decision. The importance of this definition lies in the linkages that it specifies between the collection of facts and opinions and the creation of knowledge that forms the basis of making resource management decisions. If the last phrase is included then it requires that the facts and opinions be extensive enough, and of sufficient quality so as to make appropriate, or good, resource management decisions. The judgement in this definition is thus on the quality of the decisions that can flow from the collected facts and opinions. Resource managers will

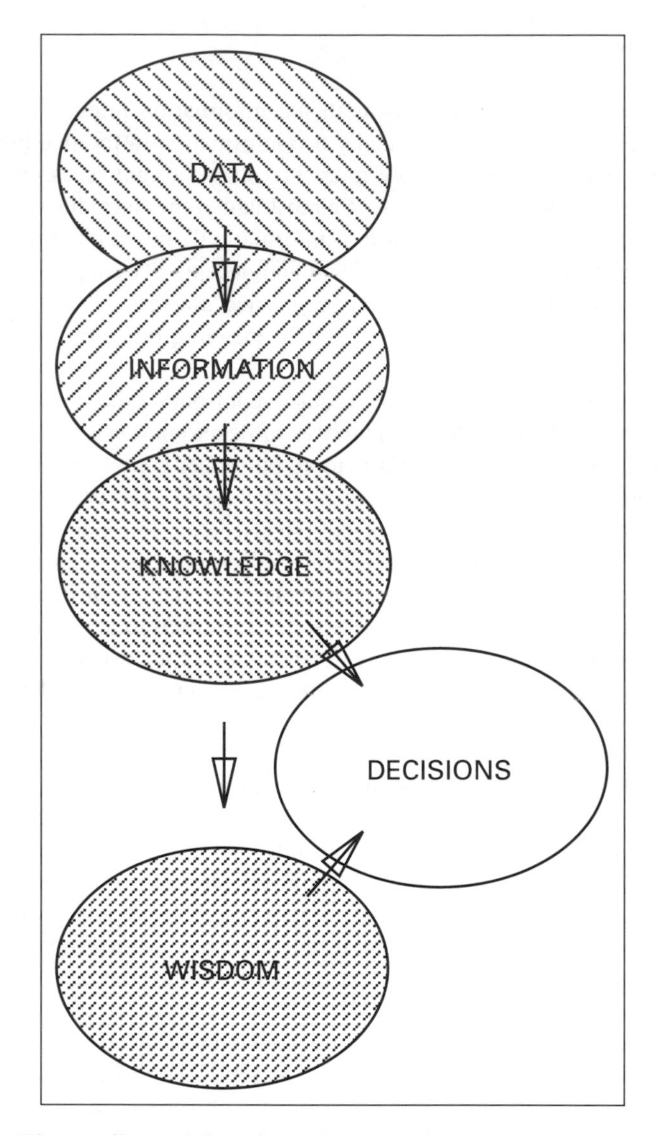

Diagram 8.2 The intellectual flow from data to information, into knowledge and ultimately to wisdom as part of the decision-making process.

only pay for information systems that facilitate their ability to make decisions, or provide some other benefit to their activities.

Information includes that material and observations extracted from data, and which increase the understanding or knowledge of the object by the recipient, thereby assisting the recipient to make a decision, or take a course of action in relation to the object. An important component of an information system is therefore the conversion of data into information (Diagram 8.2).

Information is both contextual and enabling. It is **contextual** because a particular manager, with certain attributes and experiences, requires certain information to make certain decisions. Any changes in the manager, the environment about the resource of interest, or in the nature of the decisions to be made, changes the type of information required. Thus information only has value within a specific context; information of value to one manager will be quite valueless to another. It is **enabling** because information, to have value, must either enable decisions or facilitate subsequent actions by the resource manager. If the information does not assist in either the making or the implementation of decisions then it is valueless, and it may be worse than this if, by its presence, it clouds the issue or masks the decisions that have to be taken.

Often, information derived from remotely sensed and other data are not adequate in themselves. They often need to be integrated with other data or information to create management information, so that information derived from remotely sensed data often become the data for subsequent analysis. The term, 'raw' information discriminates between data (the derived primary or raw information) and the subsequently derived secondary or management information.

In many dynamic processes the suitability of data is severely limited by its perishability over time. Thus imagery of weather conditions are of little use in weather prediction once they are more than 12–18 hours old. The time scale of events being assessed and the information required, have an impact on the changing values of the data over time.

Analysis of the base or raw information, with appropriate auxiliary information, within the context of a GIS provides management information. Management information is highly context-dependent; the raw information required, the information derived from it, and the use to which that information is put are dependent on the manager, the questions that are to be addressed, and the environment in which the questions are being posed. In consequence the extraction of management information from raw information must be done within the direction of the resource manager. Thus the process of extracting raw information from remotely sensed images, from field data or from digitization of analogue maps, may be done by a centralized agency operating under contract to the end users of the information, but the process of converting the raw information to management information will usually be done in the user agency, with the direct input of the resource manager.

It is for these reasons that institutions concerned with the management of information are more likely to establish centralized GIS, whilst those institutions concerned with the management of resources are more likely to develop distributed GIS. Information is a product made for sale to other agencies. This production is usually done most efficiently in a centralized location for the collection of data, processing into information and distribution to client agencies. The information agency may establish a distributed network of shop fronts to promote and sell their wares. Resource managers are distributed in accordance with the distribution of the resources; the GIS is to be used as a decision support tool for the management of those resources. As a tool, it needs to be located adjacent to the manager that needs to make the decision.

Consider the manager of a chemical factory, who wants to know whether to expand production of herbicides. Where are the potential markets, and when will they develop to a size sufficient to justify servicing them? If farmers are moving into no or minimum tillage agricultural practices that use herbicide to control weeds rather than tillage, then the manager wants to estimate the rate of conversion, the final percentage that can be expected to adopt these practices and when, and the distribution. The base information used to derive this management information can come from extraction of land-cover information from remotely sensed data as in the following steps.

1. Use remotely sensed data to map land-cover information.
2. Integrate the land-cover layers to give seasonal land use.
3. Analyse the land-use trends to estimate the annual rate of adoption.

4. Use this data, with questions to farmers and assessment by agronomists, to estimate both the ceiling of percentage adoption, and the time that it will take to reach this figure.
5. Repeat this analysis for each administrative area, to provide estimates of spatial distribution.

In this process the extraction of land-cover information is quite generalized; it can be used for a variety of purposes and hence can be extracted collaboratively for a number of users. However the subsequent steps are dependent on the information required by the specific resource manager and so the databases and facilities necessary for the conduct of this work need to be under the direction of the resource manager.

Just as the role of information in decision making is often difficult to define, so there is also a lack of understanding about the role of information in the management of physical resources. Many resource managers have difficulty answering the question, 'What are the characteristics of the information set that you require for the conduct of your management tasks?'. Even if a manager provides a list of information needs, it may be of little use if it is a reflection of the currently used set of information for the management of resources.

The information used to make decisions depends entirely on the technical competence and financial capability of the community within which the decisions are to be made. It is thus impossible to conceive society establishing databases on weather conditions, as a basis for predicting the likelihood of frost in an area, before the advent of computers and automated weather stations. It is similarly difficult to conceive of most under-developed countries having the resources to establish sophisticated RMIS without assistance. Whilst direct financial assistance can solve this problem, a much more constructive approach is to establish collaborative approaches to the collection of resource information since all of these information bases will ultimately strengthen the global databases that will be necessary to monitor global processes.

The advent of a new technology changes the cost structures associated with the collection of information. New sources of information may become more cost effective than old sources. Introduction of new technologies, like remote sensing and GIS, provide society with the opportunity to reassess its information requirements in relation to resource management. Not only is it important to take up this opportunity to better manage those resources, but there are likely to be other significant impacts. Introduction of better information systems may change societies' perceptions about that and other resources, changing their role and importance in society. The changing economic structure in relation to a resource may see other areas becoming economic in the use or production of that resource, changing the whole structure of society in the process. The potential impacts of a major new information system on society and its perception of its environment can be profound, and should not be underestimated.

Historically, resource managers have had access to a specific set of information types to make decisions. If this information set is sub-optimal, then the management decisions are likely to be sub-optimal. For example, in deciding when to shear sheep after winter, the farmer would dearly like to know whether there will be a frost that might chill, and kill, the sheep. As this information is not available to him, he shears as early as he thinks practical so as to get the wool to market, whilst accepting some risk of frost. Clearly there will be times when the frost does occur and some sheep die.

Not only do we need to consider the types of information that should be collected, but also the weight that will be given to each type of information in the decision-making process. Not all types of information are of equal importance in making decisions. The weight to be given to the information is important for two reasons: it indicates the priorities within the group of information types to be collected, and the accuracy required in each type of information.

The law of propagation of error states that, for a function, $w = f\{x, y, z, \ldots\}$ the variance of the dependent variable, w, can be determined from the variances of the independent variables, x, y, z, ... where these independent variables are assumed to be approximately normally distributed, from the equation:

$$\sigma_w^2 = \left(\frac{\partial f}{\partial x}\right)^2 \sigma_x^2 + \left(\frac{\partial f}{\partial x}\right)\left(\frac{\partial f}{\partial y}\right)\sigma_{xy} + \left(\frac{\partial f}{\partial y}\right)^2 \sigma_y^2 + \ldots$$

where

σ_x^2 = variance of parameter x

σ_{xy} = covariance between parameters x and y

$\left(\frac{\partial f}{\partial y}\right)$ = partial differential of function $f\{x, y, z, \ldots)$ by y

In practice the relationship between a decision, w and the information used in making that decision ($x, y, z, \ldots$) is rarely known with any precision. If the independent variables are independent of each other in the equation, the weights associated with each information type are estimated by managers and a linear model can be used. Then

$$w = A_0x + A_1y + A_2z + \ldots$$

so that the variance of w can be calculated from:

$$\sigma_w^2 = (A_0)^2\sigma_x^2 + (A_1)^2\sigma_y^2 + (A_2)^2\sigma_z^2 + \ldots \tag{8.1}$$

The contribution of each variable x, y, z, etc. to the final variance is thus proportional to the product of its constant by its standard error. Optimally, each of the product components in (8.1) should be of a similar size so that they make similar contributions to the variance. If one of the product components is significantly larger, or smaller, than the others then it will contribute most, or very little, to the variance. Variability between the contributions of the product components offers scope to either reduce the costs of collecting the information, or significantly increasing its accuracy. If one component is the major contributor to the final variance then reducing its variance will significantly increase the accuracy of the derived information. If one component is an insignificant contributor to the final variance then that information can be collected in a less accurate, and hence cheaper, manner without significantly affecting the final variance.

In practice the relationship between the independent variables x, y, z, etc. may be quite non-linear so that the choice of this sort of linear model needs to be taken with great care.

8.3.3 Knowledge and wisdom

Knowledge can be defined as an understanding of the principles and processes that influence the condition of, and changes in, an object or surface.

Knowledge of a process or situation means that the person with knowledge can create a model of that process or situation. Models are either conceptual, defined and retained by the user in the brain, or tangible where they have been described either in writing (qualitative) or mathematically (quantitative). Models are usually retained as a generalized model, i.e. they define and describe the process or condition in a general sense, but they cannot be used in specific instances as they have not been parameterized, i.e. their size or magnitude has not been set to match a specific situation. When a decision has to be made then these models need to be fleshed out with parameter values to become localized models.

Generalized models, by describing the general condition, can be applied under all conditions for which the assumptions on which the model are based can be met. For example, a model of water velocity down a slope may be expressed as follows.

Velocity of water down a slope depends on the starting velocity, the slope, frictional affects of the surface, the volume of water, and the time from the starting conditions.

Such a model may be expressed mathematically in the form

$$v_{new} = v_{old} + \text{constant} \times (\text{slope} \times \text{gravity} \times \text{time} \times \text{volume/friction})$$

However, such a model makes assumptions about the viscosity of the liquid, the gravitational pull and the ranges of values in the slope and friction. It can be applied under those conditions that meet the assumptions on which the model are based, but not under other conditions without the risk of incurring large and unpredictable errors.

Resource managers have many of these generalized models available to them; many of them are intuitive, i.e. they are built from an amalgam of knowledge and experience and retained in the mind of the manager. However, models can also be verbalized, or written down so that they can be analysed by others, and they can also be made more rigorous or quantifiable.

Generalized models can be extremely simple. For example, the question from an employee as to whether stock in one paddock should be moved to another may prompt the manager to parameterize the generalized model, 'Stock need so much fodder if they are to maintain condition' so as to respond properly to the question.

Localized models describe particular (localized) conditions so that they can be used under those conditions, but not extrapolated beyond the boundary conditions of the data used to create the model. Clearly there will be an infinite variety of localized models for each generalized model, so that in general the generalized models are retained rather than individual localized models. The process of creating a localized model from a generalized model is called **parameterization**.

In the case of the quantified model of water flow discussed earlier, values for v_{old}, slope, gravity, time, volume and friction are required to calculate v_{new}, assuming that observational data covering a range of conditions have been used to determine a value for the constant. If the water contains sediment, then another assumption will be the velocity rate below which the water will not be able to carry that sediment, so that it is deposited out of the water. The model might be used to follow stream velocity down a stream line, to determine the location at which the sediment will be deposited by the stream. In the example on stock movement, the generalized model could be parameterized by asking the question, 'How much fodder is there in the paddock?', with the answer, 'Enough for about three weeks.'

Information forms part of two processes in creating knowledge; it is important in the creation of the models themselves and it is important in parameterizing those models to make actual decisions. The information used in creating models is often different to that used in parameterizing them. In the first case the information must cover a range of conditions in a systematic manner so as to ensure that the model is applicable to a wide range of conditions. The greater the range of conditions covered, the more generally applicable will be the model. Thus the analysis may need to cover a number of seasons or years, a range of climatic conditions, soil types and other parameters. Parameterizing these generalized models requires specific information over a specific spatial and temporal domain.

Knowledge can therefore be defined as understanding specific processes sufficiently to construct or adapt generalized models of the processes, and be able to apply these models under specific conditions, i.e. be able to parameterize the models to create and use them as localized models.

Wisdom may be defined as an appreciation or understanding of the forces and factors that influence the condition of, and changes in, the various objects that constitute the whole environment. It thus considers the interactions in the whole environment, consisting of physical, social and ethical components, within the broadest global, indeed universal, context. Wisdom would thus provide the manager with an appreciation of the limitations of the different models under different conditions, and the causes of these limitations. Wisdom would thus provide the skill necessary to know whether to apply a model or not, and if to apply a model, then how it might be used.

Wisdom is a condition to which many aspire, but few attain. Ideally decisions should be based on wisdom, but rarely is this available to us. We strive to make decisions, and take actions based on adequate knowledge, but the impact of our decisions will sometimes have unexpected results due to our lack of wisdom.

To summarize these four terms are defined as a progression:

Data	Raw observations, context free
Information	Defines object characteristics; depends on context
Knowledge	Understands object dynamics; short-term predictions
Wisdom	Understands object in environment; predictions

8.3.4 Informational constraints

There are a number of constraints on the nature of information available to resource managers, and on how information can be used in the decision-making process.

Classification of data

Classification into a number of classes excludes other classes in that classification and has been discussed in Chapter 5. It is possible to construct different classifications from the one data set, to provide information classes for other purposes. A classification can only provide information on the classes retained in the classification and not on other classes. Each classification thus contains the paradox that it both creates some and ignores other information.

Order is a function of applied information in a system

Achieving a level of order in a system requires the utilization of information; increasing the order in the system will require the use of more information. It should be noted that the requirement is for application of information; it is not sufficient to have information, it must be used to contribute to order. Thus it is possible to create a higher level of order in a system with less information if the information base is better used.

Order in a system depends on internal and external information

There is a limit to the order that can be achieved without an adequate balance of internal and external information. Internal information can provide a certain level of order;

increasing the order requires external information. External information is required to ensure that the system is heading in the right direction relative to its physical, economic and social environments, so as to utilize opportunities as they arise and foresee difficulties in time to take remedial action. Internal information is necessary to ensure the proper internal functioning of the institution.

Conversion of information

Conversion of information incurs losses that decrease order. It is best to have the information required for resource managers with the managers, and not have it filtered through a number of steps. Information loss can also occur in the communication of information. This loss can be reduced by retaining a level of redundancy in messages and data files during transmission and storage. However, redundancy reduces efficiency.

The meaning of words, symbols and numbers

To unambiguously convey understanding, there must be a clear definition of the meaning of words, symbols and numbers used in information types within information systems. Thus all mapping organizations clearly define the meaning of all of the symbols displayed in their Table of Conventional Symbols on a map. For efficiency, there should also be uniformity of meaning in the words, symbols and numbers that are used as far as is practical. There are guidelines for the storage and transmission of information as well as the definition of words, symbols and numbers used in information types.

1. The words, symbols or numbers must be appropriate to expected use.
2. Significant relationships between the information types, and with other types of information, must be disclosed.
3. Environmental information, i.e. information on factors that will affect the information in some way, or how the information is used, must be included.
4. There must be uniformity of practice in the use of words, symbols and numbers, both within the one organization and between organizations.
5. There must be consistency of practices over time.
6. Information must be prioritized, by only retaining essential information, highlighting the most important information and ensuring that the information needs of users have been thoroughly examined.

Information system established to ensure proper use by management

Studies have shown that even when executives have all of the information available to them that they need to make a decision, they do not always make the correct decision. These failures are due to:

- difficulty of some managers in selecting appropriate models;
- inadequacy of the information base (either content or presentation) for parameterizing of the models;
- difficulty of managers in translating knowledge into action; and
- the hidden agendas of some managers.

Thus some of the problem lies with the managers themselves, and some lies in the nature of the information being conveyed to the manager.

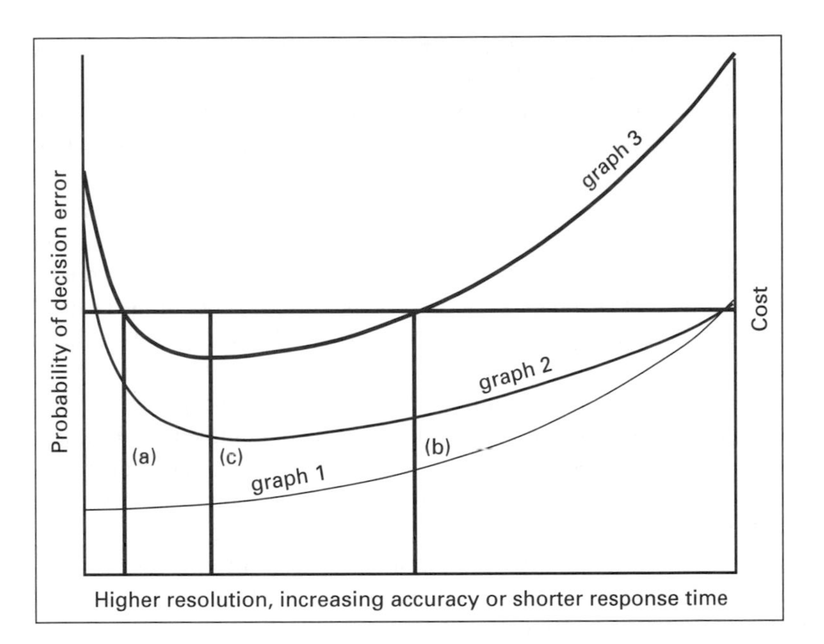

Diagram 8.3 Generalized models of the costs of acquiring information (graph 1), the probability of decision error (graph 2) and the sum of these two as a cost (graph 3).

8.3.5 Value of information

There are various approaches to assessment of the value of information. One approach centres on the minimization of the sum of two costs; the cost attributable to decision error, and the cost of assembling and analysing the required information.

The cost of assembling and analysing information is proportional to the accuracy required in the information and inversely proportional to the resolution of the information and the length of time between acquisition of the data and presentation of the information. The higher the accuracy requirements on the information, the higher the costs of the information. The finer the resolution, whether that be informational, spatial or temporal resolution, the higher the costs of the information. The tighter the time constraints in extracting information from the data, the higher the costs of the information. A generalized model of this relationship is shown as graph 1 in Diagram 8.3.

The costs of decision error depend upon certain factors.

- Management must have the ability to remedy decision error before costs are incurred. Some decisions contain considerable scope for later remedies so that the costs of decision error are relatively small. Other decisions contain negligible opportunities for future rectification, and in consequence the decision error costs can be high. Strategic planning decisions can be revised over time and so they contain low costs of decision error. As the decision moves down the management ladder, so the opportunities to modify the decision decreases and the costs of decision error increase.
- The magnitude of the impact of the decision on the institution and the resources being managed will be important. If the decision affects only a small section of the institution then the costs of decision error may be small from the perspective of the institution, whereas decisions that affect the whole of the institution can incur high costs of decision error. In general, decisions at the operational control level affect only small sections of

the institution and hence have a lower impact, whereas decisions at the strategic planning level affect the whole of the institution and hence can have a higher impact.

- The frequency with which the errors occur. Infrequent errors incur lower costs whereas frequent errors usually incur higher costs of decision error. The frequency of decision error is also affected by the resolution of the information relative to the optimum resolution to support management decisions. Information that is too coarse will increase decision error whilst information that is too detailed will also tend to increase errors since the sheer volume of information makes it difficult to see the trends from the detail. Thus graph 2 in Diagram 8.3 shows a generalized model of the costs of decision error, where the position of this curve depends on the sources of decision error as discussed.
- The contribution of that information to the decision making process will also affect cost. Decisions should be based on a number of sets of complementary information. If the contribution of an information set is small then the component of the costs of the decision error due to the information are also small. The complementary sets reduce the chances of decision error and also reduce the costs of decision error in relation to each information set. Strategic level planning tasks use many information sets whilst operational level tasks use only a few sets of information. The contribution of each set is usually small in strategic planning, but large in operational control.

The sum of these sources of error are shown as graph 3 in Diagram 8.3. These graphs are indicative of generalized models of these costs. The actual positions of the graphs will vary from application to application and between the different levels of management. For strategic planning level decisions, graph 2 will move to the left relative to graph 1; for operational control level decisions, graph 2 will move to the right. These graphs produce the following information.

1. The information that is either far too coarse or far too fine will incur significantly larger total costs. It is important to determine suitable ranges in the resolution, accuracy and temporal response specifications for the information to be used.
2. Within these specified ranges, as shown by lines (a) and (b) in Diagram 8.3, the optimum characteristics of the information (c) will be closer to the coarser information limit than the finer limit. This is important as resource managers, when their opinions are sought on information needs, will often set specifications that are tighter (i.e. further to the right in these graphs) than is necessary.
3. As graph 2 moves to the right relative to graph 1, so the optimum position moves closer to the coarser limit (line (a)).

The value of information to a manager is increased when the manager is aware of the method of generation of the information, is familiar with those who generated the information, and knows the characteristics of the information, including its accuracy, resolution, reliability and timeliness.

The second approach to assessing the value of information is based on comparing the costs of acquiring the information with the benefits gained from the use of the information. The costs of acquiring information were discussed earlier in the section. There are three basic strategies for acquiring resource information that depend in part on remotely sensed data: sampling, visual interpretation and digital image processing. All three have different cost structures (Diagram 8.4).

Sampling Low initial cost at coarse resolution, but costs become very high as the level of detail increases. At the extreme of complete data collection sampling is the same as field mapping.

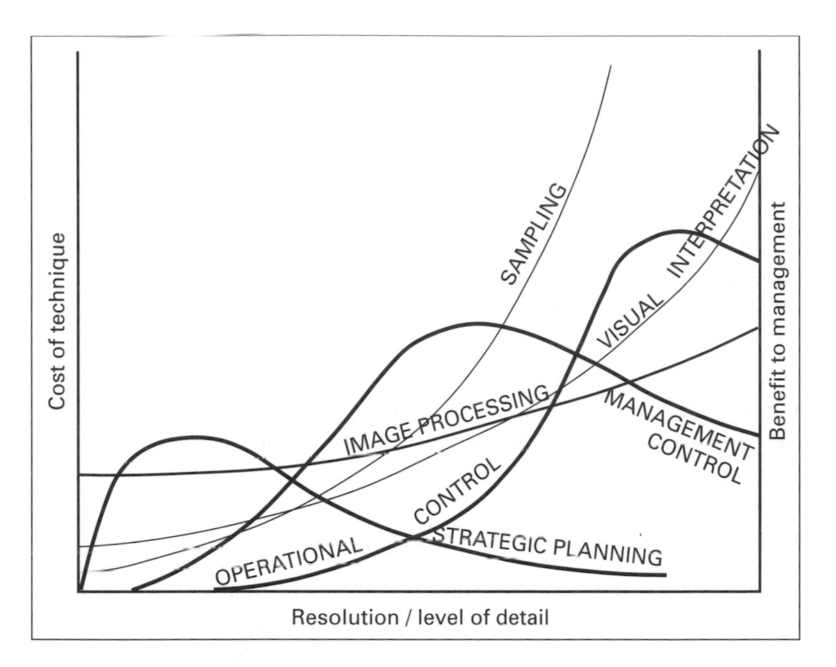

Diagram 8.4 The costs and benefits associated with using information in resource management, depicting the cost structures associated with the different information collection strategies and the benefits that can generally be expected at the different levels of management.

Visual interpretation Basic initial costing includes images and maps. Cost increases are initially low, but again become very high when the level of detail becomes very high.

Digital image processing Significantly higher initial costs reflect the higher facilities and data costs. Lower rates of increase reflect the capacity of computer processing to do large amounts of work at low cost.

The benefits derived from the use of information are much harder to assess. If the information is too coarse to be of much use in decision making then the benefits are very small. As the characteristics of the information become closer to the needs of the manager, then the benefits increase, peak and slowly decrease once the level of detail is greater than that required by the manager. This decrease occurs because the manager is faced with a plethora of information; he either incurs costs in having it reformatted to his needs, uses the information and incurs costs in time lost using the information, or the information is not used at all.

The magnitude of the benefits depend on:

1. the importance of the decision to the institution–information that contributes to significant institutional decisions has obvious high benefit value, and vice versa;
2. the contribution of the information to the decision – if the information is marginal to the making of the decision then the benefits of the information are small;
3. dependence of the decision on the information; and
4. costs of incurring decision errors, as a result of not using the information in comparison with using the information.

There are situations where certain types of information do not contribute to the quality of the decision, but they do influence its implementation. Such information often includes land-ownership information. In these situations where the decision cannot be implemented without the information, that information has a high benefit value.

8.4 The process of decision making in resource management

The process of decision making is still the subject of considerable scientific investigation. However, making a decision in resource management is generally considered to involve the following sequence of events:

- selecting an appropriate model of events;
- collecting the necessary information to parameterize the model;
- using the parameterized model to identify decision choices; and
- making the decision.

8.4.1 Selecting an appropriate model of events

Prior to any decision making, the decision maker has to be aware of the need to make a decision. Generally either of two types of reasons may trigger recognition of this need: problem detection or opportunity seeking. **Problem detection** arises when a collection of facts or evidence suggests to someone that some conditions are deviant from those expected, i.e. the facts or evidence indicate conditions different to the norm. **Opportunity seeking** arises when circumstances suggest that there may be opportunities for the decision maker.

The facts or evidence may be observed by the decision maker, brought to him by an employee in the process, or by an outsider. The resource manager will then, from the description of conditions, decide which model is appropriate for the conditions under consideration. The resource manager may need further information before selecting from amongst alternate models to select a model for use in making the decision. The model at this stage will be an unparameterized generalized model.

8.4.2 Parameterizing the model

The generalized model needs to be parameterized to create a localized model, which then needs to be tested to ensure that it is appropriate for the making of the necessary decision. Data has to be used for this purpose. This process may involve the collection of more data and it may reveal inadequacies in the model, requiring either modification of the model, or its rejection and the acceptance of some other model. The process of model selection and parameterization is thus an iterative one, in which the experience of the resource manager is a critical factor in the selection and effective parameterization of the model.

8.4.3 Identifying decision choices

The parameterized model leads to a better understanding of the conditions existing when the need to make a decision was identified. Often this understanding will require the use of other models, since the decisions made will have impacts that can only be assessed by the use of other models. These other models may also need to be parameterized before they can be applied to the specific conditions that have been met. Thus the resource manager uses the understanding generated by the model to assess whether he has adequate knowledge and understanding to make the required decision. Once this condition has been met then the alternate decision choices need to be identified. Each alternative choice must be examined under the following criteria.

- Does the choice solve the problem?
- Is it technically, economically and practically feasible?
- Does it conform to common practices and meet community legal requirements?
- Does it comply with community time constraints?
- Who will be affected and what will be these effects?
- What will be the community reaction to the solution?
- What are the environmental and social costs and benefits?

From this process of considering the advantages and disadvantages of each solution, the decision maker then selects one choice from the alternatives.

8.4.4 Making the choice

Making the choice is often a complex task.

Multi-preferences In many cases the outcome is not measured by one, but by several variables. Often some of these variables cannot be quantified, so that the choice becomes one between trade-offs that may not be easily measurable.

Uncertainty of effects Often there is a level of uncertainty as to the effects of the different choices, and again it may not be possible to quantify this level of uncertainty.

Conflicts of interest The different choices may involve conflicts of interest for the organization, the manager, or for other staff. A choice may benefit some parts of an organization, but disbenefit other parts, and the implications of these conflicts needs to be taken into account when selecting between choices.

Control It is not sufficient to make a decision unless that decision can be adequately supported. Is there sufficient information to make the follow up decisions that will need to be made? Is there enough capital to allocate to the decision, and to cope with contingencies that might arise? Can the decision be reversed if necessary, and at what cost?

8.4.5 Types of models and decisions

Models, and the decisions that flow from them, can be identified as being either structured or unstructured in form.

Unstructured models and decisions These involve intuition, common sense and heuristics; they often depend on trial and error. The relevant data will often be qualitative rather than quantitative, and it may be vague. The decisions rarely replicate previous decisions, and so they have a high degree of uniqueness, they are difficult to program. Their characteristics are that they:

1. cannot be subjected to scientific analyses;
2. are generalized with low resolution and accuracy and can be vague;
3. cannot be predictive in a rigorous, scientific sense;
4. can respond to the unexpected and outside influences – are thus flexible; and
5. can incorporate factors that are difficult to quantify.

Structured models and decisions These are based on clear logic. They are often quantitative; the factors involved and the outcomes are usually well defined. They can be routine, repetitive and are often amenable to programming. Their characteristics are that they:

1. can be subjected to scientific analyses;
2. have or require high levels of resolution, accuracy and precision;
3. can be predictive;
4. cannot work outside premises of the model – they are inflexible; and
5. cannot respond easily to outside influences.

Structured models can be either qualitative or quantitative in form. Qualitative models that have been made more rigorous through being verbalized and thus subject to analysis by others can be structured models. Quantitative models, parameterized by quantitative data, will usually be structured models. Indeed the more rigorous the data, from nominal to ratio, the more rigorous or structured the resulting model is likely to be.

8.5 The nature of resource management

8.5.1 Types of management

Management functions can be grouped into three categories: strategic planning, management control and operational control.

Strategic planning

Strategic planning is defined as planning the future directions for the corporation through establishing the corporation's long-term objectives and mobilizing its resources so as to achieve these objectives. This will normally be done by developing corporate objectives and goals, establishing the structure appropriate to the achievement of these goals, and then setting the goals of the units within the corporation. This type of activity is the primary concern of the board of directors and senior managers.

Strategic planning activities are concerned with the environments (physical, economic, social and political) within which the corporation operates. It involves assessing how these may change in the future, the potential impact of these changes on the corporation, and how the corporation needs to respond and operate under evolving conditions. Strategic planning therefore requires extensive information about environments that are external to the corporation.

Strategic planning, in considering these external environments, will need to consider many factors over which the corporation has no control, and many of which are quite unpredictable since the corporation will usually have incomplete information on the factors and their environments. Strategic planning requires the use of models that can accommodate the impact of such unexpected and unpredictable factors; it generally requires the use of unstructured models.

Generally the information used in strategic planning will be of low spatial resolution, generalized and qualitative because of the way that the information is to be used, and the extensive range of information that has to be considered.

Typical information used in strategic planning includes:

- statistical information as collected by central Statistical Bureaus providing national and regional level information on different aspects of the community – remote sensing and

GIS can improve the collection and accuracy of this information by providing sampling strata appropriate to the information being collected, and in some cases by the actual collection of the sampled information itself;
- broadscale monitoring information as collected by satellite at the global and regional level – remote sensing with field data are the basis of most of these monitoring programs, such as with weather prediction; and
- institutional information as collected from documents and publications including economic, social and political information about the institution and its relationships with other institutions, including governments.

Management control

Management control translates strategic objectives and plans into implementable programs within the units of the organization, whilst matching current operations to both current and future resource needs. It therefore involves structuring the unit to achieve the set goals, marshalling the staff resources to the achievement of these objectives, and ensuring that the unit continues to operate at peak levels of productivity. This strategic and tactical level of activity is the concern of middle management.

Management control is concerned both with optimizing productivity within the whole of the corporation, and with maintaining the resources to ensure the long-term viability of the corporation and its environment. It is in the position to influence all components of the corporation that impact on the resources of an area, and thus it is in the position to balance resource maintenance with productivity, within the general policy framework established by the corporation's strategic plan. Management control is not concerned with the overarching focus of strategic planning, but with a more direct focus on the corporation as a whole, and its interactions with its immediate environment. Management control is therefore concerned with more localized external and internal information on all aspects of the environment within which the corporation operates. Management controls concern with both maintenance and productivity means that it needs to integrate the needs of the corporation with those of others groups or corporations in the environment; it is concerned with very complex yet rigorous decision processes. It will require the necessary research and development support necessary to evaluate these decision processes.

Management control is concerned with predicting future effects of corporation activities so as to assess decision options for the corporation. This prediction must come from the use of quantitative models of processes, requiring quantitative information to drive these models. The resolution of the information must be compatible with assessing the impacts of decisions on units of management, and is a function of the size of the operational control units. (For areas of extensive units (e.g. the farm), where the unit can cover areas of over 1000 ha, then information at 100 m resolution can often be adequate. Once the unit of operational control becomes smaller then finer resolution information is required. Ten metre resolution data are often adequate for operational control units of 10–100 ha in area.) Management control requires that the quantitative and qualitative information it uses has a resolution sufficient to provide information about, or within, the smallest units of management within the corporation.

Management control is the level of management above individual units of production; it is the first level of management that can address resource degradation issues since these generally flow across these units of production, with actions in one unit having impacts on other units, often at some other time. Management control thus requires rigorous databases and predictive models to assess potential impacts and revise corporate tactical objectives so as to adjust operational control management to take these impacts

Table 8.1 Relationship between management category and the attributes of the information required for management purposes

	Management activity		
Information characteristic	Strategic planning	Management control	Operational control
Coverage	Global	Regional	Local
Accuracy	High average accuracy	High local accuracy	High local accuracy
Level of detail (resolution)	Generalized	5–100 m	0.01–2 m
Time horizon	Prediction	Prediction	Present
Sources	Primarily external	Internal and external	Internal
Data type	Qualitative	Quantitative	Quantitative and qualitative
Model type	Unstructured	Structured	Structured and unstructured

into account. Since this process will involve resolution of conflict between operational control units, it will often require accurate and detailed quantitative information.

In a very important sense, management control is the main beneficiary of these new technologies of remote sensing and GIS. For the first time, management control has the tools necessary to assess in a rigorous way how the individual units of production are operating, and respond to that information by the formulation of better tactical plans for those units of production, so as to improve utilization and maintenance of the resources of the institution or region.

Operational control

Operational control involves the direction of staff in activities directly involved in managing the resources of the unit of production, whether it be a farm, a factory or a shop, so as to achieve the short-term objectives as established by middle management. Operational control is concerned with day-to-day decisions, and requires quick decisions, often on site, using primarily on site or internal information. It requires models that can be implemented by the manager on site, and that can respond to events as they occur. Information requirements at this level are for high resolution yet simple information since the decisions are usually made quickly using empirical models.

The relationship between these categories of management and their information needs are shown in Table 8.1. As the level of management gets higher up the organizational structure, so the information needs change. The information needs become more summarized and more oriented around strategic planning issues and less around detailed management tasks (Diagram 8.5). The mix of information also changes the higher up the management structure the information is required; it becomes more dependent on external sources of information, more predictive, subjective and qualitative, simply because the number and range of factors and how they change over time both increase and become harder to predict.

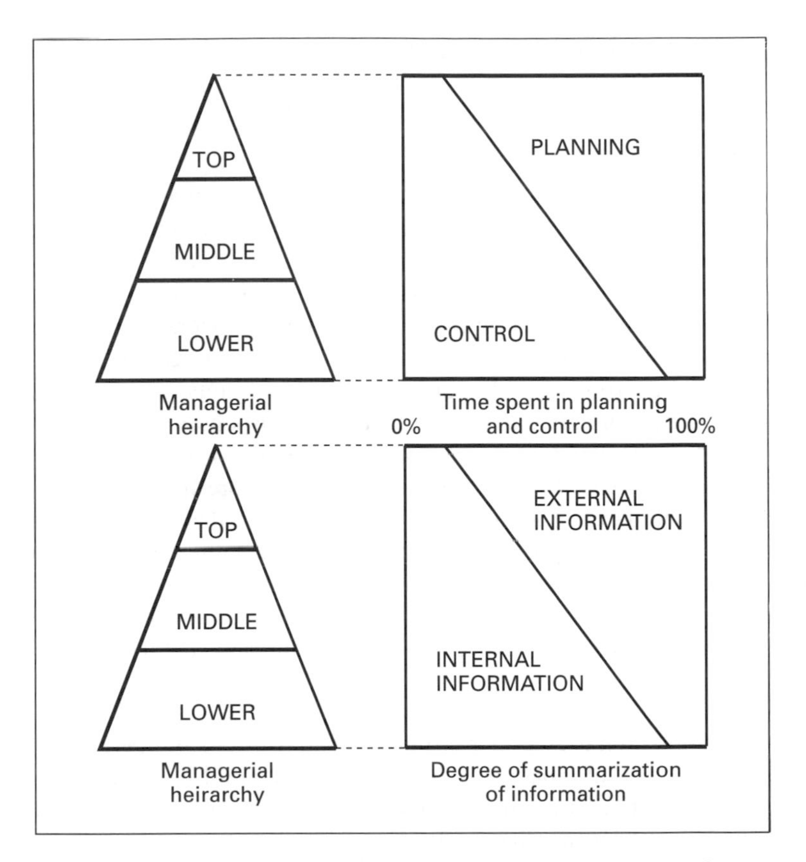

Diagram 8.5 Management levels relating to planning, control and information in (a) relation of managerial position to the time spent in planning and control and (b) relation of managerial position to summarization of information.

8.5.2 The role of remote sensing and GIS in resource management

Strategic planning

Sampling is an appropriate strategy for collecting information required for strategic planning. Remote sensing and GIS can assist in this collection and analysis.

Defining strata for sampling If the information to be collected is unevenly distributed across the area of interest then stratification of the area and random sampling within each strata will provide a better estimate of the measured parameters than will a simple random sampling. If these strata are expressed in remotely sensed images in a way that allows discrimination between them, then images are an appropriate medium for defining the strata. This is typically the case if the strata are related to land use, land-cover or land-cover condition, as applies to many forms of rural census. Typical strata could be:

- land-use intensity for agriculture and other census data collection where these strata would be identified by the visual interpretation of suitable imagery; and
- vegetative vigour as a basis for estimating yield.

Once the strata have been identified then they would be incorporated into a GIS so that they can be used with attribute data collected in some way to derive estimates of the attributes for each strata and for the whole area of interest.

Sampling within the remotely sensed data This is another way to provide strategic planning level information.

Management control

Management control is both deficient of an independent, objective source of information and yet it is the key management level in the maintenance of resources. It is deficient of suitable information sources since most of its information, in relation to resource management, comes from the operational control units that report to it. This information will tend to be subjective, self-focused and may contain biases. Operational control management perspectives are closely focused on the productivity of the unit. The impact of operational control decisions on other operational control units are of marginal interest to the operational control unit manager, particularly if responses to those impacts affect productivity. However, maintenance of resources requires a careful consideration and balancing of these impacts. Middle management is therefore in the position that it requires rigorous quantitative information if it is to balance the demands of operational control managers against the potential cost of their actions either in terms of degradation or impact on other units of production.

Management control is the level of management that can benefit most effectively from the information derived from airborne and satellite data, and incorporated in a GIS since:

- the derived information is at a finer resolution than the units of operational control so that it provides some information at the operational control level; and
- the derived information is extensive enough to cover all units of operational control that are of concern to management control.

The definition of management control should not be interpreted in too constrained a manner. Those groups in society that could typically be included in management control are:

- district agricultural extension staff and commercial agricultural consultants where they are concerned with farms as operational units;
- banks, valuers, stock and station agents and local government who deal with farms as operational units;
- forestry departments and timber millers who deal with logging concession operators as operational units;
- fishing co-operatives who deal with fishermen as operational units;
- community- and government-based integrated resource management co-operatives who deal with individual land-owners or lessees as operational units; and
- staff of commercial rural industries (herbicides, fertilizers, farm machinery) who deal with farmers and other land-users as individual units of operational control.

The information support required by management control does involve some inventory, but it will more importantly involve monitoring and prediction, to assess performance and environmental impact, and for planning and design purposes.

Operational control

Operational control can use remote sensing and GIS when the resolution in the data is appropriate to the task, and the derived information is provided on time. This requirement will often see satellite data being used for meteorological, marine and fisheries purposes since these activities cover extensive areas at the operational control level. Agriculture, forestry and other land-based activities are more likely to use large scale photography or video image data.

The demand of operational control is for current information, so as to address current management issues. This requirement places a significant demand on the system to provide the information, to an agreed level of accuracy and reliability, within a set time frame. This constraint creates an advantage for video data relative to photography because it does not lose valuable time in development and processing.

8.6 Specific characteristics of resource information

8.6.1 Quality attributes required of resource information

Resource information must be relevant, available and timely. It should also be objective, accurate, precise, of suitable sensitivity or resolution, comparable and consistent, concise, and complete.

Resource information must have the quality attributes of **relevance**, **availability** and **timeliness**, if it is to be of use to a manager. In designing an information system it is essential that each type of information included in the selected information set can be provided to managers with acceptable values in these three quality attributes. Whilst there will often be a most appropriate information set for addressing specific resource management issues, in practice this set may not be chosen because the analyst cannot guarantee acceptable values in these three quality attributes. There will always be a trade-off between the cost and benefits of information. The analyst, in selecting the information parameters to be included in the information system, must take into account the value of the information to the resource manager.

The other quality attributes are desirable and are present in varying degrees in the information used by a resource manager. To assess the importance of, and hence the values that are required in the differing quality attributes, the resource manager needs to assess the role of the information in his decision-making process, and the influence that information has on the decision. If the information has a relatively minor effect then the accuracy of that information will not need to be that great; if the information is central to the decision then the accuracy will very directly affect the quality of the management decisions made. Sometimes the costs of the information become prohibitive in the form originally envisaged by the resource manager; it may be necessary to adapt the style of management so as to operate effectively with either lower quality or different information.

For example, in monitoring forest clearing, images of clearing may be provided to the appropriate managers on time, but they may only have accuracy levels of 80–85 per cent in contrast to the 95–97 per cent accuracy levels originally sought by the forest managers. One adaption of management is to use the clearing information as a pointer to, or indicator of, probable clearing, and use some other form of verification, such as aerial flights or ground visits to confirm that the clearing information is correct or not, before using the information to make decisions. In this way the information on clearing is used to reduce the need for fieldwork, from coverage of the whole area to coverage of specific locations. If this approach is adopted then it may be very important to ensure that there are negligible errors of omission in the clearing information, since the field work focuses on identifying errors of commission.

It is not essential to have perfectly accurate information, but inaccuracies incur costs either through:

- loss of credibility;
- loss of confidence in the data; and
- costs of verifying the information before action is taken arising from the information.

The lower the accuracy, the higher the costs of establishing a supporting checking mechanism. Similar considerations apply to the other quality attributes, for example, the lower the resolution of the information, the fewer decisions that can be made using the information and hence the lower the value of the information to resource managers.

These quality attributes are favourably influenced when the information is derived from data of a higher order in the measuring scales. Qualitative information is more likely to be derived from nominal or ordinal data, whereas quantitative information is more likely to be derived from interval or ratio data. As the data increases up the scale from nominal to ratio data so the attributes of the derived information are likely to increase because the derived information is seen as being derived in a more rigorous manner.

The two attributes of conciseness and completeness include reference to the presence or absence of ambiguities in the information. **Conciseness** refers to the ability of the information to convey a lot of information in as simple and as direct a manner as possible. If the information is of land-use classes that are similar to, but not exactly what is required by the resource manager then the information loses conciseness. **Completeness** refers to the facility of the information to convey the required information without gaps or overlaps. The existence of unclassified pixels, or a miscellaneous (garbage bin) class indicates lack of completeness. It can be seen that both conciseness and completeness can be measured by a properly conducted accuracy assessment programme.

Remotely sensed images can seduce the observer into the opinion that extraction of information is easy since they suggest to the observer that, 'I can see all of the features that I am interested in'. Remotely sensed images are a record of the complex environment; but that complexity is being recorded in only a few of all the dimensions that exist in the environment. Of necessity some environmental variations are going to be recorded in a similar way in the image, and some surfaces that we see as being similar are going to be recorded in different ways. In consequence, images contain unexpected similarities and differences. The experienced analyst allows for these anomalies to occur in the analysis. The inexperienced analyst runs the great risk of ignoring this characteristic of images and rushing into conclusions without adequate cross-checking and verification.

Specification of user needs

Specification of user needs must be done in the context of:

- the individual manager;
- the information sought, i.e. the decisions to be made; and
- the environment in which the decisions have to be made.

These components are all dynamic; all three can be changing all of the time. Specification of user needs is thus a dynamic process. The components also vary from level to level within an organization, so that specification of user needs varies with the level of the manager within the organization.

8.6.2 Mapping and monitoring

Mapping is defined as the extraction of information on the values and spatial distribution of specific resource attributes for a geographical area. Mapping strategies are concerned with conditions at a time, but not with the analysis of change over time. A mapping task will be designed using either existing or new image data, the specifications for which are either known, or set, at the project design stage. The procedure is then designed around known data specifications, so that the data and the procedure can be optimized to achieve the required accuracy and reliability standards. The mapping project is then conducted without the urgency associated with the provision of timely information. This means that proper accuracy assessment and field validation are normally completed before publication of the resulting maps.

Mapping tasks are done once in any one area. There are advantages in developing highly specialized groups to conduct the mapping tasks, as occurs in topographic mapping and to a lesser degree in the mapping of other resources. These groups can be made most cost-effective if the local knowledge component of the mapping can be minimized either by the development of suitable techniques, as in photogrammetry, or in the use of highly skilled interpreters, as occurs in photogeology. In consequence, mapping tasks are usually conducted with little or no contact with local resource managers.

Monitoring is defined as maintaining a watch over, or keeping track of, the values and spatial distribution of specific resource attributes for a geographical area. Monitoring is thus concerned with the changes in resources over time and so includes an additional dimension to that of mapping. A monitoring programme is designed before implementation and before actual data availability are known. In consequence, monitoring systems cannot be designed to be optimized for the extraction of information from a specific data set. Resource managers will require acceptable results from the monitoring programme for all or most seasons. These managers will expect the monitoring to work using a wide variety of data sets that will be acquired under various conditions. In consequence, monitoring systems must operate, and produce acceptable results, using data sets that will vary between the optimum and the minimum sets as defined in the system specifications. These systems must therefore be robust, where robustness is defined as the ability of a system to maintain accuracy and reliability standards using a wide variety of data sets in the information extraction process.

If monitoring systems are designed to provide timely information then there will be considerable pressure to provide the information before accuracy assessment is complete. Monitoring systems must therefore be as independent as possible from the time-consuming task of accuracy assessment, yet users must be confident of the reliability of the information they are receiving. This is another reason why monitoring systems need to be as robust as possible.

Monitoring systems are fixed to the one geographical location, and if they are successful then resource management staff will become dependent on the information provided by them. The high level of robustness required by these systems should be improved by the integration of local field data or environmental knowledge into the information extraction component of the monitoring system. If local knowledge and data are used in this way then it will be more cost effective for resource management staff to be involved in collection of this field data. This arrangement has the advantages that the user becomes more integrated into the conduct of the monitor; they will better understand the monitoring process and its characteristics, they will take on the monitor as part of their own, and they will have more confidence in the use of the information provided by the monitor. This arrangement should not meet difficulties as the alternatives are more expensive to the user, and the resource staff will see the proper operation of the monitor to their advantage.

Resource monitoring is characterized by continuity of operation, so as to identify, quantify and analyse the significance of changes in the resources of interest. It is logical

Table 8.2 Changes in accuracy in the visual interpretation of crop fields using progressive cultivation, canopy and harvest images, from Hardy et al. *(1987)*

Shire	Image	Sample		Crop classification accuracy (%)	Errors	
		Farms	Paddocks		Omission (%)	Commision (%)
Bogan	Cult'n	12	109	84	16	3.5
	Canopy	14	121	96	4	0.0
	Harvst	10	78	100	0	2.2
Walgett	Cult'n	22	166	65	35	4.3
	Canopy	19	138	66	34	3.6
	Harvst	14	96	94	6	0.0

that monitoring information systems should likewise be characterized by a continuity of operation. In these systems individual images record land-cover at a specific time, representing a temporal sample of certain resource attributes.

The nature of this image sample relative to the characteristics of both the environmental characteristics and information required are critical to the design of monitoring systems. Land use is a discontinuous phenomenon that is usually constant over seasonal or longer periods, with potentially large breaks, sometimes indicated by changes in land-cover, at the changeover from one land use to another. Frequently, attempts to estimate land use early in the season from land-cover data will yield low accuracies, but these accuracies will usually increase as the season progresses. This characteristic of the accuracies of some information within monitoring systems is shown in Table 8.2.

The improvement in accuracy of mapping cropping by the use of a harvest image should be noted; the period of harvest has the greatest likelihood of acquiring images due to less cloud cover and the changes in radiance are the largest over the shortest period. It is possible to further improve these accuracies by incorporating agronomic information into the information extraction phase, where this agronomic information may include prior history, typical land-using patterns in the area, and the typical crop management practices adopted. All of these sources of information can be incorporated into the information extraction phase.

Once land-use information for an area is known, models may be used to estimate specific physical parameters of the land-covers associated with the different land uses. These estimation practices will often require calibration by use of field sites where the vegetation parameters, such as biomass, can be measured. Monitoring systems are therefore likely to include seasonal records of land use, as well as within-season information on vegetation canopy parameters associated with some or all of those land-use classes.

8.6.3 Accuracy and reliability

The characteristics of the data used in, and the nature of the information provided by, monitoring systems impose more severe constraints on their design than on the design of mapping systems. Considerations of accuracy and reliability can impose further constraints on monitoring systems. **Accuracy** is defined as a measure of the closeness of the estimated to the true value in the measured parameter, whereas **reliability** is a measure of the variance associated with the estimated values derived from the data, or the observations.

The accuracy and reliability of mapping has only a spatial context, whereas they both have spatial and temporal contexts in monitoring. In mapping, both accuracy and reliability can vary from class to class of land-cover and from location to location. In monitoring, the information extracted from individual images or sets of images can vary in accuracy and reliability in a manner similar to that of mapping. However, the accuracy and reliability can also vary from one classification to another so that any statements on accuracy and reliability should indicate the relevant data or information set.

The purposes for which the information are to be used should control the accuracy and reliability standards. Not only is it possible that one will be more important than the other, but their relative importance can change over time. Setting accuracy and reliability specifications is an important consideration in project design, particularly as the temporal aspects of accuracy and reliability can be arduous to assess.

Consider the case of crop monitoring to provide (a) within-season total crop area for industry and (b) a permanent diagnostic database of cropping, at field resolution, to improve productivity. The within-season information (a) places great emphasis on timeliness and reliability, but not on accuracy. If the within-season crop area information is 70 ± 4 per cent of the final crop area every year then the information is reliable but not very accurate. However the managers who use this information can use a multiplication factor to derive estimates of the final crop area that will be very close to the actual final area. The permanent database (b) places great emphasis on spatial resolution and accuracy, but not so much on reliability.

The nature of monitoring indicates that information derived at different dates, from different sources or by different techniques, will be integrated for many analyses. As the errors in the original data are propagated, so they are magnified, reducing the accuracy and hence usefulness of the processed information. If the accuracy is low then auxiliary techniques have to be instituted to protect the resource managers from making errors of judgement due to errors in the information. This raises the cost of using the information and commensurately reduces its value.

8.7 The design and implementation of resource management information systems

The purpose of project design and planning is to ensure the satisfactory completion of the project, to set quality standards, within set cost, time and other constraints. Designing the project and then planning execution is essential if problems are to be anticipated, contingency plans prepared and optimum use made of resources in the project. It will thus lead to better execution at lower cost.

There are six stages in a project life cycle:

1. analysing user requirements;
2. conceiving and designing the project;
3. detailed planning of the project;
4. implementing the project plan;
5. finalizing and evaluating the project; and
6. presenting the results to the user.

This approach is used for the following reasons.

- It provides a structure to organize a large number of inter-related activities that must be conducted to complete a project, particularly where there are constraints on the sequence of conducting these activities; some must follow others, some must be conducted in parallel with others, etc.
- It helps to predict or anticipate problems that might arise in execution of the project, so that contingency plans can be prepared.
- It provides a framework that is understood by project staff thereby providing a basis to make decisions, to understand the implications for other parts of the project, and to provide confidence that the project is being executed in a rational way. The framework also means that staff can contribute positively to the conduct of the project in ways that they could not if the plan of implementation was not known.
- It assists in resource management and reporting since managers can quantify, order and co-ordinate resource needs to minimize costs and optimize project implementation, as well as providing a framework for project assessment and reporting.

8.7.1 Specifying user needs

The objective of the designer will be to elicit from the user precise specifications of the:

- types of information required;
- spatial, temporal and informational resolutions that are acceptable and desirable to the user;
- timeliness required, specifically within season or post season;
- accuracy and reliability requirements;
- acceptable variations in these, either between classes or over time; and
- way that the information is to be used – who is going to use the information, how, and for what purposes. This will indicate the format of the required information, the facilities that will be required, time constraints, required skills levels and training needs as well as indicating potential impacts on management processes in the organization.

Often the user, being unaware of the characteristics of this technology yet being familiar with the information sets currently used to support resource management, will not be able to state the needs in these terms. The user may use a layman's terminology set in the form of management information needs, where these needs are different to the information extracted either from maps or from remotely sensed images. Having elicited user needs, the designer must transform them into user specifications. The first step is to convert the user needs into precise and quantifiable information needs. The second step is to split these needs into their components that can be acquired from remote sensing, analogue maps or field data.

For example, a user may specify that he wants erosion to be reduced below a specified level within a catchment that contains a number of land-holders. Erosion prediction depends on a number of component factors. Once the information has been derived then the user needs to discuss remedial action with land-holders; so this information is also required. Depending on the mechanisms chosen to encourage compliance, the user may also need information on other administrative units within the catchment.

The process of splitting user needs into component parts must also involve consideration of the tasks needed to integrate the components back into the management information required by the user. This integration may require the component information be collected in particular units, at particular accuracies or resolutions, or collected in particular ways.

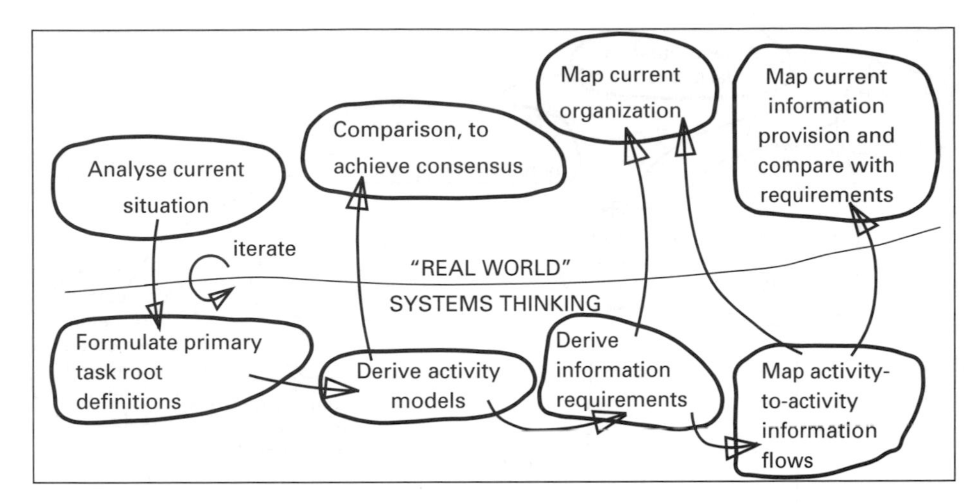

Diagram 8.6 Information requirements analysis based on soft systems methodology, from Galliers (1991).

8.7.2 Considerations in defining user needs

Precisely specifying user needs is not simply a matter of rational analysis given the nature by which decisions are made, the purposes for which information are sought, and the perceptions of manager as to the role of information in maintaining or strengthening his position. An approach that has been found to be very successful in defining information needs in other disciplines is based on the soft systems methodology described in Wilson (1990). Soft systems methodology is an appropriate basis for information systems planning, given its ability to clarify the activities required to meet objectives which may be only dimly perceived. Modifications to this methodology as discussed by Galliers (1991) have been adapted to resource management in this book.

The basic approach of Wilson and others is (Diagram 8.6) that:

1. identification of information requirements is based on modelling the primary tasks of the organization rather than addressing issues;
2. using the existing organizational structure, activity-to-activity information flows are converted into role-to-role information flows in order to identify individual managers' information requirements; and
3. the information systems required are then determined, based on performance and information needs.

The approach of Wilson has been amended to provide a better model for defining information requirements in resource management. The main changes (Diagram 8.7) are:

1. to identify environmental and community issues or activities that either affect, or are affected by the management activity sufficiently to require a response sometime in the future, and from conceptual models of these issues or activities, define information needs to create activity models and hence define information requirements; and
2. to recognize the importance of the community perception of the situation, so as to modify our conceptual models and hence amend the whole process.

As part of this process it may be appropriate to identify certain aspects.

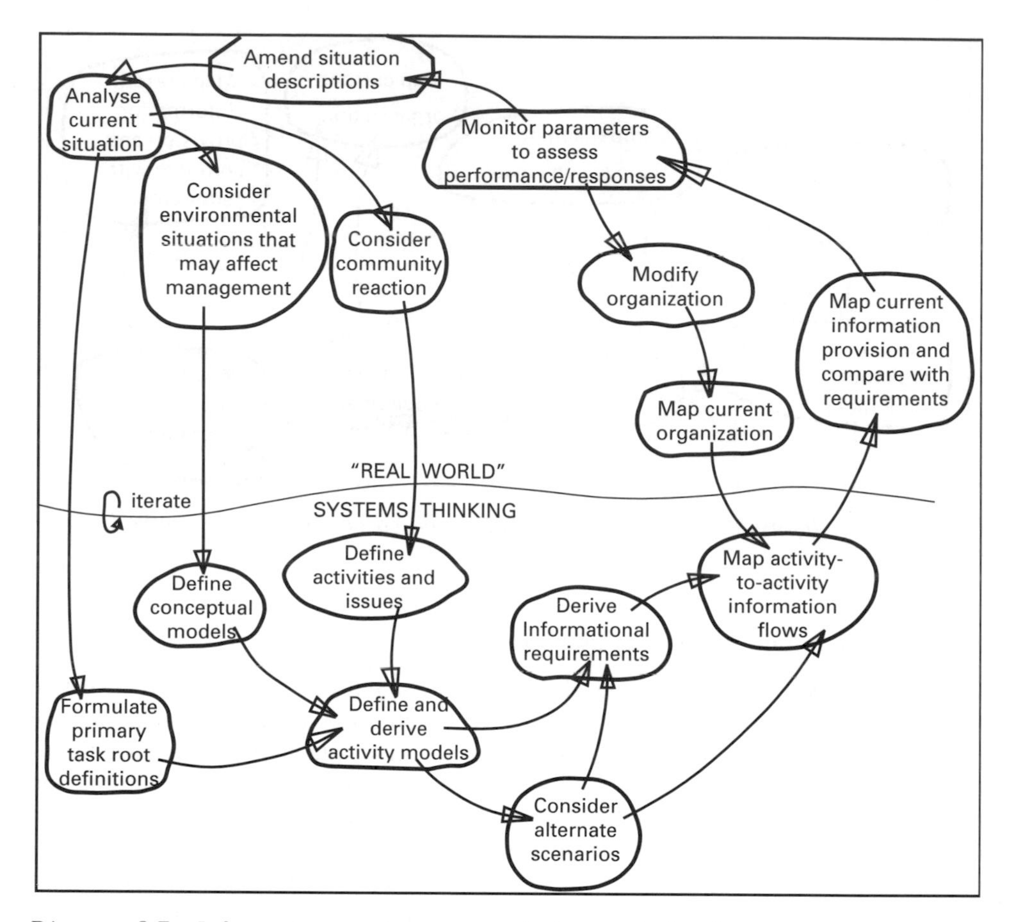

Diagram 8.7 Information requirements analysis as suitable for resource managers concerned with physical resources.

Constant factors Environmental and organizational factors that appear likely to be fairly constant over a designated period are likely to be easy to predict over the period of concern, and significant changes in them may indicate that conditions have changed sufficiently to require a reappraisal of information needs.

Trends Factors that can be expected to be subject to change over time, i.e. to exhibit trends in status or condition are often more difficult to predict than constant factors, and again significant variations from the predicted trend in these factors may indicate that a significant reappraisal of information needs is required.

Issues Factors, over which there may be considerable debate as to their status, condition or relevance, create issues because they contain either unknown or unpredictable aspects, or aspects that will have, or are seen to have, different impacts on different groups in society. The status of such factors can be highly unpredictable over the period of interest in the construction of information systems.

The design of information systems must identify and then consider appropriate responses to all three types of factors, as well as considering the implications for the monitoring programme.

8.7.3 Designing mapping and monitoring systems

Once the user needs have been specified then the analyst needs to assess fully the environmental characteristics of all land uses and land-covers so as to identify potential confusions in the mapping or monitoring process, and to identify rules that may be of use in those processes to improve the interclass discriminability. Some of this information can be effectively gleaned from the user who should be encouraged to discuss the environment of the area in the briefing. Other sources of information may also be used to gather information on the environment, including field visits to gain a better appreciation of the issues, the challenges involved, and to discuss how the information will be used with field staff. The design of the project must then consider the following aspects.

The geographic extent of the area to be monitored

It may be appropriate to consider conducting the mapping or monitoring within sub-areas that are reasonably internally consistent in their physical, climatic, and environmental characteristics. The consistency within the sub-areas may lead to simpler environmental rules and conditions being built into the monitoring process, thereby simplifying the construction of the monitoring systems.

A monitoring system should be seen as an information system for an area rather than an information system for a discipline. One of the major disadvantages of current methods of resource management is that all management is discipline-oriented; the impact of decisions on other aspects of the environment is rarely taken into account. The effect of this is to exacerbate the damage to the environment; proper environmental maintenance requires that resource management be regionally rather than discipline-oriented.

The optimum and minimum acceptable data sets

Phenological or crop calendar data (as discussed in Chapter 4) can be used to select optimum and minimum data sets that can be used in the mapping or monitoring tasks.

Designing robust systems

Monitoring systems must be as robust as possible, where robustness is a measure of the capacity of a system to continue providing the specified information under conditions of sparse or incomplete data. The robustness of the system will directly affect the reliance placed on the monitor by resource managers to routinely provide information of use for the management of resources. Robustness is increased by the use of:

- field data to supplement the remotely sensed data;
- additional images to provide a level of redundancy;
- supplementary information from various sources; and
- established techniques and processes where possible.

The monitoring as a routine updating process

To be successful, monitoring systems must be seen by the user as a reliable source of up-to-date information. To establish this reliability, use regular sequential image data, processing each image as they are acquired, to update masks of land-cover, land-use and land-cover status. Extract information in this sequence because:

- information on land-cover at a date is extracted from the individual images acquired at that date;

- then integrate this information to create land-use information for the season or period of interest; and
- use the information on land use and land-cover to control estimation of the condition of the land-cover at the time of data acquisition.

Extracting multiple sets of information

Extraction of information for a range of users has the advantage that it shares the costs of fieldwork and information extraction, thereby reducing the costs per user, whilst also providing consistent information to all of the users, a significant advantage in the maintenance of resources.

Assessing the fieldwork component of the project

Include fieldwork to improve the accuracy of the analysis or to increase the robustness of the programme, and the fieldwork becomes necessary for accuracy assessment. User agencies may be prepared to assist in the collection of field data.

Assessing training requirements

Implementation of the programme and dissemination of the information is likely to require training of staff in the user agencies. This training is likely to include initial training, then ongoing user support until staff become familiar with the information and the system used to analyse that information.

Planning implementation of the project

Having designed the project then planning implementation needs to consider resource allocation, scheduling of data, tasks, facilities and staff, staff training, conduct of the components of the project and the processes involved in reporting to the user.

8.7.4 Planning implementation of a monitoring system

Once a project has been designed then the planning phase must consider the staff, data and facilities required by the project design. The project planner must consider, for each resource used in the project, some or all of the following:

- acquisition,
- utilization in the data preparation and analysis,
- maintenance,
- coordination,
- training,
- costs, and
- termination on completion of the project.

The are a number of techniques that can be used to order the project plan so as to make the best use of resources without undue expenditure. These techniques include Gantt and PERT/CPM charts.

Gantt charts

A Gantt chart is a horizontal bar chart that graphically shows the time relationship of the tasks in a project. On the horizontal axis the time is mapped, while the vertical axis

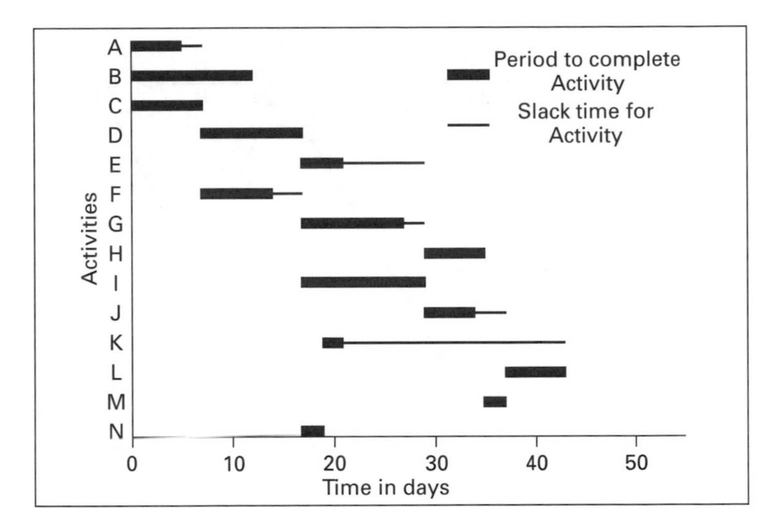

Diagram 8.8 A typical Gantt chart.

represents the list of tasks to be conducted in the project. Each task is represented by a line that extends for the period covered by the task, with solid lines representing the duration of the task and dashed lines representing longer periods over which the task can be conducted.

The Gantt chart shows the flow of activities, those that are running concurrently and those that must follow sequentially, and so readily shows when resources are required, potential clashes and bottlenecks that need to be resolved. They can thus be used to refine the planning process and to assist management in the conduct of projects by providing the basis of comparison between the planned and actual rates of completion of the tasks in the project.

PERT/CPM

PERT/CPM program evaluation and review technique critical path method is a technique for planning and controlling work which reduces a project to a series of activities and events. It is used both to control project duration and to allocate resources. As well as showing the scheduled time for each task, the PERT/CPM chart (Diagram 8.9) also shows the dependence of one task on other tasks.

Each activity in the project is represented by an arrow, so the network of arrows shows the sequence of the independent activities in the project, and their interrelationships. When time estimates are added, it can also show the critical path (the heavier arrows on Diagram 8.9) i.e. the quickest route. The free time in parallel paths, known as the slack or float, is used as the basis for analysing the plan to achieve maximum economy of time and resources.

This technique aims to answer the following questions.

- What is the shortest time in which the whole of the project can be completed and on which activities does this shortest time depend?
- What amount of free time exists in each of the other activities?
- How will the project be affected by a delay in an activity?
- What bottlenecks are likely to occur?

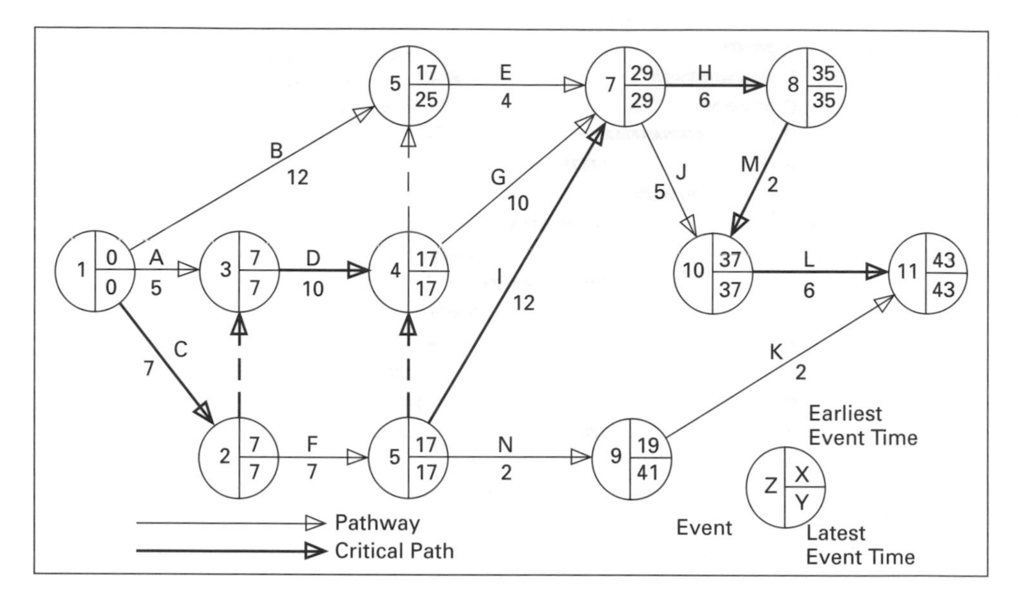

Diagram 8.9 The PERT/CPM diagram for the construction of a brick and tile house.

- How much will it cost to reduce the duration of the project and on which activities is it best to concentrate to do this as cheaply as possible?
- How could availability of resources affect the staff of the project?

The advantages of PERT/CPM are that it:

- aids planning;
- is pictorially informative;
- includes all coordination activities;
- shows problems in perspective;
- highlights decision making;
- assists decision making;
- makes best use of available information;
- displays progress situation and aids management control;
- depicts the probability of success;
- is simple to learn; and
- indicates implications of decisions.

The elements of the networks in a PERT/CPM diagram are activities, events and dummy activities. An **activity** is an element of work occupying a period of time, represented on a PERT/CPM diagram as an arrow with a description of the work to be done over the arrow. An **event** is an instant in time at which one or more activities begin or end, shown in the diagram as a circle. A **dummy activity** provides a logical link between the two types of elements in the diagram. No work is represented by a dummy activity, and the dummy takes no time. It is represented on the diagram as a broken line.

These elements are built into a PERT/CPM diagram by following a set of rules.

1. All activities start and finish in an event.
2. Activities may start only when the required preceding events have been achieved.
3. An event is achieved when all activities leading to it are complete.

Table 8.3 PERT/CPM example: construction of a house

Activity reference	Description of activity	Linking events	Estimate of days to completion
A	Obtain bricks	1 and 3	5
B	Obtain roof tiles	1 and 5	12
C	Prepare foundations	1 and 2	7
D	Erect shell of house	3 and 4	10
E	Construct roof	5 and 7	4
F	Lay drains	2 and 6	7
G	Wiring	4 and 7	10
H	Plastering	7 and 8	6
I	Plumbing	6 and 7	12
J	Flooring	7 and 10	5
K	Landscaping	9 and 11	2
L	Painting and cleaning	10 and 11	6
M	Doors and fittings	8 and 10	2
N	Lay pathways	6 and 9	2

The steps in the preparation of the PERT/CPM diagram, shown in Diagram 8.9, are as follows.

1. List all of the activities that make up the project.
2. Prepare a draft network diagram following the rules given above.
3. Test the logical progression of events and activities in the draft network diagram.
4. Draw the final network or PERT/CPM diagram.

Consider the example of the construction of a house. The activities involved in this project are listed in Table 8.3.

The events are numbered sequentially from the start of the project to the end. In each event circle is the event number, the earliest start date for this activity and the latest start date, recognizing that some activities cannot commence until others are partially or fully completed and thus placing constraints on the period that can be covered by the start date of an event. Activities that can be conducted concurrently are drafted as separate paths. The diagram not only shows the relationship between the activities, and the time duration of each, it also shows the critical path i.e. the sequence of activities that must always be started on time to ensure the earliest possible completion of the project. The critical path is found by starting with the earliest project completion time and working backwards to compute the latest start time for each activity. The latest start time for each activity is included in each circle as the second date for each event identifier.

The critical path therefore identifies the critical activities that must be completed on time to avoid delays in overall project completion, allowing the project manager to concentrate efforts to ensure that this milestone is achieved. It also identifies slack time where staff and other resources can be re-allocated to other activities.

Another important function of these task scheduling methods is to provide team leaders and project managers with a means of monitoring actual against intended progress through the project tasks, and so assist project managers to maintain schedules towards project completion.

8.8 Usage of information in resource management

Section 8.4 discussed the concepts of generalized and localized models and their role in formulating resource management decisions. Such concepts are adequate to describe the process of decision making when:

- appropriate generalized models are known and understood by the resource manager;
- methods of adequately parameterizing the models are available to the resource manager; and
- the information provided is adequate to the task, i.e. the information is at sufficient resolution, accuracy, reliability and timeliness.

In many situations not all of these requirements can be met, yet decisions may have to be made. In consequence, a strategy is required to address the need to make decisions despite these limitations. If the information base available to the resource managers is inadequate then the resource manager may not be able to parameterize the necessary generalized models, or may be able to parameterize approximately the generalized models sufficiently to give an indication of conditions.

The first situation clearly indicates that collection of the information needs to be reviewed to ensure that either the information meets the needs of the resource manager or the management style changes to use information bases that can be provided. The second situation is more likely to arise, particularly if RMIS have been properly planned. There are a number of strategies that can be implemented that use the information base to support management when this situation occurs.

The information base as an indicator

The information is used to identify areas or situations that may meet certain conditions; it is used as an indicator of those conditions. Other information is used to verify that the conditions do, or do not, exist in the indicated areas before appropriate management decisions are made. This second set of information can do things not possible with the first, so that it is likely to be more expensive. The second set of information should only be collected for those areas indicated by the first set.

Consider when information is used to monitor the felling of timber. In this type of situation enforcement of the regulations may involve legal actions, which are only going to be successful if the information used in court is accepted as being true, beyond reasonable doubt. This level of accuracy cannot always be achieved with information that has been derived from remotely sensed data. When this occurs then the primary information does not have the accuracy necessary to parameterize fully the model that illegal logging has occurred, but it would often be adequate to indicate that illegal logging may have occurred. Once areas of possible illegal logging have been identified then a second set of information must be used to determine whether logging actually occurred, e.g. by field visits. The field visits are more expensive than derivation of information from remotely sensed data, but acquisition of the primary information is justified if it reduces the overall cost, or maintains a current monitor of logging activities.

Collection of information by remote sensing to create information that will be used as an indicator to a second information collection process may be established as a deliberate strategy. When this occurs then the information collected by remote sensing can sometimes be collected at a lower accuracy or resolution than might otherwise be specified, reducing the cost of the monitoring programme.

Information base used to define the extent of more detailed monitoring programmes

A monitoring programme can be used to indicate when more detailed monitoring programmes are required, and the extent of those programmes. Thus a strategy may be adopted to monitor an extensive area at low resolution or lower accuracy, so as to identify when significant changes are occurring, and then map those areas at higher resolution or accuracy. The routine monitor is primarily used to indicate significant change and provide information so as to decide when, where and why more detailed monitoring is required.

9

Applications of digital image processing and geographic information systems

9.1 Introduction

The applications of remote sensing in Chapter 4 demonstrated the use of visual interpretation techniques in the extraction of information from remotely sensed data. Chapters 5–8 have introduced digital image processing techniques, GIS, accuracy assessment and how information interfaces with resource management, so now we are ready to consider more complex types of applications. It is necessary first to introduce some concepts in environmental monitoring as are relevant to applications of these technologies in prediction. These applications will proceed in the sequence:

1. conducting resource inventory as an application of standard classification techniques, and simple integration of the information in a GIS to provide the required management information;
2. resource monitoring as a more advanced application of classification, with simple GIS integration to provide the required management information;
3. estimation of resource status using remote sensing;
4. resource modelling by using remote sensing and GIS to predict potential outcomes, to support the management of resources; and
5. estimation of crop production with the aid of GIS data.

9.2 Application of remote sensing and GIS to resource inventory: mapping landslips in the Cordillera region of the Philippines

9.2.1 Background

The Kayapa–Bokod area of the Central Cordillera of Luzon (Philippines) is bisected by the tectonically active Philippine Fault and associated faults. The area is characterized by rugged mountainous terrain consisting primarily of extensively shattered and worked metamorphic rocks. The area varies between 900 ft and 3300 ft altitude. The ruggedness and frequent tectonic activity in the area means that landslides occur quite naturally.

The area is subjected to the normal tropical monsoonal weather, with the wet season (July–October) receiving most of the annual (2100 mm) precipitation. The area of the study is shown in Photos 9.1 and 9.2, rectified images acquired using the French SPOT sensor in January 1987 and January 1991. Both images clearly show the extensive drainage network and steep ridges that are characteristic of the area. The first image also depicts the rainshadow effect common in this area, with higher rainfall in the east than in the west.

The land use of the area is primarily subsistence or slash and burn agriculture with harvesting of timber on the hills, and paddy rice growing in the narrow valleys. The denudation of the slopes due to timber cutting and subsistence agriculture can be expected to exacerbate the natural occurrence of landslides.

On 30 June 1990 a major earthquake occurred in the complex Philippine Fault system, with the epicentre south-east of the image, causing major movement along the fault. Survey observations taken over a network of stations both prior and subsequent to the earthquake revealed up to 5 m relative lateral movement and up to 1 m vertical movement across the fault.

The earthquake was of long duration, nearly 60 seconds in Manila, and quite severe, recording 7.8 on the Richter scale. The magnitude of the earthquake could be expected to loosen rock and soil, increasing the likelihood of landslides. This tendency was exacerbated by rainfalls in the subsequent wet season.

Landslides have a number of effects on the local population. The landslides themselves destroy scrub and forest cover and both current and potential slash and burn agricultural sites, houses, roads and other infrastructure and cause fatalities to the local population. The colluvium that fills the valley floors from these landslides can destroy agricultural areas, transport corridors, houses, and cause further fatalities. The alluvium created by subsequent erosion exacerbates the filling of the valleys and drowning of agricultural areas, filling of flood works causing further damage to road and other transport corridors, and destroys sources of potable water. Many communities were isolated for months when the road network was destroyed by the earthquake.

Information on the density of landslides, as induced by the earthquake, can therefore be used to assess the level of damage that has occurred in different areas. This type of information is essential in the planning of relief works and general assistance, to assess the allocation of resources and set priorities for rehabilitation works, and hence to establish the budget for the rehabilitation programme.

9.2.2 Management needs

The objective of the project was to provide information on relative landslide impact on the different communities in the area, so as to prioritize rehabilitation work within the federal government agencies of the Republic of the Philippines, as well as programmes supported by international and local agencies.

Information for this purpose is required at the local district or similar level. However for many engineering and agricultural purposes, better statistics are those based on watershed subcatchments. These statistics are also more useful in attempting to understand the relationship between physical conditions and landslide frequency. In addition, the boundaries of watersheds could be readily defined from topographic maps of the area, but community boundaries are often unknown or vaguely defined. In consequence, the project was designed to provide information on landslide density, pre- and post-earthquake, for watershed sub-catchment areas.

These user needs are then be separated into component types of information:

- mapping of landslides/quarries prior to the earthquake;
- mapping of colluvium/alluvium prior to the earthquake;
- mapping of landslides/quarries after the earthquake;
- mapping of colluvium/alluvium after the earthquake;
- mapping of sub-watershed areas that will form the basis of the decision making; and
- navigational information to locate the information on the ground, such as roads and drainage lines.

The subsequent analysis phase revealed the need for two other types of information.

- Delineation of the valley floors was needed to discriminate the landslide/quarry class from the colluvium/alluvium class. The mapping of both classes is to be done from satellite image data, but their nature means that they may not be readily discriminated from each other in the image data. In consequence, some other information is required to achieve discrimination between them.
- Delineation of cloud affected areas was needed on both images since these areas are effectively not available for analysis.

9.2.3 Method of analysis

Aerial photographic analysis was rejected because:

- the pre-earthquake photographs were too out-of-date (1980) to represent the just-prior conditions although the 1980–81 aerial photography was used with 1:50 000 topographic maps, to identify watershed subcatchments;
- it was impossible to acquire coverage within a reasonable time period; and
- information extraction would have taken too long to be of use for planning the rehabilitation programme.

The selected approach was to map the pre-earthquake land-covers using a February 1987 SPOT multispectral scanner satellite image and a January 1991 SPOT multispectral scanner image to map the land-covers after the earthquake. The land-cover mapping task would focus on discriminating exposed rock and soil from the other land-covers. Digital image processing was used to extract the required information because of the small size of many of the landslides, and the narrow expanse of the colluvium in the narrow valleys. Visual interpretation of the classification, superimposed with topographic data in the GIS, was used to discriminate between landslides/quarries and colluvium/alluvium. The images were also used to delineated cloud cover on both images. The sub watershed areas were identified using 1:50 000 topographic maps of the area, supplemented by air photo interpretation.

The images were rectified, so that they could be registered to the other data sets of the study area. Rectification of the 1991 image proved to be straightforward as the sensor was pointing directly down during acquisition. However the 1987 image was acquired with a 13° tilt in the sensor axis, introducing displacements of up to 200 m due to height differences in the area. The resulting displacements meant that the information on pre-earthquake landslides, could not be registered to either the map, or the post-earthquake landslide data to better than ±100 m. The resulting displacements between the data sets would not introduce significant errors in statistics derived for the sub-watersheds, but it meant that the pre- and post-landslide data sets could not be compared, pixel by pixel.

Each rectified image was classified using a set of training areas as seed classes for clustering. From the clustering were derived a set of spectral class statistics that were used to classify the image into those statistical classes. The classified image was then visually analysed with the raw image, and large-scale oblique colour aerial photography, acquired in June 1991 for the purpose, to group the spectral classes into a smaller number of land-cover or informational classes. This process was done for both the 'before' and 'after' images.

Comparison of the classification of the 1991 image with the 1991 oblique aerial photographs showed that there was good discrimination between exposed rock and soil and all other land-covers. However, the classification did not provide good discrimination within this land-cover class, specifically between rockfalls and quarries on the one hand and colluvium, and fresh alluvium on the other.

The information layers were transferred to the GIS for analysis which involved:

1. creating layers that represent landslide/quarry and colluvium/alluvium on both the before and after images;
2. deriving statistics on the amount of each class within each sub-watershed;
3. extracting further statistics on the density of each class within each sub-watershed from this statistical data; and
4. mapping the sub-watersheds to depict the extent of these classes in a graduated way, from the most to the least affected.

This method provides the information necessary for the planning phase. If the information is not of sufficient accuracy for the execution of some rehabilitation works then it would show where more detailed aerial photography would be required.

9.2.4 Results achieved

The results, shown in Photo 9.2 from top-left to bottom-right, are:

Image AA Spectral classification of the 1987 (pre-earthquake image) into 25 spectral classes.

Image AB Land-cover classification by grouping the spectral classes into eight information classes, of which two exposed rock and soil classes are of interest in this study.

Image AC The two classes of interest from Image AB.

Image BA Spectral classification of the 1991 (post-earthquake image) into 43 spectral classes.

Image BB Land-cover classification into ten information classes with eight being similar to those created from the 1987 classification and two related to cloud and shadow.

Image BC The two classes of interest from Image BA.

Image CA Change in landslides and alluvium/colluvium in the study area showing: green – landslides in 1987; yellow – landslides on both images; red – landslides on the 1991 image; blue – alluvium/colluvium on the 1987 image; cyan – alluvium/colluvium on both images; magenta – alluvium/colluvium on the 1991 image; grey – areas outside the 1991 image coverage; and white – cloud.

Image CB Image CA, including the boundaries of the sub-watersheds.

Image CC Percentage cover of landslides within each sub-watershed, colour-coded to set priorities.

The GIS then contained the following layers of information:

1. landslides/quarries prior to the earthquake;
2. colluvium/alluvium prior to the earthquake;
3. landslides/quarries after the earthquake;
4. colluvium/alluvium after the earthquake;
5. sub-watershed areas that will form the basis of the decision making;

Table 9.1 Statistics on landslides (hectares) by watersheds

	WATERSHED STATISTICS			LANDSLIDE STATISTICS				
Name	GIS No.	Area 1987	Area 1991	Area 1987	Area 1991	%Area 1987	%Area 1991	%Change
Binian	1	389	389	7.2	15.9	1.85	4.10	122
Anap	4	1 800	796	3.8	28.2	0.21	3.54	1 586
Inoman	5	1 349	1 349	2.1	56.8	0.16	4.21	2 700
Benneng nth	6	2 188	2 188	6.7	118.6	0.31	5.42	1 700
Damchak sth	7	1 363	1 363	4.4	114.4	0.32	8.39	2 700
Damchak nth	8	542	542	1.0	54.8	0.18	10.10	4 950
Agat	9	1 631	1 631	4.0	110.7	0.25	6.79	2 620
Bolo	10	3 510	2 745	11.1	158.5	0.32	5.77	1 817
Bobok	11	2 878	2 046	3.8	98.0	0.13	4.79	3 100
Benneng	12	2 272	2 184	0.0	44.4	0.00	2.03	(inf)
Yabnong wst	13	2 138	1 789	1.1	5.6	0.05	0.31	500
Yabnong est	14	1 514	966	1.1	28.1	0.07	2.91	5 700
Kayapa prpr	15	3 033	133	13.1	9.3	0.43	8.20	1 722
Kayapa	17	2 783	1 567	2.9	45.9	0.10	2.93	2 800
Kiwang	18	2 574	1 497	1.2	30.8	0.05	2.06	4 000
Peuadan	19	3 483	2 027	0.3	9.4	0.01	0.46	4 500
Banao	20	3 403	1 228	3.3	74.9	0.10	6.10	6 000
Peuadan est	29	2 322	336	0.2	61.4	0.01	18.28	18 270
Darac sth	39	171	171	1.5	10.1	0.88	5.88	556
Batangbang	40	1 582	728	20.4	29.9	1.29	4.10	215
Tuduyan	41	2 808	2 558	16.8	86.4	0.60	3.38	467
Darac	43	267	267	5.6	26.3	2.08	9.82	367
Ambuklao	44	1 463	1 459	19.4	64.6	1.33	4.43	226
Agno River	45	5 726	4 026	63.8	261.9	1.11	6.51	491
Aykip	50	3 000	1 553	3.8	16.6	0.13	1.07	600
Bukig wst	52	1 508	17	0.0	0.2	0.00	1.09	(inf)

6. roads and drainage lines; and
7. cloud cover on the 1991 image, to mask classification errors that were caused by this cover.

The data in the GIS was analysed to create a table of the count of the area of the landslide/quarry class in each sub-watershed area from both the 'before' and 'after' images, making sure that the cloud cover was used as a mask to that count. This count information was then placed in a relational database file, where the data were analysed to calculate the percentage area of landslides in each sub-watershed. This information was then printed (Table 9.1). The percentage landslide data for the 'before' date was subtracted from the 'after' date, to create another column in the database file and another GIS layer representing this difference in percentage cover, with the colours coded to represent the difference ranges as shown in Table 9.2.

The areas of the sub-watersheds given in Table 9.1 are correct in the 1987 data, but are less in the 1991 results, due to cloud cover on the image over parts of the sub-watersheds. It is for this reason, and for comparison between sub-watersheds, that percentage areas have been used in the analysis.

Table 9.2 Colour scale indicating the severity of landslide impact due to the 1990 earthquake in Photo 9.1

Colour	Range of values, difference in per cent area landslides (%)
Magenta	0.00–1.00
Blue	1.00–3.00
Cyan	3.00–5.00
Green	5.00–7.00
Yellow	7.00–10.00
Red	10.00–15.00
White	15.00–20.00

The Table 9.1 shows that there was very little landslide area in 1987, with the largest being 1.2 per cent of a sub-watershed and most being less than 0.01 per cent. By contrast the 1991 results show huge increases in the landslide area as the results of the earthquake and the wet season. Image CA reinforces this observation due to the large amount of red in the image. The largest percentage area is 18 per cent, with an average of 4.5 per cent. No sub-watersheds have been exempt, with all exhibiting an increase. These percentage increases are also large, in part, because of the low area of landslides in 1987.

9.2.5 Usage by the recipient

The users of this information benefited in a number of ways. At the planning level the statistics on impact of landslides on the sub-watershed areas was important in setting work priorities and planning programmes. At the implementation level the local government officers were able to translate plans into projects and explain these to the local people.

9.3 Application of remote sensing and GIS to resource monitoring: monitoring of rice crop area

9.3.1 Background

Rice is an important export crop from Australia. In temperate Australia, rice is grown under irrigation in summer, in areas that have negligible summer rain, but long hot and dry periods that are suitable for the growth of the crop. The crop is a major consumer of water, so the managers of the upstream water resources are interested in the area and status of the crop so as to manage the volume, and staged release, of water for the crop. The rice industry is also interested in the area of the crop so as to estimate production during the season, to plan mill production runs, prepare storage, and seek sales.

The water resource managers need early season information on the area of crop within each irrigation area, and an update on changes in the area due to failures in the crop, if this occurs. The rice industry needs the same information on area, and regular updates on the condition of the crop, so as to estimate production.

The irrigation areas are located on the inland floodplains of rivers draining from the Great Dividing Range that runs down the eastern side of Australia. They are located in areas with a winter dominant rainfall of 15–25 inches per annum (380–635 mm). Temperatures range from daytime peak average values of 95°F in summer down to daytime average peak values of 10° in winter. Other annual crops, particularly wheat and vegetables, are grown in the area under irrigation. There are extensive horticultural plantings in the area.

These management information needs can be separated easily into the information layers that will be needed. These are:

- rice area across the irrigation areas;
- irrigation areas as administrative regions for the management of the supply of water and the marshalling of the grain; and
- navigational information.

9.3.2 Monitoring procedure

The monitoring procedure starts with an initial mask in which pixels are assigned either 0 (cannot be rice) or 1 (could grow rice). All horticulture areas, non-arable areas including urban development, and swamps and rivers, as well as areas outside of the study area, are assigned 0.

The procedure involves two steps for each image. The first step is an image classification, which assigns each pixel to either be within or without the rice response domain, depending upon the pixel response characteristics. Pixels that are 0 in the initial mask are not processed. The second step is a mask updating procedure that uses agronomic criteria, in conjunction with the mask, and image classification to create a current mask. Both the image classification and the mask updating criteria are different in the establishment and growing phases of the crop and could also be changed within those phases.

The image classification in the crop establishment phase (November through January in Australia) is done using the vector classifier (Chapter 5) since the rice crop has a response domain suited to this algorithm. Rice, starting from the response of the water in the rice bays, changes in response as the crop grows, until it reaches the response of the full green canopy. This change in response closely follows a linear path from the response of water to the response of the green canopy, as shown in Diagram 9.1. The vector classifier is modified by use of a variable threshold to define a conical domain. The surface of this domain is set at about the response of pixels with a 50 per cent rice cover, from analysis of scattergrams of each image.

The mask updating procedure is shown in Diagram 9.2. This procedure assumes that non-rice areas of flood irrigation are very unlikely to be in the rice response domain of a sequential pair of images during the establishment phase. Rice will only diverge from this domain when physiological changes in the plants start to affect canopy reflectance. The mask updating rules adopted in the establishment phase are as follows.

1. Pixels with mask values greater than 0 and classified as being the rice response domain are incremented one value. Pixels with a value of 2 are categorized as potentially rice, while pixels with a value of 3 or more are categorized as rice.
2. Pixels with mask values of 4 or less, and classified as not being in the rice response domain, are assigned 1 or returned to 'could grow rice'. Pixels with values greater than 4 are not changed. They could be considered to have earned the right to stay as rice, particularly as changes in response are most likely to be due to physiological changes in the crop.
3. Pixels with a value of 0 are not changed.

The system has been designed to be highly automated so as to minimize the necessary interaction and level of remote sensing skills of the operator. Operator interaction

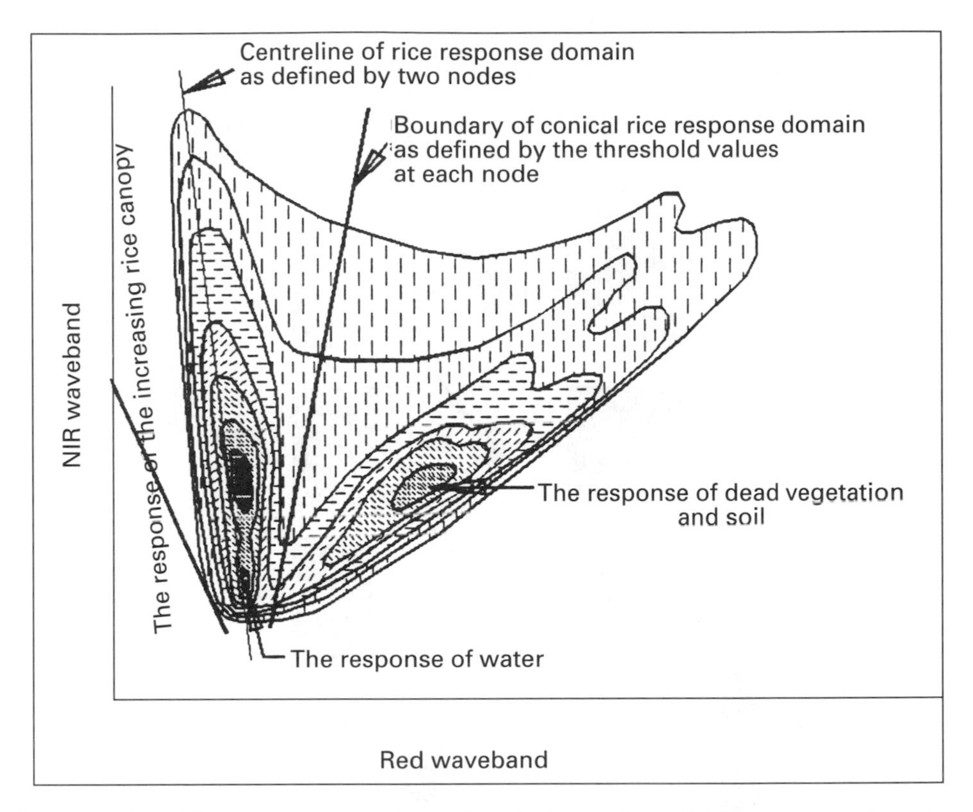

Diagram 9.1 The response characteristics of rice in the red/ NIR response domain, and the class domain for rice as defined by the vector classifier.

for each acquisition is required for setting parameters, creating a field data mask and in some cases, such as partial cloud cover, for directing the system to process different parts of the area in accordance with different rules. The parameters to be set are control points for rectification, selection of the nodes and the variable threshold for classification, and modification of the agronomic criteria used in mask updating. Results should be available within five days of receipt of the data at the image processing centre. Field data are not incorporated into the classification process because they increase both the logistic complexity of the system and the delays between data reception and information presentation and also are not needed to meet the required accuracy standards.

Field data are incorporated into the accuracy assessment by creating a field data mask that is compared to the rice mask created by the monitor. Typically, this accuracy assessment would only be done at the end of the establishment phase. The monitor was implemented in the Murrumbidgee Irrigation Area in 1983/84 and 1984/85 seasons using Landsat MSS data and in the 1986/87 season using a combination of SPOT and Landsat MSS data.

The layers of information created in the GIS are therefore:

- irrigation areas as the land administration areas within which the information is required;
- areas that cannot grow rice, such as horticulture, reservoirs, town areas, and the river floodplain;
- rice growing areas after each image acquired during the cropping season, up till development of full canopy; and
- navigational information.

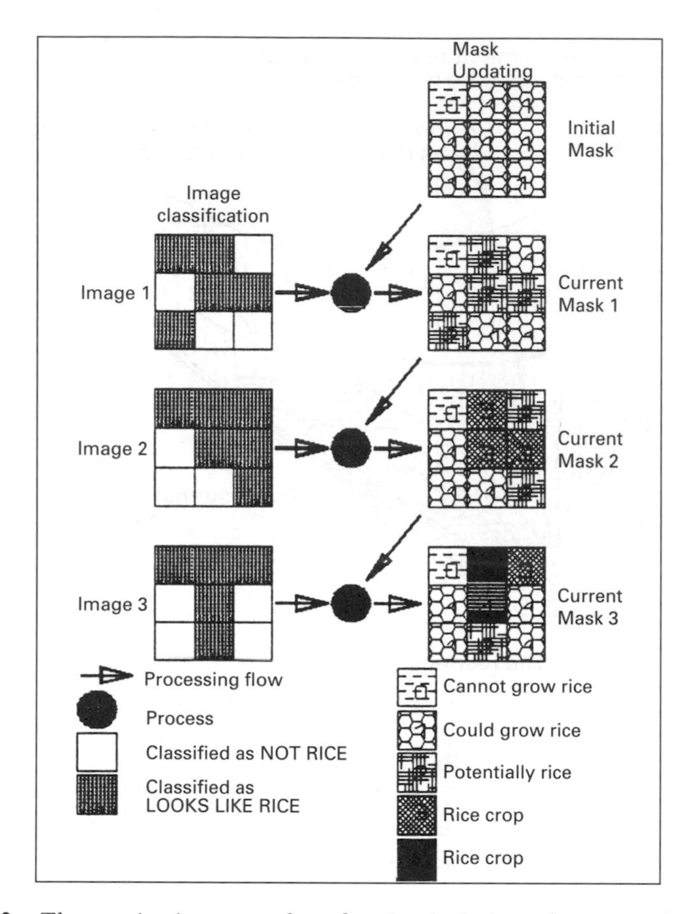

Diagram 9.2 The monitoring procedure for rice in irrigated areas, using SPOT and Landsat data, with two processes of image processing to identify the areas that 'look like rice', and the GIS procedure to update the mask using agronomic criteria to create the updating rules.

9.3.3 Results of monitoring in the 1983/84 season

Photos 9.3, 9.4 and 9.5 show the image used and the current land-cover mask created in the GIS resulting from application of the mask updating rules discussed in section 9.3.2 above. In these photographs of the GIS overlay, yellow and red represent areas that could grow rice, whilst areas that cannot grow rice are shown as black. The boundary of the Leeton Irrigation Area is shown in white. As areas are changed to rice, they change to blue (potentially rice) to green (rice).

Comparison of the land-cover mask, after the 20 January 1984 image as shown in Photo 9.4, with the field data mask, gave the results shown in Table 9.3 for total pixel counts, with the errors of omission and commission in parentheses. This comparison shows a 2 per cent error of commission and a 6 per cent error of omission, respectively, using the 9 per cent property sample.

The accuracy assessment given in Table 9.3 is a good indication of the accuracy of the mapping within the fields defined in the field data mask. However, it does not necessarily indicate the accuracy of estimating the areas of fields since it may not accurately indicate errors in classification at the boundaries of the fields. To indicate the accuracy

Table 9.3 Comparison of pixel assignments in field data and classification masks after 20 January 1984 image showing pixel counts and proportion errors of omission and commission

	Satellite results			
	Vector classifier		Maximum likelihood classifier	
Field data	Rice	Not rice	Rice	Not rice
Rice	5 077	337	4 724	690
	(0.94)	(0.06)	(0.87)	(0.13)
Not rice	249	11 672	82	11 839
	(0.02)	(0.98)	(0.01)	(0.99)

Table 9.4 Comparison of pixel assignments for the field data and land-cover masks after 14 January 1985 image showing proportion errors of omission and commission

	Satellite results	
Field data	Rice	Not rice
Rice	0.952	0.048
Not rice	0.012	0.988

of estimating the area of fields, the areas of a set of rice bays were accurately measured and compared with the areas as estimated from the rice mask.

This regression of 35 selected fields measured on aerial photographs and from the land-cover mask yielded a regression equation, with a 95.34 coefficient of determination, of

$$\text{area} = 0.579 + 1.0098 \times \text{Landsat area}$$

Both of these accuracy assessments indicate that the estimate of rice area as determined from the monitor is smaller than the actual area of rice.

9.3.4 Results of monitoring in the 1984/85 season

The same procedure was implemented in the 1984/1985 season, and the results assessed using the same methods of accuracy assessment. Comparison of the land-cover mask after 14 January 1985 image data with a field data mask is given in Table 9.4, showing the proportions, accuracy and errors of omission and commission respectively.

Regression of 30 random fields with areas measured in the aerial photographs and from the satellite analysis yielded the following regression equation, with an 84.8 coefficient of determination.

$$\text{area} = 0.97 + 1.0260 \times \text{Landsat area}$$

Table 9.5 Pixel by pixel comparison of field data layer with Landsat MSS and SPOT current rice area layers after images acquired in late January 1987, showing areas (from pixel counts) and proportion accuracy and errors of omission and commission

	Landsat MSS		SPOT	
Field data	Rice	Not rice	Rice	Not rice
Rice	693.3 (96.9%)	22.0 (3.1%)	726.6 (99.1%)	6.5 (0.9%)
Not rice	80 (0.4%)	2121.8 (99.6%)	6.9 (0.4%)	1789.6 (99.6%)

Table 9.6 Regression analysis of the areas of 25 fields as measured by planimeter against their areas as measured by the Landsat and SPOT GIS layers

	Regression		Sum of squares		
	Offset	Gain	Regressions	Residuals	R^2
Landsat (50 m)	2.97	1.05	7661.9	108.5	0.99
SPOT (25 m)	0.91	1.02	7730.8	39.6	0.99

9.3.5 Results achieved in monitoring rice in the 1986/87 season

The system was implemented in the 1986/87 season using two satellite systems: Landsat MSS on its own, and a combination of SPOT and Landsat data. The purpose of this was to compare the results achieved with the costs of production, so that the user could assess the benefits that would accrue from the higher costs of analysis of the SPOT data. Processing of the image data produced masks of areas growing rice by Landsat and SPOT, respectively, producing two estimates of rice area for the different irrigation areas. Two separate GIS databases were produced, one at 50 m resolution for the Landsat analysis, and one at 25 m resolution for the SPOT analysis. Field data were used to create 50 m and 25 m resolution field data layers for comparison with the Landsat MSS and SPOT rice land-cover layers, respectively. The pixel by pixel comparison gave the results shown in Table 9.5.

The areas of 25 fields were measured by planimeter and regressed against their areas as estimated by Landsat and by SPOT. The regression results are given in Table 9.6.

The field data have shown that monitoring of rice area by use of both Landsat MSS and SPOT data are conservative in comparison with that from aerial photography. The information derived from the SPOT data have errors of omission and commission that are of the same order as errors in the field data themselves. As experience is gained, and with more careful delineation of boundaries, it should be possible to achieve consistently accurate estimates of rice area early in the growing phase and then to monitor subsequent changes.

9.4 The use of remote sensing in estimation of rangelands tree cover

9.4.1 User needs

Extensive areas of the rangelands of Australia are being invaded by woody shrubs, typically about 3 m in height. The shrubs, by competing for moisture and light, reduce the herbage production of the country, thereby reducing stock carrying capacity. In addition, the existence of dense shrub stands increases the difficulties and reduces the efficiency of mustering stock, thereby increasing costs. Many land-holders have noticed a reduction in animal production of up to 30 per cent over the last 20 years, due primarily to reduced carrying capacity of the land.

The District Agronomy Advisors in the rangelands are encouraging land-holders to address the problem of shrub invasion. To do this they require accurate and detailed information on current cover conditions, and changes that have occurred over time periods that can be related to the management of the shrubs.

Recommended treatment of shrub infestations depends upon the density and extent of those infestations where these treatments include:

- physical destruction of the shrubs by cutting or rolling using a caterpillar tractor dragging heavy chains, designed for the purpose;
- burning, where the suitability of the area for burning depends on the availability of fuel – fuel is provided by dead herbage; areas suitable for burning occur after a good wet season produces plenty of herbage and not under high scrub densities where there is insignificant herbage growth; and
- grazing with goats on species palatable to the goats.

The advisers need detailed information on shrub density at the property level to provide advice on appropriate remedial action.

9.4.2 Technical specifications and design

The land-holdings in the rangelands vary in size between 20 000 ha and 500 000 ha, with paddock sizes typically larger than 4000 ha. Landsat MSS data can be used to provide 100 m resolution information that is quite acceptable for the management of scrub in this environment.

The response of Australian woody vegetation is quite dark in the Landsat data as shown in Diagram 9.3, in contrast to green herbage. A mixture of green herbage and darker soils could have response values similar to that of woody vegetation, or a mixture of woody vegetation and light soils. In consequence, these different cover types cannot be discriminated spectrally within image data. The solution is to select images when all of the herbage is browned off, a readily achievable situation in the rangelands, which have annual rainfalls typically less than 15 inches (380 mm) per annum, usually concentrated in a few storms.

Woody shrubs germinate and become established after good rains in two years. This sequence of events generally occurs every five years. In consequence, it was decided to map shrub cover at five-year intervals using Landsat MSS data rectified to 100 m cell sizes.

The mapping will be done using the vector classifier (section 5.5) that provides estimates of the proportion contribution to response of the parent cover types; in this

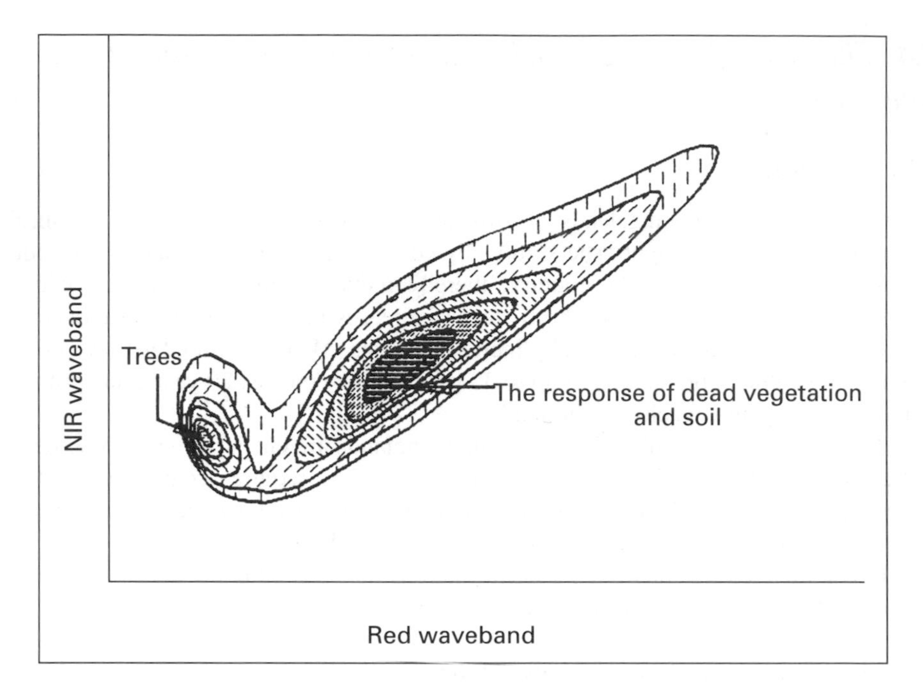

Diagram 9.3 Scattergram displaying the distribution of data typical of the rangelands of Australia in the red and NIR response domains. The 'Badge of Trees' are shown in the diagram, which represents conditions when there are no green herbage areas.

case, light and dark soil, and shrub canopy. The Landsat MSS response values for these three are selected so that all response values for mixtures of these cover types lie within either the figure defined by the nodes or a threshold distance of this figure. Since three nodes define a triangular, planar figure, very few values will lie precisely on this plane figure. A threshold distance either side of this surface is defined to capture pixels with values that are in close proximity to the plane surface.

The proportions derived by the vector classifier are not proportions of shrub cover, but rather proportion contribution to the response of the pixel by the node. The nodes themselves may be some form of mixed surface conditions. It is thus necessary to calibrate these images into proportion shrub cover images.

The calibration was conducted by the use of large scale aerial photography acquired at the time of the 1987 image data. The aerial photography was flown as a transect across the study area. The transect was chosen to be adjacent to a road so that the aerial photographs could be located in the satellite image data. The transect was also chosen so as to have a representative range of shrub densities within different samples along the transect.

A series of 63 1-ha sample cells were chosen across the transect, such that each cell could be identified in the image data and marked on the photography. The proportion contribution to response for each sample was taken from the vector classification. The proportion shrub cover for each sample was measured using a dot grid that contained about 50 dots within the 1-ha sample.

The two sets of data were analysed, and used to develop a linear regression model that was then used to transform the processed data from the vector classification into an image of per cent woody cover at the date the image was acquired. Similar processing

Table 9.7 Results of regression analysis between the Landsat MSS based estimate of the proportion contribution to response and the aerial photographic estimate of the proportion shrub cover, for a selected area in the rangelands of Australia

Area (ha)	Image date	No. of samples	Regression		R^2
			Gain	Offset	
2 706 000	2 February 1987	63	0.793	10.3	0.927

was done for each image, using the one regression model since the field data only matched the one image date.

The district advisers wish to look at management strategies within properties. The best solution is to include the cover information with property boundary information in a GIS. In this way, the advisers can consider different management strategies for paddocks within properties that have different densities of shrub cover. The advisers therefore need to be able to add existing and proposed fence lines to the property boundary layer in the GIS.

In summary, the client requires shrub density information at 100 m resolution at five-yearly intervals. This information is included in a GIS that includes:

- main roads for navigation;
- property boundary information;
- capacity to include other linear features, such as existing and proposed fence lines within the property layer of the GIS; and
- facilities to produce statistics and maps for clients, so as to measure shrub cover within polygons defined in the Property layer and produce property maps depicting shrub cover for the land-holder.

9.4.3 Conduct of the work

Images were chosen in 1975, 1982 and 1987, with the 1975 and 1987 images depicted in Photo 9.6. The images were processed using the vector classifier after identifying the three nodes and selecting an appropriate threshold distance. Large scale aerial photography flown for the purpose was used with the 1987 results to derive the regression equation given in Table 9.7 and depicted in Diagram 9.4.

This regression was then used to transform the proportion contribution to response parameter derived from the vector classification of the image data, into an estimate of shrub cover. The resulting image is shown in Photo 9.7 for the transformation of the 1987 classification. In this image, increasing greenness indicates increasing shrub cover whilst increasing redness indicates decreasing shrub cover and increasing soil and dead vegetative cover. The blue areas are of a different soil colour and are outside the threshold of the classifier, i.e. they were not classified. The regression developed from the 1987 data was used for those dates as well, recognizing that the results will include errors that arise from different placement of the nodes relative to the cover conditions. This information was then added to the GIS as a layer for each of the three dates.

The advisers were provided with the GIS layers: shrub cover, 1975; shrub cover, 1982; shrub cover, 1987; property boundaries; and main roads. The shrub cover information provided in the GIS had 100 levels of information, corresponding to 1 per cent changes in shrub cover.

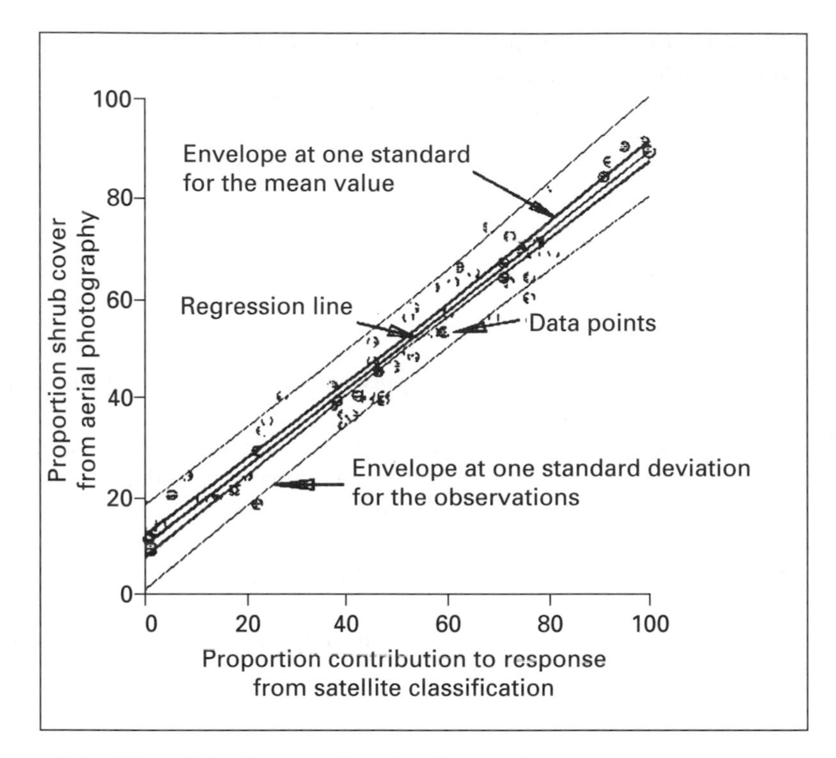

Diagram 9.4 Plot of the data used in developing the regression equation to estimate per cent of woody cover from proportion contribution to response due to woody vegetation as derived by the vector classifier and the derived regression equation.

9.5 Application in predictive modelling: prediction of water caused soil loss by means of the USLE

An important application of remote sensing and GIS is in estimation and prediction. The previous example illustrated one form of estimation that depended on processing of the image data and then converting the derived results into an estimate of shrub cover by the use of a simple regression equation. More complex modelling, involving the use of models or sub-models may need to be conducted within the GIS.

One example of a more complex model relevant to many parts of the globe concerns estimation of water-borne soil erosion. Such a model can estimate the erosion that will occur given specific input values, and it can predict the erosion that might occur given anticipated conditions. The difference between estimation and prediction is that the former uses actual data values, or data meant to represent seasons that have occurred, whilst the latter uses predicted or expected values for the parameters, for a future time or period.

The most widely used method of soil loss prediction is by the USLE:

$$A = (0.224)\mathrm{RKLSCP}$$

where

A = the soil loss, $\mathrm{kg/m^2}$ per year,
R = the rainfall erosivity factor,

K = the soil erodibility factor,
L = the slope length factor,
S = the slope gradient factor,
C = the cropping management factor, and
P = the erosion control practice factor.

The USLE was developed as a method of estimating or predicting average annual soil loss from inter-rill and rill erosion. If this estimation or prediction produces estimates of unacceptably large soil losses, then one or more of the parameters have to be changed so as to reduce the soil loss to an acceptable level. The only parameters that can be influenced are the land-using practices, i.e. the cropping management factor, C, and the erosion control practice factor, P. The model can be used with different values in these parameters, to find acceptable values, and then deduce land-using practices that will allow the land-holder to meet this requirement. Thus the USLE may properly be used to:

- estimate or predict average annual soil loss from a field with specific slope, soil and land-use conditions;
- guide the selection of cropping and management systems, and conservation practices for specific soils and slopes;
- predict the change in soil loss that would result from a change in cropping practices on a specific field;
- determine how conservation practices may be applied or altered to allow more intensive cultivation;
- estimate soil losses from all land uses, not just agricultural fields; and
- provide soil loss estimates for conservationists to use for determining conservation needs.

The equation was developed to estimate the average annual soil loss, so that its application to a specific year or storm may not be appropriate. The parameter values that are used in the equation must therefore be representative of the whole year, and not just one season, or one rainfall event. The parameters that are affected in this way are primarily those that change throughout the year, including the rainfall erosivity factor, R, the crop management factor, C and the erosion control practice factor, P.

The factors of the USLE were developed using an evaluation unit called the standard plot. A standard plot is 22.13 m long on a uniform lengthwise slope of 9 per cent. Let us consider the factors in turn.

9.5.1 The rainfall erosivity factor, *R*

The rainfall erosivity factor, R, is a measure of the erosivity of the rainfall events in an area. It is affected by two rainstorm characteristics: the kinetic energy created by the intensity of the rainfall at the surface; and the duration of this intensity. These rainfall factor values for a large area can be presented as curves of equal erosivity (iso-erodents) on a map of the area of interest, as shown for the USA in Diagram 9.5.

9.5.2 The soil erodibility factor, *K*

The soil erodibility factor, K, is a quantitative description of the inherent erodibility of a particular soil. Soils with different levels of erodibility will erode at different rates when the other factors that affect erosion are the same. The erodibility of soils are primarily affected by the per cent silt, per cent sand, organic matter content, soil structure

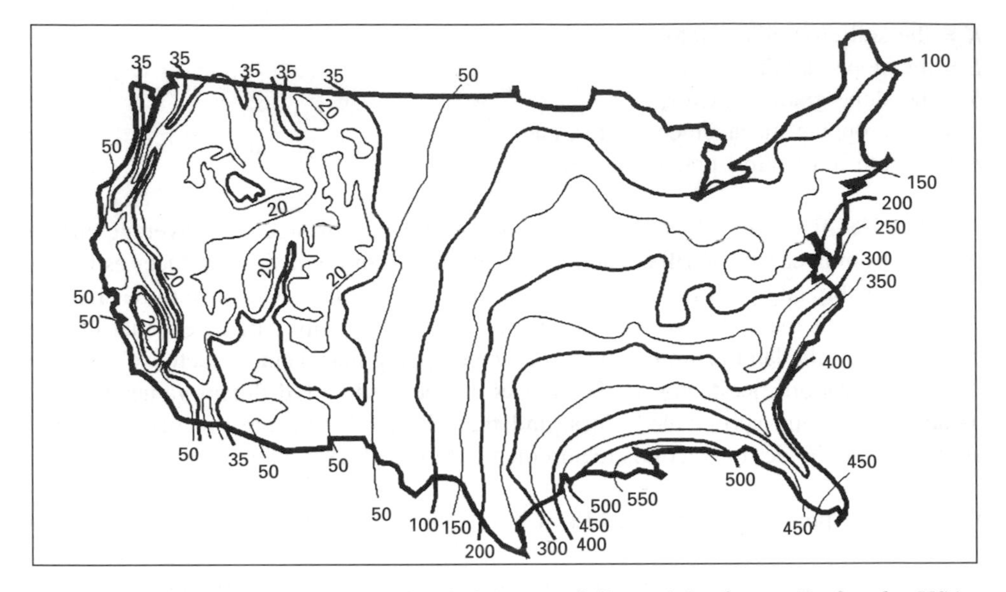

Diagram 9.5 Average annual values of the rainfall-erosivity factor, R, *for the USA.*

and soil permeability. For a particular soil, the soil erodibility factor, K, is rate of erosion per unit of erosion index from a standard plot.

The soil erodibility factor, K, can be estimated from a knowledge of the soil characteristics in the five physical parameters of per cent silt (0.002–0.05 mm) plus very fine sand (0.05–0.10 mm), per cent sand (0.10–2.0 mm), organic matter content, structure, and permeability. Once values have been obtained for these parameters, they can be fed into the nomograph shown in Diagram 9.6 to estimate the soil erodibility factor for that soil.

9.5.3 The slope length factor, *L*, and the slope gradient factor, *S*

The effects of slope length and gradient are presented in the USLE as L and S, respectively; however, they are often evaluated as a single topographic factor, LS. Slope length is defined as the distance from the point of origin of overland flow to the point where the slope decreases sufficiently for deposition to occur or to the point where runoff enters a defined channel. Irregular slopes are divided into n segments, where each segment should be uniform in gradient and soil type. The soil loss for the entire slope is then computed using

$$A = \frac{(0.224)\mathrm{RKCP}}{x_c(22.13)^m}\left\{\sum_{j=1}^{n}(S_j x_j^{m+1} - S_j x_{j-1}^{m+1})\right\}$$

where

x_j = the distance from the top of the slope to the lower end of the jth segment in meters
x_{j-1} = the distance from the top of the slope to upper end of the jth segment in meters
x_e = the overall slope length in meters
S_j = the value of the slope-gradient factor for the j segment
m is a power that is a function of the slope, and A, R, K, C, and P are as defined previously.

It would be reasonable to expect that whilst progressing down a slope, the other parameters would also change. Changes in soil type may incur changes in the soil erosivity

Table 9.8 Values of x_j^m, x_j^{m-1} *and* x_j^{m-2} *expected for the* m *values used and distances of one cell, 10 cells and 100 cells from the start of the slope*

		m											
		0.5			0.4			0.3			0.2		
		x_j^m	x_j^{m-1}	x_j^{m-2}	x_j^m	x_j^{m-1}	x_j^{m-2}	x_j^m	x_j^{m-1}	x_j^{m-2}	x_j^m	x_j^{m-1}	x_j^{m-2}
	1*p*	1.0	1.0	1.0	1.0	1.0	1.0	1.0	1.0	1.0	1.0	1.0	1.0
x_j	10*p*	3.2	0.3	0.0	2.5	0.3	0.0	2.0	0.2	0.0	1.6	0.1	0.0
	100*p*	10.0	0.1	0.0	6.3	0.1	0.0	4.0	0.0	0.0	2.5	0.0	0.0

factor, *K*, and changes in land-cover may result in changes in *C* and/or *P*. The USLE can be modified to allow for changes in all parameters down the slope, giving a modified equation:

$$A = \frac{0.224}{x_c(22.13)^m} \sum_{j=1}^{n} R_j C_j P_j K_j S_j (x_j^{m+1} - x_{j-1}^{m+1}) \qquad (9.1)$$

This equation is almost suitable for use in a GIS as the values of R_j, C_j, P_j, K_j and S_j can be calculated for each cell or polygon within the GIS. However the component $(x_j^{m+1} - x_{j-1}^{m+1})$ is dependent on both the distance of the point from the start of the slope, and the slope at the point x_j. The component $(x_j^{m+1} - x_{j-1}^{m+1})$ can be expanded

$$x_j^{m+1} - x_{j-1}^{m+1} = x_j^{m+1} - (x_j - p)^{m+1} \qquad (9.2)$$

where *p* is the size of a cell in the GIS (in meters), and the processing will be done in units of one cell size in the GIS. Expanding (9.2) gives

$$\begin{aligned} x_j^{m+1} - x_{j-1}^{m+1} &= x_j^{m+1} - x_j^{m+1} + (m+1)x_j^m\, p - \frac{1}{2}(m+1)mx_j^{m-1}p^2 \\ &\quad + \frac{1}{6}(m+1)m(m-1)x_j^{m-2}p^3 - \ldots \\ &= (m+1)x_j^m\, p - \frac{1}{2}(m+1)mx_j^{m-1}\, p^2 \\ &\quad + \frac{1}{6}(m+1)m(m-1)x_j^{m-2}\, p^3 - \ldots \end{aligned} \qquad (9.3)$$

For small sub-watersheds the distance x_j will usually be less than 100 p, so that the range of values to be expected for x_j^m, x_j^{m-1} and x_j^{m-2} are as shown in Table 9.8. In consequence the first term of (9.3) is a reasonable approximation for (9.2), giving

$$x_j^{m+1} - x_{j-1}^{m+1} = (m+1)x_j^m p$$

Now x_j^m and *p* are in units of meters, whereas distances calculated in a GIS will usually be in units of cell size, *p*. This is most easily accommodated by use of the conversion, $x_j^m = X_j^m p$ where *X* is the distance from the start of the slope in units of cell size in the GIS.

Table 9.9 The vegetal cover factor and cultural techniques (C factor) in West Africa

Practice		Annual average C factor
Bare soil		1
Forest or dense shrub, high mulch crops		0.001
Savannah, prairie in good condition		0.01
Over-grazed savannah or prairie		0.1
Crop cover of slow development of late planting: 1st year		0.3–0.8
Crop cover of rapid development or early planting: 1st year		0.01–0.1
Crop cover of slow development or late planting: 2nd year		0.01–0.1
Corn, sorghum, millet (as a function of yield)		0.4–0.9
Rice (intensive fertilization)		0.1–0.2
Cotton, tobacco (2nd cycle)		0.5–0.7
Peanuts (as a function of yield and the date of planting)		0.4–0.8
1st year cassava and yam (as a function of the date of planting)		0.2–0.8
Palm tree, coffee, cocoa with crop cover		0.1–0.3
	burned residue	0.2–0.5
Pineapple on contour (as a function of slope)	buried residue	0.1–0.3
	surface residue	0.01
Pineapple and tie-ridging (slope 7 per cent)		0.1

Equation (9.1) can now be put in a more useful form for application within a GIS:

$$A = \frac{0.224}{x_c(22.13)^m} \sum_{j=1}^{n} R_j C_j P_j K_j S_j (m+1) X_j^m p^2$$

for A in units of kg m^{-2} per year, or

$$A = \frac{0.224}{x_c(22.13)^m} \sum_{j=1}^{n} R_j C_j P_j K_j S_j (m+1) X_j^m$$

for A in units of kg/Cell area per year.

9.5.4 The cropping management factor, C

The cropping management factor represents the ratio of soil loss from a specific cropping or cover condition to the soil loss from a tilled, continuous fallow condition for the same soil and slope and for the same rainfall. This factor includes the interrelated effects of cover, crop sequence, productivity level, growing season length, cultural practices, residue management, and rainfall distribution. The evaluation of the C factor is often difficult because of the many cropping and management systems. Crops can be grown continuously or rotated with other crops; rotations are of various lengths and sequences; residues can be removed or left on the field or incorporated into the soil, and the soil may be clean tilled or one of several conservation tillage systems may be used. Each segment of the cropping and management sequence must be evaluated in determining the C factor for a field. This is done by estimating the C factor for each crop stage, and allocating a weight to it by the proportion of the year that it covers, with the weighted proportions summed to derive the annual C factor for the field.

Sets of C factor tables have been produced for various countries and regions. An example of such data are given in Table 9.9, representative of cropping conditions in

Table 9.10 *C values for undisturbed land in the USA*

Surface: Vegetal canopy, type and height	Percentage cover	Mulch or vegetation at ground Percentage cover					
		0	20	40	60	80	95–100
None		0.45	0.24	0.15	0.91	0.04	0.01
Tall weeds or short brush, 0.5 m effective height	25	0.36	0.20	0.13	0.83	0.04	0.01
	75	0.17	0.12	0.09	0.07	0.04	0.01
Brush or brushes, 2 m effective height	25	0.40	0.22	0.14	0.09	0.04	0.01
	75	0.28	0.17	0.12	0.08	0.04	0.01
Trees, 4 m effective height	25	0.42	0.23	0.14	0.09	0.04	0.01
	75	0.36	0.20	0.13	0.08	0.04	0.01
Factor to obtain *C* values with grass or compacted duff ground cover*		1.0	0.83	0.67	0.46	0.30	0.27

* To obtain *C* values for ground cover of grass or compacted duff, multiply the *C* value from the table by the factor.

Table 9.11 *Erosion control practice factor,* P

Land slope, percentage	Contouring	Contour strip cropping and irrigated furrows	Terracing
1–2	0.60	0.30	0.12
3–8	0.50	0.25	0.10
9–12	0.60	0.30	0.12
13–16	0.70	0.35	0.14
17–20	0.80	0.40	0.16
21–25	0.90	0.45	0.18

West Africa. In many situations the analyst may also be interested in estimating the soil loss from non-agricultural lands. Table 9.10 provides *C* Factor values for undisturbed lands.

9.5.5 The erosion control practice factor, *P*

The erosion control practice factor is the ratio of soil loss using the specific practice compared with the soil loss using up-and-down hill culture. The erosion control practices usually included in this factor are contouring, contour strip cropping, and terracing. Conservation tillage, crop rotations, fertility treatments, and the retention of residues are important erosion control practices. However, these cultural practices are included in the cropping management factors described earlier.

The practice factors for the three major mechanical practices of contouring, contour strip cropping and irrigated furrows, and terracing are shown in Table 9.11. Within a practice type, the *P* factor is most effective for the 3–8 per cent slope range and values increase as the slope increases. As the slope decreases below 2 per cent the practice factor value increases due to the reduced effect of the practice when compared with up-and-down-hill cultivation. The factor for terracing in Table 9.11 is for the prediction of the total off-the-field soil loss. If within-terrace interval soil loss is desired, the terrace interval distance should be used for the slope length factor, *L*, and the contouring *P* value used for the practice factor.

9.5.6 Soil loss tolerance

Soil loss tolerance is the maximum rate of soil erosion that permits a high level of productivity to be sustained. In general, deep, medium textured, moderately permeable soils that have sub-soil characteristics favourable for plant growth were assigned tolerances of 1.1 kg/m^{-2} per year. Soils with a shallow root zone or other detrimental characteristics were assigned lower tolerances. Recommended soil loss tolerance values for soils in the US may be obtained from Soil Conersvation Service handbooks.

The soil loss tolerance for a specific soil is used as a guide for soil conservation planning. The USLE is used to estimate the actual soil loss and to evaluate how changes in practices can be applied to reduce soil loss to below the tolerance level.

9.6 Application of remote sensing and GIS to environmental modelling: application of the USLE in a GIS

9.6.1 Introduction

Section 4.2 described derivation of the raw information needed to provide management information for the Harbour creek catchment in New South Wales, Australia. Those information needs include estimation of soil loss.

9.6.2 Estimation of erosion risk

Estimation of erosion risk using the USLE requires derivation of the following factors used in that equation. All of these factors vary spatially across the catchment, so that there will be unique values in each parameter, in each cell of the GIS, indicated by the subscript *j*: rainfall erosivity, R_j; soil erodibility, K_j; slope and slope length factor, LS_j; crop management factor, C_j; and erosion control practice factor, P_j.

The **rainfall erosivity** R_j of the area has rainfall erosivity index values that varied between 160–200 across the watershed, as shown in Photo 9.9. The terrain classification was used as the basis for deriving estimates of **soil erodibility**, K_j. This was done by collecting field data from a number of sites in each terrain class on the physical parameters that influence estimation of soil erodibility from the nomograph in Diagram 9.6. The values used in estimating soil erodibility are given in Table 9.12. The data in this table were used to create a soil erodibility layer in the GIS (Photo 9.9).

The slope data, was converted into classes in units of per cent slope using the data in Table 9.13. This data was then used to calculate the slope factor S_i; for each cell, as shown in Photo 9.9. The GIS was used to create watershed boundaries from the sub-watersheds identified from the topographic maps. These were used to calculate the shortest distance from these boundaries to each cell in the watershed. This distance is somewhat different to the length of slope distance required by the USLE equation.

To more closely match the slope length required by the USLE, the sub-watersheds were made small enough to approximate single valleys or basins, i.e. they could not

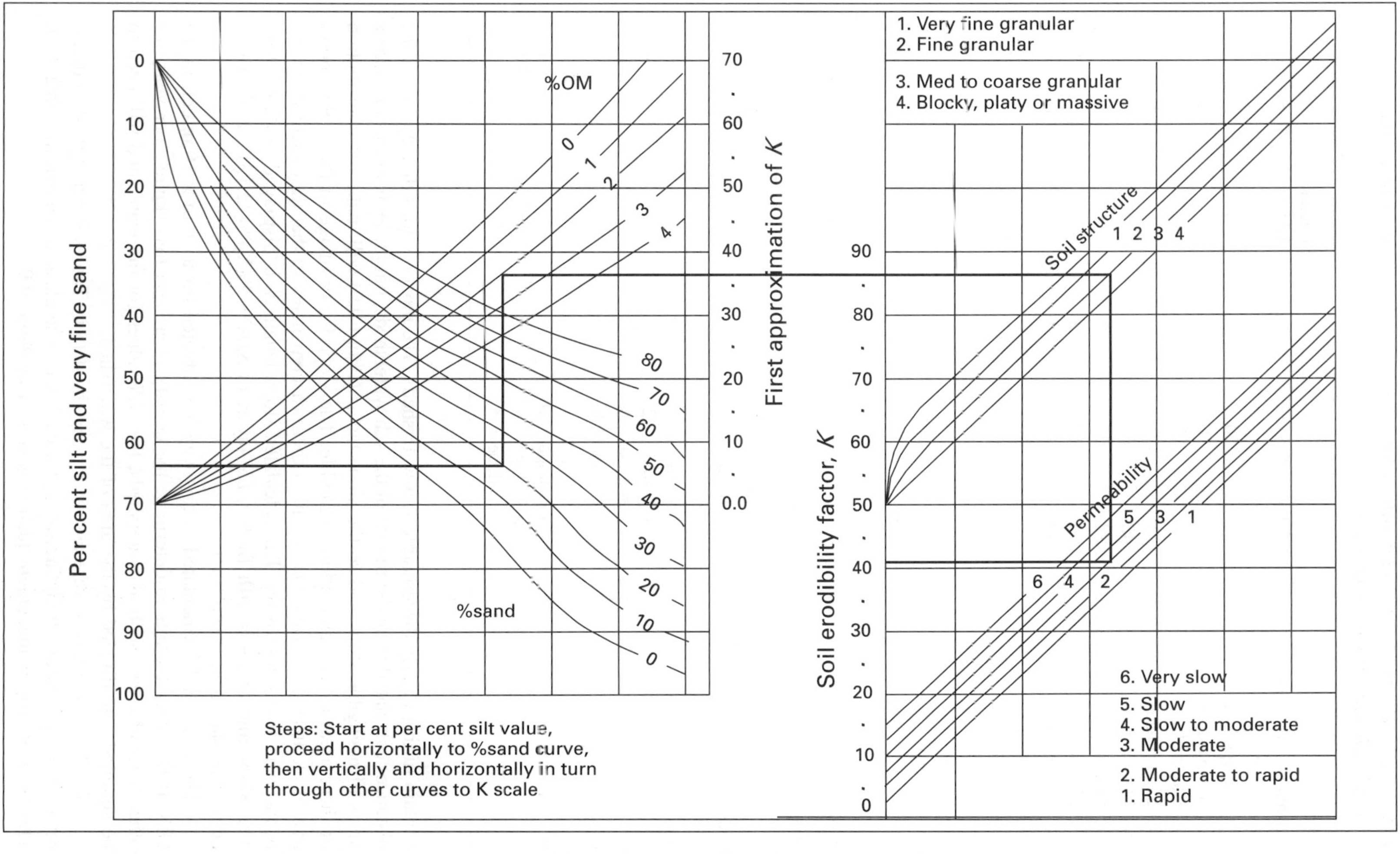

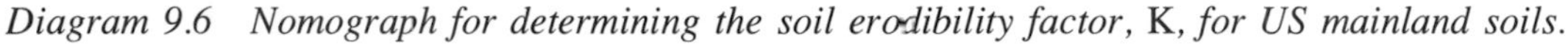
Diagram 9.6 Nomograph for determining the soil erodibility factor, K, for US mainland soils.

Table 9.12 Physical parameter values for all terrain classification units and the derived estimates of soil erodibility

	Terrain classification	%Silt	%Sand	%Org.	Structure Matter	Perm.	*K*
1	Escarpment,	30	70	1	3	2	0.24
2	Steep ridges,	50	50	2	1	4	0.35
3	Low hills,	70	30	2	1	4	0.50
4	Sand ridges	5	95	0	3	1	0.04
5	Flat plain	40	60	4	3	3	0.22
6	Slopes	80	20	2	2	3	0.52
7	Flat plain	90	10	2	1	5	0.58
8	Flat plain	90	10	3	1	6	0.54
9	Escarpment	50	50	2	2	3	0.35
10	Ridges	60	40	3	2	3	0.37
11	Low hills	70	30	3	2	4	0.45
12	Slopes	80	20	3	2	4	0.48
13	Plateau	40	60	2	3	2	0.28

Table 9.13 Conversion of slope classes in degrees into slope classes in percentage slope units

Class	Slope range (degrees)	Slope value (per cent)
1	slope 1	1.0
2	1 < slope 2	2.6
3	2 < slope 5	6.1
4	5 < slope 8	11.4
5	8 < slope 15	20.3
6	15 < slope 26	37.4
7	26 < slope 45	72.7
8	slope > 45	125.0

contain further clearly discerniable sub-watersheds. The most obvious discrepancy is the distances calculated from the mouth of the sub-watersheds, but these distances are across areas of negligible slope and so the slope factor (Photo 9.9) will make the contribution of these areas to the calculation negligible. The slope factor image in Photo 9.9 shows that black areas (zero value) cover all of these discrepancies so that they incur no errors in the estimated soil erosion. The calculated slope length X_j is displayed in Photo 9.10. The **Slope and Slope length factor**, LS_j layer, of $S_i(m + 1)(X_i)^m/(22.13)^m$, is then calculated as shown in Photo 9.10.

The value of LS_j calculated in this way is calibrated for a 10 m × 10 m cell in the GIS. If the analyst wants to determine the density at erosion loss (kg m^{-2} per year) for other types of areas, such as watersheds, then *LS* values for the watershed will need to be summed, and divided by the area of the watershed.

The land-use classes were assessed using the data in Tables 9.9 and 9.10 to determine the **crop management factor**, C_j for each land-use class, as shown in Table 9.14. The layer of crop management factors is shown in Photo 9.9.

Table 9.14 Transformation of land-use classes into crop management factors

Landcover Class.	Crop management factor, C
Pasture	0.01
Forest	0.04
Residential	0.007
Commercial	0.001
Industrial	0.004
Recreational	0.01
Open land	0.01
Horticulture	0.20
Bare sand and earth	1.00
Water surfaces	0.00
Mangroves	0.001

Table 9.15 Acceptable levels of soil loss, per annum for the Harbour Creek catchment as determined from field data for the terrain classification classes

Terrain Classification		Soil Loss Tolerance
1	Escarpment	15
2	Steep ridges	20
3	Low hills	25
4	Sand ridges	10
5	Flat plain	60
6	Slopes	20
7	Flat plain	15
8	Flat plain	20
9	Escarpment	30
10	Ridges	35
11	Low hills	30
12	Slopes	25
13	Plateau	45

In this example there has been no use of control practices so that **Erosion Control Practice Factor**, P_j has not been considered in the analysis.

Calculation of erosion risk

The product of the four factors provides the estimate of soil loss for those given conditions, as shown in Photo 9.10.

9.6.3 Analysis of the erosion risk data

Acceptable soil losses per year, as determined from field data for the terrain classification classes are shown in Photo 9.9, in accordance with the data in Table 9.15.

Table 9.16 Grouped classes indicating erosional status of the Harbour Creek catchment

Colour	Value	Description
black	(Predicted-Acceptable) < –10	Erosion significantly below acceptable levels
red	(Predicted-Acceptable) < 0	At risk from erosion
green	(Predicted-Acceptable) < +10	Serious risk of unacceptable erosion levels
blue	(Predicted-Acceptable) < +20	Significantly unacceptable erosion occurring
yellow	(Predicted-Acceptable) > = +20	Seriously unacceptable erosion is occurring

Table 9.17 Comparison of erosional status classes and land-use classes in the Harbour Creek catchment

Erosion class	Land-use class	Count	Per cent	Cumulative	Area (ha)	Legend
0						insignificant erosion
	1	44 117	15.58	15.58	441.2	Residential 110
	2	81 906	28.92	44.49	819.1	Pasture 210
	4	1 605	0.57	45.06	16.1	Beach 720
	5	3 202	1.13	46.19	32.0	Commercial 120
	6	3 257	1.15	47.34	32.6	Streams 510
	7	3 339	1.18	48.52	33.4	Recreational 170
	8	15 399	5.44	53.95	154.0	Openland 190
	9	10 445	3.69	57.64	104.5	Mangrove 610
	10	55 244	19.50	77.15	552.4	Forest 410
	11	426	0.15	77.30	4.3	Ocean 550
	12	60 593	21.39	98.69	605.9	Horticulture 220
	13	3 001	1.06	99.75	30.0	Industrial 130
	14	434	0.15	99.90	4.3	Bay 540
	15	278	0.10	100.00	2.8	Unknown
1						At risk
	10	3 117	26.72	26.72	31.2	Forest 410
	12	8 550	73.28	100.00	85.5	Horticulture 220
2						some erosion
	10	86	8.22	8.22	0.86	Forest 410
	12	960	91.78	100.00	9.6	Horticulture 220
3						significant erosion
	12	264	100.00	100.00	2.6	Horticulture 220
4						serious erosion
	12	225	100.00	100.00	2.3	Horticulture 220

Comparison of the predicted soil loss with the acceptable level of soil loss in Photo 9.10 shows areas that are currently suffering excessive soil erosion. In these photographs the information on erosion has been grouped into four classes to make it easier to understand and use, in accordance with Table 9.16.

To better understand the implications of this mapping, the information in the GIS on land use and slopes is compared with the erosional status in Tables 9.17 and 9.18. The data in Table 9.17 shows that the areas of significant and serious soil erosion are in areas of horticulture. The data in Table 9.18 shows that the areas of significant and serious soil

Table 9.18 Comparison of erosional status classes and slope classes in the Harbour Creek Catchment

Erosion class	Slope class	Count	Per cent	Cum. %	Area (ha)	Legend
0						Insignificant erosion
	1	1 067	0.38	0.38	10.67	slope < 1°
	2	103 030	36.37	36.75	1 030.30	1° < slope < 2°
	3	15 975	5.64	42.39	159.80	2° < slope < 5°
	4	18 689	6.60	48.99	186.90	5° < slope < 8°
	5	77 281	27.28	76.27	772.80	8° < slope < 15°
	6	59 630	21.05	97.32	596.30	15° < slope < 26°
	7	6 607	2.33	99.66	66.70	27° < slope < 45°
	8	971	0.34	100.00	9.71	slope > 45°
1						At risk
	6	5 858	50.21	50.21	58.58	15° < slope < 26°
	7	2 692	23.07	73.28	26.92	27° < slope < 45°
	8	3 117	26.72	100.00	31.17	slope > 45°
2						some erosion
	7	932	89.10	89.10	9.32	27° < slope < 45°
	8	114	10.90	100.00	1.14	slope > 45°
3						significant erosion
	7	210	79.55	79.55	2.10	27° < slope < 45°
	8	54	20.45	100.00	.54	slope > 45°
4						serious erosion
	8	225	100.00	100.00	2.25	slope > 45°

erosion are in areas of steep slope, 15–26° and 27°–45° slope respectively, slopes that are far too steep for horticultural pursuits under conditions in this area. Clearly there is a need to verify these analyses in the field. If this verification shows that horticulture is being practised in these areas, that the slopes are correct and that there is significant soil erosion, then the farmers will need to be approached, for the purpose of demonstrating the impact of this erosion on productivity once the soil resource becomes depleted. The purpose of this demonstration will be to influence those farmers with a current or potential erosion problem to change their land-using practices in the affected areas before the degradation becomes worse, or before it seriously affects productivity. The farmers to approach can also be determined from the GIS using the property layer that contains a matching attribute file of information on the address and owners of the properties. The farms that are affected are shown in Table 9.19.

9.7 Estimation of crop production using a GIS

9.7.1 Introduction

Section 4.4 discussed methods of estimating the density of cropping, and crop yield so as to derive an estimate of production. That section showed how both of these types of information could be used with information on the area of the polygons or strata over which the yield and crop density parameters are constant, to calculate crop production. However doing this calculation by manual techniques is very laborious. A better

Table 9.19 Comparison of erosional status classes of significant and serious erosion with property information

Erosion Class	Property	Count	Per cent	Cumulative	Area (ha)	Legend
3						Significantly eroding
	62	75	28.41	28.41	0.75	
	87	101	38.26	66.67	1.01	
	101	58	21.97	88.64	0.58	
	103	3	1.12	89.76	0.03	
	121	27	10.24	100.00	0.27	
4						Seriously eroding
	16	7	3.11	3.11	0.07	
	62	58	25.78	28.89	0.58	
	87	130	57.78	86.67	1.30	
	104	30	13.33	100.00	0.30	

Table 9.20 Relational database of mean crop density values for land-using strata, derived from field sampling of those strata

		Wheat			Oats			Barley		
Strata	Key code	Cane	Alb.	Dune	Dora	Dora2	Spain	Ibis	Ibis2	Triticale
ALLUVIUM	2	0.04	0.16	0.06	0.02	0.06	0.09	0.03	0.09	0.03
SLOPES	3	0.20	0.00	0.13	0.06	0.05	0.04	0.12	0.00	0.00
IRRIGATION	4	0.04	0.09	0.02	0.00	0.01	0.06	0.01	0.05	0.00
RIDGES	1	0.02	0.00	0.01	0.00	0.00	0.00	0.02	0.00	0.00
LOW HILLS	5	0.04	0.00	0.02	0.00	0.00	0.00	0.02	0.00	0.00
STEEP HILL	6	0.01	0.00	0.01	0.00	0.00	0.00	0.00	0.00	0.00
URBAN	13	0.00	0.00	0.00	0.00	0.00	0.00	0.00	0.00	0.00

Table 9.21 Relational database of average yield values for productivity strata, as derived from field sampling of those strata

		Wheat			Oats			Barley		
Strata	Key code	Cane	Alb.	Dune	Dora	Dora2	Spain	Ibis	Ibis2	Triticale
GOOD YD	1	2.6	2.8	2.6	2.4	2.4	2.5	2.5	2.6	2.3
MODERATE	2	1.9	2.1	2.0	1.6	1.6	1.8	1.7	1.8	2.1
POOR YD	3	1.3	1.4	1.4	1.1	1.0	1.2	1.1	1.2	1.6

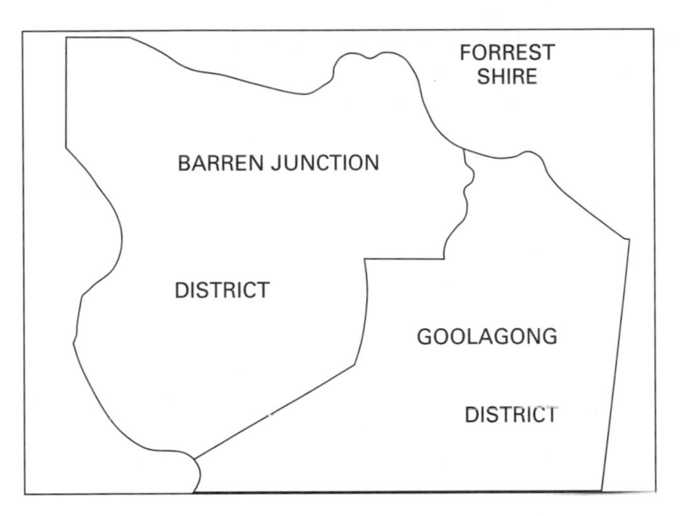

Diagram 9.7 Administrative districts for which production information is required within Forrest Shire.

approach is to use a GIS. Applications of this type depend on the interface between GIS and databases containing the sampled field data and its summarization into average values for each strata. Such data would be of the form shown in Tables 4.10 and 4.11 in section 4.4.

9.7.2 Preparation of the data in a relational database

The data on crop density and yield would form two relational databases as shown in Tables 9.20 and 9.21.

The databases have, as the second column, a key code numbered the same as that strata in the relevant GIS layers of land use and yield as discussed in section 4.4. The other columns are the proportion area and yield for each crop type within the relevant strata. The column of strata name is not necessary for the analysis, but is an advantage in managing the process.

The objective of estimating yield requires solution of

$$\text{Production} = \text{polygon area} \times \text{crop density} \times \text{yield}$$

This equation is solved within a GIS by conducting the following steps for each crop by:

1. calculating the area of crop within each cell or polygon in the GIS;
2. calculating the production within each polygon or cell as the product of the area and the yield; and
3. summing across the administrative areas for which the production data are required.

In this example a raster-based GIS is used and so the production will be calculated for each cell in the GIS and then summed across the administrative regions into a statistical table of production information. The GIS interrogates Table 9.20 to create a layer for each crop, with cell values being the proportion area values given in the table, and shown in Photo 9.11 for the nine crops of interest to the client. The process is repeated

Table 9.22 Final information on production (Tonnes) for Barren Junction and Goolagong Districts of Forrest Shire

		Wheat			Oats			Barley		
Strata	Key code	Cane	Albert	Dune	Dora	Dora2	Spain	Ibis	Ibis2	Triticale
B. Junct	1	1211.03	458.11	864.09	214.68	275.49	344.77	643.66	217.30	83.32
Goolagong	2	1917.98	763.56	1291.45	437.97	488.53	710.79	1043.94	388.34	63.18

for the yield data given in Table 9.21, to give the crop yield images shown in Photograph 9.12. The two layers per crop are then multiplied together to give the production per cell of the GIS, and shown in Photograph 9.13. This data is then summed by the GIS within the administrative layers for which production information is required. The boundaries of the two administrative districts within Forrest Shire are shown in Diagram 9.7. The final statistics are given in Table 9.22.

Appendix: Digital image data and systems

A.1 Introduction to digital image data

Definitions

Digital data are data that can be utilized by a computer, while **image data** are data recording the values for specific attributes of the surface or object, for which the relative and absolute spatial and temporal locations are known to some level of accuracy.

Digital image data are image data that are in the form of computer compatible or digital files, and digital map data are map data that are in the form of computer compatible or digital files.

Image data that has been converted into a digital form are usually structured as raster or grid cell data that covers the whole of the image. Map data that has been converted into a digital form are often created as strings of coordinates recording the locations of all lines and points on the map, or as vector data. Each type of data can be converted into the other, so that a distinction between the two is somewhat arbitrary.

A.2 The nature of digital data

A.2.1 Binary data structure

The basic building block of digital data is the *bi*nary digi*t* (bit). All computers use bits as the basis of all data processing. A bit can be thought of as either an 'ON' or 'OFF'; it has a value of either 1 or 0. If two bits are joined together then there are four possible combinations of the states of each bit: 00, 01, 10 and 11. If four bits are joined together then the possible combinations of the states of each bit are shown in Table A.1.

Two bits can thus store values in the range 0–3 or 0–(2^2–1) and four bits can store values in the range 0–15 or 0–(2^4–1). In the same way n bits can store values in the range 0–(2^n–1). Nowadays most computers group bits into bytes (8 bits = 1 byte), and have 'words' that are 1, 2 or 4 bytes long depending on the computer. Depending on the word length in the computer, then each word will consist of 8, 16 or 32 bits and can store values in the range {0–255}, {0–64 535} or {0–3 164 518 079} respectively if all values can be assumed to be either positive or negative. Inclusion of the sign uses one bit, reducing the range of values that can be stored to {0–±127}, {0–±32 267} and {0–±1582 259 039} respectively. Values that are in this range can therefore be stored in bytes in the computer. A computer word length depends on the make of computer. The way the computer's central processing unit is referred to usually indicates the word length used. For example, 32-bit machines such as Microvax use 32-bit or 4-byte words, and 16-bit machines such as the IBM PC AT use 2-byte words.

A.2.2 Storage of data in a computer

The most common ways for computers to store data are as follows.

Coded data Each character is stored in a single byte. Eight bits can store one of 256 different characters as a value in the range {0–255}. A typical assignment of the different

Table A.1 Arrangement of four bits and their equivalent octal, decimal and hexadecimal values

bit1	bit2	bit3	bit4	Value base 8 (octal)	Value base 10 (decimal)	Value base 16 (hex)
0	0	0	0	0	0	0
0	0	0	1	1	1	1
0	0	1	0	2	2	2
0	0	1	1	3	3	3
0	1	0	0	4	4	4
0	1	0	1	5	5	5
0	1	1	0	6	6	6
0	1	1	1	7	7	7
1	0	0	0	10	8	8
1	0	0	1	11	9	9
1	0	1	0	12	10	A
1	0	1	1	13	11	B
1	1	0	0	14	12	C
1	1	0	1	15	13	D
1	1	1	0	16	14	E
1	1	1	1	17	15	F

characters, including the whole set of alphanumeric characters and other symbols, to a byte is the American Standard Code for Information Interchange (ASCII) format.

When storing a message e.g. 'HELLO', this requires five bytes of storage, one for each character in the message. Coded data can be an inefficient manner to store data, particularly numerical data. An alternative is to store the data as binary values.

Binary data Numerical data can be stored in binary form, in a number of ways. In 'straight binary' or 'fixed point' form the range of numerical values that can be stored depends on the number of bytes allocated. Numbers using 1 byte can be in the range ± 0–127. With 2 bytes the range extends to ± 0–32 511, and so on. Alternatively, a 'floating point' form can be used with 2 bytes for the mantissa plus 1 for exponent giving a range of ± 0–32 511 × 10 (± 0–127). Adding another byte can be used to increase the number of significant digits in the numbers or increase the range of values possible. Binary data allows for more efficient storage of numeric data, when calculations are to be done.

Digital data are stored in records and files. Usually a set of data values that are read from, or written to a storage devicc at a time constitute a record of data. A set of records constitutes a file. A record can be considered to be like a sentence. Just as a sentence consists of a number of words, so a record consists of a number of data values. A file could be considered to be the equivalent of a complete article in a magazine, or a complete chapter in a book.

With remotely sensed data, each scanline of data is usually structured as a record, with the record containing data values for each gridcell or pixel. The set of data records creates a file for the whole image. There are standards for the structure of remotely sensed data, the most common being those developed for the storage of satellite data from the Landsat and SPOT satellites. The two forms of data files are band interleaved by line or BIL format and the band sequential or BSQ format. In the BIL format all of the data values (waveband 1, waveband 2, etc.) associated with gridcell 1 (pixel 1) are stored sequentially in the record, then the data values (waveband 1, waveband 2, etc.) for pixel 2 are stored until the whole of the scanline is included. In BSQ, the values for waveband 1 for the whole image are stored, as one record per scanline for *m* records (scanlines), then all of the values for waveband 2 are stored, until all of the data values have been stored. The BIL format allows the computer quickly to access all of the values associated with a pixel, so that it reduces time in conducting many image processing tasks. But the disadvantage is that the format creates difficulties in combining images of the same area as additional bands in the one image. The BSQ format is good for writing images to some output devices that writes the bands one at a time, and it is easy to merge bands from different images, but of the same area. The disadvantage is that there is more search time in accessing the data from the storage device during many image processing tasks.

The end of files are flagged by end of file (EOF) markers, a unique combination of bit values that allows the computer to clearly identify that word or byte as marking the end of a file. In the same way each record ends with a unique combination of bit values that will identify that word, or byte, as an end of record (EOR) marker.

Files are usually one of two types: serial or direct access.

Serial files

Data is written to the file as individual records, one after the other. The records can be of variable length, each record being completed by an EOR mark. The computer serially reads data from the file by records. To find a specific record in the file the computer counts the number of EORs to find the required record. It then reads this record before placing the data in the specified output location from the storage device, usually a location in RAM (random access memory). The next READ from that file will start after the EOR and continue until the next EOR is encountered. Serial files only use the space required to store data in the file and so are not wasteful of storage space. Because the records are not of fixed length, it is not possible to amend a record and write the new version into the original location, since the revision will usually change the length of the record. New records can only be written at the end of the file, and if the file is amended then a new version of the file will usually have to be created. They are the usual type of file for storing data on magnetic tapes. However, they are inefficient when searching for a specific data record within the file as the search has to start at the first record in the file and proceed serially through all of the records. They are also inefficient where data records have to be amended, as this will require the writing of another version of the file.

Direct access

Direct access files have records of a fixed length. Each record is uniquely labelled or numbered with a 'key', and this key is held in a table so that interrogation of the table can establish the storage location of individual records in the file. Consequently individual records can be accessed directly rather than having to search through the whole file to find a record. This means that records can be accessed as they are required, and the data in them can be changed without threatening the integrity of the whole file. The start address of each record is stored so the records do not have to be contiguous on the storage media; new records can be inserted at any point in the file just by adjusting the record index in the table.

Direct access files are more suitable for searching for a specific value in a parameter, that is stored in the records in sequential order, than are serial files because the search can start at the middle of the file, test to determine if the middle value is higher or lower in the sequence than the sought value for the attribute, and then search the middle record in either the top or bottom half as appropriate. This search technique is known as the binary search. Consider searching for an entry Little D.L. in a file of names arranged alphabetically. The search would start at the centre and find (say) Newton K.M. which is further along than the required name. Choose the first half of the file and interrogate the centre record of that half to get Gates A.Z. The required name is now later than this record, so interrogate the mid record in the second quarter of the file to get Kenny S.F. The search continues in this way until the required record is found. With serial files, in comparison, the search must start at the first record.

For a file of n items the average search of a serial file takes $(n + 1)/2$ search operations to find the required item. If the file contains 10 000 records and it takes 1 second to do one search operation then the average search will take about 80 minutes to complete. In contrast a binary search takes on average $\log_2(n + 1)$ search operations to find the desired entry. With the previous database of 10 000 entries, a search will take on average 13 seconds. This type of file must be a direct access file so that it can be ordered and then searched.

A.3 Digital image and map data

The important characteristics of digital image and map data are as follows.

- The time or period of acquisition are known.
- The image area and geometry are known so that the arrangement and size of the grid cells are known.
- The attributes being recorded are known, and the values of these attributes are recorded as different layers.
- The relative spatial locations of one grid cell to any other grid cell are known to a set level of accuracy.
- The absolute spatial locations of each grid cell on the Earth's surface are known to another level of accuracy.

Each digital data point has at least four values associated with it, either explicitly or implicitly, where these are:

- time of acquisition;
- location of each cell in x, y coordinates; and
- value in one attribute for the cell.

In practice each point of digital data usually has many more values associated with it. These values may include:

- the elevation, or z, coordinate;
- additional attribute values e.g. Landsat TM data contains seven attribute values and SPOT data contain three attribute values;
- hooks or addresses into other attribute value data sets, such as those contained within databases or as an address to a particular process or model.

Of the four or more values associated with each location or cell, some are constant either within the whole image, or across significant portions of the image. When this happens then considerable savings can be made in both storage and computer processing if the values that are constant are recorded only as necessary to reconstitute the image data set. There are various methods of reducing duplication within image data sets. The various methods have characteristics that make them more appropriate for some conditions than others. The more common situations are listed below.

1. Constant attribute value across the image. Typical examples are the time of acquisition of the image, the sensor used and the coordinates of the platform at the centre of the image. Whilst these attributes are constant for the whole image, they are still unique to that image. In consequence, the normal way of handling these types of attributes is to create a header record for these constant data values.
2. Attribute values change in a numerically known way. The typical example of this type of data are the spatial coordinates of each grid cell or pixel in raster and scanner data. Because the numerical model that describes the geometric relationships between cells or pixels is known, the actual spatial coordinates do not need to be retained as long as the key parameters in the model are known. With most satellite sensor systems the key parameter is the name of the sensor, and this is retained in the header record.
3. Attribute values are constant for contiguous groups of grid cells or pixels. Groups of contiguous pixels with the same value can occur along scanlines in the data, or across a group of scanlines. Both situations occur, and packing mechanisms for one are not necessarily the best for the other. There are various ways of reducing duplication in the image data set, and packing of the data can take various forms.

- Convert the data to difference values from a mean value, or as differences from the immediate previous value. Since differences are usually smaller than the actual data values, the differences can be stored in fewer bits thereby saving storage space. The disadvantages of this method are the risk of large differences and the processing required to reconstitute the data set.
- Contiguous runs of cells or pixels within a scanline with the same data values are stored as two numbers, the number of pixels and the attribute value, called run length encoding (Diagram 7.11). To provide consistency the data will need to include flags to show when one pixel is involved or when multiple pixels are involved. This approach is very efficient for storage of raster GIS data. It leads to variable length records.
- Contiguous groups of pixels with constant attribute values can be stored using variable cell sizes. Groups of cells with the same value are grouped into larger cells so as to save on storage. Irregular areas will consist of a number of cells, with many being of different sizes, the method, called quadtree structure, is shown in Diagram 7.10. This method of reducing duplication is usually the most efficient when areas of contiguous pixels with the same value occur, as is often the case with theme images. The disadvantage is the additional processing time and programming sophistication necessary to utilize the technique.
- Replace a set of image files with one image file that contains a key attribute linked to a relational database. The relational database will show the relationship between the key attribute and values in the other attributes. This relational key approach to reducing duplication is suitable if the number of entries in the relational database will be much less than the number of cells in the image data.

A.4 The display of digital data

The characteristics of the different methods of displaying digital data are:

- Fast and cheap output occurs when many images may need to be displayed in short time periods e.g. during image analysis. The normal solution in this situation is to display the images on a colour video display unit (VDU).
- Cheap, multiple copies can be used in the field. Whilst good colour rendition and resolution is important, the most important criteria are cost and ease of production of multiple copies. Inkjet plotters and electrostatic plotters fall into this category of providing suitable material for field staff and for the collection of field data.
- Good quality, high resolution and good colour fidelity are required in hard copy images to be used for photo-interpretation.

There are a variety of different output devices, their characteristics making them suitable for satisfying different subsets of these needs. The main types of interest in remote sensing are: the lineprinter; colour monitor; filmwriter; inkjet plotters; thermal transfer plotters; cathode ray tube filmwriter; and cameras.

A.4.1 The lineprinter

Most lineprinters can display either 80 or 132 characters per line, with a character size and spacing that usually differs in the x and y directions. If this is the case then scale will be different in the x and y directions unless the pixels in the data are generated so that they are at a similar height to base ratio to the lineprinter character spacing in the output. Some lineprinters can have their line and character interval adjusted, thereby eliminating the gap between lines of characters and effectively producing square pixels in the output image.

A.4.2 Colour monitor

Colour monitors with high resolution and good stability can be used to display image data. As with a television set, the image on the monitor needs to be refreshed because the phosphor coating on the screen does not retain the image for very long. The refreshing process is controlled by a dedicated CPU called the display CPU. The role of the display CPU is to continuously access a video memory containing the digital image data to be displayed. The display CPU sends the digital data to a digital-to-analogue converter (DAC) that directs the resulting analogue signal to the monitor. The display CPU repeats this process at the refresh rate, normally about 50 times per second (50 cycles per second). Many systems require that the display CPU conduct simple transforms on the data before it is sent to the DAC. These transforms are particularly convenient as the changes caused by them appear to happen instantaneously on the screen. The complexity of the transforms that can be processed in this way is dependent on the refresh rate and the speed of the CPU.

The data are placed in the video memory by the main CPU in the image processing system. The memory used by the display CPU needs to be large as it stores the data for each pixel in the display at a particular location. If the displayed image is to have $(2^n - 1)$ shades in each of the three primary colours, then n bits of storage are required for each colour, or $3n$ bits per pixel in total. An additional bit is required for the cursor display on the monitor, and further bits would be required if graphics or other information is to

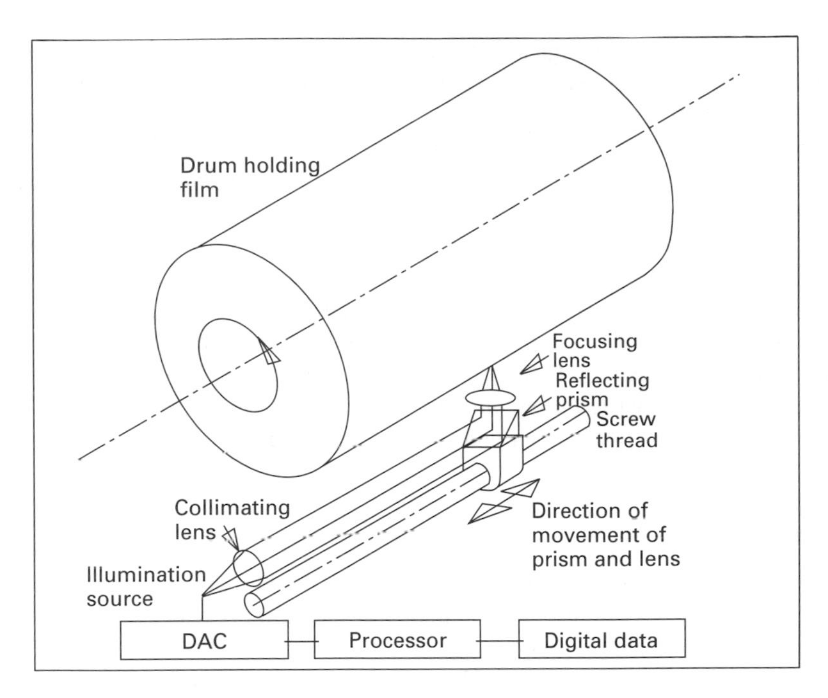

Diagram A.1 Schematic layout of a photowrite.

be displayed in addition to the image data. A system with a display of 512 pixels by 512 lines would require a minimum of 512 × 512 × (3n + 1) bits of storage. The recommended minimum is 16 shades per primary colour, or 4 bits, requiring a minimum of 3 407 872 bits of storage. Most systems support 8 bits per colour per pixel, a cursor bit and at least 7 bits for graphics data, or 32 bits per pixel. This capacity would require 1 mbytes of memory to refresh the video (video memory) at 512 by 512 resolution, or 4 mbytes at 1024 by 1024 resolution.

Most display monitors have resolutions of better than 512 by 512, and resolutions of 1024 by 1024 are common in more powerful systems. The memory should support the display of images of at least the size of the display unit that forms part of the image processing system. With some systems the video memory is larger. The advantages of this are in:

- creating a large video image and viewing a window of this on the monitor, with the ability to rove about this larger image area;
- storage of multiple images in video memory, for fast display of different images and comparison between images; and
- storage of the original data, that is in the 8 or more bits per data value, so that the video memory could be used to supplement the main memory in image processing by the main computer, or to provide more flexibility when transforming the data using the display CPU.

A.4.3 The filmwriter

The photowrite or colourwrite creates either black and white or color photographic images from digital data. Both instruments consist of a drum (Diagram A.1) securing the film, an illumination source and optical train to project the illumination onto the film and electronics to convert digital signals into an optical signal. Illumination from the source is projected through a variable intensity filter as a parallel beam onto a mirror that can traverse the length of the drum. The mirror reflects the beam into a lens that focuses the illumination onto the film. As the mirror traverses a scanline the digital values for the pixels in the scanline control the setting on the filter, thereby controlling the amount of energy transmitted through the filter and onto the film. As the mirror traverses across the film a scanline of imagery is created by exposing the film using the illumination that varies in intensity as a function of the digital data values for the pixel at that location in the scanline. The drum rotates by the scanline distance after writing of the scanline in preparation for writing the next line. The next line is written by the mirror traversing back, in the opposite direction to the earlier line. The process is repeated until the image is complete. The colourwrite differs only in that the image writing is repeated, with each recording being for the blue, green and red primary colours in turn, using appropriate filters.

A.4.4 The inkjet plotter

The inkjet plotter consists of a plotting base, either a flat bed or roller depending on the machine, a battery of inkjets that spray a fine jet of ink onto the paper secured on the bed and a carriage drive assembly to carry the inkjet heads precisely across the base. Inkjet plotters typically contain four differently coloured inkjets: cyan, magenta, yellow and black. The cyan, magenta and yellow are used to produce six colours: cyan, magenta, yellow, blue (magenta + cyan), green (cyan + yellow) and red (magenta + yellow). White is produced from the paper itself, and black is produced by the fourth inkjet. The instrument sends the battery of jets on the printing head across the bed, squirting ink onto the paper so as to produce the desired colour, at a resolution of about 200 dots per inch depending on the instrument, creating a scanline of image data for each pass across the paper. For flatbed plotters the printing head is incremented by a scanline width after each traverse, and before scanning across to create the next traverse or scanline. With roller plotters the roller bed is rotated by an increment equivalent to the width of a scanline on the surface of the roller after each scanline and before initiating the next traverse.

A.4.5 The thermal transfer printer

The thermal transfer printer consists of a ribbon cassette that contains either the three primary colour dyes of cyan, magenta and yellow or these three dyes as well as black, a thermal print head and paper. Both the thermal print head and ribbon are as wide as the paper. The thermal print head transfers dye from the ribbon to the paper for the whole image in each primary colour in turn. At each point dye is either transferred or not in each of the primary colours, providing a range of eight colours. However, the printer uses groups of dots (2 by 2), (3 by 3), (2 by 3) and (4 by 3) to provide a greater range of tones in each colour, but sacrificing spatial resolution. Thus a 2 by 2 dither can have the following combinations;

No Colour	Pale Colour	Mod. Colour	Intense Colour	Saturated Colour
00	01	01	10	11
00	00	10	11	11

providing five shades of intensity. The three primaries can thus provide 125 shades using a 2 by 2 dither, and more shades using larger dither patterns. Thermal transfer printers have resolutions of about 300 dots per inch, produce glossy images that have better intensities than inkjet plotters and are very robust in operation since little can go wrong with the printing process.

A.4.6 Cathode ray tube filmwriter

The cathode ray tube filmwriter consists of:

- a cathode ray tube which is used as the light source for exposing the film;
- microprocessor to control exposure and internal operations of the instrument;
- film cassette and transport module; and
- filters and focusing optics.

Images are created by scanning colour film with an intensity modulated raster produced on the surface of the cathode ray tube and focused by the lens onto the film. The raster is produced by a moving spot of light generated by scanning an electron beam across the face of the cathode ray tube. The intensity in the raster is controlled by the microprocessor as a function of the digital data value at that location in the image. The film is scanned once for each of the three primary colours, with filters controlling exposure of the film in each scan.

A.4.7 Cameras

The simplest and cheapest way of photographically recording digital data is to take either a photograph or slide of the colour VDU or screen using a portable camera. Quite satisfactory photographs can be achieved as long as the camera axis is perpendicular to the screen surface, and normal speed films are used with long exposure to integrate over a number of refreshes of the screen. (The screen is usually refreshed at 50 cycles per second or similar depending on the system being used.) Exposures that are shorter than this will capture part of the screen not refreshed, that is a strip across the screen will be dark. These dark strips are avoided by using an exposure longer than the refresh rate, i.e. longer than 1/50th seconds.

A.5 Image processing equipment

The following functions need to be performed by image processing equipment.

- The ability to read, write and store large data files. An unrectified Landsat MSS image requires about 30 mbytes of storage, and a Landsat TM image takes about 270 mbytes of storage. In operations that involve creating a new image by processing an old image, such as rectification, twice this storage will be required as mass storage to the system. If the processing involves multiple images then even more than this would be required. Generally high cost image processing systems can process whole image data sets, both interactively and in batch mode. Mid-range image processing systems can usually handle sub-image data sets interactively, but can process whole image data sets in batch mode. Low cost image processing systems can usually handle only sub-image data sets, both interactively and in batch mode.
- A central processor unit optimized for the processing of large numerical data sets.
- A Colour Visual Display device for the rapid and cheap display of processed data.
- Devices available to create hard-copy images from digital image data.

- The capacity to create files for use in GIS, and being able to use GIS files in image processing.
- The software with the capacity to manage data files, conduct interactive and batch processing tasks, display and store derived information, support the development of specialized processing tasks, and manage the system and operations conducted on that system.

A.5.1 Data storage

Image data is generally stored on 1600 or 6250 bits per inch storage density magnetic tapes although data are becoming available on optical discs and other media. Subsets of this data can be purchased on floppy discs.

A.5.2 Central processor

The Central Processor controls the flow of data throughout the system, as well as processing the data in accordance with the program instructions. Image processing involves short periods of very intensive 'number crunching' during which the CPU is fully utilized, separated by relatively long periods of inactivity by the CPU. During these periods the CPU could be productively used for other background tasks to image processing. When the operator initiates CPU action, he usually is waiting for a result, so that delay causes inconvenience and frustration. Therefore interactive image processing tasks should have a high priority.

The central processor requires memory adequate to hold the sophisticated image processing software system, and very large data arrays that may have 6000 values per array or per band. The processor needs to be optimized for the numerical processing of data, where this processing may be speeded up by the use of parallel processors or mathematical coprocessors.

A.5.3 Display devices

The current best solution for the rapid and cheap display of images is by means of a video monitor. The video monitor is driven by a CPU (the display CPU) accessing data in a large memory called the video memory.

A.5.4 Hard-copy devices

The choice of hard-copy devices depends on the needs of the user. Organizations that use the image processing equipment primarily to prepare the data for visual analysis will require high quality photographic images as can be provided by film writing devices. If the main function of the image processing facilities is to classify or transform image data into estimates of physical parameters then greater emphasis may be placed on means of quickly and cheaply distributing processed imagery to users in the field by means of inkjet or electrostatic printer images.

A.5.5 Linkage with GISs

Image processing will often provide information as input to GIS, and may use information in such systems to control or influence image processing. The linkage to such systems is therefore an important consideration in the development of image processing facilities.

A.5.6 Image processing software

Good software is essential to the satisfactory use of image processing equipment. Often users will wish to build specialized systems designed to conduct specific tasks in the most cost effective way. Building of these tasks will often be based on using modules and software libraries provided by the system. In consequence the software must be satisfactory in five main ways: documentation; user interface; applications functions; expandability; and system services.

Documentation

Documentation should be in three forms: as manuals, on-line help and as application remarks in the software. **Manuals** have the advantage that they can provide detailed explanations of a function or application task including the options and implications for each user interaction with the system. They can also be perused away from the system. **On-line documentation** provides a quick ready reference as prompts or explanations during use of the workstation. The two, and the actual processing stage in the task, should be linked by appropriate cross-reference mechanisms for ease in accessing either documentation during processing.

Hard-copy documentation should include the following as standard.

1. System overview documentation
   A synopsis of the required system hardware
   An overview of the software packages, capabilities and interfaces
   How to initiate (log on) entry to the facility and start to use the system
   Use of user modifiable parameters affecting overall processing
   Command syntax
   Modes of system operation (interactive, batch, multi-user, multi-tasking)
   System processing limits (e.g. number of bands)
   File types, characteristics and naming conventions
   Applications programmer's reference guide
2. Applications function documentation
   Function name and brief description of the task
   Processing limitations, e.g. number of bands, number of classes
   Parameter names, descriptions, default options and valid values
   Input and output image and data file types
   Examples of running the function
   Typical processing times
   Detailed description of the algorithm used
   Error messages and appropriate user response
   Information that will provide insights into the use of the function
   Listing of the hardware used
   Reference material
   Date the function was released

The level of on-line help required depends on the experience of the user with the software system. One solution is to allow users to categorize their own level of experience from a set of offered categories in the software. This level of help is then provided until the user changes their category. The amount of on-line help provided is a function of the categories of experience. Thus novices will get a lot of help, much of it unprompted;

moderately experienced users will get a lot of help, but little of it unprompted, whilst experienced users will get little help.

User interface

Image processing systems are complex software packages. Novice users will require considerable assistance in using these packages whereas experienced users will want to save time by avoiding all of the help provided to the novice. A menu approach is particularly useful to the novice, whereas the experienced user may wish to input the critical, non-default parameters, as a command, thereby speeding up use of the facilities. Systems should have both options available to the user, and have them linked so that one pass through the menu can be retained, modified, and used in the command form. For example, users may wish to run the same application a number of times, varying one, or a few parameters, by some amount. In this situation, it would be an advantage if the software automatically logged the tasks done as a processing history, and allowed previous tasks to be resurrected and rerun after modifying the required parameters.

User interfaces must consider:

- proper error and bound handling by the software;
- consistent interface between all application tasks and the user;
- ease of interfacing between programs, for the interchange of parameters and files; and
- recall of previous commands and parameters.

Applications functions

The software capabilities of the system will often be limited in a number of very important ways that must be known to the user. These will often include:

- number of bands that can be processed;
- number of classes that can be mapped;
- size of image that can be processed and displayed; and
- restrictions on the display of images.

System services

System services include reading/writing data to/from storage devices, display of image data on the video monitor, cataloguing of data sets and output to hard-copy devices. Data is provided in various formats. It is important that the system be able to read the data to be used on the system, and it is an advantage to be able to read all data formats of data that can be provided. Since many data sets may be larger than the capacity of the system, it is an advantage if the system can read subscts of the data from tape for storage on a mass storage device.

Bibliography

Carlson, N.L., 1967, Dielectric constant of vegetation at 8.5Ghz, Technical Report 1903–5, Electro-Science Laboratory, Ohio State University, Columbus.

Carswell, 1965.

Condit, H.R., 1970, The spectral reflectance of American soils, *Photogrammetric Engineering*, **36**, 955–66.

Galliers, R.D. (Ed.), 1987, *Information Analysis, selected readings*, Sydney: Addison-Wesley Publishing Company.

Galliers, R.D., 1991, A scenario based approach to strategic information systems planning, in Jackson *et al.* (Eds) *Systems Thinking in Europe*, New York: Plenum Press.

Gates, D.M., 1965, Spectral properties of plants, *Applied Optics*, **4**, 11–20.

Gausman, H.W., Allen, W.A., Wiegand, C.L., Escobar, D.E. and Rodriguez, R.R., 1970, Leaf light reflectance, transmittance, absorptance and optical and geometrical parameters for eleven plant genera with different leaf mesophyll arrangements, *Technical Monograph* (7), Texas A&M University, College Station, Texas.

Grove, C.I., Hook, S.J. and Paylor II, E.D., 1992, Laboratory reflectance spectra of 160 minerals, 0.4 to 2.5 micrometers, *Jet Propulsion Laboratory Publication* Pasadena, pp. 92–2.

Kauth, R.J. and Thomas, G.S., 1976, The tasselled cap – A graphic description of the spectral-temporal development of agricultural crops as seen by Landsat, *Proceedings, Symposium on Machine Processing of Remotely Sensed data*, Purdue University, West Lafayette, Indianna.

Lintz, Jr., J. and Simonett, S.D., 1976, *Remote Sensing of Environment*, Reading, Mass: Addison-Wesley Publishing Company.

Lundien, J.R., 1966, Terrain analysis by electromagnetic means: radar responses to laboratory prepared samples, US Army Waterways Experimental Station Technical Report No. 3–639, Report 2.

McCloy, K.R., Schoneveldt, R. and Kemp, D., 1992, Measurement of pasture parameters from reflectance data, submitted to *Int. Journal of Remote Sensing*.

O'Neill, A.L., 1990, Personal communication.

Sabins, F.F., 1987, Remote Sensing.

Stoner, E.R. and Baumgardner, M.F., 1981, Characteristic variations in reflectance of surface soils, *Soil Science of America Journal*, 45, 1161–65.

Wiegand, C.L., 1972, Physiological factors and optical parameters as bases of vegetation discrimination and stress analysis, *Proceedings*, Seminar on Operational.

Index